STATISTICAL ANALYSIS
OF GEOLOGICAL DATA

George S. Koch, Jr.
and
Richard F. Link

Two Volumes Bound as One
Volume I

Dover Publications, Inc.
New York

Published in Canada by General Publishing Company, Ltd., 30 Lesmill Road, Don Mills, Toronto, Ontario.

Published in the United Kingdom by Constable and Company, Ltd., 10 Orange Street, London WC2H 7EG.

This Dover edition, first published in 1980, is an unabridged and corrected republication of the work originally published as two separate volumes in 1970 (Volume I) and 1971 (Volume II) by John Wiley & Sons, Inc.

International Standard Book Number: 0-486-64040-X
Library of Congress Catalog Card Number: 80-66468

Manufactured in the United States of America
Dover Publications, Inc.
180 Varick Street
New York, N.Y. 10014

Preface

All geologists who work with numbers use statistical methods, whether or not formally, and can profit from a knowledge of applied statistics. The purpose of this book is to explain some effective statistical procedures for the analysis of geological data and to discuss methods to obtain reliable data that are worth analyzing.

We write for the person who has numerical data from which he wants to draw conclusions or who has a problem whose solution will require obtaining and interpreting numerical data. We stress basic statistical methods and emphasize that thoughtful application of these methods will yield valid results. No statistical methods are introduced for their own sake but only because they have proved useful for data analysis. Because geology is a complicated and diverse science, we have purposely included some involved geological arguments. The mathematics is relatively simple, however, and requires for its understanding only elementary algebra and geometry.

This book is for readers with some geological training, who may or may not have professional experience. Although most readers may be geologists, the book will also interest mining and petroleum engineers, geochemists and geophysicists, mineral economists, and others.

The scope and arrangement of the book, outlined in Chapter 1, reflect our selection of the statistical methods that are most useful now, given the present state of geology, statistics, and electronic computing. The book is divided into two volumes. The first chapters, constituting most of Volume 1, place statistics first and geology second because the same statistical methods serve for varied geological problems. Most of the example analyses in this volume are chosen for numerical properties that illustrate specific statistical methods and are fictitious. Volume 2 places geology first and statistics second, and most of the examples are real ones for which the statistical principles introduced in Volume 1 are put to work, extended, and refined if necessary. Nowhere in the book is a large body of geological data analyzed completely because to do so would require a monographic presentation, too long for a textbook.

Although data for examples come from many fields of geology, most are from economic geology, and many are data that we ourselves have analyzed. We think that the reader will profit more from our experience with real data than from analyses of fictitious data or the second-hand data of others. This opportunity to learn from real analyses should offset the two disadvantages

iii

that readers relatively unfamiliar with economic geology may need to invest a small amount of thought in order to make a transfer to the fields of geology that interest them and that our own work is mentioned more often than we would like.

We intend this book for the practicing geologist as well as for the student in a formal course. We have closely tied it to J. C. R. Li's book on applied statistics, which is excellent for self-instruction, and have adopted Li's notation. A page of notation follows the table of contents; we hope that readers familiar with other mathematical symbols will not be seriously inconvenienced.

Our collaboration on the statistical analysis of geological data began at Oregon State University in 1956 and has been continued since 1962 in the Mine Systems Engineering Group of the U.S. Bureau of Mines. This book was outlined and partly written in the spring of 1966, when we were visiting research statisticians at Princeton University.

We are grateful for financial support to Oregon State and Princeton universities, the American Philosophical Society, and the National Science Foundation.

Several mining companies, cited in the text, have generously supplied data. We thank in particular two Mexican companies which have, at some expense and inconvenience, furnished data. Henry B. Hanson, General Manager, Minera Frisco, S.A., has permitted us access to assay records, and his staff has been most cooperative in assembling the data. J. B. Stone, formerly General Manager, Compañia Fresnillo, S.A., and his successor Ing. Luis Villaseñor, also allowed us to copy assay records; G. K. Lowther, Chief Geologist, discussed Fresnillo geology with Koch on many occasions.

Many of the statistical analyses in the book were first made for the U.S. Bureau of Mines and published in the Bureau reports cited in the text. S. W. Hazen, Jr., formerly chief of the Bureau's Mine Systems Engineering Group, encouraged and aided us in this work. Many past and present members of this group have also helped us, particularly J. H. Schuenemeyer, G. W. Gladfelter, and Velma Sturgis in computer programing. This book is not an official statement or representation of the U.S. Bureau of Mines, however.

Many of the geologists and statisticians cited in the text have helped us by discussions as well as by their published works. We have also profited from the criticisms of the following men who read all or part of Volume 1: J. C. Davis, R. C. Flemal, David Hoaglin, D. F. Merriam, A. T. Miesch, D. B. Morris, C. W. Ondrick, G. S. Watson, and Alfred Weiss. The errors and misjudgments that remain are ours alone.

We have also been helped by librarians of the U.S. Geological Survey, Denver, Colorado; the Colorado School of Mines, the Denver Public Library, and especially by Samuel Shephard, librarian of the U.S. Bureau of Mines library in Denver. G. W. Johnson and Sally Konnak improved the English expression and organization of the book by careful editing. The book was typed by Vicky Yen Contreras and Verna Bertrand.

Finally, GSK would like to acknowledge a personal debt of gratitude to L. C. Graton, who taught him the importance of careful observation and verification of scientific data, to Ernst Cloos, who taught him to draw conclusions from numerical measurements, and to the late H. E. McKinstry, who taught him that information of fundamental geologic interest resides in the numerical data of the mining industry and urged him, in 1952, to begin the investigations that eventually led to this book.

Similarly RFL would like to acknowledge the help and inspiration of W. J. Dixon, who introduced him to statistics and computing, and to J. W. Tukey, who contributed materially to his development in statistics and data analysis.

George S. Koch, Jr.
Richard F. Link

Denver, Colorado
New York, New York
February 1970

Contents

Chapter 5. ANALYSIS OF VARIANCE

Chapter 6. DISTRIBUTIONS AND TRANSFORMATIONS

PART III. SAMPLING AND VARIABILITY IN GEOLOGY

Chapter 7. GEOLOGICAL SAMPLING

INTRODUCTION

Chapter 1

Introduction

Every geologist obtains numerical data—the number of hand specimens collected from a formation; strike and dip of a bed or fault; chemical rock analyses; porosity measurements; assays of ore, coal, or oil; etc. And every geologist summarizes these numerical data when he prepares the report that, as an essential part of the scientific process, communicates his findings to others. He summarizes because it is usual in the discipline of geology to obtain many more data than can be reported verbatim; for instance, a geologist may omit strike and dip measurements on a quadrangle map because they are too crowded, or he may average chemical or modal rock analyses. Exactly what he summarizes, and how, will determine much of the value of the report.

The geologist who has substantial amounts of data must use statistics—whether formally or informally. Our thesis is that he can sharpen his thinking and improve the reliability of his conclusions through purposefully devised statistical methods that yield incisive results. He may thus avoid the end effect too often produced today—tables of uninterpreted or misinterpreted numerical data, attached only as ornaments to his report.

Although an increasingly large number of geologists study statistics, a gap remains between academic study and useful practical application. Most textbooks on statistics stress such problems as drawing black and white marbles from urns, measuring the life of light bulbs, and counting beer bottles that leak. Problems of time and space relations, problems that are at the heart of geological thinking, are seldom found. Therefore it is difficult for geologists to make the translation from statistical theory to their immediate problems.

In this book we emphasize data analysis rather than the application of formal statistics. Although the geologist interested in applying statistics to geology must learn some formal techniques, it is even more important for him to develop taste and judgment. We would rather explain the effective allocation of effort to real problems than develop complicated analyses that are likely to be mathematically unstable and computationally unreliable. Our point of view owes much to J. W. Tukey (1962, p. 2), who has written the following:

> I have come to feel that my central interest is in *data analysis*, which I take to include, among other things: procedures for analyzing data, techniques for interpreting the results of such procedures, ways of planning the gathering of data to make its analysis easier, more precise or more accurate, and all the machinery and results of (mathematical) statistics which apply to analyzing data. Large parts of data analysis are inferential . . . but these are only parts, not the whole. Large parts of data analysis are incisive, laying bare indications which we could not perceive by simple and direct examination of the raw data, but these too are only parts, not the whole. Some parts of data analysis, as the term is here stretched beyond its philology, are allocation, in the sense that they guide us in the distribution of effort and other valuable considerations in observation, experimentation, or analysis. Data analysis is a larger and more varied field than inference, or incisive procedures, or allocation.

1.1 SCOPE AND ARRANGEMENT OF THE BOOK

Because we write for those who want to apply statistics to geology, we assume that our readers know some geology, but not necessarily any statistics. The book starts from elementary statistical principles that are sufficiently developed to be put to work on geological data. No attempt is made to survey applied statistics or to review the literature on statistics applied to geology.

Although the discussion of statistical methods starts with first principles, it sometimes ends in specialized procedures not treated in elementary statistics books. Only those statistical methods are introduced that advance interpretation of geological data and are needed to explain a geological problem. Many numerical examples, most of them based on real geological data, are calculated. Simple algebraic arguments are given but complicated proofs requiring calculus are omitted. Even the reader who knows statistics will find it helpful to scan the first, elementary chapters to become acquainted with our viewpoint and notation.

All of the above is within the scope of the book. In the next paragraphs, some related subjects that are not treated in the book are mentioned, and works that explain them are suggested.

Because we are concerned with selected statistical subjects, a broad exposition of statistics, even on an elementary level, is not attempted. Of the many excellent textbooks on statistics, two of the best for geologists are by J. C. R. Li (the two volumes are referenced as 1964, I; and 1964, II) and Dixon and Massey (1969). Li's book (especially vol. I) is well suited to self-instruction. It is written for scientists and engineers and explains statistics in great detail for readers whose mathematics is limited or rusty. Anyone who reads one or both of these books and works the problems will obtain a good grounding in the basic statistics useful to a geologist.

We do not discuss mathematical statistics, a subject explained in many books, for example, the introductory books by Hoel (1962 and 1966) and an advanced book by Wilks (1962). Probability, a central subject in theoretical statistics, is discussed here only briefly; the reader interested in this subject is referred to an elementary text by Mosteller and others (1961) and an advanced book by Feller (vol. I, 1957 and vol. II, 1966).

Five excellent books on statistics in geology cover some additional topics not touched upon in this book. Miller and Kahn (1962) provide an excellent summary of the literature up to 1962, after which so many papers have appeared that a full review volume could contain little else. Besides reviewing the literature, Miller and Kahn discuss such subjects as probability, probability density functions, and paleobiometrics, to all of which we devote little attention. Krumbein and Graybill (1965) stress model formulation and draw most of their examples from sedimentation and oil geology. In *U.S. Bureau of Mines Bulletin 621*, Hazen (1967) covers mining technology, special kinds of sampling, and size distributions of particles in ores and rocks. Griffiths (1967) discusses statistical methods of studying sediments, but his book has wider application, contains statistical methods not mentioned in our book, and offers perceptive comments on the science of geology. Finally, Harbaugh and Merriam (1968) emphasize statistical methods implemented by electronic computers for studying stratigraphy.

This book is divided into two volumes and six parts, three in each volume. Part I is this first introductory chapter. In Part II, which covers Chapters 2 to 6, univariate statistical methods for the analysis of single variables are explained and illustrated through examples; only enough geological data are introduced to help explain the statistics. Part III, which is Chapters 7 and 8, takes up geological sampling and variability. In Part III the formidable problems of data interpretation as it pertains to geology are stressed, and attention is given, therefore, to geological as well as statistical problems of data analysis. Part IV, comprising Chapters 9 to 11 in Volume 2, develops multivariate statistical methods for the analysis of two or more variables. Part V, which includes Chapters 12 to 16, is about the statistical analysis of data from applied geology, mainly mining geology; the chapters pertain to exploration for natural resources and their valuation, to decision making through operations research, and to some specialized methods of sampling and data analysis. Finally, Part VI, or Chapter 17, is about the use of electronic computers to implement the statistical methods.

1.2 PROBLEM SOLVING

The geological investigation that statistics is likely to benefit is that which focuses on solving a specific problem. Good advice, simply summarized, is as follows: have a goal, plan ahead, and use any valid geological and statistical methods, but only those methods that will lead to that goal.

First and foremost, the investigator should think carefully about the problem he is about to pursue: the scope, limitations, ramifications, and objectives. Then through appropriate methodology he may construct that type of hypothesis known as a model (sec. 1.5), purposefully collect data to test the hypothesis, and solve the problem within stated limits of reliability—perhaps in the process revising his model in the light of the preliminary results and then, if necessary, collecting additional data to refine and to verify his conclusions. Thus, he will neither become overwhelmed with enormous amounts of data, nor fail to collect that which is pertinent. He will be able to reach an effective solution to a specific problem while keeping within limits of cost—whether reckoned in money or in time, or in both.

An example of the success to be achieved by striving for one objective is the exploration program that resulted in the discovery in the United States after World War II of large new uranium deposits by government and industry geologists. Much of the new uranium was found by systematic search in environments deemed unfavorable by theoretical geology. It is interesting to note that later, when the data collected primarily for finding ore were used for secondary purposes, such as stratigraphic and geochemical studies, they were found to be not wholly suitable. Thinking in retrospect, some of the investigators wished that, for the new studies directed at different problems, they had collected new data.

In problem solving, rather than being tied only to traditional geology, one should accept the imaginative, integrated use of any and all sciences and engineering fields that have anything to contribute. It is in this spirit that this book includes approaches from a number of disciplines. Cloud (1964) writes in a similar vein:

What is most characteristic philosophically, and most gratifying to me personally, about the earth sciences today is their blending of the useful parts of classical science with the most exciting aspects of advancing science. . . . All of the systematic sciences, and I use *systematic* in the broad sense of classifying *and* explaining, are in a state of ferment as new equipment, new measurements, and improved computer facilities provide different and in some instances more fundamental bases for classification and rapid quantitative methods of evaluation—this is true, not only of mineralogy but also of paleontology and petrology. . . .

If we were to tabulate the things that most generally characterize the earth sciences in the modern world we might include:

1. A growing restiveness with traditional methods of investigation.

2. An increasing tendency to express observations and conclusions quantitatively wherever it is possible to do so.

3. An increasing degree of interaction with other sciences.

4. The assumption or requirement of an increasing degree of familiarity with mathematics as a form of communication, and with physics, chemistry, and biology. . . .

5. A high degree of sophistication of instrumentation that is increasing our resolving power in all fields and permitting us to make new observations and discoveries, both in new fields and in fields that once appeared on the verge of foreclosure.

William Hambleton (1966), associate director of the Kansas Geological Survey, takes a similar view toward the work of state geological surveys in particular. He writes as follows:

As to the program of a modern survey, one might say that it is characterized by change, urgency, and involvement in the social, economic, and political problems of the state and region which it influences. . . . One hundred years ago, the purpose of our organization was to survey the mineral resources of the state. Today, it is an instrument of economic development; its research serves to catalyze activity in the mineral industries; and its plans, purposes, and programs are characterized by innovation. For a number of years we have engaged in projects involving computer techniques for fundamental geologic problems. . . . For the past several years we have involved people from geophysics, statistics, petroleum engineering, mining engineering, geology, and econometrics in studies to develop methodologies for regional economic analysis.

In other words, we are dealing increasingly with the whole field of systems analysis and operations research in the economics of the mineral industries. We are launching urban development and environmental geology studies that relate to the problems of environmental health, transportation, land use, and urban and regional planning. . . .

Most of the studies that I have mentioned emphasize change and new directions. The methods used involve transference of ideas from one discipline to another. The systems approach of the engineer is evident; we have drawn heavily upon management science, biometrics, the correlation techniques of psychology, mathematics and statistics, as well as chemistry and physics.

1.3   THE NATURE OF GEOLOGICAL DATA

Geology differs from the experimental sciences in that most geological data are fragmentary and are derived from the surface manifestations of natural processes that are uncontrolled by the investigator. When a geologist inspects a particular place on the earth, he finds a unique situation developed over a long time through more processes than he can take into account. He cannot erase the natural processes that produced this environment and do it over with a simpler, controlled laboratory experiment. Nor can he observe the natural processes; most of them are finished, and the results are fixed. Furthermore, he finds that the natural processes worked to destroy or remove part of the evidence. And most of the remaining evidence is buried inaccessibly deep in the ground, while the surface outcrops are contaminated by water, weather, and the works of mankind.

So the geologist must make do with the data available, which are seldom those with which he would prefer to work. It is here that statistics may enable him to plan data collection and deduce inferences that are not readily discernible from the raw fragmentary observations that he collects.

Despite these difficulties the investigator will find before him ten thousand times more potential data than he can collect, most of it useless to his purpose. He must, first, be aware that this is the situation and then be selective in choosing the data he will collect. Sampling guided by statistical design will serve him well in choosing these data. For this reason, designed sampling and deliberate selection of data are the major concerns of this book. The available material and the purpose govern the design, the design governs the sampling, and sampling provides the data from which valid conclusions may be inferred.

The sampling results—raw data, derived observations, and conclusions— should be verified before being accepted as valid information on which to base an analysis. Verification, an essential requirement of the scientific method, is too often done informally or not at all in geology. The reader has undoubtedly visited areas whose geology does not correspond to the published maps and has searched for fossil or mineral localities whose positions are inadequately described. Even "quantitative" data such as rock analyses are often suspect, as the well-known silicate rock studies of the U.S. Geological Survey showed ("B-1" and "G-1" rocks, R. E. Stevens and others, 1960). Yet with reasonable care verification is possible. Duplicate or replicate samples or specimens can be sent to two or more analysts, petrographers, or paleontologists. If ore is sold, the mine's assay is verified by the smelter or by an umpire if buyer and seller disagree.

Classes of Geological Data

Of the many ways to classify geological data, one of the most suitable for data analysis is according to the method of collection. Four classes of data can be distinguished by the *operations* used to collect them. They are: *measurement, counting, identification,* and *ranking.*

One class is primarily derived from *measurement,* involving such operations as measuring the deflection of a needle on a dial, the width of an x-ray line, or the thickness of a sedimentary bed. Examples are compass measurements of the strike and dip of a plane or of the bearing and inclination of a line; distance measurements by scale, tape, or alidade; and microscopic measurements, including optic angle, extinction cleavage angle, and index of refraction.

A second class of data is derived primarily from *counting.* Examples are the number of zircon grains recognized in a microscope traverse, the number of right- and left-handed clam shells found in a study of life-and-death fossil assemblages, and the number of oil fields in a state.

In a third class of geological data the interest is in *identification.* For instance, in the discovery of a fossil, such as the first brachiopods found in metamorphic rocks in New Hampshire (Billings, 1935), the fact of discovery and the veracity of identification are important rather than the number of specimens or other numerical data. Many data about rocks and minerals are recognized intuitively much as one recognizes the face of a friend, rather than by a formal procedure. In this stage, common in field work, counting and measurement may be unnecessary. Just as the physician will diagnose at the bedside and later confirm his diagnosis through laboratory tests, so will the geologist confirm field identifications through tests, such as laboratory measurements on thin sections or on powders.

To these three classes of data may be added a fourth: data that can be *ranked,* where a scale of measurement is difficult if not impossible to assign. Examples of data in this category are color descriptions, "favorability" of rocks for oil or ore, and a ranking of oil fields or mineral prospects in the order of which to drill first.

Thus, as Philip Frank (1957, pp. 311 ff.) explains in his excellent book, *Philosophy of Science,* when he discusses operational definitions, scientific data are derived from many kinds of human activity and cannot be divorced from the methods of collection. Regardless of the philosophical concept of the geologist and the elegance of his scientific theory, the data to support a study are obtained by some person reading a needle on a compass dial, counting pebbles, or performing some other physical operation. The choice of operations and the details of how they are performed may be of as much or more consequence than the philosophy and theory of the geologist.

Sources of Data

Many of the difficulties in statistical analysis of geological data stem from the data sources. In science, data are derived ideally from controlled experiments rather than from uncontrolled natural processes. The monitoring of seismic quakes resulting from nuclear explosions at the Nevada test site is an example of a new and controlled experiment. In geology most data are obtained either from uncontrolled natural processes such as volcanic gases and earthquakes, or, more often, from the results of events concluded in the geologic past, for instance from basalt flows and granitized rocks. The data collected may therefore be regarded as coming from already completed as well as uncontrolled experiments. Furthermore, the data that must be used may have been collected years ago under circumstances that today are often unclear. It is a tribute to the geologist that he can make any sense at all from such inferior data. Yet such an investigation may yield valuable results, as when Chayes (1963) studied analyses of basalt made in the past from which he was able to draw valid conclusions.

Because the scientific method demands that data be verified and because the inferences of the geologist working in construction, or in oil or mineral extraction will become apparent during implementation of his decisions, the reliability of data must always be considered. This is a familiar exercise for geologists, who customarily distinguish observed, inferred, and indicated contacts in geologic mapping; differentiate proved, inferred, and indicated ore; and predict volcanic eruptions and earthquakes.

Although the geologist who uses old existing data is restricted in the breadth of his statistical analysis and is often doubtful of the validity of his conclusions, the geologist who plans to collect new data is in a better situation. He can choose those kinds of data that will illuminate a problem. He can also seek kinds of data that are easy to obtain and as useful as data that are difficult to collect. This subject is further developed in section 5.11 of this book. Whitten (1964) has also written about it.

Still another distinction about geological data has been made by Krumbein (1962, p. 1088), who contrasts *observational* data of geology, comprised of the qualitative or quantitative data of natural objects obtained in the field or laboratory, with *experimental* data obtained only in a laboratory. In the laboratory, with data obtained under controlled experimental conditions, a geological problem must be highly simplified. Indeed one experimentalist has told us, in a pessimistic moment, that the only chemical systems about which he feels confident are those of no geological interest! Krumbein (1962, p. 1088) provides an instructive table contrasting observational and experimental data.

Data Analysis

We are often approached by a geologist with a thick pile of paper representing the results of a large drilling campaign and are asked to interpret them statistically. Often we can offer very little help because the data are diffuse and the problem ill-defined. With the advent of new methods for gathering data rapidly—by electronic well logging, rapid rock analyses, and aerial geophysics—a geologist can collect data very quickly. In a recent year, one laboratory alone (the Exploration Research Laboratory of the U.S. Geological Survey in Denver, Colorado) made one and a half million chemical analyses of geological samples. Unless a geologist takes the time and has the training to plan data collection and interpretation, he is soon submerged in a mass of information, and cataloging alone will take most of his time.

For data analysis to be effective the geologist must be willing to manage the data gathering and analysis from start to finish. He must be alert for the odd findings that signal the unusual; that is, he must be serendipitous. Nothing is worse than a too-systematized data collection, such as providing printed forms with no space for extra comments, so that valuable data are neglected only because they do not fit a preconceived classification. This point is well made by Stephan and McCarthy (1958), who, in discussing the social sciences, stress that the principal investigator or project leader should do some of the data collecting himself.

Many geologists are making excellent analyses of data, with and without formal statistics. Any one of several books will give good examples of proficient analysis of geological data. Besides the books primarily on statistics in geology (sec. 1.1), three recent books are outstanding in their analysis of geological data: Irving (1964) on paleomagnetism, Ager (1963) on paleoecology, and Leopold and others (1964) on geomorphology. Among the best of the older books is one on mining geology by McKinstry (1948).

1.4 SOURCES OF DATA FOR EXAMPLES OF ANALYSES

In principle, numerical data for examples of analyses are available from most if not all fields of geology. In practice, suitable published data are hard to find, although more appear every year. Numerical data are common in hydrology and in special studies in paleontology and geomorphology. From sedimentology come data on size and shape characteristics of particles. There are many data in petroleum geology, although those most interesting are locked up in company files. Petrology also furnishes many data, although those potentially most interesting are rock analyses difficult to compare because of marked inconsistencies from one laboratory to another.

Although some example data in this book come from these fields, most are from mining geology, for two reasons. First, we have been analyzing mining data for long enough to deal with them with some assurance. Second, numerical data are abundant in mining geology. Mining geologists are concerned with sampling in order to determine if a mineral deposit is rich enough to mine and, if so, how the mine should be designed. Because of the large variability within a set of mine data and the substantial amounts of money that will be spent as a result of their conclusions, mining geologists require more data than other geologists and usually more data than investigators in other fields of the natural and social sciences.

Although many of our example data come from mining geology, the reader can readily interpret the statistical analyses in terms of any field of geology in which he is interested; for instance, our studies of distribution of metals in ore deposits are, in principle, like those in quantitative petrology. To make the transition easier, data are introduced from other fields of geology. Made-up data are designated "fictitious illustrative data" to separate them from real data.

Several examples are drawn from investigations based on assay data from the Frisco and Fresnillo mines in Mexico. Because these studies are repeatedly referred to, the geology of the ore deposits and some of their problems are summarized for convenience in this section. Other sources of data for examples are characterized briefly the first time they are mentioned.

The Frisco Mine

The Frisco mine, located in San Francisco del Oro, Chihuahua, Mexico (fig. 1.1), is 550 kilometers south of El Paso, Texas. The mine yields monthly some 70,000 metric tons of ore with a grade of 0.4 gram (0.01 ounce) gold per metric ton, 180 grams (5 ounces) silver, 5 percent lead, 0.6 percent copper, and 8 percent zinc. The primary ore minerals visible without a microscope are galena, chalcopyrite, sphalerite, and fluorite. The ore is produced from about 70 typical fissure veins that cut calcareous argillite of Mesozoic age. The mine occupies a block of ground about 3 kilometers long, 1 kilometer across, and 800 meters deep. Details are given in a general account of the geology (Koch, 1956).

The data that we have studied are assay results from 19,050 chip samples cut at 2-meter intervals in all drifts on mine levels 7, 10, 13, 14, and 15 and from 7676 mine samples cut at the same intervals in all drifts, raises, and

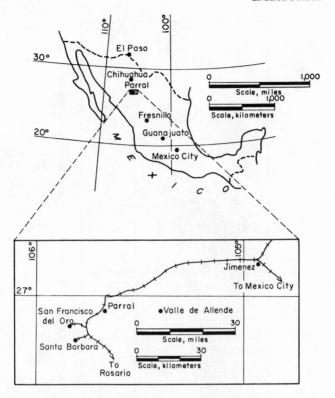

Fig. 1.1. Location map. Frisco and Fresnillo mines.

stopes on the Don Tomás and Brown veins. These levels were selected to provide information on changes in mineralization in depth, from level 7, one of the upper levels, to level 15, the lowest developed level. The Don Tomás and Brown veins were chosen as typical of large veins in this mine.

The general pattern of veins in the Frisco mine is shown in composite plan in figure 1.2 and in cross section in figure 1.3. A longitudinal section of the Don Tomás vein appears as figure 4.11.

The mining company assayed each sample for gold, silver, lead, copper, and zinc; the variables that we calculated and analyzed statistically are vein widths, metal contents (calculated by multiplying grades by vein widths), logarithms of metal contents, and logarithms of metal ratios (except those formed with gold, which are mostly indeterminate because most gold assay results were zero). Metal contents are measured in meter-grams per metric ton for precious metals and in meter-percent for base metals. The reasons

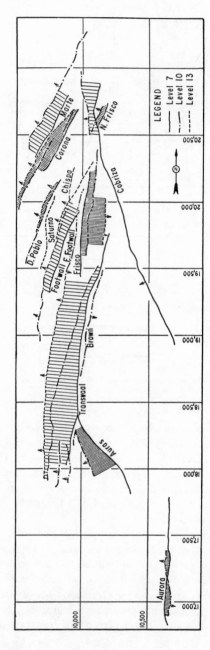

Fig. 1.2. Frisco mine. Composite plan of drifts following veins on levels 7, 10, and 13. (For clarity most short drifts and a few long ones are omitted.)

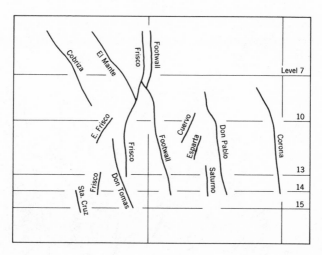

Fig. 1.3. Frisco mine. Vertical, east-west cross section at N-20,000. Observer looking south.

for working with metal contents rather than with unweighted assays are reviewed in section 13.2.

The Fresnillo Mine

The Fresnillo mine is at the city of Fresnillo (fig. 1.1), near the center of the state of Zacatecas, Mexico. Fresnillo is on the Pan-American Highway, 766 kilometers (476 miles) north of Mexico City, and 1316 kilometers (818 miles) south of El Paso, Texas.

The Fresnillo mine has been developed below vein outcrops on Proaño hill, which rises 100 meters above the general level of the plain. For the most part, the Fresnillo veins strike northwestward and dip northeast or southwest at angles ranging from 50 degrees to nearly vertical. There are 6 major and more than 50 minor veins. The veins cut graywackes, carbonaceous and calcareous shales of Mesozoic age, and sediments of Tertiary age. The primary ore minerals are galena, chalcopyrite, sphalerite, pyrargyrite, and other silver sulfides. The general geology of the mine has been described by Stone and McCarthy (1948).

The Fresnillo data that we have studied are assay results from 16,400 chip samples cut at 2-meter intervals in all drifts on the 2137, 2137 Footwall, 2200, 2630, and Esperanza veins. We treat the data in the same manner as the Frisco data described in the preceding subsection. The general pattern of these five veins is shown in composite plan in figure 1.4, in cross section in

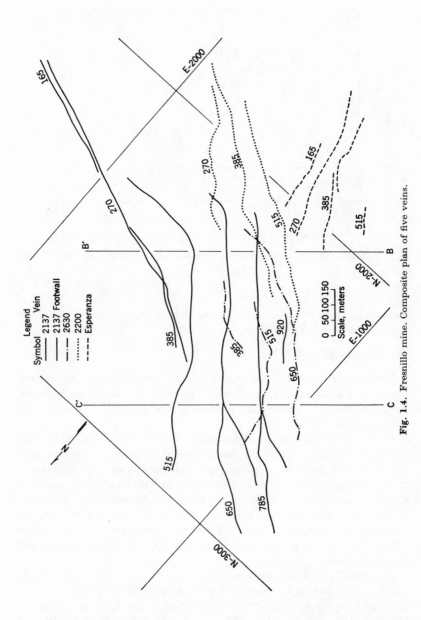

Fig. 1.4. Fresnillo mine. Composite plan of five veins.

Legend

Symbol	Vein
	2137
	2137 Footwall
	2630
	2200
	Esperanza

Scale, meters
0 50 100 150

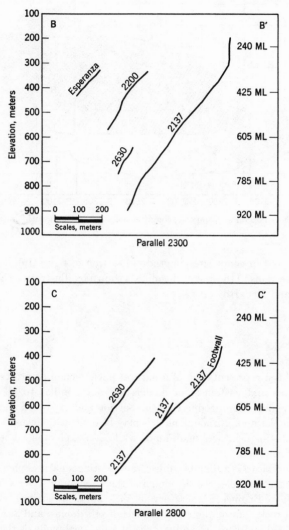

Fig. 1.5. Fresnillo mine. Vertical cross sections at Parallels 2300 and 2800. Observer looking northwest.

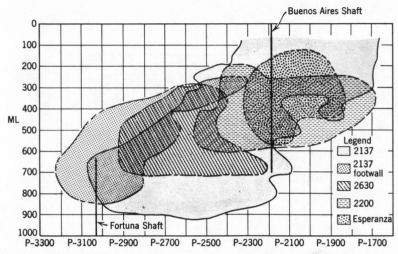

Fig. 1.6. Fresnillo mine. Composite longitudinal section. Observer looking northeast.

figure 1.5, and in composite longitudinal section in figure 1.6. The relation of these five veins to the rest of the mine is shown by illustrations in the paper by Stone and McCarthy (1948).

1.5   MODELS

A *model* is a representation of a natural phenomenon or process. Because the concepts and usefulness of models are not familiar to all geologists, elementary ideas are introduced in this section and expanded and refined in subsequent chapters. Although models may be defined and classified in many ways, we distinguish three main kinds while also explaining some principles of model formation.

A *physical model* is a tangible object representing a natural phenomenon or process. Some examples are the model of the Mississippi River system in the U.S. Corps of Engineers laboratory in Vicksburg, Mississippi; clay models simulating rock deformation; wax models of salt domes; and atomic models made of wood and wire. The advantage of a physical model is that a part of nature whose elements can be manipulated is isolated for study. Providing that the proper part of nature is selected, the principal difficulty in using a physical model is that the model is usually only a scaled representation of nature. For physical models in geology, the most troublesome scale changes are liable to be in size and time. Other scale changes, such as in temperature and pressure, may also be large but are usually easier to control.

A *geologic model* is an abstract formulation of a geologic concept that may be tested by collecting geologic data. For example, the model has been postulated (Gross and Nelson, 1966) that the change in radioactivity of present-day sediments with increasing distance from the mouth of the Columbia River is due solely to radioactive decay and that there is no mixing of material from the Hanford, Washington, nuclear reactor with previously deposited sediments. Another geologic model is the concept that every oil field consists of three essential elements: a trap, a seal, and a reservoir rock. Still another is the hypothesis that metals in some hydrothermal ore deposits, such as that at Butte, Montana, were deposited under a temperature gradient decreasing outward from a hot center.

Physical and geological models are the bases for devising mathematical models. A *mathematical model* is a set of formal rules defining exact relationships among variables in order to describe the essential elements of a physical process for a specified range of conditions. Often the relationship can be expressed in equations with mathematical symbols. An example of a mathematical model is Boyle's gas law that pressure multiplied by volume is equal to a constant, for a certain pressure range. Another example is Gross and Nelson's (1966) assertion that the relationship between radioactive concentration in sediments and the distance from the Hanford reactor is defined by the linear equation

$$y = a + bx,$$

where y is the activity ratio of Zn^{65} and Co^{60}, x is distance, and a and b are experimentally determined constants.

From the several subtypes of mathematical models, it is convenient to define two. One is the *deterministic model*, in which the relationship among variables is completely predictable so that, if one or more are known, the other or others can be exactly calculated. An example is the previous equation for radioactive concentration in sediments. The other is the *statistical* or *stochastic model*, in which the relationship among variables is not entirely predictable because a *random* or *chance element* is added to an otherwise deterministic model. An example of a random element is the "experimental error" introduced in measuring distances in surveying or in making weighings on an analytical balance. The formal definition of a random element and its relation to a mathematical model requires complicated reasoning that is explained as the subject is developed.

The purpose of a model, whether physical, geological, or mathematical, is to abstract, simplify, and organize reality in order to focus attention on one or a few factors in a geological situation. Thus the geologist, rather than trying to describe every small detail, can use a model to define a specific problem for which a reliable solution can be found.

Models, particularly geological and physical ones, have been used in geology since earliest times. However, they have been used less than in the simpler physical sciences, because to be of much value, they must deal with conceptually complex natural phenomena. Nonetheless, modern geology is making the first steps toward wide application of models, particularly mathematical ones.

To clarify the definitions and to illustrate the sequence of models that may be used in a geologic problem we discuss a simple example which utilizes three types of models, one after another. The example comes from a study of shearing stress in soil. The aim of this investigation was to prevent the failure of engineering structures caused by soil movement. The example uses both a physical apparatus and a mathematical formula of the sort developed by Terzaghi (1948; the book includes references to the original works) and other pioneer students of soil mechanics.

The first step is to devise a geological model. A list of variables and some tentative relationships among them is drawn up. Some of the variables that may affect shearing stress in soils and cause soil failure are to be found in a list of questions such as these:

1. Is the type of soil important?
2. Is the pressure on the soil a factor?
3. What is the role of water and other natural fluids?
4. Must the force applied to the soil reach a definite value before failure occurs?
5. At what displacement of the soil does rupture take place?
6. Is time a factor?
7. Does the temperature make a difference?

From these questions one may develop a geological model. Terzaghi's model was the hypothesis that shearing stress in soils is determined mainly by pressure, water content of the soil, displacement of the soil, and time. These factors can be regulated with Terzaghi's (1948, p. 79) direct-shear apparatus, one of the many devices, or physical models, used to study soil failure.

In the direct-shear device (fig. 1.7) the soil sample is placed in an apparatus consisting of a movable upper frame and a fixed lower frame. The sample is contained on the sides by the two frames and on the top and bottom by two porous, grooved stones which function to prevent slippage of the soil and to allow water to be drained from moist soils. The shearing force is applied by pulling the upper frame of the shear box. As the displacement of the upper

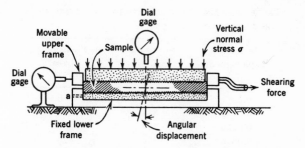

Fig. 1.7. Direct-shear apparatus (after Terzaghi, 1948, p. 79).

frame increases the force required to increase the displacement increases and approaches a maximum, as shown by the graph in figure 1.8.

By varying the experimental conditions one may use the physical model to collect data like that graphed in figure 1.8. Different types of soils with various moisture contents can be used. Different pressures can be applied by varying the load on the upper stone. Shearing force can be applied in various ways, either by a steady pull or by increments, and the time taken to apply it can be fast or slow. Temperature can be changed.

Consider the results of one series of tests (Akroyd, 1957). Horizontal pull was applied in increments, each corresponding to a horizontal displacement

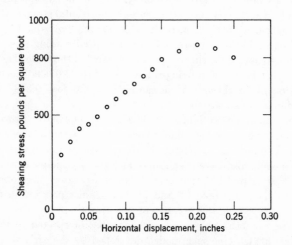

Fig. 1.8. Graph of shearing stress versus horizontal displacement in direct-shear apparatus (data from Akroyd, 1959).

TABLE 1.1. EXPERIMENTAL DATA OBTAINED IN A
DIRECT-SHEAR APPARATUS *

Horizontal displacement (in.)	Stress dial reading (1/10,000 in.)	Shearing stress (lb/ft^2)
0.0	300	0
0.0125	334	289
0.0250	342	357
0.0375	350	425
0.0500	353	451
0.0625	358	493
0.0750	364	544
0.0875	369	587
0.1000	373	621
0.1125	378	663
0.1250	383	706
0.1375	387	740
0.1500	393	791
0.1750	398	833
0.2000	402	867
0.2250	400	850
0.2500	394	799

* Data from Akroyd, 1957.

of 0.05 inch per minute. A stress dial on an instrument called a proving ring performed like a spring balance to measure the horizontal stress in the system at each displacement interval, as recorded in column 2 of table 1.1. After the initial reading of 300 was subtracted, the stress dial reading was multiplied by a constant to obtain the shear stress in column 3 of the table. The constant was found by dividing the proving-ring constant, 0.33/0.00001 inch, specific to the instrument, by the area of the sample in square feet, 0.3882, to yield a result in pounds per square foot.

When the experimental data in table 1.1 were plotted in figure 1.8, the maximum shearing stress was found to occur at a horizontal displacement of 0.2000 inch. The purpose of plotting figure 1.8 was to obtain this maximum point, which could also have been found, at least in principle, by means of a maximum reading dial like that on a maximum reading thermometer. The shear stresses listed in table 1.2 and plotted against pressure in figure 1.9 were found by repeating the experiment four times.

These data from the physical model may be interpreted in terms of a mathematical model. One such model, developed by the early investigators in soil mechanics, may be written simply as

$$w = (\tan \theta)x,$$

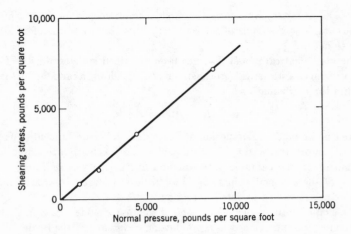

Fig. 1.9. Shear stress versus pressure (data from Akroyd, 1957).

where w is shear strength, x is pressure, and θ, named the angle of internal friction, is the angle between a straight line through the data points and the x axis (fig. 1.9). From similar experiments on cohesive rather than non-cohesive soils, a more general model can be formulated as

$$w = K_1 + (\tan \theta)x,$$

where K_1, a constant for a particular soil, is named the coefficient of cohesion.

The formulation of the mathematical model in the preceding paragraph is rather arbitrary. An alternative formulation is

$$w = K_1 + K_2 x,$$

TABLE 1.2. SHEAR STRESSES OBTAINED BY REPEATING THE EXPERIMENT OF TABLE 1.1 FOUR TIMES *

Experiment number	Normal pressure (lb/ft²)	Shear stress (lb/ft²)	Bulk density (lb/ft²)	Moisture content (%)
1	1100	869	115	29.0
2	2200	1603	116	26.8
3	4400	3580	118	29.1
4	8800	7090	118	27.8

* Data from Akroyd, 1957.

where $K_2 = \tan \theta$. However, the first formulation is more meaningful because $\tan \theta$ is the angle of repose of dry, loose sand and therefore affords a tie to nature.

The mathematical model that has been described is deterministic. It can be rewritten as a statistical (stochastic) model by adding a random fluctuation to form the new equation

$$w = K_1 + (\tan \theta)x + e,$$

where e is the random fluctuation or "experimental error," including factors such as temperature and type of soil packing not explicitly accounted for in the model. In the example data, random fluctuation was reflected by the failure of the graphed points (fig. 1.9) to lie exactly on a straight line, a discrepancy discussed in detail in section 9.2.

The simple shear device is only one of the many models used in geology. Other models are described in the following chapters and the reader will be able to think of many more, even though they may not have been formally or explicitly defined; for instance, the method of multiple working hypotheses of T. C. Chamberlain (1897) is essentially a geologic model.

In choosing between a deterministic or a statistical model for a mathematical study of a geological problem, one relies on taste and judgment, influenced above all by one's view of the real nature of the world. An example of a deterministic model is that for the accumulation of sand in dunes, the geometry of which may sometimes be rather well specified if wind direction, velocity, and distribution of sand-grain sizes are measured. Another example is the formation of coral reefs, which grow at a fairly constant rate if water temperature, salinity, etc., are constant. Examples of statistical models are those that describe the distribution of gold in conglomerate at Witwatersrand, South Africa, the dispersal of fossil populations, and the occurrence of earthquakes in California. These examples are each in a sensible category, but any one of the deterministic models may be regarded as statistical, and vice versa, to suit the purpose of the investigator.

Once a model is formulated, its correspondence to the real world is tested by gathering and analyzing data. Usually, the model turns out to be less than ideal, and the process of data collection itself suggests a better model. The sequence is like that followed by the field geologist who gathers data during the field season, interprets it the following winter, and in the next field season collects new data to test a revised hypothesis.

Models are discussed throughout this book. The first extensive and explicit treatment is in section 5.11 on linear models, and the next is in section 8.6 on experimental designs.

An outstanding proponent of using models in geology is W. C. Krumbein, whose book, *An Introduction to Statistical Models in Geology* (1965), coauthored by F. A. Graybill, summarizes thinking developed through many years. In Chapter 2 of their book models in geology are introduced with a somewhat different set of definitions and with many references to the literature.

REFERENCES

Ager, D. V., 1963, Principles of paleoecology: New York, McGraw-Hill, 371 p.

Akroyd, T. N. W., 1957, Laboratory testing in soil engineering: London, G. T. Foulis & Co., 233 p.

Billings, M. P., and Cleaves, A. B., 1935, Brachiopods from mica schist, Mt. Clough, New Hampshire: Am. Jour. Sci., 5th ser., v. 30, no. 180, p. 530–536.

Chamberlin, T. C., 1897, Studies for students: the method of multiple working hypotheses: Jour. Geology, v. 5, p. 837–848.

Chayes, Felix, 1963, Relative abundance of intermediate members of the oceanic basalt-trachyte association: Jour. Geophys. Research, v. 68, no. 5, p. 1519–1534.

Cloud, P. E., Jr., 1964, Earth science today: Science, v. 144, p. 1428–1431.

Dixon, W. J., and Massey, F. J., Jr., 1969, Introduction to statistical analysis: New York, McGraw-Hill, 638 p.

Feller, William, 1968, An introduction to probability theory and its applications, v. 1: New York, John Wiley & Sons, 509 p.

———, 1966, An introduction to probability theory and its applications, v. 2: New York, John Wiley & Sons, 626 p.

Frank, Philipp, 1957, Philosophy of science: Englewood Cliffs, N.J., Prentice-Hall, 360 p.

Griffiths, J. C., 1967, Scientific method in analysis of sediments: New York, McGraw-Hill, 508 p.

Gross, M. G., and Nelson, J. L., 1966, Sediment movement on the continental shelf near Washington and Oregon: Science, v. 154, p. 879–885.

Hambleton, W. W., 1966, Education of geologists for geological surveys: Jour. Geol. Education, v. 14, no. 3, p. 83–86.

Harbaugh, J. W., and Merriam, D. F., 1968, Computer applications in stratigraphic analysis: New York, John Wiley & Sons, 259 p.

Hazen, S. W., Jr., 1967, Some statistical techniques for analyzing mine and mineral-deposit sample and assay data: U.S. Bur. Mines Bull. 621, 223 p.

Hoel, P. G., 1962, Introduction to mathematical statistics: New York, John Wiley & Sons, 427 p.

———, 1966, Elementary statistics: New York, John Wiley & Sons, 351 p.

Irving, E., 1964, Paleomagnetism: New York, John Wiley & Sons, 399 p.

Koch, G. S., Jr., 1956, The Frisco mine, Chihuahua, Mexico: Econ. Geology, v. 51, p. 1–40.

Krumbein, W. C., 1962, The computer in geology: Science, v. 136, no. 3522, p. 1087–1092.

Krumbein, W. C., and Graybill, F. A., 1965, An introduction to statistical models in geology: New York, McGraw-Hill, 475 p.

Leopold, L. B., Wolman, M. G., and Miller, J. P., 1964, Fluvial processes in geomorphology: San Francisco, W. H. Freeman, 503 p.

Li, J. C. R., 1964, Statistical inference: Ann Arbor, Mich., Edwards Bros., v. 1, 658 p.; v. 2, 575 p.

McKinstry, H. E., 1948, Mining geology: Englewood Cliffs, N.J., Prentice-Hall, 680 p.

Miller, R. L., and Kahn, J. S., 1962, Statistical analysis in the geological sciences: New York, John Wiley & Sons, 483 p.

Mosteller, Frederick, Rourke, E. K., and Thomas, G. B., Jr., 1961, Probability with statistical applications: Reading, Mass., Addison-Wesley, 478 p.

Stephan, F. F., and McCarthy, P. J., 1958, Sampling opinions, an analysis of survey procedure: New York, John Wiley & Sons, 451 p.

Stevens, R. E., and others, 1960, Second report on a cooperative investigation of the composition of two silicate rocks: U.S. Geol. Survey Bull. 1113, 126 p.

Stone, J. B., and McCarthy, J. C., 1948, Mineral and metal variations in the veins of Fresnillo, Zacatecas, Mexico: Am. Inst. Mining Engineers Trans., v. 178, p. 91–106.

Terzaghi, Karl, 1948, Soil mechanics in engineering practice: New York, John Wiley & Sons, 566 p.

Tukey, J. W., 1962, The future of data analysis: Annals Math. Statistics, v. 33, p. 1–67.

Whitten, E. H. T., 1964, Process-response models in geology: Geol. Soc. Amer. Bull., v. 75, p. 455–464.

Wilks, S. S., 1962, Mathematical statistics: New York, John Wiley & Sons, 644 p.

UNIVARIATE STATISTICAL METHODS

Because the same statistical methods are useful for different geological problems, Part II, which contains Chapters 2 to 6, is organized statistically rather than geologically. The methods given are for the analysis of single variables.

In Chapters 2 to 5 is introduced the minimum number of statistical methods essential for conducting a serious study. The reader prepared in statistics may decide to skim over these chapters. Chapter 2 explains frequency distributions of observations, for instance, grain-size measurements or a count of oil wells per square mile, and relates empirical distributions to theoretical distributions and to probability. Chapter 3 is about statistical principles of sampling, and Chapter 4 is about statistical inferences that can be drawn from samples. Chapter 5, on the analysis of variance and the sources of variability, concludes the introduction to the statistics that must be understood before real geological data can be effectively studied.

Distributions of observations are examined in more detail in Chapter 6, and transformations are introduced, which are devices to change original observations into distributions that can be more easily handled with mathematics.

Chapter 2

Distributions

Chapter 2 introduces some methods to organize numerical information about geological phenomena and relates the resulting empirical arrangements of data to some theoretical principles of statistics and probability. Additional methods to organize geological data are given later in the book, particularly in Chapters 6, 9, 10, and 11.

2.1 OBSERVATIONS

In principle almost any geological phenomenon or item can be represented by a numerical expression. The numerical expression may be obtained by various means, including measuring, counting, or ranking. The expression may be a single number defining copper grade in a stope, the number of zircon grains in a microscope slide, or a series of numbers designating, as a vector, the plunge of a fold axis. Whatever its origin, the numerical expression is named an *observation*.

In this chapter only the technical details for manipulating observations are considered. At this point, it is not necessary to consider the important matter that the assignment of a number may be of no interest, as, for example, in the discovery of the first trilobite in Oregon, where the fact of discovery rather than the number of fossils is the point of interest, nor is it necessary to consider that the assignment of a numerical expression may be very difficult, as in describing earthquake intensity. To simplify we restrict the discussion to *univariate observations*, that is, observations that can be represented by a single number. In Chapter 9 (Vol. 2) *multivariate observations*, those represented by more than one number, are taken up for the first time. However, essentially all of the discussion of univariate observations is also required for the analysis of multivariate data.

2.2   EMPIRICAL FREQUENCY DISTRIBUTIONS

A person cannot comprehend a large number of observations, say 10,000, by reading the list of individual observations. To make sense of the numbers, one must organize them by sorting, grouping, and averaging. In this section, the use of *frequency distributions* to organize observations is explained. The object is to summarize numerical information—no profound reasoning is involved, nor are there any optimum methods to employ.

Table 2.1 is a frequency distribution for 224 furnace-shale phosphate assays from the Phosphoria formation (Permian) near Fort Hall, Idaho (Hazen, 1964, p. 6). In column 1, assay intervals of 2 percent P_2O_5 are tabulated to include the range of assays from lowest to highest. In column 2 the midpoints of the assay intervals are tabulated, and in column 3 the number of assays in each interval, named the *frequency*, is recorded. To prepare the entries for column 3 one reads the list of assays and keeps a count for each interval. In column 4 the frequency is converted to a percentage of the 224 total assays. In column 5 the cumulative frequencies obtained by summing

TABLE 2.1. Example frequency distribution of 224 phosphate assays[*]

Assay interval ($\%$ P_2O_5)	Interval midpoint, w	Frequency, f	Relative frequency ($\%$)	Cumulative frequency, c.f.	Relative cumulative frequency, r.c.f. ($\%$)
14–16	15	1	0.45	1	0.45
16–18	17	1	0.45	2	0.90
18–20	19	8	3.57	10	4.47
20–22	21	21	9.37	31	13.84
22–24	23	44	19.64	75	33.48
24–26	25	54	24.12	129	57.60
26–28	27	56	25.00	185	82.60
28–30	29	30	13.39	215	95.99
30–32	31	7	3.12	222	99.11
32–34	33	2	0.89	224	100.00

[*] After Hazen, 1964, p. 6.

the successive lines in column 3 up to the final total of 224 are recorded, and in column 6 the values in column 5 are converted to percents.

In order to prepare the frequency distribution in table 2.1, the choice had to be made of interval width and the related number of groups. By use of an interval width of 2 percent, 10 groups were obtained. In general, both practical experience and theory (Tukey, 1948) have shown that, if the observations in any set are grouped into 10 to 50 intervals, all of the essential information is preserved. Because the interval width is generally taken to be the same for all groups, the more groups chosen, the narrower the interval width. However, if a set of data contains extreme values, it is sometimes convenient to widen the widths for the intervals with small or large midpoints.

2.3 HISTOGRAMS

Because it is even easier to look at a picture than to read a table, it is generally useful to draw a *histogram* to represent the frequency table pictorially. Figure 2.1 is a histogram of the phosphate data represented by the frequency distribution of table 2.1. The diagram is made by plotting a series of contiguous rectangles whose base length corresponds to the interval width from the frequency table (all are equal if the interval widths are equal) and whose height corresponds to the number of observations in each interval.

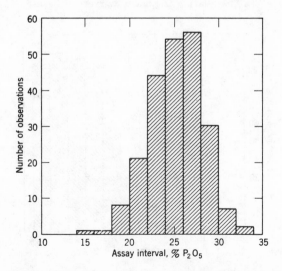

Fig. 2.1. Histogram, showing frequency distribution of 224 phosphate assays.

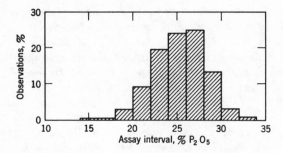

Fig. 2.2. Histogram, showing frequency distribution of 224 phosphate assays, recalculated to 100 percent area.

In figure 2.1, the height corresponds directly to the number of observations. In figure 2.2, for the same data, the height corresponds to the percentage of observations, so that the total area is 100 percent, or unity. The purpose of a unit area plot is to allow ready comparison of histograms from two sets of observations that differ in number. Because the total area is 100 percent, it is also possible to see at once from the histogram the percentage area of observations in a given range; for example, figure 2.2 shows that about 25 percent of the observations, corresponding to one-fourth the area of the histogram, are between 26 and 28 percent P_2O_5.

Figure 2.3 is a graph of the relative frequency curve from the phosphate data in table 2.1. The relative frequency curve can be used to determine

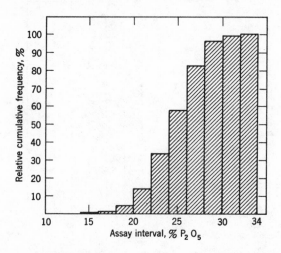

Fig. 2.3. Percent relative cumulative frequency plot for 224 phosphate assays.

directly the proportion of observations that are larger or smaller than an observation of given size. For example, figure 2.3 shows that only 4.5 percent of the observations are smaller than 20 percent P_2O_5.

2.4 THEORETICAL FREQUENCY DISTRIBUTIONS

In order to progress in interpreting observations beyond the summary afforded by the frequency table and histogram, the abstraction of the *theoretical frequency distribution* must be introduced. The starting point is the histogram of the phosphate data (fig. 2.1) which is for only 224 observations grouped into only 10 classes. If more phosphate observations were made and if more class intervals were taken at narrower widths, the rectangle heights in the histogram would take on more and more the appearance of a smooth curve. The smooth curve would resemble the dashed line on figure 2.4. Because the transition from a histogram to a smooth curve implies an infinite number of observations, it is essential to recognize that a smooth curve represents all *potential* observations of a specified kind, rather than actual observations.

Such a smooth curve is a theoretical representation, for which a mathematical model can be found, of potential observations. Thus a theoretical frequency distribution is a mathematical model which represents potential

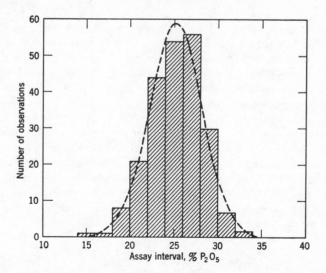

Fig. 2.4. Smooth curve fitted to histogram in figure 2.1.

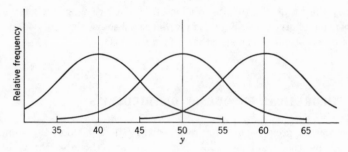

Fig. 2.5. Three normal curves with different means but the same variance (after Li).

observations. The value of a theoretical frequency distribution, in contrast to an empirical frequency distribution, is that the theoretical frequency distribution can be manipulated mathematically to develop statistical methods. Therefore a theoretical frequency distribution is essential, whether one's interest is in theoretical statistics or in applied statistics.

By way of illustration the *normal distribution*, the most important theoretical frequency distribution, is introduced. The equation for the normal distribution is †

$$f(w) = \frac{1}{\sigma\sqrt{2\pi}} \exp\left[-\frac{1}{2}\left(\frac{w-\mu}{\sigma}\right)^2\right],$$

where w is an observation, and μ and σ are variable constants, which must be evaluated to specify the distribution uniquely. Constants like μ and σ, whose values are different for different normal distributions, are called *parameters*, to distinguish them from fixed constants, which are 2, e, and π in the above formula. Regardless of the parameter values, the normal curve is always symmetrical and bell-shaped, but its form changes as the parameters change, as illustrated by figures 2.5 and 2.6 for normal curves with different parameter values. The normal distribution is also known by other names, for example, *Gaussian* and *error*. In present usage, the name is arbitrary and does not mean that other theoretical frequency distributions are abnormal. The normal curve is fully described in textbooks on mathematical statistics, for example, that by Hoel (1962, p. 98).

Not all symmetric bell-shaped distributions are normal; for example, the Cauchy distribution is bell-shaped but its equation is

$$f(w) = \frac{k}{\pi[k^2 + (w-\eta)^2]},$$

† The notation exp (x) is another form for the exponent e^x.

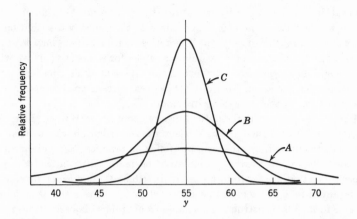

Fig. 2.6. Three normal curves with different variances but the same mean (after Li).

where k and η are parameters. Figure 2.7, a plot of a Cauchy distribution and a normal distribution, is presented to show that two distributions with different mathematical functions can look much alike.

There is, in principle, an infinite number of theoretical frequency distributions, many of which may under some conditions have frequency curves that look alike, just as in figure 2.7, which compares the normal and Cauchy distributions. From an empirical frequency distribution, there is no way to find a theoretical frequency distribution that is a *unique* representation of a set of actual observations. Nevertheless, because the detailed behavior of

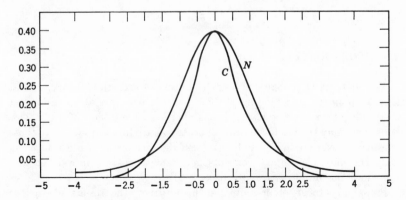

Fig. 2.7. Comparison of curves from normal and Cauchy distributions.

observations, as developed by statistical theory, depends very much on the assumed distribution, the choice of a theoretical frequency distribution has always been a matter of concern. Theoretical considerations may lead to a choice; for example, one may postulate a model for radioactive decay which yields a specific theoretical distribution, the exponential distribution, for the time of particle emission. Most often, however. there is no choice, much less a clear choice, for the theoretical distribution.

Mainly because it is mathematically convenient to work with, the normal distribution is by far the most used theoretical distribution. Fortunately, the normal distribution is one that can be applied in practice. Many observations are normally, or at least approximately normally, distributed. Even for observations that are far from normally distributed, the normal distribution can be applied if a problem can be answered by considering the behavior of a variable formed by constructing statistics from groups of observations rather than by considering the behavior of the original observations. Or, we may use a mathematical function to transform each observation into a new observation that follows the normal distribution. Moreover, many theoretical frequency distributions that are not followed by observations are themselves derived from the normal distribution, and these derived distributions may be useful even if the original observations are not normal.

Chapters 3 to 5 take up properties of the normal distribution and other distributions derived from it. Only that material necessary to supply essential tools for use on geological data is presented. Then, in Chapter 6, statistical methods are presented for observations that do not follow or even approximately follow the normal distribution. Also in Chapter 6, other distributions that have been useful in the analysis of geological data are presented, as well as mathematical transformations than can be used to change observations from other distributions into observations that follow the normal distribution.

2.5 PROBABILITY

The concepts of probability lie at the heart of statistical theory, and at least a minimal understanding of them is necessary to be able to interpret the results of statistical analyses. Probability is a large subject of study. This book gives only the briefest outline in order to introduce the subject to those who have never studied it and to refresh the memory of those who have. Only the most elementary concepts and calculations are introduced and some fictitious problems are solved.

Probability theory represents, in the abstract, the results of various physical occurrences, called *experiments*. A mathematical *point* is defined for each possible outcome, or *event*, of an experiment. Each mathematical point is therefore associated with an event. The collection or *set* of points associated with all possible events of the experiment is named the *sample space* for the experiment.

If there are two sets, named A and B, their *logical* sum, designated $A + B$, is the collection of the points in either set A or set B. Their *logical product*, designated AB, is the collection of the common points in both set A and set B. If sets A and B have no points in common, the logical product is an *empty set*, and sets A and B are said to be *mutually exclusive*.

The meaning of these terms will become clearer as the explanation of probability progresses.

Some Probability Calculations

These definitions are illustrated in the two idealized experiments of tossing two coins and tossing two dice. Two formulas for calculating probabilities, the *addition formula* and the *multiplication formula*, are introduced and illustrated. These experiments are then related to three fictitious geological illustrations.

Suppose that a coin is tossed. It may fall heads or tails. If these two outcomes are represented conceptually by two mathematical points, the points describe all possible results of the experiment and define a sample space. The phrase "all possible results" is an oversimplification, of course, since the points do not reflect such real happenings as a coin landing on edge or being lost under the table. The simplifications will help the reader to understand probability theory, which can then be applied to practical statistical problems from real-life situations.

If two coins are tossed, the sample space contains four points: HH, HT, TH, and TT (H means heads and T means tails). The event HH means that both coins landed heads, the event HT means that the first landed heads but the second landed tails, the event TH means that the first landed tails but the second landed heads, and the event TT means that both landed tails. If five coins are tossed, the sample space contains 32 points.

Suppose that two ordinary dice, each a cube whose sides are numbered 1 to 6, are tossed. The sample space has 36 points, as shown in figure 2.8, corresponding to the 36 possible combinations of points on the two dice. Thus $(1, 1)$ means that both dice are one's, $(1, 2)$ that the first die is 1 and the second die is 2, $(2, 1)$ that the first die is 2 and the second die is 1, etc. As before, the experiment has been idealized in that one die could have fallen under the table or leaned up against the wall so that the experimenter could not decide which side was up.

In these experiments with coins and dice, and in real experiments, such as drilling oil wells or sending prospecting teams to the field, numbers named

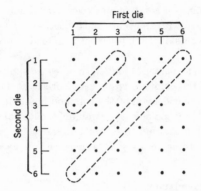

Fig. 2.8. Sample space for tossing two dice.

probabilities may be associated with each of the points of the sample space. Probabilities must be positive or zero numbers and their sum taken over the whole sample space must be 1. Since each event is represented by one or more points in the sample space, the probability of any one event is the sum of the probabilities of the points for that event. If only one point is associated with an event, the probability associated with the point is the probability of the event.

The *addition formula* allows the probabilities for some events to be calculated when the events are represented by two or more points. Consider the sample space associated with the two coins. Let A designate the event that coin 1 lands heads and B designate the event that coin 2 lands heads. What is the probability of the event, designated $(A + B)$, that *at least* one of the two coins lands heads? This situation is represented by figure 2.9, in which

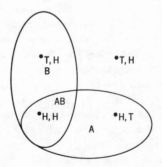

Fig. 2.9. Sample space for two coins, A and B.

events A and B are circled. The event $(A + B)$ is the logical sum of the events A and B. The addition formula, which relates these probabilities, can be written

$$P(A + B) = P(A) + P(B) - P(AB),$$

where $P(A + B)$ is the probability of the event $(A + B)$, $P(A)$ is the probability of event A, $P(B)$ is the probability of event B, and $P(AB)$ is the probability of both events. The event $(A + B)$ is that at least one coin lands heads, the event A is that the first coin lands heads, the event B is that the second coin lands heads, and the event AB is that both coins land heads.

If the events A and B are *mutually exclusive*, $P(AB)$ is zero, because no points are associated with AB, and the addition formula becomes

$$P(A + B) = P(A) + P(B).$$

The probability of the sum of any number of mutually exclusive events is simply the sum of the probabilities associated with the events; that is, by applying the addition formula,

$$P\left(\sum C_i\right) = \sum P(C_i),$$

where C_i denotes the events from C_1 to C_n.

The *multiplication formula* is used to calculate the probabilities of two events which occur simultaneously. Thus AB is the probability that both coins, tossed simultaneously, will land heads. The multiplication formula introduces the concept of *conditional probability*, which is the probability of an event, *given* the fact that another event has already occurred. The probability that coin 1 lands heads $[P(A|B)]$, *given* that coin 2 has already landed heads, is

$$P(A|B) = \frac{P(AB)}{P(B)},$$

which, rewritten, becomes

$$P(AB) = P(A|B)\ P(B),$$

the *multiplication formula*, and, by symmetry,

$$P(AB) = P(B|A)\ P(A).$$

If $P(A|B) = P(A)$, the events A and B are said to be *independent* in a probability sense, and

$$P(AB) = P(A)\ P(B).$$

Both the addition and multiplication formulas are illustrated by the following example. When two dice are tossed, it is reasonable from symmetry to believe that each of the 36 events (fig. 2.8) is equally probable, so that the probability associated with each of the 36 points is 1/36. Through the addition formula, the probability of the sum of the two dice being 7 can readily be calculated; the sample space is encircled in figure 2.9. A 7 corresponds to one of the six mutually exclusive events (1, 6), (2, 5), (3, 4), (4, 3), (5, 2), and (6, 1). Thus the probability of a sum of 7 is simply

$$P(7) = 6(1/36) = 1/6.$$

Next consider the calculation of the probability of the event (1, 6) through the multiplication formula, by considering separately the two events of getting a 1 on the first die and a 6 on the second die. For each die the probability associated with the event of getting a 1 or a 6 is 1/6 because there are six faces of each die and therefore six points in the sample space. Into the multiplication formula

$$P(AB) = P(A) \, P(B),$$

are substituted numerical values to yield

$$P(1, 6) = P(1) \, P(6) = (1/6)(1/6) = 1/36.$$

The concept of independence is intuitively apparent in this result since the one die is obviously not controlled by the other.

In geological applications of probability one of the most important concepts is that of *odds*, for instance, the odds that a fossil will be found in a certain outcrop or that an exploration program will discover a profitable oil field. The word odds in probability has the same meaning as in everyday life and may always be applied to the probability associated with two events, one of which is certain to happen. In the previous illustration, the probability of getting a 1 or a 6 was 1/6; therefore the odds of getting a 1 and a 6 are

$$P(1, 6) + P(6, 1) = 1/36 + 1/36 = 1/18,$$

or 5.56 percent. In general, if the odds of event A to event B are x to y, the probability of event A is $x/(x + y)$ and that of event B is $y/(x + y)$. If the probabilities of two events are known, the corresponding odds are simply the ratio of the two probabilities.

The experiments with coins and dice may readily be related to a fictitious geological illustration. Suppose that instead of a coin being tossed, a wildcat oil well is drilled. It can be classified as a discovery well or a dry hole, just as the coin can be classified as heads or tails. As in the coin example, it may not be possible in practice to classify the oil well since the records may have been lost or because only traces of oil were found. Similarly, instead of two coins being tossed, two wildcat oil wells might be drilled, or, instead of two dice being tossed, two states might be explored for iron ore, the value of which could be referred to six classes, each with an equal chance of occurring. The calculations made for the coins and dice apply equally well to these fictitious geological situations. The reader will find it instructive to rework the calculations in terms of these or other geological situations of interest to him.

Finally, *Bayes' theorem*, a formula for the calculation of conditional probabilities, is introduced. (An application for it is described in sec. 4.8.) The theorem may be stated as follows: If A_n is a series of mutually exclusive events such that $\sum P(A_n) = 1$, and if B is any other event,

$$P(A_n|B) = \frac{P(B|A_n) \, P(A_n)}{\sum P(B|A_i) \, P(A_i)}.$$

Bayes' theorem, fully explained by Feller (1957, p. 114) and Lindley (1965, p. 114) may be illustrated by a simple fictitious illustration.

Table 2.2 represents the situation that a geologist collected 60 fossils that are divided between two species, B and C, from three outcrops, designated A_1, A_2, and A_3. There were 10 fossils at A_1, 20 at A_2, and 30 at A_3. If one of the 60 fossils is picked at random, the probability $P(A_1)$ that it came from outcrop A_1 is $10/60 = 1/6$. Similarly, $P(A_2) = 1/3$, and $P(A_3) = 1/2$. Moreover, given that a fossil is from outcrop A_1, the conditional probability $P(B|A_1)$ that it is of species B is $7/10$. Similarly, $P(B|A_i)$ can be calculated for outcrops A_2 and A_3. The probability statements are listed in table 2.2. Now, by using Bayes' theorem, the probability that a given fossil of species B came from outcrop A_1 can be calculated. The arithmetic is shown in the table and yields the result $7/27$. This development was given to illustrate Bayes'

TABLE 2.2. FICTITIOUS ILLUSTRATION OF COLLECTING FOSSILS, TO ILLUSTRATE BAYES' THEOREM

Outcrop names	Number of observations	Number of fossils	
		Species B	Species C
A_1	10	7	3
A_2	20	10	10
A_3	30	10	20

Probability statements:

$P(A_1) = 1/6$	$P(A_2) = 1/3$	$P(A_3) = 1/2$
$P(B|A_1) = 7/10$	$P(B|A_2) = 1/2$	$P(B|A_3) = 1/3$

Calculation of probability $P(A_1|B)$:

$$P(A_1|B) = \frac{7/10 \times 1/6}{7/10 \times 1/6 + 1/3 \times 1/2 + 1/2 \times 1/3} = 7/27 = 0.259$$

theorem; for this fictitious illustration, the probability 7/27 can be obtained directly, because there are 27 fossils of species B altogether, and 7 of these were found at outcrop A_1.

REFERENCES

Feller, William, 1968, An introduction to probability theory and its applications, v. 1: New York, John Wiley & Sons, 509 p.

Hazen, S. W., Jr., 1964, Statistical analysis of some sample and assay data from the bedded deposits of the Phosphoria Formation in Idaho: U.S. Bur. Mines Rept. Inv. 6401, 29 p.

Hoel, P. G., 1962, Introduction to mathematical statistics: New York, John Wiley & Sons, 427 p.

Lindley, D. V., 1965, Introduction to probability and statistics from a Bayesian viewpoint, pt. 2, inference: Cambridge Univ. Press, 286 p.

Tukey, J. W., 1948, Approximate weights: Annals Math. Statistics, v. 19, p. 91–92.

Chapter 3

Sampling

Chapter 3 is about statistical principles of sampling. First some statistical principles that can help the geologist devise a sampling plan are explained in sections 3.1 and 3.2. In the remainder of the chapter organization and preliminary interpretation of the data obtained from sampling are discussed.

3.1 POPULATIONS AND SAMPLES

In Chapter 2 some simple methods to organize and summarize a set of numerical geological data are explained. Generally, a set of data represents only a small fraction of the total number of observations that might in principle be obtained about the subject of interest. The total set of potential observations is named a *population*, and the set of actual observations in hand is named a *sample* from this population.

As explained in detail in section 3.2 and throughout the book, a population is just what the investigator chooses to make it through a definition that is arbitrary, although it should be thoughtful and purposeful. If analyses of quartz percentage are made for several hand specimens from a batholith, the total population of observations may be the set of hand specimens that could be taken from the rock body; the sample of quartz percentage values consists of the small proportion of values actually obtained. If microfossil counts are made for several specimens of sediment, the total population may be defined to be the set of specimens that could conceivably be collected; the sample consists of the small number of counts actually obtained.

Unfortunately the statistical definition of *a sample* as a group of observations does not agree with the geological usage of sample as a single hand specimen or bag of rock chips yielding one observation on chemical analysis, assay, or petrographic study. Therefore, whenever there is a possibility of confusion, the clumsy terminology *statistical sample* or *geological sample* is adopted.

Now, several questions arise. If the set of observations at hand is only a sample of the total information of interest, what conceptual framework encompasses all of the information? Moreover, if the set of observations is only a sample, what can be inferred from the sample about the population? When the behavior of all possible samples from a population is discovered, then it is possible to learn what information about a population is conveyed by one particular statistical sample.

This chapter is mainly statistical and the mathematics is purely deductive. Geology is almost always concerned with the reverse process of making inductive inferences about populations from a single statistical sample. Many or all possible statistical samples in this chapter are considered in order to learn how they behave, not to devise a direct guide for action. In later chapters the geological consequences of the material in this chapter are discussed. If, as rarely if ever happens, an entire population is available for study, population *parameters* can be calculated directly from the available observations. Moreover, if this population is described by a theoretical frequency distribution, the values of the population parameters are known.

The normal distribution has two parameters, the mean and the variance, which are defined in section 2.4. Almost always, however, the set of observations under study constitutes only a sample of the entire population that interests the investigator, and it is therefore necessary to estimate the parameters by corresponding numbers named *statistics* that are derived from the sample. In this book the common convention of designating parameters by Greek letters and statistics by corresponding English letters is followed; for example, s^2 designates a statistic derived from a sample that estimates the parameter σ^2. The notation is consistent with that of Li (1964).

3.2   TARGET AND SAMPLED POPULATIONS

In section 3.1 the concept of a population comprising the total number of potential observations is introduced and contrasted with the statistical sample of observations actually taken. In this section, two kinds of populations are distinguished, the one from which the sample is actually drawn, named the *sampled population*, and the generally larger one of geological interest, named the *target population*. These two types of populations were distinguished and named by Cochran et al. (1954, p. 18), who writes as follows:

> We have found it helpful in our thinking to make a clear distinction between two population concepts. The *target* population is the population of interest, about which we wish to make inferences or draw conclusions. It is the population which we are trying to study. The *sampled* population requires a more careful definition but, speaking popularly, it is the population which we actually succeed in sampling.
>
> The sampled population is an important concept because by statistical theory we can make quantitative inferential statements, with known chances of error, from sample to sampled population. It must be carefully distinguished from the target population, the population of interest, about which we are tempted to make similar inferential statements.
>
> Insofar as we make statistical inferences beyond the sample to a larger body of individuals, we make them to the sampled population. The step from sampled population to target population is based on subject-matter knowledge and skill, general information, and intuition—but not on statistical methodology.

The terms target and sampled populations were introduced into geology by Rosenfeld (1954), and the concept has been extensively used, particularly by Krumbein (1960; 1965, pp. 147–169) and his students. Krumbein's discussions, which are particularly clear and thoughtful, merit close attention.

We compare target and sampled populations here by presenting examples. Because of the importance of the subject and the complicated nature of the argument, the discussion is purposely made detailed. As an example of a target population, we choose the Jurassic Morrison Formation, which contains the source beds for many uranium deposits in the western United States. Assume that the problem is to sample the formation for uranium. To make the problem more definite, assume further that the formation is to be drilled to obtain NX core that will be cut transversely to yield geological samples consisting of rock cylinders each 3 inches long by about 2 inches in diameter. However the population is defined, the Morrison Formation clearly contains an essentially infinite number of such cylinders.

The first task is to define precisely the target population. Although the definition might appear easy because the entire formation is of interest, it at once becomes necessary to choose among alternatives, including:

1. The entire Morrison Formation as originally deposited in Jurassic time.

2. The entire Morrison Formation excluding that part removed by erosion or stoped out by magma since Jurassic time.

3. The Morrison Formation excluding the part covered by younger rocks and colluvium.

4. The Morrison Formation excluding the part covered by younger rocks.

5. The Morrison Formation available for mining under certain specifications, e.g., covered by at most 500 feet of younger rocks, in areas not withdrawn from mining by law, or in areas not owned by competing mining companies.

The list is not intended to be complete, and it is at once clear that there are additional alternative definitions of the target population, as well as combinations among the listed alternatives. Decisions among some alternative definitions may require geological judgment; for example, has a certain rock formation been dissolved by magma or is the rock found today a granitized sedimentary rock formation? Other decisions may require information that can be obtained only in the course of field sampling, for example, the depth below the surface of a formation. These complications are purposely introduced to emphasize that the definition of the target population may be difficult. Yet in the appropriate definition may lie the best way of posing a problem to obtain an interesting or useful answer.

By and large the definition of a target population relies on geological judgment and knowledge, which often must be applied with considerable skill. In specifying an appropriate target population, one can often formulate in geological terms a good question, the answer to which will illuminate the science. The purpose of the sampling must be kept in mind. In order to indicate that the problem is a real one, a real example may be mentioned. We once wrote (Link and Koch, 1962) that, in sampling a granite body, "if there were extraneous material at a grid point, for example a dike, there is a real question as to whether the dike should in fact be sampled. . . ." Another worker (Exley, 1963, p. 650) commented that "the question as to whether or not a dike should be sampled in a granite survey simply does not arise; the dike is not the granite and is excluded," a comment that indicates how misunderstanding can stem from different concepts of target population. If a granite is to be sampled as a potential uranium source and if the dikes would be mined along with the granite, the dikes certainly should be included, although perhaps the samples should be kept separate. If a granite is to be sampled to determine its origin and if the decision was made before sampling that the dikes were post-granite, the dikes certainly should be excluded. If a granite is to be sampled to determine its origin and if the relative age of granite and dike were unknown (possible post-dike granitization), probably the dikes should be included, although the samples should be kept separate.

Thus, after some travail, a target population can be defined precisely. The definition should never be made in a routine way. If a target population is not defined explicitly and precisely, it will be defined implicitly and imprecisely, both by the worker and by those concerned with the result; and if the definitions are hazy, they will lead to conflict. Therefore the geologist should think first—rather than just set off, Brunton compass and topographic map in hand, to collect hand specimens.

Once a target population is defined the next step is to define the sampled population. Some of the target population may clearly be unavailable for sampling, for example, because of time. It is too late to sample the part of the Morrison Formation that has been eroded away, and it is too early to sample the gas from next year's volcanic eruption in Guatemala. Some of the target population may be expensive or inconvenient to sample. For the example of the Morrison Formation the sampled population might be defined in one of the following ways:

1. The entire formation exposed in outcrop, in outcrop below colluvium, and in suboutcrop to a depth of 500 feet.
2. The entire formation exposed in outcrop and in outcrop below colluvium.
3. All outcrops in the state of Utah.
4. All outcrops in roadcuts.
5. All outcrops within 1 mile of a road.
6. All outcrops below 10,000 feet elevation.

The connection between the sampled and the target population must be made for geological reasons. Particularly because of the time element, in geology the sampled population can seldom if ever correspond to the target population. The connection must be made on geological rather than on statistical grounds, although the statistician may be able to offer some help from his general experience. However, geological problems tend to be more complicated and the populations less accessible for sampling than in such simpler situations as sampling cars of coal delivered to a steam generating station, sampling light bulbs, or quality control in a brewery.

Thus, in defining a sampled population for the Morrison Formation, such questions as the following would have to be faced:

1. Is the Morrison Formation different at higher topographic elevations, because it is more resistant to erosion, or for another reason?
2. Even though it may be almost impossible to sample in river valleys, is the formation different there because of zones of weakness, or fractures that may carry mineralization?
3. How deep does weathering extend? It will certainly be absent if the formation is sampled deep enough, but will such sampling be prohibitively expensive?
4. If the sampled population is restricted to outcrops in Utah, perhaps because the work is done by the Utah Geological Survey, how applicable will

the results be to Colorado ? Can a larger scale subsampling scheme be set up to extend the validity of results into Colorado ?

All these questions are basically geological, whether posed by a geologist or by some other person such as a mining engineer or geophysicist.

Definitions of target and sampled populations and distinctions between them are taken up repeatedly in this book. Exploration of these subjects will sharpen one's geological thinking and improve his proficiency as a geologist. In Chapter 7 on geological sampling the two kinds of populations are treated further in some detail.

3.3   MEAN, VARIANCE, STANDARD DEVIATION, AND COEFFICIENT OF VARIATION

In Chapter 2 ways to organize and summarize data by a frequency table and a histogram are explained. Here, methods of formulating numerical expressions that summarize information about observations in mathematical terms, which can be manipulated, are introduced. For many types of observations, particularly those that follow the normal distribution, the important summary expressions are the *mean* (a measure of central tendency), the *variance* (a measure of spread of a distribution), the *standard deviation* (the square root of the variance), and the *coefficient of variation* (the ratio of standard deviation to mean).

These summary expressions are only a few of the many that have been devised by statisticians; however, they constitute those that are generally of most use for geological data. One additional expression, the *correlation coefficient*, is introduced in section 9.1. Information on other more specialized expressions, such as measures of skewness, are given by Hoel (1962) and in other books on mathematical statistics.

If all the observations of a population are available, the population mean is simply the arithmetic average calculated by dividing the sum of the values of all observations by the number of observations; that is,

$$\mu = \frac{\sum w}{n},$$

where μ is the population mean, w is an observation, and n is the number of observations. If the population is represented by a theoretical frequency distribution, n is indeterminate, because the population is conceptually infinite in size, and therefore the simple formula above cannot be applied. In this case the population mean may be obtained by the formula

$$\mu = \int_{-\infty}^{\infty} w f(w) \, dw,$$

where $f(w)$ is the expression for the particular theoretical distribution. For example, for the normal distribution introduced in section 2.4, the formula obtained is

$$\mu = \int_{-\infty}^{\infty} \frac{w}{\sigma\sqrt{2\pi}} \exp\left[\frac{1}{2}\left(\frac{w - \mu_0}{\sigma}\right)^2\right] dw.$$

When this integral is evaluated, the result obtained is

$$\mu = \mu_0.$$

Thus the parameter μ in the normal distribution is in fact the population mean.

If, as is almost always the case, the set of observations is a sample rather than a population, the mean is computed by the parallel formula

$$\overline{w} = \frac{\sum w}{n}$$

where the mean is designated by the English letter $\overline{w}$ rather than by the Greek letter μ. The relation of $\overline{w}$ to μ depends on many things, and investigation of this relation is an important subject that reappears repeatedly in this book. In table 3.1, the calculation of the sample mean $\overline{w}$ is illustrated for a sample with $n = 5$, $\sum w = 25$, and $\overline{w} = 5$.

TABLE 3.1. CALCULATION OF MEAN AND VARIANCE FROM A FICTITIOUS SAMPLE OF SIZE 5

	Observations, w	Observations squared, w^2	Deviation from mean, $(w - \overline{w})$	Squared deviation from mean, $(w - \overline{w})^2$
	2	4	-3	9
	4	16	-1	1
	6	36	1	1
	6	36	1	1
	7	49	2	4
Sum	25	141	0	16

Calculations:

$$\overline{w} = 25/5 \qquad s^2 = 16/4$$
$$= 5 \qquad\qquad = 4$$
$$s = \sqrt{4}$$
$$= 2$$

Although the calculation of a mean is commonplace, the calculation of a variance may be less so, although the concept of variation is familiar. The idea that observations vary is familiar, for no one expects all men to be of the same height, all eggs to be identical in size, or all molybdenum assays to be of the same value. In order to measure the spread of observations, some calculation is required of how close on the average the observations are to the mean. One measure of variation might be obtained by subtracting the mean from each observation and summing according to the expression

$$\sum (w - \mu).$$

However, this formula is useless because the sum is equal to 0 for all popalations. In order to overcome this difficulty, each term $(w - \mu)$ is squared to make its value positive and to yield the expression

$$\sum (w - \mu)^2.$$

The sum of squared deviations from the mean is used rather than the sum of absolute values, $\sum |w - \mu|$, because it is a more convenient quantity to manipulate mathematically. To obtain an average measure of variation, one divides the expression $\sum (w - \mu)^2$ by the sample size n according to the formula

$$\sigma^2 = \frac{\sum (w - \mu)^2}{n},$$

where σ^2 is named the population variance. Thus, the population variance σ^2 is the average squared deviation of observations from the population mean. The square root of the variance σ^2 is the standard deviation σ.

If the population is represented by a theoretical frequency distribution, n is indeterminate, and the formula for the variance becomes

$$\sigma^2 = \int (w - \mu)^2 f(w) \, dw,$$

where, again, $f(w)$ is the expression for the particular theoretical frequency distribution, and the integration extends from $-\infty$ to $+\infty$. For the normal distribution, the formula obtained is

$$\sigma^2 = \int_{-\infty}^{\infty} \frac{(w - \mu)^2}{\sigma_0 \sqrt{2\pi}} \exp \left[-\frac{1}{2} \left(\frac{w - \mu}{\sigma_0} \right)^2 \right] dw,$$

and when the integral is evaluated, the result obtained is

$$\sigma^2 = \sigma_0^2.$$

Thus the parameter σ^2 in the normal distribution is in fact the population variance.

If the observations constitute only a sample rather than a population and if, as almost always happens, the population mean μ is unknown, the variance is computed as

$$s^2 = \frac{\sum (w - \overline{w})^2}{n - 1},$$

where s^2 is the sample variance, and its square root s is the sample standard deviation. The term $(n - 1)$ rather than n is used as the divisor, as explained in section 3.7. In table 3.1 the sample variance is calculated for the sample of five observations. The mean deviation, tabulated in column 3, is 0 and is therefore useless for measuring variation. Dividing the squared mean deviation, which is 16, by $(n - 1)$ which is 4, yields an s^2 value of 4. The numerator $\sum (w - \overline{w})^2$ in the formula for sample variance is called the *sum of squares*, which is a short form for "sum of squared deviations from the sample mean" and is designated SS. The method of calculating sample variance in table 3.1, given to illustrate the principle, is inconvenient to use in practice, because the sample mean must be calculated before the sum of squares can be calculated. In table 3.2 a short-cut procedure is given that is well suited for use with a

TABLE 3.2. CALCULATION OF VARIANCE FROM THE FICTITIOUS
SAMPLE OF TABLE 3.1 BY SHORT-CUT COMPUTING FORMULA

$$
\begin{aligned}
\text{SS} &= \sum (w - \overline{w})^2 \\
&= \sum w^2 - (\sum w)^2/n \\
&= 141 - (25)^2/5 \\
&= 141 - 125 \\
&= 16 \\
s^2 &= 16/4 \\
&= 4
\end{aligned}
$$

desk calculator (but not with a digital computer, Chap. 17). The short-cut procedure depends on the algebraic identity

$$\text{SS} = \sum (w - \overline{w})^2 = \sum w^2 - \frac{(\sum w)^2}{n},$$

which is proved in statistical textbooks (Li, 1964, I, p. 89). In table 3.2, this formula, which is applied to the data of table 3.1, yields an SS value of 16 and the same value of 4 for s^2.

The mean, variance, and standard deviation of a sample are calculated from the observations. If the observations are changed to new ones by adding a constant to each observation or by multiplying each observation by a constant, the mean, variance, and standard deviation, in general, also change.

If all of the original observations w are changed to new observations u by adding a constant a,

$$u = w + a,$$

the new mean is equal to the original mean plus the constant,

$$\overline{u} = \overline{w} + a,$$

and the variance and standard deviation are unchanged,

$$s_u^2 = s_w^2,$$
$$s_u = s_w.$$

If all of the original observations w are changed to new observations u by multiplying by a constant b,

$$u = bw,$$

the new mean is equal to the original mean multiplied by the constant,

$$\overline{u} = b\overline{w},$$

the new variance is equal to the original variance multiplied by the constant squared,

$$s_u^2 = b^2 s_w^2,$$

and the new standard deviation is equal to the original standard deviation multiplied by the constant,

$$s_u = b s_w.$$

The change in an original observation by introducing one or more constants by addition or multiplication is an example of a *linear transformation* (Chap. 6).

An illustration of linear transformations applied to the data of table 3.1 is given by table 3.3. In column 2, the constant 5 is added to each observation; in column 3, each observation is multiplied by the constant 4; and in column 4, each observation is multiplied by the constant 4, and the constant 5 is also added. The calculations verify the remarks in the preceding paragraph.

An especially important linear transformation is to subtract the population mean μ, which is a constant, from the observations w, and divide the result by the population standard deviation σ, which is also a constant, according to the formula

$$u = \frac{w - \mu}{\sigma}.$$

This transformation converts a set of w values into u values whose mean is 0 and whose variance is 1. In the first part of this transformation,

$$u' = w - \mu,$$

TABLE 3.3. EFFECT ON MEAN AND VARIANCE CAUSED BY CHANGING THE OBSERVA-
TIONS IN THE FICTITIOUS SAMPLE OF TABLE 3.1 BY ADDING OR MULTIPLYING BY A
CONSTANT

	Original observations, w		Observations + 5, $u = w + 5$	Observations × 4, $u = 4w$	Observations × 4 and + 5, $u = 5 + 4w$
	2		7	8	13
	4		9	16	21
	6		11	24	29
	6		11	24	29
	7		12	28	33
$\sum w$	25	$\sum u$	50	100	125
n	5		5	5	5
$\overline{w}$	5	$\overline{u}$	10	20	25
$(\sum w)^2$	625	$(\sum u)^2$	2,500	10,000	15,625
$(\sum w)^2/n$	125	$(\sum u)^2/n$	500	2,000	3,125
$\sum w^2$	141	$\sum u^2$	516	2,256	3,381
SS	16		16	256	256
d.f.	4		4	4	4
s^2	4		4	64	64
s	2		2	8	8

where u' is the intermediate u value. The mean of u', designated $\mu_{u'}$, is 0 because

$$\mu_w = \mu,$$

and the variance of u', designated $\sigma_{u'}^2$, is still σ^2. In the second part of the transformation, u' is multiplied by $1/\sigma$, to yield the final result, that is,

$$u = u' \frac{1}{\sigma}.$$

Then, because the new variance is equal to the original variance multiplied by the constant squared, the variance of the final result u is

$$\sigma_u^2 = \frac{1}{\sigma^2} \sigma_{u'}^2,$$

but, because $\sigma_{u'}^2$ is equal to σ^2,

$$\sigma_u^2 = \frac{1}{\sigma^2} \sigma^2 = 1.$$

This important transformation is applied in section 3.6.

The *coefficient of variation*, the ratio of standard deviation to mean, is a useful measure of relative variability of observations, provided that they are all either positive or negative. For a population, the coefficient of variation is found by the formula

$$\gamma = \frac{\sigma}{\mu},$$

and, for a sample, the coefficient of variation is found by the formula

$$C = \frac{s}{\bar{w}}.$$

For geological data, most of which are positive, and whose variability tends to be larger than that of many other kinds of data, the coefficient of variation is an important measure that is used repeatedly in this book, particularly in Chapter 6, as a guide to whether to transform observations nonlinearly, and in Chapter 8, to evaluate variability in geological data.

If a frequency distribution of observations has been made, an alternative calculation of the elementary statistics discussed in this section may be made. As an example, phosphate assays from section 2.2 have been used. When the calculations are performed on the original data, the results obtained are

$$\bar{w} = 25.13$$
$$s = 2.95$$
$$C = 0.117.$$

The calculations can also be performed on the classed data, as shown in table 3.4, which is part of table 2.1 with two additional columns to give the results of multiplying the frequency in each class by the midpoint of the interval and

TABLE 3.4. Calculation of mean and variance for classed data from figure 2.1

Assay interval	Interval midpoint, w	Frequency, f	fw	fw^2
14–16	15	1	15	225
16–18	17	1	17	289
18–20	19	8	152	2,888
20–22	21	21	441	9,261
22–24	23	44	1,012	23,276
24–26	25	54	1,350	33,750
26–28	27	56	1,512	40,824
28–30	29	30	870	25,230
30–32	31	7	217	6,727
32–34	33	2	66	2,178

Calculations:

$$\overline{w} = \sum fw / \sum f$$
$$\sum fw = 5652$$
$$\sum f = 224$$
$$\overline{w} = 25.23$$

$$s^2 = \frac{SS}{\sum f - 1} = \frac{\sum fw^2 - (\sum fw)^2 / \sum f}{\sum f - 1}$$

$$\sum fw^2 = 144{,}648$$
$$(\sum fw)^2 = 31{,}945{,}104$$
$$(\sum fw)^2 / n = 142{,}612$$
$$SS = 2{,}036$$
$$s^2 = 9.13$$
$$s = 3.02$$
$$C = 0.119$$

by the squared midpoint of the interval. When the calculations are performed as indicated in the table, the new results obtained are

$$\overline{w} = 25.23$$
$$s = 3.02$$
$$C = 0.119.$$

It can be seen that the two methods of calculation yield essentially the same results. The classed method is useful for desk-calculator computations or if only classed data are available; otherwise, with a digital computer it is generally easier to work with the individual observations.

This section will now be summarized. For any set of observations, the concepts of average value and variation of individual observations from average value are familiar ones. Based on these concepts are two specific measures that are particularly useful, the mean and the variance. For a population, the mean is μ and the variance is σ^2, corresponding to the same parameters in the theoretical normal distribution. For a sample, the mean is $\overline{w}$ and the variance is s^2. The standard deviation, the square root of the variance, is σ or s. The coefficient of variation, the ratio of standard deviation to mean, is γ for a population and C for a sample.

3.4   RANDOM SAMPLES

Before considering the behavior of samples from a population, in this section we first explain the method used to obtain a *random sample*, sometimes named a probability sample. The sampled population is a collection of potential observations, generally defined to be infinite in number through appropriate definition of the sampled population. If a scheme is devised to take a fraction of these observations so that each potential observation has an equal chance of being selected, the sample obtained is named a *random sample*. All statistical theory depends essentially on the behavior of random samples. When samples have been selected by nonrandom processes, the validity of the conclusions is always indeterminate, although the consequences, as discussed later (especially in secs. 7.2 and 8.6), may or may not be serious.

In order to take a random sample from a population, unambiguous rules must be thoughtfully made and carefully followed. One useful device is to assign to each item in the population a unique serial number in addition to the number or numbers representing the observation; for example, everyone eligible to collect a social security payment in the United States has a social security number in addition to the monthly payment, if any, that he receives. The social security numbers are serial numbers, and the monthly payments, including the zero ones, are observations. Although many individuals may be eligible to receive the same payment each month, so that the observations are the same, the individuals may be differentiated by the serial numbers. In order to take a random sample from this population, it is only necessary to find a method of specifying the serial numbers for the individuals to be selected.

The serial numbers may be specified and the random samples then selected by methods of chance familiar to everyone. For example, a coin may be tossed, a card may be drawn, a number may be chosen in a lottery, or dice may be thrown. In practice, it is more convenient to use a table of random numbers, because the serial numbers can be selected more rapidly and because any simple mechanical process is liable to be biased. Random numbers are generated by random processes, which are devised so that each of the digits from 0 to 9 has an equal chance of occurring. Because there are ten digits in the range 0 to 9, the chance or probability of one occurring each time is 10 percent, or 0.1. The mathematical process is such that there is no memory of which digit was generated last time, just as the toss of an unbiased coin does not depend on the result of the previous toss. In technical terms, the digits are generated with statistical independence.

Tables of random numbers are available, for example, the Rand table (Rand Corporation, 1955) of one million digits. Table A-1, in the appendix to this volume, is a small table of random digits from the Rand table. The table is arranged with serial numbers in the far-left column and with 10 five-digit columns of random numbers to the page. To use the table, choose any starting point, for example, by picking a page, column, and row at random. Read in any consistent direction, for example, down columns or across rows, and record numbers of the desired size. For example, in order to obtain two-digit numbers, the first or last two digits of a group of five could be read.

Because random numbers may be assembled to form numbers of any required size, tables of random numbers may be used to obtain random samples from populations with serially numbered observations. If serial numbers are not explicitly attached, it is sometimes possible to attach them implicitly. If serial numbers cannot be attached, the sampling becomes more or less obscure, and it may be difficult to determine whether a sample is indeed random. For example, pulverized rock might be passed through a Jones splitter three times to obtain eight fractions. If the eight fractions are serially numbered, and if one is selected at random for an analysis, then this phase at least of the sampling is indeed random. On the other hand, if the eight fractions are not serially numbered and preserved for a final random selection, there is always the question whether the sample splitting was indeed random, or whether some bias was introduced, perhaps by the splitter not doing an unbiased job of dividing a powdered sample.

In order to illustrate a specific method of taking a random sample, an example of sampling diamond-drill core from the Homestake mine, Lead, South Dakota, is described here in detail. Because our study (Koch and Link, 1967) supplies other examples for this book, pertinent details of the mine geology are reviewed here. In the Homestake mine, the gold is localized in the Precambrian Homestake Formation. The Homestake Formation, originally a low-grade iron formation, has been metamorphosed and intensely and intricately folded. Its present composition is mainly chlorite, cummingtonite, and quartz—with subordinate ankerite, garnet, iron sulfides, and other minerals. Typically, the ore contains 0.33 troy ounce of gold per ton, 8 to 10 percent of sulfide minerals (mainly pyrrhotite), and a little silver. The Homestake Formation has been folded into steeply dipping ore bodies of banded and foliated rock. The ore is developed on each level by crosscuts and between levels by diamond-drill holes from sublevel stations.

The purpose of our study was to investigate the distribution of gold in gold ore of the world. Such distribution is of use both for theoretical and applied geology and for mining. Thus our initial target population was gold distribution in gold ore throughout the world. The gold-bearing Homestake Formation was the second target population. Because we wanted to compare variability in gold in this formation at short range as well as at longer range, we put the further requirement on our target population that we would sample in two locations some tens of feet apart, with more than one borehole at each location.

Having specified the target population, we could next obtain a potential sampled population corresponding to the specifications. The actual sampled population obtained was ore at two locations (fig. 3.1) selected as typical. Thus, a long chain of decisions and events extended from our initial vague concept through to the actual sampling, connected by geological reasoning within a framework of geological assumptions. To sample the ore at the two

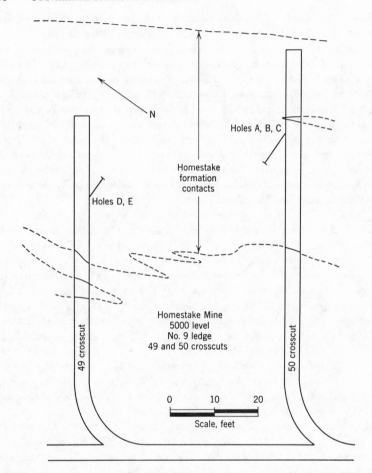

Fig. 3.1. Locations of diamond drill holes for experimental sampling in the Homestake Mine.

chosen locations, we bored five drill holes, three designated A, B, C, in the 50 crosscut; and two, designated D and E, in the 49 crosscut. Because the total combined hole length was 56 feet and the core was sawed transversely into 1-inch-long cylinders, our original sample consisted of 672 (56 × 12) cylinders. We prefer to think of the sample as composed of the actual rock cylinders rather than as a collection of 672 potential gold assay values, because other observations could have been made on the cylinders, such as mineralogical composition or silver assay. This usage puts our samples into a more geological framework.

The purpose of choosing 1 inch for the length of the rock cylinders was to obtain geological samples for assaying, each weighing about 1 assay ton (29.1667 grams), so that no variation would be introduced by splitting pulverized material or by handling coarse gold that could not be ground. To save time and expense, we decided to sample one-third of the original sample of 672 cylinders, under the restriction that two pairs of adjacent cylinders, or a total of four cylinders would be taken from each foot. This restriction is a *stratification* of the sampling, a method, discussed in section 7.2, that is not considered here.

In order to select randomly the four rock cylinders per foot required by our scheme, the cylinders were serially numbered as shown in table 3.5 and

TABLE 3.5. RANDOM SAMPLING SCHEME FOR DIAMOND-DRILL CORE FROM THE HOMESTAKE MINE

Foot interval	Core pair numbers	Random number from table	Constant	Random number plus constant
1	1			
	2			
	3	3	0	3
	4			
	5	5	0	5
	6			
2	7	1	6	7
	8			
	9	3	6	9
	10			
	11			
	12			
3	13	1	12	13
	14			
	15			
	16			
	17	5	12	17
	18			
4	19			
	20	2	18	20
	21			
	22			
	23	5	18	23
	24			

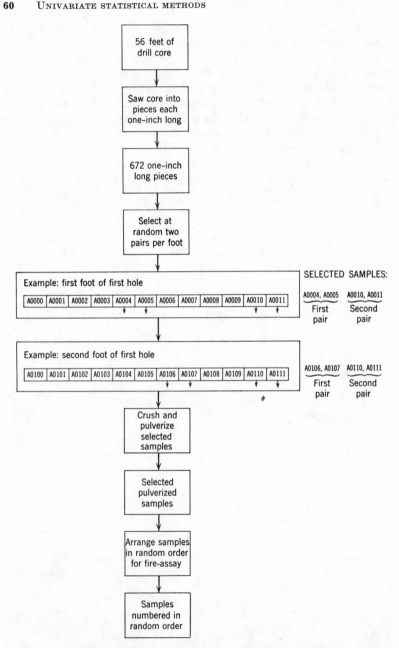

Fig. 3.2. Flow chart for sampling drill core from the Homestake mine.

figure 3.2. Because paired samples were required, for the first foot it was necessary to choose from a random number table the first two numbers between 1 and 6. For the second foot, the next two numbers between 1 and 6 were chosen, and 6 was added to each. Through this process, 224, or one-third of the total 672 cylinders, were selected. The rock from one selected inch was lost in drilling; 223 cylinders and 111 pairs were assayed.

3.5 SAMPLING DISTRIBUTIONS

In this chapter the relation of sample statistics to population parameters has been introduced, and the calculation of some parameters and statistics has been explained. From the observations in a sample, various statistics can be calculated, for instance, the sample mean and the sample variance. Consider, in particular, one statistic and a particular sample size. For repeated samples from the population the calculated value of the statistic fluctuates from sample to sample because the observations in the various samples are different. The calculated values can be summarized in a frequency distribution. The *sampling distribution* of a statistic is simply the theoretical frequency distribution of the values of a statistic calculated for all possible samples of a particular size from a population. The values used to form the sampling distribution are new observations that together compose a new population.

A sampling distribution is valuable because it relates a statistic to an underlying population parameter. In practice only one statistical sample of a population is taken at a time, as we took only one random sample at the Homestake mine. But one must know the behavior of statistics from all possible samples in order to relate the statistic from the single sample to the population parameter that we wish to estimate.

For a small population it is easy to illustrate the sampling distribution by enumerating all possible samples of a given size. Two types of sampling must be distinguished: sampling with replacement and sampling without replacement. Although in practice it is almost never necessary to distinguish between these two methods of sampling, both should be understood.

Consider a population with only three observations—2, 4, and 6—from which a sample of two observations, that is a sample of size 2, is to be drawn. In table 3.6, the population is listed, and the mean μ and variance σ^2 are calculated. The population may be thought of as three playing cards of the appropriate values. When the cards are shuffled and one is drawn at random, the first observation must be either 2, 4, or 6. In *sampling with replacement*, the playing card first selected is put back in the deck before the second is

TABLE 3.6. CALCULATION OF MEAN AND VARIANCE OF FICTITIOUS POPULATION OF SIZE 3

Observation, w	Deviation, $w - \mu$	Squared deviation, $(w - \mu)^2$
2	-2	4
4	0	0
6	2	4
Sum 12		8

Calculations:

$$\mu = \tfrac{12}{3} = 4 \qquad \sigma^2 = \tfrac{8}{3} = 2\tfrac{2}{3} = 2.667$$

drawn; accordingly, the second observation may again be 2, 4, or 6. The total number of possible samples is 3^2 or 9, according to the scheme of table 3.7 (Li, 1964, I). The order in which observations are selected counts—the sample with observation 2 followed by 6 is different from that with observation 6 followed by 2, even though a particular statistic such as the mean is the same for both samples, because other statistics, such as one that gives more weight to the first observation than to the second, are different. In general, the sample size in sampling with replacement is equal to N^n, where N is the population size and n is the sample size. Thus for the example population of size 3 there are three possible samples of size 1 $(3^1 = 3)$, namely, 2, 4, and 6; and there are 27 of size 3 $(3^3 = 27)$.

TABLE 3.7. SCHEME FOR SAMPLING WITH REPLACEMENT FROM POPULATION OF SIZE 3*

First observation	Second observation	Sample
2	2	2, 2
	4	2, 4
	6	2, 6
4	2	4, 2
	4	4, 4
	6	4, 6
6	2	6, 2
	4	6, 4
	6	6, 6

* After Li, 1964, I, p. 37.

TABLE 3.8. SCHEME FOR SAMPLING WITHOUT REPLACEMENT
FROM POPULATION OF SIZE 3

First observation	Second observation	Sample
2	4	2, 4
	6	2, 6
4	2	4, 2
	6	4, 6
6	2	6, 2
	4	6, 4

From the same population of size 3, next consider *sampling without replacement*. The first observation is obtained as before; the cards are shuffled and one is drawn at random. Again, the first observation may be 2, 4, or 6. But the first card is not replaced in the deck as before; therefore the second observation cannot be the same as the first. Thus, as table 3.8 shows, there are fewer possible samples, 6 instead of 9. In general the sample size is equal to $N!/(N - n)!$. Thus for the example population of size 3, in sampling without replacement, there are 3 possible samples of size 1, 6 of size 2 $[3!/(3 - 2)!]$, and 6 again of size 3 $[3!/(3 - 3)!]$.

For each sample of a given size from a population, the value of a particular statistic can be calculated. When the means are calculated for each sample of size 2 from the population of size 3, the results obtained are those listed in table 3.9 for sampling with replacement and in table 3.10 for sampling

TABLE 3.9. CALCULATION OF SAMPLE MEANS FROM SAMPLES OF SIZE 2
SELECTED WITH REPLACEMENT *

First observation	Second observation	Sample	$\sum w$	$\overline{w}$
2	2	2, 2	4	2
	4	2, 4	6	3
	6	2, 6	8	4
4	2	4, 2	6	3
	4	4, 4	8	4
	6	4, 6	10	5
6	2	6, 2	8	4
	4	6, 4	10	5
	6	6, 6	12	6

* From Li, 1964, I, p. 37.

Table 3.10. Calculation of sample means from samples of size 2 selected without replacement

First observation	Second observation	Sample	$\sum w$	$\overline{w}$
2	4	2, 4	6	3
	6	2, 6	8	4
4	2	4, 2	6	3
	6	4, 6	10	5
6	2	6, 2	8	4
	4	6, 4	10	5

without replacement. The frequency distributions of the sample means are graphed as histograms in figures 3.3 and 3.4 to compare with the original distribution of observations in the population.

In sampling with replacement, the sampling is said to be *statistically independent*, because no observation is influenced by the previous one. Among many familiar examples of sampling with statistical independence are tossing a fair coin, where the observations are 1 or 0, depending on whether the coin comes up heads or tails; throwing fair dice, where the sum of the numbers is the observation; and making repeated potassium analyses on a specimen by a nondestructive method such as x-ray fluorescence.

In contrast, in sampling without replacement, the sampling is said to be *statistically dependent*, because each observation is influenced by the preceding one. A familiar example of sampling with statistical dependence is drawing cards from a deck, when the cards are not replaced after each draw.

In general, in sampling, the probabilities change with each observation taken, so that all stages of sampling from complete dependence to complete

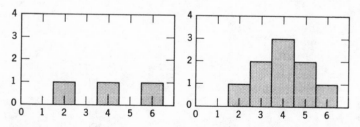

Fig. 3.3. Histogram of original population of size 3 and samples of size 2; sampling with replacement.

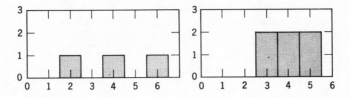

Fig. 3.4. Histogram of original population of size 3 and samples of size 2; sampling without replacement.

independence are possible. For example, if successive cards are drawn from a 52-card deck, the value of the card being the observation, the probability of the first card being a 10 is 4/52 (7.7 percent). The probability of the second card being a 10 is not 4/52, because it depends on the value of the first card: it is 4/51 (7.8 percent) if the first card is not a 10, and 3/51 (5.9 percent) if the first card is a 10. Such probabilities are known, intuitively if not formally, to all poker and blackjack players.

Appropriate mathematical theory has been devised both for sampling with replacement (corresponding to statistical independence) and sampling without replacement (corresponding ·to statistical dependence). However, as suggested by the changing probabilities in the card-drawing example, the mathematical theory of sampling with replacement is simpler. Fortunately, most if not all geological sampling may be regarded as sampling with replacement, by making an appropriate definition of the sampled population. If the sampled population is regarded as being infinite in size, the theoretical considerations are identical, whether the sampling is done with or without replacement, because drawing 1 or even 100,000 observations from a conceptually infinite number of observations does not change the probabilities.

Although the population of tables 3.6 to 3.8 contains only three observations, the total number of samples of size 2 is 9 in sampling with replacement and is 6 in sampling without replacement. For populations and samples of sizes ordinarily of interest, enumeration of the samples is impractical. Table 3.11 gives the number of all possible samples of sizes 2, 5, and 10, drawn from populations of sizes 10 and 100; even for these small sizes, the number of samples is formidable. Fortunately, it is unnecessary to enumerate the samples in the manner of tables 3.7 and 3.8 because the same results can be obtained through mathematical theory. The details of the mathematics, which are given in books on mathematical statistics and probability, for example, those by Hoel (1962) and Cochran (1963), are not reviewed here. In sections 3.6 and 3.7, the results for the sample mean and the sample variance

TABLE 3.11. NUMBER OF ALL POSSIBLE SAMPLES OF SIZES 2, 5, AND 10 FROM POPULATIONS OF SIZES 10 AND 100, SAMPLING WITH AND WITHOUT REPLACEMENT

Population size	Sample size	Number of samples	
		With replacement	Without replacement
10	2	100	90
	5	100,000	30,240
	10	10^{10}	$10! = 3.6288 \times 10^6$
100	2	10^4	9900
	5	10^{10}	$100!/95! = 9.0345 \times 10^9$
	10	10^{20}	$100!/90! = 6.2815 \times 10^{19}$

Note: "$n!$" denotes factorial n; for example, "$10!$" denotes the product of integers from 1 to 10.

are presented. Detailed discussions based on the population 2, 4, and 6 are given to illustrate the similarities and differences of sampling with and without replacement.

3.6   DISTRIBUTION OF THE SAMPLE MEAN

From sampling distributions in general attention is now turned to the sampling distribution of a specific statistic, the sample mean $\bar{w}$. The *sampling distribution of the sample mean* is the theoretical frequency distribution of all sample means calculated from all possible samples of the same size. The purpose of this discussion is to relate the mean and variance of the distribution of the sample mean to the mean and variance of the population. The discussion is introduced with the same simple example of table 3.7 for sampling with replacement from the population of three observations 2, 4, and 6, which defines the theoretical frequency distribution plotted in figure 3.3. From calculations in table 3.6, the mean μ of this population is 4, and the variance σ^2 is $2\frac{2}{3}$. The nine sample means define a new secondary population, whose observations are listed in table 3.9, and graphed in figure 3.3 as a secondary theoretical frequency distribution called the *distribution of the sample mean*. From the nine sample means a mean of sample means $\mu_{\bar{w}}$ and a variance of sample means $\sigma_{\bar{w}}^2$ can be calculated (see table 3.12). Regardless of the initial theoretical frequency distribution of w, comparison of the two kinds of means and variances illustrates two important results:

TABLE 3.12. DISTRIBUTION OF SAMPLE MEANS FROM SAMPLES OF SIZE 2 SELECTED
WITH REPLACEMENT

First observation	Second observation	Sample	$\sum w$	$\overline{w}$	$(\overline{w} - \mu)$	$(\overline{w} - \mu)^2$
	2	2, 2	4	2	-2	4
2	4	2, 4	6	3	-1	1
	6	2, 6	8	4	0	0
	2	4, 2	6	3	-1	1
4	4	4, 4	8	4	0	0
	6	4, 6	10	5	1	1
	2	6, 2	8	4	0	0
6	4	6, 4	10	5	1	1
	6	6, 6	12	6	2	4
Sum				36		12

Calculations:

$$\mu_{\overline{w}} = \tfrac{36}{9} = 4 \qquad\qquad \sigma_{\overline{w}}^2 = \tfrac{12}{9} = 1.333$$

1. The mean of sample means is equal to the population mean; that is,

$$\mu_{\overline{w}} = \mu.$$

For the numerical example, the two quantities are equal to 4.

2. The variance of sample means is equal to the population variance divided by the sample size; that is,

$$\sigma_{\overline{w}}^2 = \frac{\sigma^2}{n}.$$

For the numerical example, $\sigma_{\overline{w}}^2$ is equal to $1\tfrac{1}{3}$, and σ^2 is equal to $2\tfrac{2}{3}$.

In table 3.13 the calculations of table 3.12 are repeated for sampling without replacement to illustrate the following results:

1. As before, the mean of sample means is equal to the population mean; that is,

$$\mu_{\overline{w}} = \mu.$$

2. A new formula,

$$\sigma_{\overline{w}}^2 = \frac{\sigma^2}{n} \left(1 - \frac{n}{N}\right)\left(\frac{N}{N - 1}\right),$$

relates the sample variance to the population variance. Because sampling without replacement is seldom if ever important in geological situations, the difference between the two variance formulas is not discussed.

Table 3.13. Distribution of sample means from samples of size 2 selected without replacement

First observation	Second observation	Sample	$\sum w$	$\overline{w}$	$(\overline{w} - \mu)$	$(\overline{w} - \mu)^2$
2	4	2, 4	6	3	-1	1
	6	2, 6	8	4	0	0
4	2	4, 2	6	3	-1	1
	6	4, 6	10	5	1	1
6	2	6, 2	8	4	0	0
	4	6, 4	10	5	1	1
Sum				24		4

Calculations:

$$\mu_{\overline{w}} = \tfrac{24}{6} = 4 \qquad \sigma_{\overline{w}}^2 = \frac{4}{6} = 0.6667 = \frac{\sigma^2}{n}\left(1 - \frac{n}{N}\right)\left(\frac{N}{N-1}\right)$$

The method of calculating the parameters of the distribution of sample means from the population parameters has been stated. If the theoretical frequency distribution followed by the observations is known, the distribution of the sample means can be calculated. It can be shown mathematically (Wilks, 1962, p. 206) that, if the theoretical frequency distribution of the observations is normal, the theoretical frequency distribution of the sample mean is also normal for all sample sizes. It can also be shown mathematically that, even if the theoretical distribution of the observations is not normal, the distribution of the sample mean tends to be nearly normal for moderate sample sizes and becomes more nearly normal as the sample size increases. Even for a sample of size 2 from the far-from-normal population of figure 3.3, the distribution of the sample mean is symmetrical and somewhat bell-shaped. For a sample of size 4 from the same population, the distribution of the sample mean is almost normal.

The preceding facts are based on a remarkable theorem named the *central limit theorem*, which may be stated as follows: *If random samples of fixed size are drawn from a population whose theoretical distribution is of arbitrary shape, but with a finite mean and variance, the distribution of the sample mean tends more and more toward a normal frequency distribution as the size of the sample increases.* The theorem provides no guidance on how large the sample size must be for the distribution of the sample mean to be nearly normal in shape. The size depends on the shape of the original frequency distribution of the observations. For most geological distributions a sample size of 50 to 100 is adequate for the distribution of the sample mean to be nearly normal. The outstanding exceptions to this rule are some distributions of gold assays and other trace elements.

Because of the central limit theorem, whenever the observations are normally distributed, or whenever there is a moderate number of observations even though they are not normally distributed, a discussion of the distribution of the sample mean reduces to a discussion of the normal distribution. Of course, many distributions depend on the values of the two parameters: the mean, which is the population mean, equal to the mean of sample means; and the variance, which is the variance of sample means, equal to σ^2/n. The different distributions may be expressed in terms of the single normal distribution obtained by the linear transformation in section 3.3 because, when the statistic u is defined as

$$u = \frac{\overline{w} - \mu}{\sigma/\sqrt{n}},$$

it is clear from section 3.3 that u has a mean of 0 and a variance of 1. When this equation is solved for $\overline{w}$, the result is

$$\overline{w} = u\frac{\sigma}{\sqrt{n}} + \mu.$$

In table A-2 the normal distribution with μ equal to 0, and σ equal to 1, is tabulated. The first column gives percentage points, and the second gives the standardized normal distribution. The entries in the first line mean that 99.9 percent of calculated values of the statistic

$$\frac{\overline{w} - \mu}{\sigma}$$

will be larger than -3.0902.

The importance of random sampling now becomes apparent. The standardized normal distribution gives the distribution of all possible sample means. If a sample is random, any particular sample has the same chance of being drawn as any other; hence, the chance of its lying between any two percentage points, for example, the 0 and 95 percent points, can be specified. If the sample is not random, no such probability calculation can be made. Although the sample need not be a simple random sample, because restricted or extended sampling methods can be used, the element of randomness must somehow be introduced.

Finally, the variability of the sample mean depends both upon the variability of the original population and upon the sample size; the variability decreases as the sample size increases. Specifically, the standard deviation of the sample mean, $\sigma/\sqrt{n}$, also named the *standard error of the mean*, decreases

TABLE 3.14. FREQUENCY DISTRIBUTIONS OF MEANS OF SAMPLES
OF VARIOUS SIZES DRAWN AT RANDOM FROM A POPULATION OF
900 GOLD ASSAYS FROM THE HOMESTAKE MINE

Class interval (ppm Au)	Sample size			
	1	5	25	100
0–1	439	175	0	0
1–2	120	121	3	0
2–3	67	105	36	1
3–4	44	88	82	5
4–5	35	69	124	38
5–6	26	66	131	149
6–7	23	57	146	222
7–8	14	40	114	201
8–9	23	45	83	198
9–10	14	35	73	98
10–11	16	31	56	49
11–12	13	16	38	22
12–13	13	12	25	10
13–14	13	23	34	5
14–15	9	15	15	2
15–16	10	19	15	
16–17	8	11	7	
17–18	7	5	5	
18–19	9	4	7	
19–20	1	1	1	
20–21	6	5	0	
21–22	7	7	4	
22–23	1	5	2	
23–24	6	7	0	
24–25	1	1	0	
25–26	3	4	0	
26–27	3	5	0	
27–28	4	1	0	
28–29	2	3	0	
29–30	6	6	1	
30–31	2	4		
31–32	6	1		
32–33	7	0		
33–34	1	1		
34–35	0	1		
35–49	14	7		
50–99	21	3		
100–	8			

as the square root of the sample size increases. Furthermore, unless a significantly large fraction of the population is sampled, as seldom if ever happens in geological sampling, the fraction taken has no effect on the reliability of the sample mean, except as reflected in the sample size. Thus, even though the fraction of the population sampled is minute, precision can be attained by taking sufficiently large statistical samples.

The central limit theorem may be illustrated by randomly drawing samples of various sizes from 900 gold values obtained by assaying 1-foot-long EX diamond-drill cores from the Homestake mine. The original frequency distribution, defined for the present purpose as a population of size 900 and normalized in column 1 of table 3.14 to a basis of 1000 observations, is extremely skewed to the right because there are a few very high observations. From the population of size 900, we drew 1000 independent random samples with replacement of sizes 5, 25, and 100; the frequency distributions of the means of these samples are listed in table 3.14. With increasing sample size, the distributions become less and less skewed; the distribution for sample size 100 is skewed slightly and looks nearly normal.

3.7 DISTRIBUTION OF THE SAMPLE VARIANCE AND THE CHI-SQUARE DISTRIBUTION

The discussion of sampling distributions is continued with that of the sample variance s^2. In section 3.6, the distribution of the sample mean is shown to be symmetrical and to approach the normal distribution rapidly as the sample size increases. However, the distribution of the sample variance is more complicated and, for small sample sizes, is noticeably skewed. This discussion of sampling distributions is somewhat different from that in the previous sections of this chapter in that sampling with replacement is discussed first in detail, then sampling without replacement is discussed, at the end of the section.

The distribution of the sample variance is of interest for two reasons: first, the sample variance can be used to estimate the population variance; second, and usually more important for geological problems, the sample variance is required to estimate the reliability of a sample mean, a reliability that is nearly always of great interest.

The distribution of the sample variance is first considered for the samples of size 2 drawn with replacement from the population 2, 4, 6. In table 3.15 the sample variances and standard deviations are calculated for each sample. Figure 3.5, a histogram of distribution of sample variances which can be compared with figure 3.3 for the distribution of population and sample

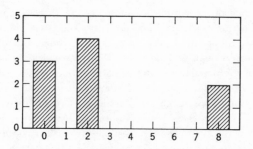

Fig. 3.5. Histogram of sample variances from samples of size 2; sampling with replacement.

means, shows that the distribution of sample variances is skewed in comparison with that of sample means. However, the mean of the sample variances,

$$\frac{\sum s^2}{n} = \frac{24}{9} = 2\tfrac{2}{3} = \sigma^2,$$

is equal to the population variance. The calculation explains why s^2 is calculated by using $(n - 1)$ rather than n as the divisor; the purpose is to make the sample variance equal on the average to the population variance. Thus table 3.15 illustrates the relation, always true, that the mean of sample

Table 3.15. Distribution of sample variances from samples of size 2 selected with replacement *

		Sample	Sample variance, s^2	Sample standard deviation, s
		2, 2	0	0.000
		2, 4	2	1.414
		2, 6	8	2.828
		4, 2	2	1.414
		4, 4	0	0.000
		4, 6	2	1.414
		6, 2	8	2.828
		6, 4	2	1.414
		6, 6	0	0.000
	Total		24	11.312
	Mean		$\frac{24}{9} = 2\tfrac{2}{3}$	1.257
Corresponding	Parameter		$\sigma^2 = 2\tfrac{2}{3}$	$\sigma = 1.633$

* From Li, 1964, I, p. 70.

variances calculated from all possible samples of a given size drawn with replacement from any population is equal to the population variance. However, the mean of sample standard deviations is not equal to the population standard deviation, although it becomes more nearly equal as the sample size increases. Table 3.15 shows the considerable discrepancy for a sample size of 2 for these observations.

Although the relation of sample variance to population variance could be conveniently introduced by the simple example of table 3.15, the most useful means of relating these two kinds of variance is by a new statistic, *chi-square* defined by the formula

$$\chi^2 = \frac{(n-1)s^2}{\sigma^2}.$$

Because, as illustrated in table 3.15, the mean value of the ratio s^2/σ^2 is 1, the mean value of the statistic chi-square must be $(n-1)$; that is, the mean value depends only on the sample size. The quantity $(n-1)$ is named the *degrees of freedom*, abbreviated d.f.; the meaning of the name is revealed as the subject develops. Because the chi-square distribution has a single parameter, the mean, equal to the number of degrees of freedom, there is a separate curve with tabled percentage points for each different number of degrees of freedom. Relating the sample and population variances by the chi-square formula rather than by the simple ratio s^2/σ^2 may seem unnecessarily complicated, but it is done just as the standardized normal deviate u is defined as

$$u = \frac{w - \mu}{\sigma},$$

because table construction is facilitated.

If the original observations come from a normal distribution, the chi-square statistic follows a distribution named the *chi-square distribution*. For the chi-square distribution, as the degrees of freedom increase, the basic shape of the distribution changes. Figure 3.6 gives frequency curves for the chi-square distribution for degrees of freedom from 1 to 6. Although all of the curves are skewed to the right, the skewness is less for the larger values. Although s^2 can never be smaller than 0, the skewness occurs because there is no upper limit to its size, and inclusion of a few extreme values of w in a small sample can lead to a very large value of s^2.

Table A-3 gives percentage points for the chi-square distribution. Each row of the table is for a specific number of degrees of freedom, and each column is for a certain percentage point. Any entry in the table gives the value of the statistic $(n-1)s^2/\sigma^2$ that will be exceeded by a certain percentage of values calculated from a random sample with a certain sample size of number of degrees of freedom. For example, the entry 10.8508 in the column for 95

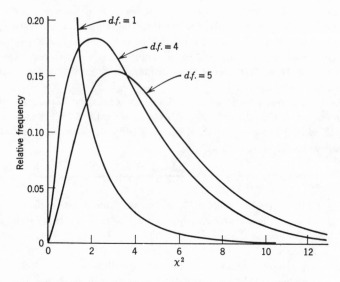

Fig. 3.6. Chi-square distributions for 1, 4, and 5 degrees of freedom.

percent and 20 d.f. means that 95 percent of the values calculated from random samples of size 21 will be *larger* than the tabulated value.

The chi-square table is used in applied statistics, in situations when the variance of a sample needs to be controlled. An example is in the manufacture of lamp bulbs, where the variation in life needs to be controlled within limits so that the bulbs will burn long enough on the average to meet specifications. In this book several applications to geology are given in later chapters for the construction of interval estimates for the variance in section 4.6, and for the curve fitting in section 6.1.

Figure 3.6 indicates that, as the degrees of freedom increase, the skewness of the chi-square distribution decreases. Because the sample variance s^2 is a kind of average, the central limit theorem, that averages tend to follow the normal distribution with increasing sample size, should apply. Table 3.16 presents results and calculations for 50 degrees of freedom to illustrate the good correspondence of percentage points calculated by using the normal distribution to the exact values in the chi-square table. As shown, the central limit theorem indeed applies even to this originally highly skewed chi-square distribution. The calculations are based on the relation we have presented— that the mean of the chi-square distribution is equal to the number of degrees of freedom—and on the relation, not illustrated in this book, that the standard deviation of the chi-square distribution with $n - 1$ degrees of freedom is

TABLE 3.16. APPLICATION OF THE CENTRAL LIMIT THEOREM TO THE CHI-SQUARE DISTRIBUTION

Source of value	95% point	5% point
Exact values from table A-3	34.8	67.5
Values from normal approximation, calculations below	33.5	66.5

Calculations:

$$\text{d.f.} = 50$$
$$n - 1 = 50$$
$$\mu = n - 1$$
$$= 50$$
$$\sigma = \sqrt{2(n - 1)}$$
$$= \sqrt{2(50)}$$
$$= \sqrt{100}$$
$$= 10$$
$$\text{Percentage point} = 1.645$$
$$95\% \text{ point} = 50 - 1.645 \times 10$$
$$5\% \text{ point} = 50 + 1.645 \times 10$$

$\sqrt{2(n - 1)}$. According to normal theory (see table A-2) the value for the 95-percent point is 1.645.

The name *degrees of freedom* is somewhat troublesome because its usage is partly descriptive. It is related in a sense to the usage in the phase rule of physical chemistry or the mineralogical phase rule, but it is not exactly comparable. As this book progresses, the usage will become clearer. In general, the number of degrees of freedom is simply the number used to divide the sum of squared deviations from the mean to get a sample variance s^2 that is an unbiased estimate of the population variance σ^2. Because

$$s^2 = \frac{\text{SS}}{n - 1},$$

the chi-square statistic,

$$\chi^2 = \frac{(n - 1)s^2}{\sigma^2},$$

can be rewritten, substituting SS for $(n - 1)s^2$, to obtain

$$\chi^2 = \frac{\text{SS}}{\sigma^2}.$$

Indeed, the chi-square statistic is often introduced in the form SS/σ^2. This formulation has been postponed to this point to emphasize the key importance of the degrees of freedom. Although thus far in this book the sample variance has been obtained by dividing the SS by $(n - 1)$, some more complicated calculations to be presented later require division of SS by some different quantity to get the sample variance. Then the mean value of the chi-square statistic is not $(n - 1)$ but equal to whatever divisor is used.

In many older books, s^2 is defined as exactly analogous to the sample mean by using a divisor of n, as used for the sample mean calculation, rather than $(n - 1)$. In most modern books, including this one, only $(n - 1)$ is used for several reasons, perhaps the most important being that, because all statistical tables involving the variance are constructed on the basis of degrees of freedom, they cannot be applied exactly if s^2 has been calculated by using n. Also, although the numerical difference between n and $(n - 1)$ is small for large sample sizes, numerically large discrepancies may arise even for large sample sizes if complicated estimates of s^2 are required.

The preceding discussion entirely concerns variances of random samples drawn from a normal population. But even for samples drawn from a nonnormal population, the mean of sample variances for all possible samples is equal to the population variance, as is illustrated by variances calculated for the samples previously drawn for the 900 Homestake gold values (sec. 3.6). In table 3.17 are recorded the population mean μ equal to 7.59 ppm Au and the variance σ^2 equal to 327.21 for the 900 observations. For the samples of sizes 5, 25, and 100, the calculated mean of sample means and the mean of sample variances are tabulated. As expected, the mean of sample means is nearly equal to the population mean, that is,

$$\mu_{\bar{w}} \approx \mu,$$

TABLE 3.17. DISTRIBUTION OF SAMPLE VARIANCES FOR GOLD DATA FROM THE HOMESTAKE MINE

$\mu = 7.59$ ppm Au, $\sigma^2 = 327.21$

| Sample size | Mean | | Variance, $s_{\bar{w}}^2$ | Ratio of population variance to sample variance, $\sigma^2/s_{\bar{w}}^2$ | |
	$\bar{w}$	s^2		Predicted	Observed
5	6.67	300	57.40	5	6
25	7.59	313	12.60	25	26
100	7.56	324	3.08	100	106

and the mean of sample variances is nearly equal to the population variance, that is,

$$\mu_s^2 \approx \sigma^2.$$

The fourth column lists the variances of the sample means, which should be equal to the population variance divided by the sample size, according to the formula

$$\sigma_{\bar{w}}^2 = \frac{\sigma^2}{n}.$$

The ratios of population variance to predicted and observed variances of sample means are given in columns 5 and 6. If all possible samples had been taken, this ratio would be exactly equal to the sample size; because only 1000 were taken, it is only approximately equal to the sample size, but close enough to verify the relationship. However, if the observations are not normally distributed, the distribution of the quantity SS/σ^2 is not necessarily close to the chi-square distribution because the distribution of this quantity is quite sensitive to the normal assumption being met. Fortunately, because the sample variance seldom need be estimated for itself, this problem is not serious.

Finally, the distribution of the sample variance in sampling without replacement is illustrated through table 3.18, which is constructed like table

TABLE 3.18. DISTRIBUTION OF SAMPLE VARIANCES FROM SAMPLES OF SIZE 2 SELECTED WITHOUT REPLACEMENT

Sample	Sample variance, s^2	Sample standard deviation, s
2, 4	2	1.414
2, 6	8	2.828
4, 2	2	1.414
4, 6	2	1.414
6, 2	8	2.828
6, 4	2	1.414
Sum	24	11.312

Mean	$s^2 = \frac{24}{6} = 4$
	$s = 1.885$
Corresponding parameter	$\sigma^{*2} = \left(\dfrac{N}{N-1}\right)\sigma^2$
	$= \frac{3}{2} \times \frac{8}{3} = 4$
	$\sigma^* = 2$

3.15. Table 3.18 shows that for sampling without replacement, the mean of sample variances is 4 rather than the $2\frac{2}{3}$ calculated for sampling with replacement. Therefore, when sampling without replacement, either the sample variance or the population variance must be redefined to make the sample variance an unbiased estimate of the population variance. It is conventional to redefine the population variance σ^2 as a new variance σ^{*2} according to the formula,

$$\sigma^{*2} = \frac{N}{N-1}\sigma^2,$$

where N, as before, is the population size. When this is done, the table illustrates that the sample variance, $s^2 = 4$, is an unbiased estimate of the redefined population variance $\sigma^{*2} = 4$. As in sampling with replacement, the sample standard deviation is not an unbiased estimate of the population standard deviation.

REFERENCES

Cochran, W. G., Mosteller, Frederick, and Tukey, J. W., 1954, A report on sexual behavior in the human male: Am. Statistical Assoc., p. 18.

Cochran, W. G., 1963, Sampling techniques: New York, John Wiley & Sons, 413 p.

Exley, C. S., 1963, Quantitative areal modal analysis of granitic complexes: a further contribution: Geol. Soc. Am. Bull., v. 74, p. 649–654.

Hoel, P. G., 1962, Introduction to mathematical statistics: New York, John Wiley & Sons, 427 p.

Koch, G. S., Jr., and Link, R. F., 1967, Gold distribution in diamond-drill core from the Homestake mine, Lead, South Dakota: U.S. Bur. Mines Rept. Inv. 6897, 27 p.

Krumbein, W. C., 1960, Some problems in applying statistics to geology: Applied Statistics, v. 9, no. 2, p. 82–91.

Krumbein, W. C. and Graybill, F. A., 1965, An introduction to statistical models in geology: New York, McGraw-Hill, 475 p.

Li, J. C. R., 1964, Statistical inference: Ann Arbor, Mich., Edwards Bros, v. 1, 658 p.

Link, R. F., and Koch, G. S., Jr., 1962, Quantitative areal modal analysis of granitic complexes: discussion of article by E. H. T. Whitten: Geol. Soc. Am. Bull., v. 73, p. 411–414.

Rand Corporation, The, 1955, A million random digits with 100,000 normal deviates: Glencoe, Illinois, Free Press, 200 p.

Rosenfeld, M. A., 1954, Petrographic variation in the Oriskany Sandstone (abs.): Geol. Soc. Am. Bull., v. 65, p. 1298–1299.

Wilks, S. S., 1962, Mathematical statistics: New York, John Wiley & Sons, 644 p.

Chapter 4

Inference

Chapters 2 and 3 are about descriptive statistics. Chapter 2 explains how observations can be summarized by an empirical frequency distribution and introduces theoretical frequency distributions and probability. Chapter 3 explains the sampling distributions of two statistics, the mean and the variance, for observations that follow the normal distribution. In this chapter, the material in Chapters 2 and 3 is extended for making *inferences* about the population mean and the population variance from the information contained in a single statistical sample. By *inference* is meant the drawing of a conclusion or conclusions about a population from a sample by inductive reasoning.

Two methods of inference are explained. The first is interval estimation to determine the range of parameter values that are consistent with the information in a sample. The second is the hypothesis test to learn whether the information in a sample is consistent with a postulated value or range of values for a parameter. In making both kinds of inferences, mistakes are inevitably made because the source of information is one or more statistics whose values fluctuate from sample to sample. Evaluation of these mistakes and their control within acceptable bounds constitutes the basic contribution of statistical theory to the analysis of data.

4.1 ESTIMATION OF POPULATION PARAMETERS

In Chapters 2 and 3, we point out that many kinds of numerical data may be represented by a theoretical frequency distribution. However, to represent data completely, we must not only choose a mathematical function but also specify parameter values. Thus, for the normal distribution, several different distributions (figs. 2.5 and 2.6) are plotted for different values of the parameters μ and σ. Now, in this chapter one of the central problems of statistics is introduced: How, assuming a theoretical frequency distribution, can the population parameters best be estimated from a sample?

A population parameter is estimated by calculating the value of a statistic from the observations in a statistical sample. Because more than one statistic can always be calculated to estimate a given parameter, it is necessary to choose among the statistics. Statistical criteria may aid in making a choice, but application of the statistical criteria depends on correspondence of data to underlying assumptions about the theoretical frequency distribution. Thus, to estimate a given parameter, no one statistic is always best. This important fact is often forgotten.

The two most important statistical criteria for choosing a statistic to estimate a given population parameter are now explained. One criterion is that a statistic should be an *unbiased estimate* of a population parameter. An unbiased estimate is one with a sampling distribution whose mean is equal to the estimated parameter. For example, the sample mean is an unbiased estimate of the population mean, but the sample standard deviation is a biased estimate of the population standard deviation (secs. 3.6 and 3.7). The *bias* δ is equal to the difference between the average value of the statistic and the value of the parameter being estimated. Thus, for the examples,

$$\delta = \mu_{\bar{w}} - \mu = 0,$$

but

$$\delta = \mu_s - \sigma \neq 0.$$

Although unbiased estimates are desirable, estimates with a small bias, for example, the sample standard deviation, may be perfectly acceptable and useful for some purposes.

The second criterion is that an estimate should be *efficient*. An *efficient estimate* is a statistic that, for a given sample size, has the smallest mean-square error of all the statistics that might be calculated. The mean-square error (M.S.E.) may be written as

$$\text{M.S.E.} = \sigma_g^2 + \delta^2,$$

where σ_g^2 is the variance of the statistic g calculated from the statistical sampling distribution of g, and δ is the bias of the statistic g. If the statistic is an unbiased estimate, δ is 0, and the preceding equation becomes

$$\text{M.S.E.} = \sigma_g^2.$$

Table 4.1. Relative efficiencies of three statistics for estimating the population mean

Name of statistic	Value of statistic	Variance of statistic	Efficiency of estimate (%)
Mean, $\overline{w}$	5	$\sigma^2/n = 0.2\sigma^2$	100
Midrange	4.5	$0.261\sigma^2$	77
Median	6	$0.287\sigma^2$	70

If the sampling distribution of the statistic is symmetrical and more or less normal and if the statistic is an unbiased estimate, the efficient estimate is likely to be the best that can be made under any circumstances.

The estimation of a parameter by several statistics is illustrated by considering three statistics that can be used to estimate the population mean μ. In section 3.3 the population mean is estimated by the sample mean $\overline{w}$, but the estimation can be made by other statistics, for example, the midrange and the median. How does one decide which estimate to use? In table 4.1 these three statistics are calculated for the fictitious data of table 3.1. In column 1 the value of the statistics is given. From table 3.1 the mean is 5. The midrange, the average of the highest and lowest observations, is $(2 + 7)/2 = 4.5$. The median, the value with half the observations larger and half smaller, is 6. The variance for the mean is equal to the population variance divided by the sample size of 5. For the midrange and the median, the variances of their sampling distributions can be obtained from the population variance by multiplying by constants tabulated by Dixon and Massey (1969, table A-8b(4), p. 488). Because from other considerations the mean is known to be the best estimate, the last column shows that the other estimates are only 70 to 77 percent as efficient.

Departures of data from the assumptions made about the theoretical frequency distribution can affect both the lack of bias and the efficiency of estimates of parameters. Sometimes a choice of estimate may be made. For example, one may choose to lose some efficiency to guarantee not to introduce bias. On the other hand, in another situation, one might choose to keep efficiency up, even though a moderate bias would be introduced. If one can find estimation procedures that are relatively insensitive to departures of the data from the underlying assumptions, he has a high degree of confidence in the conclusions drawn. But if the procedures are very sensitive to departures of data from the assumptions, then one must be wary of the results. We attempt to choose and present procedures that do not depend sensitively on departures of data from the underlying assumptions; such procedures are named *robust*. Because no procedure is completely robust, it is difficult to state quantitatively how insensitive a particular procedure is, but procedures in this book that are clearly very sensitive are pointed out.

So far only statistical criteria have been given for choosing among statistics. Actually there are other criteria based on such matters as the calculation time required, the costs of obtaining different kinds of observations, and geological judgments about how well data correspond to assumptions. Thus, when there are several statistics available to estimate a parameter, it is necessary to consider carefully which is most appropriate for a particular purpose. We sometimes present several methods, as, for example, in section 6.2, in which four methods are given for calculating a coefficient of variation for data that follow the lognormal frequency distribution. Each of these four methods has good and bad points, and to choose among them one must think not only about what the data are like internally and how they are related to the geological environment but also about how much time is to be spent in calculating and what use is to be made of the final statistic.

The choice of statistic may also be complicated by lack of the appropriate statistical theory to give criteria for choosing among different statistics. Theory is particularly lacking for the effects of departures from normality, although this field is one of active statistical research today in which better understanding can be anticipated in the next decade.

4.2 POINT AND INTERVAL ESTIMATES

In section 3.3 the methods of calculating the mean $\overline{w}$ and the variance s^2 of a statistical sample are explained. For five observations with the values 2, 4, 6, 6, and 7 the mean is 5 and the variance is 4. Both $\overline{w}$ and s^2 are point estimates, each with a single value. In sections 3.6 and 3.7 it is also shown that these statistics are unbiased estimates of the corresponding population parameters because, if all possible samples are taken, the mean of the corresponding sampling distribution is equal to the value of the parameter being estimated.

There are two disadvantages to a point estimate. The first is that a point estimate is almost sure to be wrong. Of the 9 samples of size 2 drawn from the population 2, 4, 6 (table 3.9), only three give the population mean of 4, even though the mean of all sample means is an unbiased estimate of the population mean. In the real world an estimate of driving time between two points will not be accurate to the minute; an estimate of grade of ore in a mine will not exactly check with the grade as finally mined; an estimate that a granite contains 65.1 percent quartz will be questioned by the first new analysis. The second disadvantage to a point estimate arises from the first:

the point estimate conveys no information about how wrong it is liable to be.

Because of the disadvantages of point estimates, the *confidence interval* estimate was devised to incorporate three basic attributes of a sample: the mean, the variability as measured by the standard deviation, and the sample size. In interval estimates the parameter being estimated is specified to lie between two values, called *confidence limits*, for a specified percentage of the intervals so calculated. Instead of a point estimate that a granite contains 65.1 percent quartz, a confidence interval estimate is made of the form: with a 90-percent confidence, the granite contains not less than 60.1 percent nor more than 70.1 percent quartz. Confidence intervals are of two kinds: two-sided and one-sided. A two-sided interval, like the illustration, is bounded on both sides by a calculated value. In a one-sided interval, one side is bounded by a noncalculated value, such as 0 or 100 percent, about which there is no risk of being wrong. A statement for a one-sided confidence interval is of the form: with a 90-percent confidence, the granite contains not less than 62.8 percent quartz.

The two procedures most used today for calculating interval estimates were introduced by Fisher in 1935 [1950, (reprinted) paper 25, p. 391] and by Neyman (1937). Fisher coined the name *fiducial intervals*, and Neyman used the name *confidence intervals*. At present confidence intervals are more used, although fiducial intervals are also used, particularly by British workers. In practice, although the definitions and calculations differ for some multivariate cases, the two methods usually give the same results, and in this book only confidence intervals are calculated. The differences between the two kinds of intervals are discussed by Wilks (1962, p. 370).

In order to calculate confidence limits about a mean, the standard deviation and the sample size are used. It seems to be intuitively obvious that, if the variability among observations as measured by the standard deviation is small, an estimate of the mean within narrow limits can be obtained; conversely, if the variability is large, an estimate of the mean within only broad limits is available. Moreover, it seems clear that a better estimate of the mean can be obtained with many observations than with only a few.

These concepts can now be put into formal terms. In section 3.6 on the distribution of the sample mean, it is shown that, if $\overline{w}$ is the mean of a random sample from a normal distribution with mean μ and variance σ^2, the probability is 90 percent (9 chances out of 10) that $\overline{w}$ is not more than 1.645 standard deviations larger or smaller than μ. In other words, in nine random samples out of 10, the sample mean is in the calculated interval. This relation may be written as the algebraic inequality

$$\mu + 1.645 \frac{\sigma}{\sqrt{n}} > \overline{w} > \mu - 1.645 \frac{\sigma}{\sqrt{n}}.$$

Subtracting the quantity $(\mu + \overline{w})$ from this inequality yields

$$-\overline{w} + 1.645\,\frac{\sigma}{\sqrt{n}} > -\mu > -\overline{w} - 1.645\,\frac{\sigma}{\sqrt{n}},$$

and multiplying through by -1 gives the desired result, that

$$\overline{w} - 1.645\,\frac{\sigma}{\sqrt{n}} < \mu < \overline{w} + 1.645\,\frac{\sigma}{\sqrt{n}}.$$

This interval is presented graphically for two different values of σ in figure 4.1. If σ is known, the interval may be evaluated to give a 90-percent confidence interval for μ. Such an interval is said to have a 90-percent *confidence coefficient*. If such intervals are calculated repeatedly for random samples

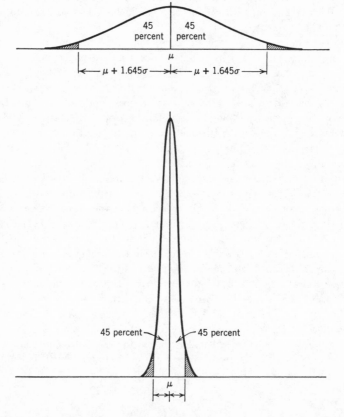

Fig. 4.1. Confidence intervals for two values of σ^2 but the same value of μ.

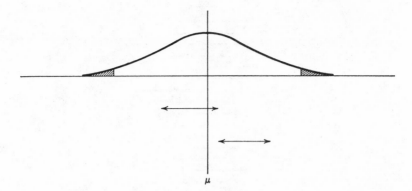

Fig. 4.2. Two confidence intervals, one containing μ and one not containing μ.

from a normal distribution, 9 out of 10 will be correct (that is, they will contain μ); whereas 1 out of 10 will be incorrect (that is, they will not contain μ). In figure 4.2, two intervals are graphed, one containing μ and one not containing μ. In figure 4.3 fifty 90-percent confidence intervals are plotted for samples of size 10 drawn from a table of random numbers from a population with a mean of $\mu = 50$, and a standard deviation of $\sigma = 100$. For these 50 intervals, 5 (or 10 percent) are incorrect as expected; they do not contain μ. Of course, in practice, whenever a confidence interval is calculated, it is unknown whether it is one of the correct ones, but the interval does give some feel for the parameter values, and the interval width gives an idea of the closeness of the estimate.

The method of calculating confidence limits outlined so far is correct, but generally it is useless because σ is seldom known. Therefore σ must usually be replaced by its estimate s. First, however, it is necessary to introduce another sampling distribution, Student's t-distribution.

4.3   STUDENT'S t-DISTRIBUTION

In section 4.2 it is pointed out that interval estimates can be calculated from the standardized normal distribution,

$$\text{s.n.d.} = \frac{\overline{w} - \mu}{\sigma_{\overline{w}}} = \frac{\overline{w} - \mu}{\sigma/\sqrt{n}},$$

with μ equal to 0 and σ equal to 1, provided that σ is known. Because σ is

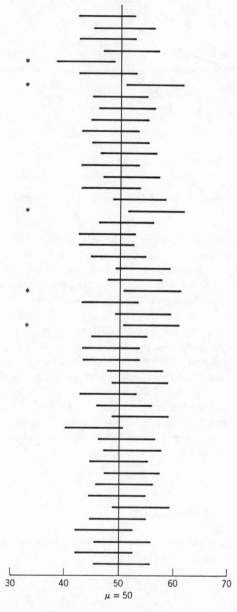

Fig. 4.3. Fifty 90-percent confidence intervals for samples of size 10 from a population with $\mu = 50, \sigma^2 = 100$; σ assumed to be known.

seldom if ever known, however, it must be replaced by its estimate s, the sample standard deviation, to form the new statistic,

$$t = \frac{\overline{w} - \mu}{s/\sqrt{n}},$$

named *Student's t* after its originator, W. S. Gosset, who used the pseudonym "Student."

The frequency distribution of the statistic t is a new sampling distribution. In sections 3.6 and 3.7 the sampling distributions for the sample mean and the sample variance are illustrated for all possible samples of size 2 from a population of size 3. A similar illustration for the t-statistic is given by Li (1964, I, p. 100). The distribution of the t-statistic for normally distributed observations has been calculated mathematically.

Unlike the standardized normal distribution, which is one curve, the t-distribution is a family of curves, one for each number of degrees of freedom, as illustrated in figure 4.4 for selected numbers of degrees of freedom. Selected percentage points for different numbers of degrees of freedom are given in table A-4. To find t values for degrees of freedom not tabulated it is often accurate enough to use the value for the next smallest tabulated number of degrees of freedom. A more exact value may be obtained by linear interpolation by using the reciprocal of the number of degrees of freedom; for example, interpolation for 240 degrees of freedom can be done as follows:

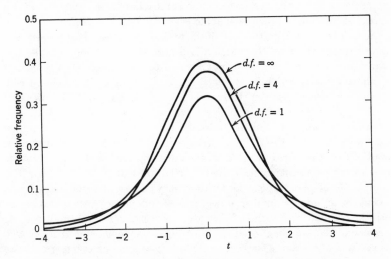

Fig. 4.4. Student's t-distributions with various degrees of freedom.

between the tabulated 5 percentage points for 120 and infinite degrees of freedom

$$1/120 = 0.0083333 \qquad t = 1.980$$
$$1/240 = 0.00416667 \qquad t = 1.970 \quad \text{(by interpolation)}$$
$$1/\infty \;= 0.00000000 \qquad t = 1.960.$$

It is convenient to explain the t-table, starting with the bottom row of numbers for an infinite number of degrees of freedom, for which s^2 is equal to σ^2, because of the following simple algebraic argument. As stated in section 3.7, the mean of the statistic

$$\frac{(n-1)s^2}{\sigma^2}$$

is $n-1$, and the variance is $2(n-1)$.

If each of these statistics is divided by the constant $(n-1)$, it is known by the rule for dividing a statistic by a constant (sec. 3.3), that the new mean is equal to the original mean divided by $(n-1)$, and the new variance is equal to the original variance divided by $(n-1)^2$. The results are that the mean of the statistic

$$\frac{s^2}{\sigma^2}$$

is 1 and the variance is $2/(n-1)$. As the number of degrees of freedom approaches infinity, the variance of s^2/σ^2 approaches 0; therefore for infinity degrees of freedom s^2 becomes equal to σ^2.

The numbers in the bottom row of the t-table are the same for corresponding percentage points of the standardized normal distribution in table A-2; for instance, for the 5-percent point t equals 1.645, meaning that 5 percent of calculated t-values with infinity degrees of freedom will be larger than 1.645. The only difference between the bottom row of table A-4 and the lower half of column 2 of table A-2 is that fewer percentage points are listed in table A-4. Only the upper *tail* of the t-distribution is tabulated. The percentage points of the lower tail are found from the symmetry of the t-distribution about 0. Thus the 95-percent point of the t-distribution has the same numeric value as the 5-percent point of the t-distribution with a negative sign. For example, the 95-percent point for t with infinity degrees of freedom is -1.645.

Thus for an infinite number of degrees of freedom the t-distribution is, in fact, the standardized normal distribution. For smaller numbers of degrees of freedom, the sampling distribution of t is still symmetrical about 0; but the variance of t becomes larger than 1. The variance becomes larger because, for an infinite number of degrees of freedom, only the numerator $\overline{w}$ varies from sample to sample; but for smaller degrees of freedom, both the numerator $\overline{w}$ and the denominator $s/\sqrt{n}$ vary from sample to sample. Whenever s as calculated from a particular sample is smaller than σ, the term $s/\sqrt{n}$ is smaller than $\sigma/\sqrt{n}$, a condition that leads to a spread-out distribution.

In the t-table the numbers become smaller in the lower part of each column because, as the number of degrees of freedom increases, the stability of the estimate s of σ in the denominator increases. The numbers in each row become larger from right to left because, to get in a larger and larger percentage of the t-distribution, we must go farther outward from the mean.

4.4 CONFIDENCE INTERVALS FOR THE POPULATION MEAN

In this section the t-statistic is used to calculate confidence intervals for the population mean μ and for the difference between two population means. The expression for a two-sided confidence interval for the population mean with σ unknown,

$$\overline{w} - t_{5\%} \frac{s}{\sqrt{n}} < \mu < \overline{w} + t_{5\%} \frac{s}{\sqrt{n}},$$

is like that with σ known, presented in section 4.2, except that σ is replaced by its estimate s, and the 5-percent point of the standardized normal distribution, 1.645, is replaced by the corresponding t-value with the appropriate number of degrees of freedom from table A-4. In table 4.2 calculations of

TABLE 4.2. EXAMPLE CALCULATIONS OF TWO-SIDED CONFIDENCE INTERVALS

	Source of data		
Calculations	Fictitious data from table 3.1	Phosphate data from table 3.4	Fresnillo mine, 2200 vein, 305-m level ($\%$ Pb)
$\overline{w}$	5	25.23	9.42
s	2	3.02	8.47
n	5	224	218
$\sqrt{n}$	2.236	14.967	14.731
$s/\sqrt{n}$	0.8945	0.2018	0.5750
d.f.	4	223	217
$t_{5\%}$	2.132	1.651	1.651
$t(s/\sqrt{n})$	1.9	0.33	0.95
μ_L	3.1	24.90	8.47
μ_U	6.9	25.56	10.37

Formulas:

$$\mu_L = \overline{w} - t_{5\%} \frac{s}{\sqrt{n}}$$

$$\mu_U = \overline{w} + t_{5\%} \frac{s}{\sqrt{n}}$$

confidence intervals are presented for three sets of data. The first two sets are the fictitious data used to introduce the calculation of mean and variance in table 3.1 and the phosphate data from table 3.4. The third set is 218 lead values from assays of mine samples taken at 2-meter intervals in the Fresnillo mine. The lead-assay data are typical of base-metal assay data in that the coefficient of variation is 0.90 in contrast to the phosphate data with a coefficient of variation of only 0.12. Although the number of observations is the same in both sets of data, the confidence intervals for the lead assays are almost three times as wide, as we note when we compare the values of $t(s/\sqrt{n})$ for the two cases.

In figure 4.3 fifty 90-percent confidence intervals for random samples of size 10 from a population with a mean of 50 and a known population variance of 100 are plotted. In figure 4.5, a comparison plot is presented, the single change being the different assumption that the population variance is unknown so that its estimate s is calculated for each sample. In figure 4.5, unlike figure 4.3, the confidence intervals are of different lengths because the value of s changes from sample to sample. Although the mean of s is nearly equal to σ, on the average the intervals are wider because the $t_{5\%}$-value for 9 degrees of freedom, 1.833, replaces the corresponding standard-normal-deviate value of 1.645. However, on the average 90 out of 100 intervals still contain the population mean μ. In the case of figure 4.5, 46 (92 percent) contain the population mean, showing good correspondence to theory. The first three misses are the same as those for the case of figure 4.3 with σ assumed known, but the fourth is different. Because s is calculated separately for each sample, exact correspondence with figure 4.3 is not to be expected, although there is some relationship because the means are the same in both cases.

In calculating the confidence intervals, a confidence coefficient of 90 percent was used. If another confidence coefficient is adopted, the value of t is different, and the confidence interval has a different width for the same data. In table 4.3, confidence intervals are calculated for the Fresnillo data using different confidence coefficients. For the confidence coefficient of 50 percent, the width of half the interval, $t(s/\sqrt{n})$, is only 0.39, but there is a 50-percent chance of making a mistake. However, for a confidence coefficient of 98 percent the chance of making a mistake is only 2 percent, but the width of half

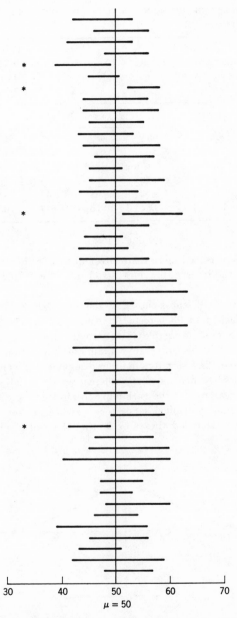

Fig. 4.5. Fifty 90-percent confidence intervals for samples of size 10 from a population with $\mu = 50, \sigma^2 = 100$; σ assumed to be unknown.

Table 4.3. Effect of changing confidence coefficient on length of confidence interval

	Item	Value
Summary data from table 4.2	$\overline{w}$	9.42
for the Fresnillo mine	$s/\sqrt{n}$	0.5750
	d.f.	217

Item	Values				
Confidence coefficient	50%	80%	90%	95%	98%
Percentile of t	$t_{25\%}$	$t_{10\%}$	$t_{5\%}$	$t_{2.5\%}$	$t_{1\%}$
Value of t	0.676	1.286	1.651	1.970	2.342
$t(s/\sqrt{n})$	0.39	0.74	0.95	1.13	1.35
Lower confidence limit, μ_L	9.03	8.68	8.47	8.29	8.07
Upper confidence limit, μ_U	9.81	10.16	10.37	10.55	10.77

the interval has increased more than threefold to 1.35. Figure 4.6 illustrates that narrowing the interval provides less chance of catching the population mean.

The confidence coefficient is chosen by the investigator. We usually use a confidence coefficient of 90 percent for confidence intervals involving the mean, so that, for random samples from normally distributed populations, 90 percent of the confidence intervals contain the population mean μ. Some geologists, and most investigators in other disciplines, conventionally choose higher confidence coefficients. Our preference for the 90-percent level is made for three reasons: (a) the variability in most geological data is larger than in data from controlled laboratory experiments or manufacturing processes, (b) the likelihood always exists that the geological body or other

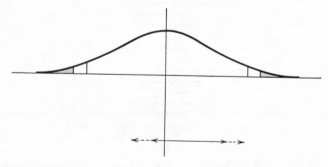

Fig. 4.6. Relation of width of confidence interval to confidence coefficient.

entity under investigation is terminated by an unexpected structural discontinuity, and (c) in applications involving valuation and other subjects of mineral economics, use of the 90-percent level is consistent with the higher risks associated with the mineral industries, risks that are illustrated by the higher interest rates in mineral industries than in manufacturing industries.

From two-sided confidence intervals we now turn to the actual calculation of one-sided confidence intervals by using the t-statistic, where the confidence statement is of the form "with a specified confidence of being correct, the population mean is not lower than a specified value," or of the form "with a specified confidence of being correct, the population mean is not higher than a specified value." The first kind of interval is named a *one-sided lower confidence interval*, and the second kind a *one-sided upper confidence interval*. In figure 4.7 a one-sided lower confidence interval is illustrated schematically. The words "one-sided" refer to the fact that only one side is calculated; actually the other side is present but is known, so there is no chance of its being incorrect. It is known to be either plus or minus infinity—or the extreme value possible, such as 0 or 100 percent or the content of the metal being estimated that is present in the pure mineral. A one-sided 90-percent lower confidence interval is calculated from the relation

$$\overline{w} - t_{10\%} \frac{s}{\sqrt{n}} < \mu < \infty,$$

and a one-sided 90-percent upper confidence interval is calculated from the relation

$$-\infty < \mu < \overline{w} + t_{10\%} \frac{s}{\sqrt{n}}.$$

In mining geology, in which it is usually more important that the ore be above cutoff grade than to set an upper limit to the grade, the one-sided

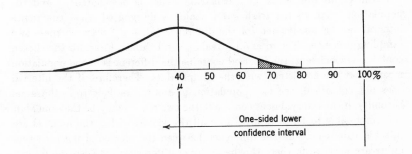

Fig. 4.7. Schematic representation of one-sided confidence interval.

Table 4.4. Example calculations of one-sided lower confidence intervals

		Source of data		
Calculations	Fictitious data from table 3.1	Phosphate data from table 3.4	Fresnillo mine, 2200 vein, 305-m level (% Pb)	Remarks
$\overline{w}$	5	25.23	9.42	From table 4.2
$s/\sqrt{n}$	0.8945	0.2018	0.5750	From table 4.2
d.f.	4	223	217	
$t_{10\%}$	1.533	1.286	1.286	From table A-4
$t(s/\sqrt{n})$	1.4	0.26	0.74	
μ_L	3.6	24.97	8.68	

confidence interval is especially pertinent. In table 4.4 one-sided confidence limits are calculated for the data in table 4.2. Comparison of the one-sided confidence limits in table 4.4 to the two-sided confidence limits in table 4.2 shows that the one-sided limits are substantially higher. In estimating grade of ore, the differences may well be enough to put an estimate above rather than below a cutoff grade.

One more kind of confidence interval for the mean, the confidence interval for the difference between two population means, is now explained. The calculation is made for a two-sided interval in the fictitious illustration of table 4.5. Assume that in a study comparing average compositions of marine and fresh-water shales, a geologist has obtained the Fe analyses in table 4.5. The six marine shales (group 1) yield a mean Fe analysis of 4.883. The seven fresh-water shales (group 2) yield a mean Fe analysis of 4.857. The question is whether the population means for the two kinds of shales are evidently the same or different.

If the population mean of the marine shales is designated μ_1 and the population mean of the fresh-water shales is designated μ_2, a confidence interval can be constructed for the difference $(\mu_1 - \mu_2)$ between these two sampling processes. The interval reveals several things. First, for the chosen confidence coefficient, the interval contains the differences in the population means that are consistent with the sample data. Therefore, if the interval does not contain 0, the two population means are unlikely to be the same. Secondly, if the interval does contain 0, the two means may be the same, but it is impossible to be certain because all the values within the interval are suitable values for the true difference between the means. If both interval limits are close to 0, there can be no large differences between the means. How close is "close" depends on the geologist's judgment.

TABLE 4.5. EXAMPLE CALCULATION OF TWO-SIDED CONFIDENCE INTERVAL FOR DIFFERENCE OF TWO POPULATION MEANS *

Item	Statistical sample number		Combination	
	(1)	(2)		
Observations:	6.9 6.5	4.8 4.7		
	3.8 3.5	3.9 6.0		
	4.9 3.7	5.9 3.7		
		5.0		
$\sum w$	29.3	34.0		
n	6	7		
$\overline{w}$	4.883	4.857	$\overline{w}_1 - \overline{w}_2$	0.026
$(\sum w)^2$	858.49	1156.00		
$(\sum w)^2/n$	143.08	165.14		
$\sum w^2$	154.25	169.84		
SS	11.17	4.70	pooled SS	15.87
d.f.	5	6	pooled d.f.	11
			s_p^2	1.4427
$1/n$	0.1667	0.1429	$1/n_1 + 1/n_2$	0.3096
			$s_p^2(1/n_1 + 1/n_2)$	0.4467
			$\sqrt{s_p^2(1/n_1 + 1/n_2)}$	0.6683
			$t_{5\%}$, with 11 d.f.	1.796
			$t_{5\%}\sqrt{s_p^2(1/n_1 + 1/n_2)}$	1.200
			μ_L	-1.174
			μ_U	1.226

Formulas:

$$s_p^2 = \frac{SS_1 + SS_2}{(n_1 - 1) + (n_2 - 1)}$$

$$\mu_L = (\overline{w}_1 - \overline{w}_2) - t_{5\%}\sqrt{s_p^2(1/n_1 + 1/n_2)}$$

$$\mu_U = (\overline{w}_1 - \overline{w}_2) + t_{5\%}\sqrt{s_p^2(1/n_1 + 1/n_2)}$$

* From Li, 1964, I, p. 151.

In geological terms, there is no evident difference in the Fe analyses of the marine and fresh-water shales. [Although fictitious data are used for illustration, the geological situation is copied from a real investigation of marine and fresh-water shales by Keith and Bystrom (1959).]

In table 4.5 a confidence interval with a 90-percent confidence coefficient is calculated for the example data. The first step is to estimate the difference $(\mu_1 - \mu_2)$ by the difference between the two sample means $(\overline{w}_1 - \overline{w}_2)$; the second is to estimate the variance of the difference $(\mu_1 - \mu_2)$. It is shown later (sec. 5.8) that if both sampling procedures are assumed to have a common variance, the variance of the difference is

$$\sigma^2\left(\frac{1}{n_1} + \frac{1}{n_2}\right).$$

Because the sample sizes are different, it is necessary to give different weights to the estimates s_1^2 and s_2^2 from the two individual samples, which are to be combined to form a *pooled* estimate of σ^2. The appropriate weights are the respective numbers of degrees of freedom, $(n_1 - 1)$ and $(n_2 - 1)$. The pooled estimate s_p^2 is calculated by adding the two SS values and dividing by the sum of the numbers of degrees of freedom. Because the number of degrees of freedom of the pooled estimate is the sum of the individual numbers of degrees of freedom, the appropriate t-value has 11 degrees of freedom. As the last step, the upper and lower confidence limits for the difference $(\mu_1 - \mu_2)$ are calculated. The conclusion may be drawn that there are no important differences between the means because the limits include 0, and moreover the upper and lower limits are not far from 0.

One- and two-sided confidence intervals are now compared and contrasted. If a two-sided confidence interval does not contain the parameter (or function of the parameter) being estimated, the parameter may be either above the upper limit of the interval or below the lower limit of the interval. However, if a one-sided confidence interval does not contain the parameter being estimated, the parameter is known to be either above or below the interval limits. With a two-sided confidence interval, it is conventional to distribute the risk so that one-half of the intervals have their high limits set too low, and the other one-half have their low limits set too high. However, in principle the risk can be distributed in any way desired, although with symmetric frequency distributions, putting one-half of the risk at either end of the interval leads to the shortest intervals. With a one-sided confidence interval, all the risk has been put into one side of the interval. Consequently, the calculated limit for a one-sided confidence interval is closer to the observed statistic than the corresponding limit for a two-sided interval. The two-sided interval is also shorter on the average than the one-sided interval, but if the investigator's interest is really in only one side, this fact is of no interest.

The distribution of risk in confidence intervals is simple for the cases given, but it can become complex. More complicated situations in apportionment of risk are considered in section 5.12 on multiple comparisons.

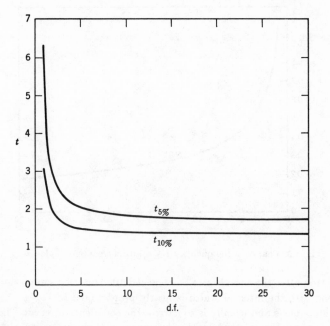

Fig. 4.8. Change in value of selected percentage points of the t-statistic with increasing degrees of freedom.

In conclusion it is important to note that the width of the two-sided confidence interval for the mean depends on the term

$$t \frac{s}{\sqrt{n}}.$$

Normally, it is desirable to make this term small so that the confidence interval will be narrow. The change in this term as the sample size increases is illustrated by figures 4.8 to 4.10. Figure 4.8 shows that, as the sample size increases, t decreases, thus making the term smaller. However, for degrees of freedom larger than about 10, the decrease is relatively small, and the value of t soon becomes very close to that of the standardized normal deviate. Figure 4.9 shows that, as the sample size increases, $1/\sqrt{n}$ decreases, also making the term smaller. The rate of decrease slows as the sample size gets larger, as shown by the fact that the sample size must be multiplied by 4 to cut the factor $1/\sqrt{n}$ in half. Combining these factors (fig. 4.10), one can see that the shrinkage of the average half-interval length is dominated by the

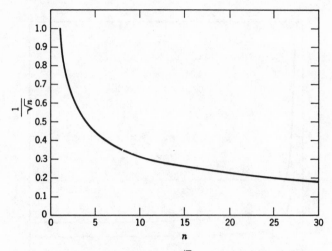

Fig. 4.9. Change in the quantity $1/\sqrt{n}$ with increasing sample size.

factor $1/\sqrt{n}$ after the sample size has reached 10. Thus, for samples of sizes larger than 10, the interval width is nearly proportional to $1/\sqrt{n}$.

As stated, one's aim usually is to narrow the confidence interval. This can be done in several ways. As shown by the graphs, n can be increased, although the advantage does not increase very fast above a sample size of about 10.

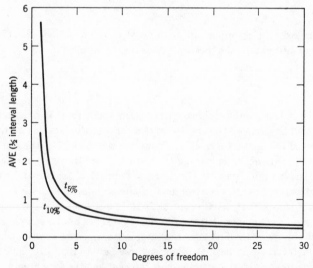

Fig. 4.10. Average of half-interval length divided by σ for increasing degrees of freedom.

More risk of being wrong can also be accepted by choosing a lower confidence coefficient. The problem can also be formulated so that a one-sided rather than a two-sided confidence interval is calculated. A final way is to decrease the size of s. Part of the variation expressed by s is natural variation inherent in the rock body or other material sampled. But part is introduced by the procedures of sampling, sample preparation, and assaying or chemical analysis. This part can be reduced by more careful methodology, as discussed in Chapter 8.

2/ 3044

4.5   EXAMPLE ESTIMATION OF ORE GRADE WITH CONFIDENCE INTERVALS

Confidence intervals are one of the most useful statistical concepts for geology. Geologists have traditionally been inclined toward interval rather than point estimates because their data are characterized by large and changing variability. Through the confidence interval, the variability, sample size, and mean can be incorporated in a single quantitative statement. To illustrate the value of confidence limits in a specific geological situation, we turn to the problem of ore estimation from diamond-drill-hole cores from a typical vein. The procedures are readily transferable to other geological situations. As usual when one is dealing with the real world, complications arise in this analysis, complications that are treated in section 4.9.

Can useful estimates of metal content be made by boring diamond-drill holes through a vein? That is, can information from drill holes provide usable quantitative results, or are the drill holes useful only to locate vein structures? Our analysis (Koch and Link, 1964), using confidence limits calculated for typical data to answer this question, is summarized here.

The data for the example analysis are assay values from diamond-drill cores from 18 boreholes through the Don Tomás vein of the Frisco mine (sec. 1.6). The Don Tomás vein (fig. 4.11) is a well-defined geologic structure that was first found on the lower levels of the mine and pinches out above level 9. When the data were obtained, development was essentially complete to level 15; drifts on each level had been driven to where the vein became too narrow or the grade became too low to warrant additional development. The entire vein is below the zones of oxidation and supergene enrichment. The Don Tomás vein is cut by the Frisco fault whose displacement is 35 meters or more. As discussed in section 5.4, because the distribution of metals is markedly different on either side, we believe this fault to be of pre-ore age. Two post-ore faults (100-S and 480-N faults) displace the vein horizontally about 14 and 11 meters, respectively, but the components of displacement in the plane of the longitudinal section are negligible.

Table 4.6. Confidence interval estimates for the Don Tomás vein based on assays from 18 diamond-drill core samples and 1829 drift samples

| | Estimate from 18 core samples | | | Estimate from 1829 drift samples | | |
| | | Confidence limits | | | Confidence limits | |
Item	Mean	Lower	Upper	Mean	Lower	Upper
Gold content	0.27	0.09	0.45	0.28	0.26	0.30
Silver content	278.0	158.0	398.0	329.0	316.0	342.0
Lead content	6.6	3.5	9.7	8.6	8.3	8.9
Copper content	0.36	0.23	0.49	0.49	0.46	0.52
Zinc content	8.6	4.9	12.3	11.2	10.8	11.6

For this study metal content was measured in meter-percent per metric ton for base metals and in meter-grams per metric ton for precious metals. Reasons for use of these units, obtained by multiplying assay values by vein widths, are taken up in section 13.2. All drifts had been sampled by chip samples at 2-meter intervals (sec. 1.4). In all, 18 holes were bored between 1940 and 1958 at locations plotted on figure 4.11. Cores of either AX or EX size were recovered and split; half of the core in vein was assayed and the other half kept for reference. Although core recoveries were not available for all holes, they were generally about 90 percent, judging from five holes logged by one of us (GSK). Sludge was not assayed. Most holes were bored at essentially right angles to the vein, but measured widths were reduced for three holes that were not.

In table 4.6 two kinds of estimates for metal content in the Don Tomás vein are presented, one based on 1829 drift samples and the other based on 18 diamond-drill holes. The table emphasizes that, although the estimated standard deviations in each case are similar, the confidence intervals based on 1829 samples are much narrower because the sample size is so much greater. In fact, as expected, the intervals are about 10 times as wide for the data from the 18 diamond-drill holes, because

$$\sqrt{18} \approx \frac{\sqrt{1829}}{10}.$$

Thus the estimate from the large sample of drift assay values forms a standard for comparison of the less precise estimates made from assays of the drill-hole samples. Nonetheless, the less precise estimates are successful in that, for all five metals, the confidence limits include the standard means from the assays of drift samples.

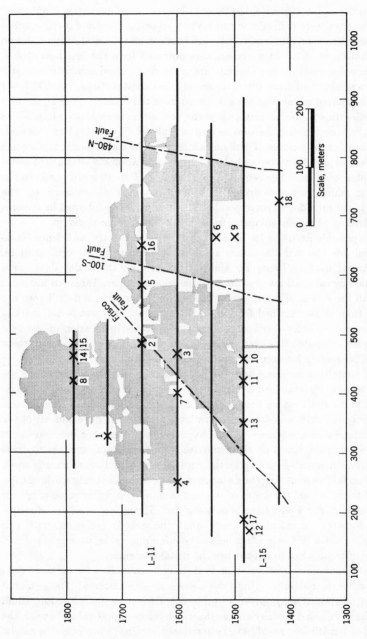

Fig. 4.11. Don Tomás vein, vertical longitudinal section. Observer looking west. Drifts shown by solid lines, other workings by gray pattern, faults by dashed line, locations of diamond drill holes by "x's" with hole numbers.

Considering, in sequence, the means obtained from increasing numbers of holes as they were drilled, one can make a second estimate of metal content for the entire vein. Obviously, little information about metal content and no information at all about variation were obtained from the first hole drilled, but more information was obtained from the first-second combinations; still more was obtained from the first-second-third combinations; etc. Table 4.7 gives estimated metal contents calculated from the first hole (line 1), the first two holes (line 2), and so forth, up to the best estimate calculated from all 18 holes (line 18), which is the same as that of table 4.6. The table illustrates that, in general, as the number of drill holes increases, the confidence limits become narrower or, in other words, that an increasingly accurate estimate of metal content is obtained. The sequence of holes from 1 to 18 corresponds to the order in which they were drilled. This order is somewhat arbitrary, and the order is not intended to correspond to one that would be followed in exploration drilling of a vein unknown from development mine workings.

It is possible from the 18 diamond-drill holes to specify confidence limits that include the metal contents actually obtained from the drift samples. Similar satisfactory limits for tonnages, based on the width data, were obtained by calculations presented in the original paper. Table 4.6 indicates that all the means of metal content calculated from diamond-drill core are lower than those obtained from the drift samples. This fact is not surprising because the holes were bored where the existence or location of the vein was most in doubt. However, when the sample variance is considered, there is no discrepancy between the two methods of sampling.

The confidence intervals based on data from the 18 diamond-drill holes are wide because the standard deviations are large, as is characteristic for base-metal veins and because the number of observations is small. Now assume that only the drill-hole data were available prior to the development of the vein through mine workings. In this hypothetical situation, a decision would have been made, based on the estimates from these data, whether to drive development workings to expose the vein. Through analysis of mining costs, based on factors including the distance of the vein from existing workings, the dip of the vein, and the width of the vein, a minimum dollar value of ore for profitable mining would have been computed. Through application of metal-price factors, the calculated grade would have been compared with this minimum value. If the grade is below the minimum value, we might proceed in one of three ways to re-estimate the metal contents.

First, a one-sided confidence interval could be computed for the same data by putting the risk entirely into the lower side of the interval. The results of such a computation are given in table 4.8. Of course this procedure could have been followed initially and might well have been preferable. Second, the confidence coefficient could have been reduced, say, to 75 percent—the results

TABLE 4.7. ESTIMATES OF METAL CONTENT FOR THE DON TOMÁS VEIN BASED ON CUMULATIVE ASSAYS OF CORES FROM 1 TO 18 DRILL HOLES BORED IN SEQUENCE

| | Point estimates of metal content | | | | | 90-percent confidence intervals | | | | | | | | | |
| | | | | | | Gold content (ppm) | | Silver content (ppm) | | Lead content (%) | | Copper content (%) | | Zinc content (%) | |
Hole no.	Gold content (ppm)	Silver content (ppm)	Lead content (%)	Copper content (%)	Zinc content (%)	Lower	Upper	Lower	Upper	Lower	Upper	Lower	Upper	Lower	Upper
1	0.15	17	1.8	0.09	1.4										
2	0.20	66	2.9	0.08	7.3	0	0.48	0	377	0	10.2	0.02	0.14	0	44.4
3	0.53	172	4.2	0.09	6.8	0	1.51	0	492	0	8.4	0.05	0.13	0	16.9
4	0.40	172	4.4	0.13	6.4	0	1.04	0	355	2.0	6.9	0.04	0.22	0.5	12.2
5	0.32	316	6.2	0.24	6.1	0	0.80	0	649	2.0	10.4	0	0.50	2.0	10.2
6	0.27	277	5.4	0.22	5.3	0	0.65	8	546	1.7	9.1	0.02	0.42	1.7	8.9
7	0.41	272	6.2	0.25	5.4	0	0.84	52	491	2.8	9.5	0.07	0.42	2.4	8.3
8	0.37	242	5.7	0.24	5.3	0	0.74	49	436	2.7	8.7	0.10	0.39	2.8	7.8
9	0.33	294	6.1	0.26	5.7	0	0.66	101	487	3.4	8.8	0.13	0.39	3.4	7.9
10	0.30	265	5.5	0.24	5.1	0	0.59	87	443	2.9	8.1	0.12	0.36	2.9	7.4
11	0.27	249	5.6	0.23	5.8	0	0.54	87	411	3.2	7.9	0.13	0.34	3.5	8.1
12	0.34	249	5.4	0.26	6.1	0.06	0.61	103	396	3.3	7.5	0.15	0.37	3.9	8.3
13	0.31	250	5.8	0.26	7.6	0.06	0.56	116	384	3.7	7.9	0.16	0.36	4.3	10.9
14	0.29	253	7.7	0.33	9.6	0.05	0.52	130	377	3.9	11.6	0.18	0.48	4.9	14.3
15	0.27	239	7.3	0.33	9.3	0.05	0.49	123	356	3.7	11.0	0.19	0.47	4.9	13.7
16	0.28	241	7.0	0.32	8.9	0.08	0.49	132	349	3.6	10.5	0.18	0.45	4.7	13.1
17	0.27	236	6.7	0.35	9.0	0.07	0.46	134	338	3.4	10.0	0.21	0.49	5.1	12.9
18	0.27	278	6.6	0.36	8.6	0.09	0.46	158	398	3.5	9.7	0.23	0.49	4.9	12.3

TABLE 4.8. ONE-SIDED LOWER CONFIDENCE INTERVAL ESTIMATES FOR THE DON TOMÁS VEIN, BASED ON ASSAYS OF CORES FROM 18 DIAMOND-DRILL HOLES

Item	Mean	One-sided lower confidence limit, 90%	One-sided lower confidence limit, 75%
Gold content	0.27	0.13	0.20
Silver content	278.0	186.0	230.0
Lead content	6.6	4.3	5.4
Copper content	0.36	0.26	0.31
Zinc content	8.6	5.8	7.1

are given in table 4.8. Reduction in the confidence coefficient could also have been made for the two-sided interval. Third, we could improve the quality of the data, or obtain more data. A small reduction might be made in the estimated standard deviations by assaying all of the core instead of only half (sec. 7.4). More holes could be drilled, although this procedure would be expensive because a total of about 72 holes would be needed to cut the interval widths in half through the increase in the sample size (because $1/\sqrt{72} = 1/2\sqrt{18}$). The 72 holes would have only a small effect on the reliability of s for the purpose of constructing one-sided confidence limits for the mean because, with a 90-percent confidence coefficient, the t-value for 17 degrees of freedom is 1.333, and the corresponding value for 71 degrees of freedom decreases to 1.294.

4.6   CONFIDENCE INTERVALS FOR THE POPULATION VARIANCE

The previous sections of this chapter mainly concern the calculation of confidence intervals for a population mean. Confidence intervals can be calculated for any other parameter for which a method has been devised. This section briefly describes a method for calculating confidence intervals for the population variance. One way to use confidence intervals for the variance is in estimating the variability of a chemical element in a geological body; another way is in estimating variability in a fossil population.

In section 3.7 it is shown that the distribution of the statistic $(n-1)s^2/\sigma^2$ follows the chi-square distribution with $(n-1)$ degrees of freedom. If the value of this statistic is calculated for repeated random samples of a given size from a population with a given variance σ^2, 95 percent of the calculated values will be larger than the chi-square 95-percent point from table A-3, and 5 percent will be larger than the chi-square 5-percent point; that is, 90 percent of the calculated values will fall between these two tabulated values, satisfying, for nine samples out of 10, the inequality

$$\chi^2_{95\%} < \frac{(n-1)s^2}{\sigma^2} < \chi^2_{5\%}.$$

If the reciprocal of the above inequality,

$$\frac{1}{\chi^2_{95\%}} > \frac{\sigma^2}{(n-1)s^2} > \frac{1}{\chi^2_{5\%}},$$

is multiplied by $(n-1)s^2$, the result is

$$\frac{(n-1)s^2}{\chi^2_{95\%}} > \sigma^2 > \frac{(n-1)s^2}{\chi^2_{5\%}},$$

which is a 90-percent confidence interval for the population variance σ^2.

A lower one-sided confidence interval for the population variance may be calculated, starting with the relation

$$\chi^9_{90\%} < \frac{(n-1)s^2}{\sigma^2} < \infty.$$

If the reciprocal of the above inequality,

$$\frac{1}{\chi^2_{90\%}} > \frac{\sigma^2}{(n-1)s^2} > 0,$$

is multiplied by $(n-1)s^2$, the result is

$$\frac{(n-1)s^2}{\chi^2_{90\%}} > \sigma^2 > 0.$$

As with confidence intervals for the population mean, the confidence coefficient is chosen by the investigator. We again adopt the value of 90 percent.

4.7 HYPOTHESIS TESTS

Sections 4.4 to 4.6 on confidence intervals explain ways to estimate a reasonable range of parameter values from information contained in a sample. Now, in this section on hypothesis tests a statistical method is explained to answer such questions as: Is the information contained in a sample consistent with a postulated parameter value? At the end of the section, statistical inference and hypothesis tests are contrasted with inference by confidence limits. Although confidence limits are more useful than hypothesis tests for most geological purposes, the discourse on hypothesis tests has some application and also will strengthen the reader's understanding of sampling distributions.

A hypothesis test decides which of two alternatives is correct. Such a dichotomy is often encountered in everyday life: a jury finds a defendant innocent or guilty; a doctor decides that his patient is sick or well. In each of these situations, a decision is made between only two alternatives. The decision is either right or wrong. In a trial, the decision is right if a guilty person is convicted or if an innocent person is acquitted; the decision is wrong if an innocent person is convicted or if a guilty person is acquitted. In a doctor's office the decision is right if a sick person is diagnosed to be sick or if a well person is diagnosed to be well; the decision is wrong if a well person is diagnosed to be sick or if a sick person is diagnosed to be well. Although the alternative may be made more complicated, for example, by considering the degree of guilt of a convicted person, or the degree of sickness of a person diagnosed sick, the two-choice situation is the easiest to think about and to discuss. As applied in statistics, the hypothesis test defines a more or less artificial framework for making inferences about data and arrives at the necessary dichotomy by introducing some rather artificial restraints into the problem. Just as in the practice of law or medicine, complicated social or medical situations can be forced into a simple pattern and useful results obtained, so in the practice of statistics, useful decisions can be made.

Principles and Nomenclature

In order to explain hypothesis test procedures, a highly artificial situation is introduced in which the problem is to choose between two population means, the assumption being that one is correct. An analysis is made of a set of fictitious data consisting of 25 observations drawn at random from a population with a known mean μ of 25 and a known variance σ^2 of 25. The first step is to introduce the terminology of hypothesis testing. Suppose that the problem is to decide whether a sample comes from a population with a mean μ of 25 or a mean μ of 27. The observations are assumed to follow a normal distribution with a variance σ^2 of 25, but, of course, the population mean is assumed to be unknown. Because the population mean μ is to be investigated, it is natural to base the decision on the value of the sample mean $\bar{w}$. Clearly, if the sample mean is large, the decision will tend to favor a population mean of 27, whereas, if the sample mean is small, the decision will tend to favor a population mean of 25. The problem thus resolves itself into defining a single value so that, if the sample mean is larger, the population mean of 27 is chosen; whereas if the sample mean is smaller, the population mean of 25 is chosen. Because the decision must be based on a statistic whose value fluctuates according to its sampling distribution, a wrong decision may be made. A wrong decision corresponds to one of two mistakes: the decision can be for a population mean of 25 when the true mean is 27, or the decision can be for a population mean of 27 when the true mean is 25.

The preceding general remarks are now restated formally in the technical language of hypothesis testing. The hypothesis chosen is that μ is equal to 25, which may be written

$$H_0: \mu = 25.$$

The alternative hypothesis is that μ is equal to 27, which may be written

$$H_1: \mu = 27.$$

When a choice is made between the hypotheses H_0 and H_1 on the basis of a sample, there is always a risk of making an incorrect decision. If the alternative hypothesis H_1 that μ is equal to 27 is chosen, and if the hypothesis H_0 is actually true, a true hypothesis has been rejected, and the mistake made is named a *type I error*. On the other hand, if the hypothesis H_0 that μ is equal to 25 is chosen, and if the alternative hypothesis H_1 is actually true, a false hypothesis has been accepted, and the mistake made is named a *type II error*. The names of the two types of mistakes are arbitrary and conventional.

Because the decision in a hypothesis test is based on information from a sample, there is always the probability of making a type I error. This probability, named the α *risk level* (also designated *level of significance* in some other books), is the probability of rejecting a true hypothesis H_0 and accepting the false alternative hypothesis H_1. The α risk level is chosen by the investigator; we choose an α value of 10 percent. Once the α risk level is set, the probability of making a type II error, named the β *risk level*, can be found. The β risk level is the probability of accepting a false hypothesis H_0 and rejecting the true alternative hypothesis H_1. In table 4.9, the two hypotheses and the two kinds of outcomes of a hypothesis test are listed in a table with two rows and two columns.

TABLE 4.9. HYPOTHESIS-TEST OUTCOMES

Decision of investigator	Actual status of hypothesis	
	(H_0 true; H_1 false)	(H_0 false; H_1 true)
Accept hypothesis H_0	Correct decision	Incorrect decision Type II error Percentage risk $= \beta$
Reject hypothesis H_0; accept alternative hypothesis H_1	Incorrect decision Type I error Percentage risk $= \alpha$	Correct decision

In terms of the legal analogy, in a criminal trial the jury accepts either the hypothesis H_0 that the defendant is innocent or the alternative hypothesis H_1 that the defendant is guilty. If the jury accepts either hypothesis when it is in fact true, no mistake is made. But if the jury accepts the hypothesis H_0 that the defendant is innocent, when in fact he is guilty, it has committed a type II error. On the other hand, if the jury accepts the hypothesis H_1 that the defendant is guilty, when in fact he is innocent, it has committed a type I error. Of course, in the United States and in many other countries, the intent in criminal law is to choose a low α risk level so that few innocent persons will be sent to prison—at the expense of a high β risk level, even though some criminals will thereby escape conviction.

To make a hypothesis test one must choose a statistic. For the example problem, the sample mean is a natural choice that also is statistically valid, because the sample mean is a good estimate of the population mean under a wide variety of circumstances and is the best estimate if the observations follow a normal distribution. The value of the sample mean $\bar{w}$ may be plotted as a point on a line extending from minus infinity to plus infinity. This line must be partitioned into segments, one group containing values of $\bar{w}$ for which the hypothesis H_0 is to be accepted, and the other group containing values of $\bar{w}$ for which the alternative hypothesis H_1 is to be accepted. In statistical nomenclature, these two groups of line segments are named *regions*, although in all cases discussed in this book the regions are segments of one straight line, and there is no connotation of area. In particular, the group of line segments for which the alternative hypothesis H_1 is to be accepted is named the *critical region*, and a statistic is said to lie *inside* or *outside* a critical region.

In principle the line segment could be partitioned into many pieces, some inside the critical region and some outside. What guide then can be used to define the critical region? For one thing, it is clear that the size of the β risk level should be as small as possible. Although sometimes this goal is difficult to accomplish, for the specific case under discussion it can be shown that a critical region consisting of the line segment to the right of a *cutting point*, designated Ω, minimizes the size of β, that is, the probability of making a type II error. Therefore the hypothesis H_0 is accepted if the sample mean is less than Ω, and the alternative hypothesis H_1 is accepted if the sample mean is more than Ω.

Now that the critical region has been generally defined, all that remains is to evaluate Ω. Associated with the line (fig. 4.12) on which the sample means lie is a frequency distribution representing the sampling distribution of the sample mean. Therefore associated with any line segment there is an area under the frequency curve for the *true* population mean and variance, an area corresponding to a probability. Accordingly, the evaluation of the

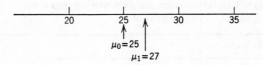

Fig. 4.12. Hypothetical means for example test of hypothesis H_0: $\mu = 25$, H_1: $\mu = 27$.

cutting point Ω is made by a calculation such that, if the hypothesis H_0 is true, the probability of making a type I error is equal to α, the area under the frequency curve that applies in this case.

Because α is chosen equal to 10 percent, in the example problem, the location of the cutting point Ω can be easily calculated since, if the hypothesis H_0 that μ is equal to 25 is true, the distribution of the sample mean has a mean of 25 and a variance equal to the population variance divided by the sample size n. Thus, if the mean is normally distributed, by using the 10-percent point from the standardized normal distribution (table A-2), the result is that

$$\Omega = 25 + 1.282 \frac{\sigma}{\sqrt{n}}.$$

The probability that the mean of a random sample will be larger than Ω if the hypothesis H_0 is true is 10 percent. Notably, the decisions about the hypothesis to be tested, the alternative hypothesis, and the α risk level of making a type I error can all be made before any observations are obtained, or even before a sample size is decided upon.

In order to illustrate the initial test of the hypothesis H_0 that μ equals 25 against the alternative hypothesis H_1 that μ equals 27, and to illustrate subsequent hypothesis tests, an example set of fictitious data consisting of 25 observations is randomly drawn from a population with a known mean μ of 25 and a known variance σ^2, also of 25. The fictitious data may be regarded as phosphate assays, alumina determinations from igneous rocks, or other variables, as the reader prefers. As the subject is developed, it will be pretended that one or both of the two parameters are unknown and that their estimates $\bar{w}$ of 24.36 and s^2 of 16.49 are to be used in their places. In table 4.10 the original observations are listed, and the sample means are calculated.

Example Hypothesis Test. H_0: $\mu = 25$, H_1: $\mu = 27$

The first example is to perform the actual calculations for the hypothesis test introduced in the preceding subsection on principles and nomenclature. The test is of the hypothesis H_0 that μ equals 25 against the alternative hypothesis H_1 that μ equals 27. Now that the observations in table 4.10 are presumed to be at hand, the calculations for the hypothesis test can be

TABLE 4.10. OBSERVATIONS AND SAMPLE MEAN FOR
EXAMPLE HYPOTHESIS TESTS *

Observations:

29	20	21	26	32
19	29	26	27	33
22	21	21	23	25
26	20	25	27	19
30	21	21	24	22

Computations:

Item	Value
n	25
$\sum w$	609
$\overline{w}$	24.36

* From Li, 1964, I, pp. 65 and 575.

performed, the pretension being that the population variance is known but
that the population mean is unknown. In table 4.11 a rather formal procedure
is set up in a format that is followed throughout this section. The hypothesis
and alternative hypothesis are listed first. The statistic to be computed to
test the hypothesis is the sample mean $\overline{w}$. The chosen α risk level, the size of

TABLE 4.11. EXAMPLE TEST OF HYPOTHESIS H_0: $\mu = 25$, H_1: $\mu = 27$

Hypothesis:	H_0: $\mu = 25$
Alternative hypothesis:	H_1: $\mu = 27$
Statistic:	$\overline{w}$
Risk of type I error:	$\alpha = 10\%$
Critical region:	$\overline{w} > \Omega$

Computation:

Line	Item	Value	Remarks
1	σ	5	
2	n	25	
3	$\sqrt{n}$	5	
4	$\sigma/\sqrt{n}$	1	Standard error of the mean
5		1.282	10% point from table A-2
6	Ω	26.282	$25 + 1.282\sigma/\sqrt{n}$
7	$\overline{w}$	24.36	

Conclusion.   Accept hypothesis H_0 that $\mu = 25$.

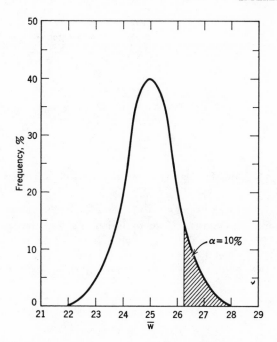

Fig. 4.13. Sampling distribution of $\bar{w}$, showing size of α risk level if H_0 is true for example test of hypothesis in table 4.11.

type I error, is 10 percent. The population standard deviation σ is assumed to be known to be equal to 5. Two other assumptions are made in this test, as in all other tests of significance in this section: (a) the major one that the observations are a random sample from the population of interest, and (b) the minor one that the theoretical frequency distribution of the observations is normal. In section 4.9 the consequences that follow when data do not fit these assumptions are considered.

If the hypothesis H_0 that μ equals 25 is true, the sampling distribution of $\bar{w}$ is that plotted in figure 4.13, because σ is known to be 5, and n is known to be 25. On computation, the value of Ω is found to be 26.282 (line 6). Because the sample mean of 24.26 is smaller than 26.282 (fig. 4.14), it is *outside* the critical region, and the conclusion is to accept the hypothesis H_0.

For the example of table 4.11 the β risk level, the size of the type II error, can be readily calculated. A type II error can be made only if the alternative hypothesis H_1 that μ equals 27 is true; whereupon the statistic $\bar{w}$ follows the sampling distribution with mean equal to 27, graphed as a solid line in

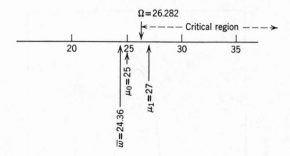

Fig. 4.14. Critical region for example test of hypothesis in table 4.11.

figure 4.15. The probability of a type II error is equal to the size of the area under the curve with mean at 27 and to the left of 26.282, the cutting point Ω. The calculations to obtain this area are given in table 4.12. In line 2, Ω, found to be 26.282 in table 4.11, is calculated to be -0.718 unit from 27. In line 3, -0.718 is divided by the standard error of the mean, 1, from line 1, to

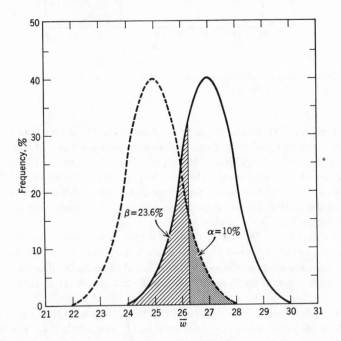

Fig. 4.15. Sampling distribution of $\bar{w}$, showing sizes of α and β risk levels for example test of hypothesis in table 4.11.

TABLE 4.12. EXAMPLE CALCULATION OF β RISK LEVEL

Alternative hypothesis: $H_1: \mu = 27$
Statistic: $\overline{w}$
Ω: 26.282

Computation:

Line	Item	Value	Remarks
1	$\sigma/\sqrt{n}$	1	From table 4.11
2	$\Omega - 27$	-0.718	Boundary of critical region minus μ, $26.282 - 27$
3	$(\Omega - 27)/(\sigma/\sqrt{n})$	-0.718	s.n.d.
4	$100\% - \beta$	76.4%	From table A-2
5	β	23.6%	

yield -0.718. From table A-2 by interpolation, or directly from a larger table, -0.718 is found to correspond to 76.4 percent; in other words, 76.4 percent of sample means lead to the correct conclusion if the hypothesis H_1 that μ equals 27 is true. Because 76.4 percent give the correct result, $(100 - 76.4)$ or 23.6 percent give the wrong result if H_1 is true, thus giving the size of the type II error or the β risk level.

Example Hypothesis Test. $H_0: \mu = 25$, $H_1: \mu < 25$ or > 25

The test of the hypothesis H_0 that μ equals 25 against the alternative hypothesis H_1 that μ equals 27 has been discussed in detail. But usually the alternative hypothesis is less specific, and sometimes the hypothesis itself is also less specific. In the remaining examples three of the less specific cases are explained. In the first the case of a specific hypothesis with a less specific alternative hypothesis is considered. In the second, the case of a less specific hypothesis and a less specific alternative hypothesis is considered. In the last, the hypothesis test with the population variance unknown is considered.

The test of the hypothesis H_0 that μ equals 25 against the alternative hypothesis H_1 that μ is smaller or larger than 25 is now considered. Clearly, the hypothesis H_0 should be rejected if the sample mean $\overline{w}$ is too much larger or smaller than 25. Thus the critical region should consist of two line segments, one to the right of 25 and one to the left of 25, with the segment of the line in the vicinity of 25 being outside the critical region. The 10-percent α risk of making a type I error must be divided between the two line segments constituting the critical region. Although the risk need not be divided evenly, there is no reason in this example not to do so, and therefore it is divided evenly. Because the critical region is divided into two line segments that lie

Table 4.13. Example test of hypothesis H_0: $\mu = 25$, H_1: μ is smaller or larger than 25

Hypothesis:	H_0: $\mu = 25$
Alternative hypothesis:	H_1: $\mu < 25$ or $\mu > 25$
Statistic:	$\bar{w}$
Risk of type I error:	$\alpha = 10\%$
Critical region:	$\bar{w} < \Omega_L = 25 - 1.645\sigma/\sqrt{n}$ or
	$\bar{w} > \Omega_U = 25 + 1.645\sigma/\sqrt{n}$

Computation:

Line	Item	Value	Remarks
1	$\sigma/\sqrt{n}$	1	Standard error of the mean
2		1.645	5% point from table A-2
3	Ω_L	23.355	$25 - 1.645\sigma/\sqrt{n}$
4	Ω_U	26.645	$25 + 1.645\sigma/\sqrt{n}$
5	$\bar{w}$	24.36	

Conclusion. Accept hypothesis H_0 that $\mu = 25$.

under the tails of the assumed H_0 frequency distribution, this type of hypothesis test is named a *two-tailed* test, in contrast to the hypothesis test of the first example which is named a *one-tailed* test.

In table 4.13 the hypothesis-test calculations are performed by using the set of observations from table 4.10. The α risk level is still chosen equal to 10 percent; this area is divided equally between the two tails of the postulated distribution, as shown in figure 4.16. Each half of the critical region is defined with the use of the 5-percent (or 95-percent) point from the standardized normal distribution (table A-2). Because $\bar{w}$ lies outside the critical region, the hypothesis H_0 that μ equals 25 is accepted.

Example Hypothesis Test. H_0: $\mu \leq 25$, H_1: $\mu > 25$

The next example of a hypothesis test is a generalized version of the one-tailed test in table 4.11, in which the hypothesis H_0 that μ equals 25 is tested against the alternative hypothesis H_1 that μ equals 27. In the new example, table 4.14, the hypothesis H_0 that $\mu \leq 25$ is tested against the alternative hypothesis H_1 that $\mu > 25$. Again, the critical region is chosen to maximize the chance of rejecting the hypothesis H_0 if it is false. All the calculations are the same as those for table 4.11. However, although the α risk level is still 10 percent, the size of the α area now is 10 percent only if μ is exactly 25; otherwise the size of the α area is between 0 and 10 percent if μ is smaller than 25. The value of Ω is still 26.282 and is calculated for the case in which μ is

Table 4.14. Example test of hypothesis H_0: μ is equal to or less than 25, H_1: μ is larger than 25

Hypothesis:	H_0: $\mu \leq 25$
Alternative hypothesis:	H_1: $\mu > 25$
Statistic:	$\overline{w}$
Risk of type I error:	$\alpha = 10\%$
Critical region:	$\overline{w} > \Omega$
Boundary of critical region:	26.282
$\overline{w}$:	24.36

Conclusion. Accept hypothesis H_0 that $\mu \leq 25$.

exactly 25, which corresponds to the cutting point between acceptance of the hypothesis H_0 and the alternative hypothesis H_1.

Example Hypothesis Test. H_0: $\mu \leq 22$, H_1: $\mu > 22$, σ unknown

In the preceding examples σ is assumed to be known. Actually, in practice σ must usually be replaced by its estimate s derived from the sample. An

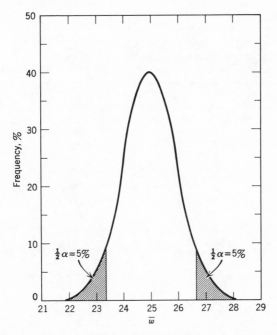

Fig. 4.16. Sampling distribution of $\overline{w}$ if H_0 is true for example test of hypothesis H_0: $\mu = 25$, H_1: μ is smaller or larger than 25.

example of the procedure is given in the test of table 4.15, where the data from table 4.10 are used, but the assumption is made that both σ and μ are unknown. To make a realistic problem, assume that the observations in table 4.10 are phosphate assays from a mine with a cutoff grade of 22 percent. Because several assays in table 4.10 are below 22 percent, there is a question whether the true grade is above 22 percent. The appropriate test is to set up the hypothesis H_0 that $\mu \leq 22$ percent and compare it with the alternative hypothesis H_1 that $\mu > 22$ percent. The reason for stating the hypothesis in this form is that the risk of mining nonpaying phosphate rock is controlled by specifying the α risk level. In this test, the t-statistic rather than the sample mean $\overline{w}$ is chosen for the test statistic because it is necessary to introduce the t-sampling distribution for 24 degrees of freedom in place of the standardized normal distribution. The value of Ω is the tabulated t-value of 1.318; calculated t-values larger than 1.318 are inside the critical region. The only basis

TABLE 4.15. EXAMPLE TEST OF HYPOTHESIS H_0: μ IS EQUAL TO OR LESS THAN 22, H_1: μ IS LARGER THAN 22, FOR THE FIRST SET OF DATA

Hypothesis:	H_0: $\mu \leq 22$
Alternative hypothesis:	H_1: $\mu > 22$
Statistic:	t
Risk of type I error:	$\alpha = 10\%$
Critical region:	$t > \Omega$ ($t_{10\%} = 1.318$, with 24 d.f.)

Computations:

Line	Item	Value
1	$\sum w$	609
2	n	25
3	$\overline{w}$	24.36
4	$(\sum w)^2$	370,881
5	$(\sum w)^2/n$	14,835.24
6	$\sum w^2$	15,231
7	SS	395.76
8	d.f.	24
9	s^2	16.49
10	s	4.06
11	$\sqrt{n}$	5
12	$s/\sqrt{n}$	0.812
13	$\overline{w} - 22$	2.36
14	t	2.91

Conclusion. Reject hypothesis H_0 that $\mu \leq 22$. Accept alternative hypothesis H_1 that $\mu > 22$.

for rejecting the hypothesis H_0 is by calculated values that are too large, never by ones that are too small. Because the computed value of t, 2.91, is larger than Ω, 1.318, and is therefore inside the critical region, the hypothesis H_0 is rejected, and the alternative hypothesis that $\mu > 22$ is accepted. In this case, a correct decision has been made because the sample is known to have been drawn from a population with μ equal to 25.

In the fictitious example of phosphate cutoff grade of table 4.15 the hypothesis test reaches a correct decision. Because the sample is actually drawn from a population with μ equal to 25, a type I error can never be made, but because the α risk level is 10 percent a type II error is necessarily made in 10 percent of the cases in which H_0 is true. An example of how type II error can be made is provided by the random sample of table 4.16, which is

TABLE 4.16. EXAMPLE TEST OF HYPOTHESIS H_0: μ IS EQUAL TO OR LESS THAN 22, H_1: μ IS LARGER THAN 22, FOR THE SECOND SET OF DATA

Observations:

22	21	14	19	23
20	15	19	21	21
22	21	35	28	23
16	23	14	27	18
20	21	24	23	22

Computations:

Line	Item	Value
1	$\sum w$	532
2	n	25
3	$\overline{w}$	21.28
4	$(\sum w)^2$	283,024
5	$(\sum w)^2/n$	11,320.96
6	$\sum w^2$	11,806
7	SS	485.04
8	d.f.	24
9	s^2	20.21
10	s	4.49
11	$\sqrt{n}$	5
12	$s/\sqrt{n}$	0.898
13	$\overline{w} - 22$	-0.72
14	t	-0.802

Conclusion. Accept hypothesis H_0 that $\mu \leq 22$. (Type II error is made.)

also drawn from the population with μ equal to 25 and σ^2 equal to 25. The same hypothesis test is made as that in table 4.15, and when the calculations are performed, a t-value of -0.802 is obtained. Because this t-value is outside the critical region, the hypothesis H_0 that $\mu \leq 22$ is accepted. A false hypothesis H_0 has been accepted, and a type II error has been made. Translated into the geological situation, the mistake made has been to infer that the sampled grade is below the cutoff grade.

The Power of a Test

The *power* of a test is the probability that a hypothesis H_0 will be rejected. Because the concept of power explains how the β risk level varies and also is important for decision making (Chap. 14), an outline is given here; a particularly clear and full account is presented by Dixon and Massey (1969, pp. 263–273). The *power* of the test is of interest only when the hypothesis H_0 is in fact false (hence its name). Because in practice it is not known whether H_0 is false, the power is defined for all parameter values—both consistent with H_0, i.e., H_0 true, and inconsistent with H_0, i.e., H_0 false. If the hypothesis H_0 is in fact false, the power of a test is equal to 100 percent minus the β risk level. If the hypothesis H_0 is in fact true, the power of a test depends on whether the parameter hypothesized under H_0 is a single value or a range of values. If it is a single value, the power is equal to the α risk level; if it is a range of values, the power is equal to the α risk level only for the extreme value at the end of the range; otherwise the power is between 0 percent and the α risk level. The use of power of a test is that, if there is a choice among two or more otherwise appropriate tests, the one with the highest power is selected in order to minimize the probability of making a type II error. These rather abstract and difficult definitions are illustrated and clarified here by reference to some of the previous examples of hypothesis testing.

When the simple hypothesis H_0 that μ equals 25 is tested against the simple alternative hypothesis H_1 that μ equals 27 (table 4.11), the power is computed to be (table 4.12, line 4) 76.4 percent when H_0 is false. For the more complicated two-tailed test of the hypothesis H_0 that μ equals 25, against the alternative hypothesis H_1 that μ is less than or greater than 25, the power depends on the actual value of μ. When the power is calculated for all values of μ, it may be plotted against the values of μ to obtain a *power curve*. In figure 4.17 the solid line on the graph is the power curve for this two-tailed test for a sample size of 25. The graph shows that, if the hypothesis H_0 that μ equals 25 is true, the probability or power of rejecting a mean of 25 is 10 percent, which corresponds to the α risk level. For true means increasingly different from 25, both arms of the power curve rise because the probability β of rejecting a false hypothesis H_0, which is increasingly different

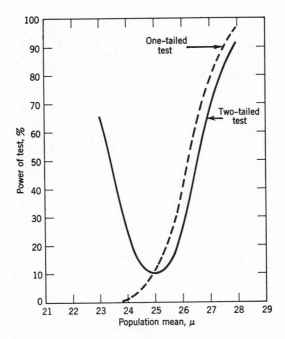

Fig. 4.17. Example power curves for one- and two-tailed test, $\mu = 25$, $\alpha = 5$ percent, $n = 25$.

from the true hypothesis, decreases. In figure 4.17 the dashed line on the graph is the power curve (also for sample size of 25) for the corresponding one-tailed test of the hypothesis H_0 that μ is equal to or less than 25 against the alternative hypothesis H_1 that μ is larger than 25. Notably, if H_0 is false, the one-tailed test is more powerful than the two-tailed test, because none of the α risk level is "spent" in the wrong tail. Therefore, if a one-tailed test is appropriate, it is always preferable to a two-tailed test.

The power of a test depends upon the sample size as well as on the type of test. Both examples in figure 4.17 are for a sample size of 25. In figure 4.18 are plots of the power curves for the second test in figure 4.17 of the hypothesis H_0 that μ is less than or equal to 25 against the alternative hypothesis H_1 that μ is greater than 25, for three different sample sizes, 10, 25, and 100. As the sample size increases, so does the power, which depends on the standard error of the sample mean, $\sigma/\sqrt{n}$. The increase from a sample size of 10 to one of 100 is striking. The importance of this relation in deciding how many mine samples to take of a given deposit is considered in Chapter 14.

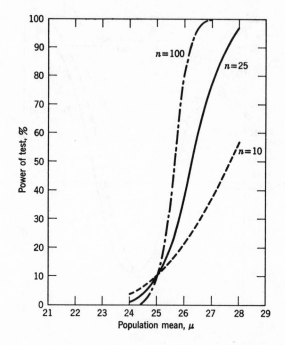

Fig. 4.18. Example power for one-tailed test, $\mu = 25$, $\alpha = 5$ percent, $u = 10$, 25, and 100.

Relation between Hypothesis Tests and Confidence Intervals

The point of view of the investigator making a hypothesis test is different from that of the investigator constructing a confidence interval, although most of the arithmetic is similar. Consider, as an example, hypothesis tests regarding the population mean contrasted with confidence intervals for the population mean. In a hypothesis test, the hypothetical mean and the estimated standard error of the mean are used to construct a critical region, and then a decision is made depending on whether the sample mean is inside or outside the critical region. On the other hand, in constructing a confidence interval, the sample mean and the estimated standard error of the mean are used to construct an interval that sets upper and lower bounds for the population mean.

In a sense, constructing a confidence interval corresponds to making an infinite number of hypothesis tests. Thus, if a confidence interval for the mean, with a certain confidence coefficient, is constructed from a set of data, it is known that, if a hypothesis test with an α risk level equal to 100 percent

TABLE 4.17. EXAMPLE TEST OF HYPOTHESIS OF EQUALITY
OF TWO POPULATION MEANS*

Hypothesis:	H_0: $\mu_1 = \mu_2$
Alternative hypothesis:	H_1: $\mu_1 < \mu_2$ or $\mu_1 > \mu_2$
Statistic:	t
Risk of type I error:	$\alpha = 10\%$
Critical region:	$|t| > t_{5\%}$ ($t_{5\%} = 1.796$, with 11 d.f.)

Computations:

Line	Item	Value	Remarks
1	$\overline{w}_1 - \overline{w}_2$	0.026	From table 4.5
2	$\sqrt{s_p^2(1/n_1 + 1/n_2)}$	0.6683	From table 4.5
3	t	0.039	

Conclusion. Accept hypothesis, $\mu_1 = \mu_2$.

* From Li, 1964, I, p. 151.

minus the confidence coefficient is made for the same set of data, all hypotheses about means falling inside the confidence interval are accepted, and all hypotheses about means falling outside the confidence interval are rejected. For example, in table 4.5 a 90-percent confidence interval is constructed for the difference between two population means of fictitious marine and freshwater shale data. The hypothesis is accepted when, in table 4.17, a test of the hypothesis H_0 that μ_1 equals μ_2 at an α risk level of 10 percent is made for the same data, because the conclusion is that μ_1 minus μ_2 equals 0, with 0 falling inside the confidence interval.

Uses for both confidence intervals and hypothesis tests in analysis of geological data are given throughout the book. In general the confidence interval is most useful when the aim is to summarize the information in a set of data to report what has been found out and how well it is known. The length of the confidence interval gives a direct indication of the precision of estimate. On the other hand, a hypothesis test is generally more useful if a specific decision is required between two alternatives. In the hypothesis test the precision is not directly indicated, although it can be obtained from the power curve.

4.8 BAYESIAN STATISTICS

The preceding sections of this book present the orthodox, classical kind of statistical inference that has dominated the thinking of most statisticians in this century. During the last few years, a different approach has been developed named Bayesian statistics, after Thomas Bayes who published a famous statistical paper in 1763. Statisticians of the Bayesian school argue that little experimental work, data gathering, and statistical analysis are performed in a vacuum and that prior information is usually available before an investigation is begun. For instance, if an investigator assays gold ore, he expects the gold content to be a few parts per million rather than a few parts per hundred; if he searches rocks for fossil birds, he expects to find only a few, not thousands. The Bayesian school argues that advantage should be taken of whatever relevant prior information is available, and it has developed formal methods to do this based on Bayes' theorem (sec. 2.5).

The principal difference between classical and Bayesian statistics is in the interpretation of the parameter under investigation. Classical statistics assumes that this parameter has a single value to be estimated, either by forming a confidence interval or by making a hypothesis test concerning a postulated single value. In contrast, Bayesian statistics assumes that the parameter has a probability distribution rather than a single value. The assumed probability distribution of the parameter is combined with the sampling distribution of the corresponding statistic to form a new modified distribution for the parameter. Thus, in Bayesian statistics, the information in the sample as summarized by a statistic modifies the investigator's postulated distribution of the parameter; whereas in the classical approach, the statistic influences the investigator's opinion about a single postulated value of a parameter or about the range of reasonable values for a parameter.

In classical statistics every distribution has a frequency interpretation, but in Bayesian statistics the distribution of the parameter may or may not be subject to a frequency interpretation. Two kinds of distributions illustrate this point. Records may show that a gold mine which has been in production for several years has had during this time an average weekly production of 20,000 ounces, with a standard deviation of 3000 ounces and a roughly normal distribution; for this distribution a clear-cut frequency interpretation is evident. On the other hand, suppose that 10 geologists hunt for fossils at an outcrop. If one of them offers an opinion that the time needed to locate the first fossil has a distribution that is normal with a mean of 3 hours and a standard deviation of 1 hour, the distribution, arrived at informally and subjectively, is not subject to a frequency interpretation. Either of these two kinds of distributions is acceptable to a Bayesian statistician.

The Use of Bayes' Theorem

As previously explained, Bayesian statistics assumes that a parameter has a probability distribution. The distribution of the parameter assumed before data have been taken is named the *prior probability distribution of the parameter*, or simply the *prior distribution*; and the distribution of the parameter calculated after the data have been collected is named the *posterior probability distribution of the parameter*, or simply the *posterior distribution*. The posterior distribution is obtained from the prior distribution and the distribution of the sample statistic by applying Bayes' theorem (sec. 2.5) in the version for continuous rather than discrete distributions.

For continuous distributions Bayes' theorem states that, *if $D_0(\theta)$ is the distribution of the parameter and if $f(w, \theta)$ is the distribution of the statistic w for a given value of θ, the posterior distribution $D_1(\theta|w)$ is a conditional distribution dependent upon the observed value of the statistic that is calculated by the formula*

$$D_1(\theta|w) = \frac{f(w, \theta)\, D_0(\theta)}{\int f(w, \theta)\, D_0(\theta)\, d\theta}.$$

Suppose that the prior distribution of a parameter θ is normal with a mean equal to ϕ_0 and a variance equal to σ_0^2. If a random sample is then taken from another normal distribution with a mean of θ and a variance of σ^2, the distribution of the sample mean $\overline{w}$ is normal with a mean of θ and a variance of σ^2/n, where n is the sample size. Bayes' theorem may be used to calculate the posterior distribution of the parameter θ, given $\overline{w}$. The calculation indicates that the posterior distribution of θ is normal with the parameters

$$\text{Posterior mean} = \phi_1 = \frac{n\overline{w}/\sigma^2 + \phi_0/\sigma_0^2}{n/\sigma^2 + 1/\sigma_0^2},$$

and

$$\text{Posterior variance} = \sigma_1^2 = \frac{1}{n/\sigma^2 + 1/\sigma_0^2}.$$

The weighting factors to combine the prior mean ϕ_0 and the sample mean $\overline{w}$ to obtain the posterior mean ϕ_1 are the reciprocals of the variances of these two quantities, and the variance σ_1^2 of ϕ_1 is the reciprocal of the sum of the weighting factors. Such weighting by reciprocals of variances is used again and again in statistics in many contexts (e.g., sec. 5.3).

The use of Bayes' theorem can be shown numerically by considering three fictitious cases of sampling from normal populations with known variances (table 4.18). In case 1 assume that, based on long experience, the weekly production of a gold mine is known to have a mean of ϕ_0 of 20,000 ounces, with a standard deviation σ_0 of 3000 ounces. Each week the mill superintendent forecasts current production before the final returns are in with a standard deviation σ of 1000. (This standard deviation measures how closely his forecast production agrees with the production calculated from the final returns; based on many years' forecasts, it may be considered to be a parameter.) In the week in question his estimate of w is 22,000 ounces, and

Table 4.18. Three fictitious cases of sampling to illustrate Bayes' theorem

Line	Item	Case 1	Case 2	Case 3
1	ϕ_0	20,000	20,000	20,000
2	σ_0	3,000	6,000	100,000
3	w	22,000	22,000	22,000
4	σ	1,000	1,000	1,000
5	$1/\sigma_0^2$	0.1111×10^{-6}	0.0278×10^{-6}	0.0001×10^{-6}
6	$1/\sigma^2$	1×10^{-6}	1×10^{-6}	1×10^{-6}
7	ϕ_0/σ_0^2	0.2222×10^{-2}	0.0555×10^{-2}	0.0002×10^{-2}
8	w/σ^2	2.2×10^{-2}	2.2×10^{-2}	2.2×10^{-2}
9	$\phi_0/\sigma_0^2 + w/\sigma^2$	2.4222×10^{-2}	2.2555×10^{-2}	2.2002×10^{-2}
10	$1/\sigma_1^2 = 1/\sigma_0^2 + 1/\sigma^2$	1.1111×10^{-6}	1.0278×10^{-6}	1.0001×10^{-6}
11	ϕ_1	21,800	21,945	21,999.8
12	σ_1	949	986	999.95

the problem is to estimate the value of the week's production that will be established when the final returns are in. In lines 1 to 4 of table 4.18, the parameters already introduced are listed, and in lines 5 to 10 are calculated the quantities required in the preceding equations to estimate the mean ϕ_1 and standard deviation σ_1 of this week's production. By taking into account prior information, the superintendent changes his estimate of ϕ_1 by about 10 percent.

In case 2 the initial four parameters are the same except that the standard deviation σ_0 of the prior distribution is assumed to be larger, with the value of 6000 instead of 3000. When the calculations are repeated, because the standard deviation of the previous production is large, that is, the prior production fluctuates more widely from week to week, there is little information available in the previous production to change the superintendent's estimate; thus, his estimate of 22,000 ounces is only altered to 21,945. Finally, in case 3 assume that the mine has entered a new orebody so that the variance of the prior distribution, instead of being calculated, is arbitrarily set at the very high value of 100,000, implying that it is practically unknown. Thus the superintendent's estimate is altered insignificantly to 21,999.8.

For most real problems the variance of the prior distribution is infinite or very large, and almost all of the information about the posterior distribution is contributed by the sample, as the equations and numerical results demonstrate. Thus in practice, almost no one, except the most ardent Bayesians, considers prior information to be relevant when testing hypotheses or constructing confidence intervals, although interpretation of the results of these procedures is somewhat modified.

Hypothesis testing has little meaning in a Bayesian context, as can be demonstrated by considering a Bayesian test about means. Assume that the postulated value of the population mean is the mean of the prior distribution and that the variance of the prior distribution is infinite. The posterior distribution is then centered about the observed sample mean, with a variance equal to the variance of the sample mean. The distance in standard deviation units between the prior and posterior means and the probability of an observation at a greater distance can be calculated. If this probability is small, the conclusion is that the prior mean is inconsistent with the posterior distribution; if the probability is large, the conclusion is that the prior mean is consistent. There are at least two ways to assess the meaning of a probability of an observation at a greater distance: (a) set a fixed significance level, say 5 percent, or (b) calculate the exact probability from the prior mean to infinity. Because a hypothesis test depends on a logical framework in which type I and type II errors are considered for single values for parameters rather than for a frequency distribution of parameters, this Bayesian procedure is clearly very different from a hypothesis test.

When Bayesian estimates are appropriate, some statisticians, for instance Mosteller and Wallace (1964), advocate calculating odds rather than using classical significance tests. Prior and posterior odds can be calculated to choose between two candidate sets of parameter values, and if the odds in favor of one set are very high, that set is accepted. Even though this approach has not found wide favor, it should not be ignored.

Seen from a Bayesian point of view, the interpretation of confidence intervals is changed like that of hypothesis tests. Instead of regarding an interval as either containing or not containing a single-valued parameter, the interval is interpreted as containing the central part of the posterior distribution. For instance, for a 90-percent confidence interval, the classical interpretation is that 90 percent of the intervals constructed contain the parameter being estimated, and 10 percent do not. The Bayesian interpretation is that, if the prior distribution is assumed to have an infinite variance, a 90-percent interval contains the central 90 percent of the posterior distribution.

If the variance of the prior distribution is not infinite, the percentage of the posterior distribution is not 90. Consider once again the estimated weekly production of the fictitious gold mine. If one ignores any prior information, the estimate is 22,000 ounces, with a standard deviation of 1000. A two-sided 90-percent confidence interval for θ has the bounds [22,000 − 1.645 (1000), 22,000 + 1.645 (1000)], or (20,355, 23,645). The constant 1.645 is the 5-percent point of the standardized normal distribution. However, if one takes for case 1 the prior information that ϕ_0 is equal to 20,000 and σ_0 is equal to 3000, the posterior distribution has as parameters ϕ_1 equal to 21,800 and σ_1 equal to 949. Between the confidence limits, the amount of area in this distribution is 91 percent. Thus 91 percent of the posterior distribution lies within the confidence interval. If this calculation is repeated for the prior distribution of case 3 with σ equal to 100,000, the amount of the posterior distribution in the confidence interval is essentially 90 percent because the variance of the prior distribution is so large that it has a negligible effect.

Unless appropriate assumptions are made about the shape of the prior distribution, the calculation of the posterior distribution can be very difficult. Therefore most Bayesians use prior distributions that yield easy calculations for the posterior distribution, as explained, for instance, by Schlaifer (1961).

Concluding Remarks

If data are fairly extensive and if the prior distribution has a large variance, classical and Bayesian statistics differ but little. Because results are only slightly modified if Bayesian theory is applied, such prior distributions are named *gentle priors*, and although an investigator's attitude toward the results is somewhat different in theory, his course of action is likely to be the same. However, if data are sparse and the prior distribution has a relatively small variance, considerably more accurate results can be obtained from a Bayesian approach—if the prior information is relevant.

Deciding whether a prior distribution is indeed relevant is liable to be the stickiest point. Different people have different prior distributions in mind because of differences in their background and position. Thus it is unlikely that the president, vice president, comptroller, and production manager of a company would have the same set of priors, and it is even improbable that their priors would agree enough to be reconciled.

4.9 INDICATIONS

Two important assumptions are made whenever a confidence interval is formed or a hypothesis test is made. The first is that the sample is random, and the second is that the observations are drawn from a normal population. These assumptions, introduced in sections 2.4 and 3.4, are repeatedly examined throughout the book. In this section their relation to *indications*, a name given to statistical results whose precise inferential validity is in doubt, is considered.

Failure of real data, even when obtained in a designed experiment, to correspond exactly to the assumptions of randomness and normality is the rule rather than the exception. Moreover, almost always the relation is uncertain between the target population and the sampled population for which the randomization is performed. Because the investigator draws inferences about the target population, these inferences must rest on geological as well as on statistical reasoning. These problems arise even in carefully designed experiments; for data obtained in undesigned experiments or from the literature, there is seldom if ever a reliable way to test their correspondence to the assumptions.

When the results of a statistical analysis, though uncertain, appear to be suggesting something about a problem, the statistical inferences may be named *indications* rather than decisions or conclusions. The term indications was first used by Tukey (1968), who provides a thoughtful discussion.

If the first assumption of randomly distributed observations is not met, great trouble can result, and even at best the reliability of any statistical analysis is always uncertain. Lack of randomness causes many problems; some interesting ones are outlined by Wallis and Roberts (1956, pp. 337–340). There is no unequivocal way to prove the randomness of statistical samples obtained in undesigned experiments. Although various tests can be applied, at best they can indicate that samples appear to behave like random samples; such tests can never prove that samples are in fact random samples.

Many geological observations, including most of the example data in this book, are demonstrably not random; yet we must use them, for they are the best we have. Even if the observations are randomly drawn from a sampled population, they are seldom if ever random with respect to the target population. Yet, because of the importance of randomness in making any statistical inferences, it is not an exaggeration to state that the two most important statistical decisions that an investigator makes are the relating of target and sampled populations and the devising of an appropriate scheme for random sampling. These matters are explored further in section 7.2.

If the second assumption, that observations are normally distributed, is not met, the exact risk levels for confidence limits and hypothesis tests are unknown. For instance, if observations actually come from the skewed distribution of figure 4.19, instead of 10 percent of the observations being more than 1.65 standard deviations greater than the mean and another 10 percent being more than 1.65 standard deviations smaller than the mean, 8 percent are actually in the left tail and 12 percent in the right. Thus a "10 percent" confidence interval based on the right tail is actually a 12-percent interval. For a hypothesis test with the critical region in one tail, the α risk level is 8 or 12 percent rather than the specified 10 percent. Of course, the investigator does not know that his probability levels are changed. Although troublesome and untidy, this distortion of the normal curve does not cause the geologist too much trouble, because with a problem in the real world the risk level is chosen arbitrarily anyway.

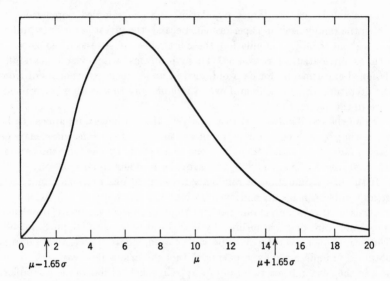

Fig. 4.19. Skewed quasi-normal distribution.

A potentially more serious problem, if the assumption of normality is not met, is that statistical procedures designed for the normal distribution may be inefficient for nonnormal distributions. Then *transformations* (Chap. 6) may be advantageously applied to observations. The attitude that one should take toward possible loss of efficiency is difficult to assess; sometimes one should be very worried, other times one can be unconcerned. The need for efficient methods is closely related to the number of observations. If observations are few and inferences are uncertain, efficiency may be very important; if observations are numerous, conclusions may be unmistakable, and efficiency may be unimportant.

As the preceding remarks suggest, failure to satisfy the assumption of randomness is much more serious than failure to satisfy that of normality. Fortunately, for a new investigation it is always possible, though not always easy, to sample randomly the sampled population. And, although few if any real data come from a population that is normal or even quasi-normal, the only consequences of failure to meet the normality assumption are distortion of the theoretical risk levels and reduction in the efficiency of estimation. As data of a certain kind accumulate, an investigator can always try to make more exact statements as he learns how skewed a distribution actually is. Even for a highly skewed distribution, such as the lognormal one (sec. 6.2), the problem is not too great.

Where to draw the line and call an inference an indication rather than a conclusion or a decision is a troublesome problem, for which judgment is the only guide.

REFERENCES

Dixon, W. J., and Massey, F. J., Jr., 1969, Introduction to statistical analysis: New York, McGraw-Hill, 638 p.

Fisher, R. A., 1950, The fiducial argument in statistical inference, *in* Contributions to Mathematical Statistics: New York, John Wiley & Sons, paper 25.

Keith, M. L., and Bystrom, A. M., 1959, Comparative analyses of marine and fresh-water shales: Penn. State Univ., Mineral Industries Expt. Sta. Bull.

Koch, G. S., Jr., and Link, R. F., 1964, Accuracy in estimating metal content and tonnage of an ore body from diamond-drill-hole data: U.S. Bur. Mines Rept. Inv. 6380, 24 p.

Li, J. C. R., 1964, Statistical inference: Ann Arbor, Mich., Edwards Bros., v. 1, 658 p.

Mosteller, Frederick, and Wallace, D. L., 1964, Inference and disputed authorship: The Federalist: Reading, Mass., Addison-Wesley, 285 p.

Mosteller, Frederick, and Tukey, J. W., 1968, Data analysis, including statistics: *in* Handbook of Social Psychology, 2nd. ed., Reading, Mass., Addison-Wesley, p. 80–203.

Neyman, J., 1937, Outline of a theory of statistical estimation based on the classical theory of probability: Royal Soc. London Philos. Trans., ser. A, v. 236, p. 333–380.

Schlaifer, Robert, 1961, Introduction to statistics for business decisions: New York, McGraw-Hill, 374 p.

Wallis, W. A., and Roberts, H. V., 1956, Statistics: a new approach: Glencoe, Illinois, Free Press, 619 p.

Wilks, S. S., 1962, Mathematical statistics: New York, John Wiley & Sons, 644 p.

Chapter 5

Analysis of Variance

In Chapters 2 to 4 selected basic statistical methods and calculation procedures are explained in detail. Comprehension of the subjects in these chapters is essential for the geologist who wishes to use statistics effectively and is also necessary for understanding the rest of this book. However, from Chapter 5 onward, only a few statistical methods are explained in detail; for other methods the reader is referred to statistics books and research papers.

Chapter 5, an introduction to the analysis of variance, is about the sources of variability in data and about the purposeful partitioning of the sum of squares of the observations into pieces, each of which can be given an interpretation in terms of geology or other discipline to which statistics are applied. Space permits only an introduction to a few of the many topics in the analysis of variance.

The following topics are taken up in this chapter. In sections 5.1 to 5.6 the one-way analysis of variance is explained, together with one modification—the nested analysis of variance. In section 5.7 the analysis of variance in randomized blocks is discussed. Section 5.8 explains linear combinations, a mathematical device necessary for the single-degree-of-freedom analysis, which is treated in sections 5.9 and 5.10. In section 5.11 the preceding sections are tied together through a discussion of linear models. Section 5.12 is about multiple comparisons of data, and section 5.13 summarizes the statistical methods introduced up to this point in the book.

Many traditional subjects in the analysis of variance are not even mentioned in this book, partly because they would expand it too much, but mainly because they pertain to the analysis of *designed experiments* (sec. 8.6), in which data are collected according to a formal plan that is carefully conceived and executed. Such designed experiments are common in agriculture, in some of the physical sciences, and in some industries. For instance, a herd of 16 cows may be randomly partitioned into four different groups; each group may be fed different rations and the weights of the 16 cows may be periodically compared. Or several different brands of light bulbs may be used in a number of factories and the lives of the bulbs then contrasted. Although some geological studies are equally formal (e.g., four different rock formations might be compared by measuring paleomagnetism at six outcrops in each), most geologists do not today seem to work in this way. There are many methods for the statistical analysis of designed experiments; Cochran and Cox (1957) and Davies (1957) give excellent accounts; and computer programs are readily available for the more popular ones.

5.1   COMPARISON OF MORE THAN TWO MEANS

In sections 4.4 and 4.7 the problem of comparing agreement among means is introduced. The means are compared by constructing confidence intervals and by making a hypothesis test. When. such a comparison is made, the question at once arises: How would the check for agreement in sampling be made if three or more groups of specimens had been taken? In this chapter this problem is investigated for means. Although agreement in data may also be compared by studying other parameters, the comparison of means is usually of greatest interest for geological problems.

The ideas and techniques that have been developed for comparing means have many uses and provide the basis for much of the applied statistics in any discipline. In this chapter these ideas and techniques are developed through the explanation of some highly simplified geological problems. Later, particularly in Chapters 7, 8, and 11, more interesting geological applications are taken up.

Suppose that in area A a geologist has mapped five limestone outcrops which he is uncertain whether to assign to a single formation. Suppose, further, that he believes the key indicator is the sand content of the rock, so that if he concludes that this content is the same in the five outcrops he will decide that they belong to the same formation. To investigate the sand content the geologist might take four hand specimens from each outcrop, or 20 altogether, and measure the sand content in each to obtain the data in table 5.1. In the table some statistics are calculated for each of the five samples.

TABLE 5.1. FICTITIOUS OBSERVATIONS OF SAND CONTENT CONSTITUTING STATIS-
TICAL SAMPLES FROM FIVE OUTCROPS IN AREA A

| Line | Item | Statistical sample number | | | | |
		(1)	(2)	(3)	(4)	(5)
1	Observations	9.8	9.5	9.3	15.5	7.6
2	(sand content,	7.5	7.2	10.6	8.9	11.0
3	%)	10.1	10.4	9.6	12.4	9.8
4		10.9	11.3	13.2	10.0	10.8
5	$\sum w$	38.3	38.4	42.7	46.8	39.2
6	n	4	4	4	4	4
7	$\overline{w}$	9.575	9.600	10.675	11.700	9.800
8	$(\sum w)^2$	1466.89	1474.56	1823.29	2190.24	1536.64
9	$(\sum w)^2/n$	366.72	368.64	455.82	547.56	384.16
10	$\sum w^2$	373.11	377.94	465.25	573.22	391.44
11	SS	6.39	9.30	9.43	25.66	7.28
12	d.f.	3	3	3	3	3
13	s^2	2.13	3.10	3.14	8.55	2.43

To compare k means corresponding in the fictitious illustration to the five sample means of sand percentage, one might be tempted to compare each mean with every other by two-sample confidence intervals or the t-hypothesis test. There are two difficulties. One is that as the number of comparisons becomes large the probability of making a type I error, by rejecting the true hypothesis that two population means are equal, also becomes large; for instance, if k is 10, 45 pairwise comparisons are possible. In a test with a 10-percent significance level four or five of the comparisons would on the average lead to rejecting the hypothesis, through chance alone, even though all were true. The second difficulty is that there may be too few observations in each statistical sample to yield good estimates of the population variance σ^2. These difficulties can be overcome by the method of *multiple comparisons*, an explanation of which is deferred until section 5.12, in order to introduce first the basic procedures involved in the analysis of variance.

Mathematical Model

Before solutions to the specific problem are discussed, a mathematical model (sec. 1.5) is introduced. For each outcrop, the four determinations of sand percentage are defined as a sample of four observations from a population corresponding to that outcrop. Thus there is a single sample of four observations from each of five populations. Each population has a mean μ_i

and a common variance σ^2. If the geologist concludes that the means of the five populations are equal,

$$\mu_1 = \mu_2 = \mu_3 = \mu_4 = \mu_5,$$

he would conclude that the five outcrops belong to the same formation. However, if he concludes that the means are unequal, he might be unwilling to conclude that the five outcrops belong to different formations without a further investigation of the data with a more complicated model (sec. 9.2). Thus his interpretation of the data depends on the mathematical model that he uses.

The problem, then, is to estimate each of the μ_i values and the common variance σ^2 and to devise a test of the hypothesis H_0 that

$$\mu_1 = \mu_2 = \mu_3 = \mu_4 = \mu_5.$$

The jth observation from the ith population may be represented as

$$w_{ij} = \mu_i + e_{ij},$$

where μ_i represents the mean of the population from which the observation is taken and e_{ij} represents the random fluctuation of the observation (e.g., the fourth observation from the third population is $w_{34} = 13.2$). The assumption is made for all the random fluctuations e_{ij} that their mean is 0 and that their variance is equal to the common population variance σ^2, or,

$$\mu_e = 0,$$
$$\sigma_e^2 = \sigma^2.$$

The expression for an observation may be rewritten

$$w_{ij} = \mu + (\mu_i - \mu) + e_{ij},$$

where μ is the average of the five μ_i values. If the μ_i values are all equal, the terms in parentheses $(\mu_i - \mu)$ are all 0. If they are not all equal, one measure of their difference is the variance of population means

$$\sigma_\mu^2 = \frac{(\mu_1 - \mu)^2 + (\mu_2 - \mu)^2 + (\mu_3 - \mu)^2 + (\mu_4 - \mu)^2 + (\mu_5 - \mu)^2}{4},$$

where the number 4 in the denominator is equal to the number of means minus 1.

The hypothesis H_0 that

$$\mu_1 = \mu_2 = \mu_3 = \mu_4 = \mu_5$$

may be rewritten

$$\sigma_\mu^2 = 0,$$

for, if the variance of population means is equal to zero, the means must also be equal to each other.

All of the quantities needed to test this hypothesis have now been introduced. They are the population means μ_i, the mean of population means μ, the common population variance of the random fluctuations σ^2, and the variance of population means σ_μ^2. Statistics to estimate the first three of these parameters are intuitively clear. The sample mean 9.58 from population 1 is an estimate of μ_1 (table 5.1), and similarly for the other populations. The mean of sample means 10.27 is an estimate of μ (table 5.2). The third parameter σ^2 may be estimated by the combination of individual sample variances, which is the pooled sample variance s_p^2 (sec. 4.4), equal to 3.87 (table 5.2).

If the fourth parameter, the variance of population means σ_μ^2, is equal to 0, the average value of the variance of the five sample means is $\sigma^2/5$. However,

TABLE 5.2. ONE-WAY ANALYSIS OF VARIANCE TO TEST THE HYPOTHESIS THAT THE POPULATION MEANS OF FICTITIOUS STATISTICAL SAND CONTENT FROM FIVE OUTCROPS ARE EQUAL

Hypothesis:	$H_0: \mu_1 = \mu_2 = \mu_3 = \mu_4 = \mu_5$, or $\sigma_\mu^2 = 0$
Alternative hypothesis:	$H_1:$ means unequal, or $\sigma_\mu^2 > 0$
Statistic:	F
Risk of type I error:	$\alpha = 10\%$
Critical region:	$F > F_{10\%}$, with 4 and 15 d.f. ($= 2.36$)

Computations:

Line	Item	Value	Remarks
1	$\overline{w}$	9.575, 9.600, 10.675, 11.700, 9.800	From table 5.1
2	n	4, 4, 4, 4, 4	,,
3	s^2	2.13, 3.10, 3.14, 8.55, 2.43	,,
4	$\sum \overline{w}$	51.350	
5	k	5	
6	$\overline{\overline{w}}$	10.27	
7	$(\sum \overline{w})^2$	2636.8225	
8	$(\sum \overline{w})^2/k$	527.36450	
9	$\sum \overline{w}^2$	530.72625	
10	$\mathrm{SS}_{\overline{w}}$	3.36175	
11	d.f.	4	
12	$s_{\overline{w}}^2$	0.84044	
13	n	4	
14	$ns_{\overline{w}}^2$	3.3616	
15	$\sum \mathrm{SS}$	58.05	
16	$\sum (\mathrm{d.f.})$	15	
17	s_p^2	3.87	
18	F	0.87	

Conclusion. Accept hypothesis $H_0: \mu_1 = \mu_2 = \mu_3 = \mu_4 = \mu_5$, or $\sigma_\mu^2 = 0$.

if this quantity is larger than 0, because the five population means are unequal, the average value of the variance of the five sample means must be
larger than $\sigma^2/5$. In fact, this average value is equal to $\sigma^2/5 + \sigma_\mu^2$. For the
fictitious data, the variance of sample means $s_{\bar{w}}^2$ is equal to 0.8404 (table 5.2),
which is nearly equal to the pooled sample variance $s_p^2/5$ of 0.7740 (table 5.2).

Thus the criterion of comparison for the population means is the ratio of
two variances calculated from the k samples, both of which estimate the
population variance, *provided* that the k means are equal. The first one is
calculated as follows: Consider k samples, each with the same number of
observations n. For each sample a sample mean $\bar{w}_i$ and a sample variance s_i^2
are calculated. Then the common population variance σ^2 can be estimated
by the pooled sample variance s_p^2 obtained by the formula

$$s_p^2 = \frac{\sum (n-1)s_i^2}{k(n-1)}$$

which, to simplify notation, may be rewritten

$$s_p^2 = \frac{\sum \mathrm{SS}_i}{k(n-1)},$$

where the summation extends over the k samples. The statistic s_p^2 has
$k(n-1)$ degrees of freedom.

The second of these variances may be introduced as follows: The variance
of the sample mean $s_{\bar{w}}^2$ is

$$s_{\bar{w}}^2 = \frac{\sum (\bar{w}_i - \bar{\bar{w}})^2}{k-1}$$

with $(k-1)$ degrees of freedom, where $\bar{\bar{w}}$ is the mean of sample means. This
variance is related to the underlying population variance of observations σ^2
because the variance of a sample mean $\bar{w}$ is equal to the population variance
divided by the sample size, that is, to σ^2/n (sec. 3.6). Thus $s_{\bar{w}}^2$ is an unbiased
estimate of σ^2/n, and $ns_{\bar{w}}^2$ is an unbiased estimate of σ^2, *provided that the k
population means are equal.*

Therefore, if the population means are the same, the ratio of the two
sample variances,

$$\frac{ns_{\bar{w}}^2}{s_p^2},$$

should be nearly 1. On the other hand, if the population means are different,
both the value of $s_{\bar{w}}^2$ and the ratio will be larger, on the average—how much
larger depends on the sampling distribution of the ratio

$$F = \frac{ns_{\bar{w}}^2}{s_p^2}.$$

The F-sampling distribution, one of the most useful in statistics, is explained in section 5.2 before the actual calculations for the analysis of variance are introduced in section 5.3.

5.2   THE F-DISTRIBUTION

The F-distribution is a family of theoretical frequency distributions for the ratio of two variances, calculated from independent random samples drawn from two populations whose observations are normally distributed and whose variances are equal. Conceptually, the distribution is obtained as follows. From the first population, with variance σ^2, all possible samples of a given size n_1 are drawn, and, for each sample, the sample variance s_1^2 with $(n_1 - 1)$ degrees of freedom is calculated. Likewise, from the second population, also with variance σ^2, all possible samples of a given size n_2 are drawn; and, for each sample, the sample variance s_2^2 with $(n_2 - 1)$ degrees of freedom is calculated. Then, every sample variance s_1^2 is divided by every sample variance s_2^2 to yield a ratio

$$F = \frac{s_1^2}{s_2^2},$$

the number of ratios being equal to the number of possible samples in the first population multiplied by the number in the second population. The frequency distribution of all possible F-ratios is the F-distribution.

In figure 5.1 frequency curves of several F-distributions are graphed. The figure illustrates several points about the F-distribution. First there is a different F-distribution for every combination of number of degrees of freedom in the numerator and denominator of the F-ratio. Second, because the F-ratio is obtained by dividing two positive numbers, its minimum value is 0, and its maximum value is infinity. Third, as the number of degrees of freedom in both numerator and denominator increase, the frequency curve becomes more nearly symmetrical because the distributions of both s_1^2 and s_2^2 become nearly symmetrical and tightly clustered around 1 because of the central limit theorem. Thus, since there are fewer large values of s_1^2 and fewer small values of s_2^2, the distribution of the ratio itself becomes nearly symmetrical. Fourth, the mean of every F-distribution with fairly large n_2 is nearly 1 and is exactly equal to $(n_2 - 1)/(n_2 - 3)$.

The F-distribution may be related to the example of section 5.1. If the ratio

$$F = \frac{ns_{\bar{w}}^2}{s_p^2},$$

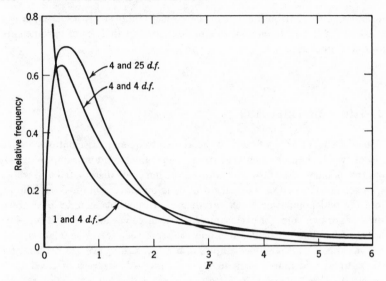

Fig. 5.1. F-curves with various degrees of freedom.

with $(k - 1)$ and $k(n - 1)$ degrees of freedom is too large, the hypothesis that the k sample means are equal is rejected. How large is too large depends upon the risk level chosen by the investigator. In figure 5.2 the F-distribution with 3 and 24 degrees of freedom is graphed. The concept is like that used for the t-test (sec. 4.7), except that the critical region is entirely in the right-hand tail. There is no reason to reject the hypothesis if the calculated F-value is in the left tail because this fact indicates that the within-sample variance is large compared with the among-sample variance. However, if F is very small (less than 0.1 for typical problems), the data are suspect because with proper randomization an F-value this small is unlikely.

The critical region of the F-test for selected α risk levels and degrees of freedom is given in table A-5. There is a separate subtable for each α risk level; the subtable for the 10-percent risk level is described, and then the differences in the other tables are mentioned. Each column corresponds to a different number of degrees of freedom in the numerator; each row corresponds to a different number of degrees of freedom in the denominator. The table entries decrease downward in each column, because the average variance of the denominator s_2^2 decreases. Likewise, except for very small degrees of freedom in the denominator, the table entries decrease toward the right, with increasing degrees of freedom in the numerator. For an infinitely large number of degrees of freedom in both numerator and denominator the value

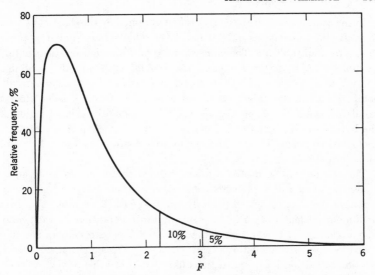

Fig. 5.2. F-curve with 3 and 24 degrees of freedom to illustrate critical regions for 5 and 10 percent risk levels.

of the F-ratio is 1, recorded in the lower right-hand corner entry, because both s_1^2 and s_2^2 become equal to the common variance σ^2 for infinitely large samples. Except for this entry, which is the same for all subtables, the subtables for the 5- and 1-percent risk levels are like those for the 10-percent subtable, except that each entry gets larger as illustrated for one case by figure 5.2.

5.3   ONE-WAY ANALYSIS OF VARIANCE

Now that the F-distribution, which is essential for interpreting the sample-variance ratio calculated in the analysis of variance, has been introduced, the one-way analysis of variance for the fictitious data of table 5.1 can be presented. In table 5.2 the hypothesis test and calculations for the data of table 5.1 are outlined as those for the t-test in table 4.15. The hypothesis H_0 that the four population means are equal is tested against the alternative hypothesis H_1 that they are unequal. If the hypothesis H_0 is true, the variance of population means σ_μ^2 is equal to 0; if the alternative hypothesis H_1 is true, σ_μ^2 is greater than 0.

In the computations (table 5.2) s_w^2 is first obtained by the short-cut variance formula; the calculations in lines 1 to 12 of table 5.2 are similar to those in lines 5 to 13 of table 5.1. The numerator ns_w^2 of the variance ratio is obtained by multiplying by n; the denominator s_p^2 is obtained by dividing $\sum$ SS in line 15 by $\sum$ (d.f.) in line 16. The F-ratio, obtained in line 18 as the final result, is compared with the tabulated F-value of 2.36. Being smaller than the tabulated value, the computed value 0.87 is *outside* the critical region, and the hypothesis H_0 that the five population means are equal is accepted. In geological language the five outcrops in area A are judged to belong to the same formation.

The calculations in table 5.2 are performed to demonstrate the algebra. However, when a desk calculator is used, it is generally more convenient to organize the calculations differently, in the format of table 5.3. In the preliminary calculations T is the sum total of observations in each sample. The advantage of this format is that $\sum w$ and $\sum w^2$ can be obtained in one operation by desk calculator, as can $\sum T$ and $\sum T^2$ as a check. As a demonstration of calculations the fictitious illustration of table 5.1 is reworked in table 5.4. If the sample sizes are unequal, the format of table 5.4 is generalized to that of table 5.5. In table 5.5 n_0 is equal to $k/\sum (1/n_i)$.

TABLE 5.3. SIMPLIFIED COMPUTING METHOD FOR ONE-WAY ANALYSIS OF VARIANCE

Type of sum	Preliminary calculations		
	(1) Total of squares	(2) Number of observations in sum	(3) = (1)/(2) Total of squares per observation
Grand	$(\sum T)^2$	kn	I
Sample	$\sum T^2$	n	II
Observation	$\sum w^2$	1	III

Source of variation	Analysis of variance			
	Sum of squares	Degrees of freedom	Mean square	F
Among-sample means	II − I	$k - 1$	ns_w^2	ns_w^2/s_p^2
Within samples	III − II	$k(n - 1)$	s_p^2	
Total	III − I	$kn - 1$		

Table 5.4. Simplified computing method for one-way analysis of variance applied to data of table 5.1

	Preliminary calculations		
Type of sum	(1) Total of squares	(2) Number of observations in sum	(3) = (1)/(2) Total of squares per observation
Grand	42,189.16	20	2109.458
Sample	8,491.62	4	2122.91
Observation	2,180.96	1	2180.96

	Analysis of variance				
Source of variation	Sum of squares	Degrees of freedom	Mean square	F	$F_{10\%}$
Among-sample means	13.45	4	3.36	0.87	2.36
Within samples	58.05	15	3.87		
Total	71.50	19			

Table 5.5. Simplified computing method for one-way analysis of variance, generalized for unequal sample sizes

	Preliminary calculations		
Type of sum	(1) Total of squares	(2) Number of observations in sum	(3) = (1)/(2) Total of squares per observation
Grand	$(\sum T)^2$	kn	I
Sample			$II = \sum (T^2/n)$
Observation	$\sum w^2$	1	III

	Analysis of variance			
Source of variation	Sum of squares	Degrees of freedom	Mean square	F
Among-sample means	II − I	$k - 1$	$n_0 s_{\bar{w}}^2$	$n_0 s_{\bar{w}}^2 / s_p^2$
Within samples	III − II	$\sum n - k$	s_p^2	
Total	III − I	$\sum n - 1$		

Table 5.6. Fictitious observations of sand content constituting statistical samples from five outcrops in area B

		Statistical sample number				
Line	Item	(1)	(2)	(3)	(4)	(5)
1	Observations	14.0	4.8	9.4	3.1	15.8
2	(sand content,	16.2	7.2	13.2	8.8	16.3
3	%)	16.9	1.8	9.2	4.2	13.6
4		15.4	1.6	9.0	5.5	14.5
5	$\sum w$	62.5	15.4	40.8	21.6	60.2
6	n	4	4	4	4	4
7	$\overline{w}$	15.625	3.850	10.200	5.400	15.050
8	$(\sum w)^2$	3906.25	237.16	1664.64	466.56	3624.04
9	$(\sum w)^2/n$	976.56	59.29	416.16	116.64	906.010
10	$\sum w^2$	981.21	80.68	428.24	134.94	910.540
11	SS	4.65	21.39	12.08	18.30	4.53
12	d.f.	3	3	3	3	3
13	s^2	1.55	7.13	4.03	6.10	1.51

A Second Fictitious Illustration

For the fictitious data from area A (table 5.1) the conclusion is drawn that the five population means are equal. Consider now the fictitious data from five outcrops in area B presented in table 5.6, for which the same hypothesis H_0 that the five population means are equal is to be tested. For area B the within-sample variances are of about the same sizes as those for area A, but the variance of the five sample means $s_{\overline{w}}^2$ is 116.2, which indicates a larger variability among sample means in area B. For area B an analysis of variance (table 5.7) yields a computed F-value of 28.59, which, being larger than the tabulated F-value of 2.36, leads to accepting the alternative hypothesis H_1. In geological language the five outcrops in area B are judged to belong to different formations.

Table 5.7. One-way analysis of data from fictitious area B

Source of variation	Sum of squares	Degrees of freedom	Mean square	F	$F_{10\%}$
Among-sample means	464.65	4	116.16	28.59	2.36
Within samples	60.95	15	4.06		
Total	525.60	19			

Remarks

In hypothesis tests with the analysis of variance, as in any hypothesis tests, whenever a hypothesis H_0 is accepted, the investigator runs the risk of making a type II error through accepting a hypothesis H_0 when it is in fact false. Sometimes this risk is verbalized by a weaseling double-negative statement, such as, "The data do not contradict the hypothesis H_0." The chance of making a type II error depends on the ratio of the population variance of the means to the population variance of the random fluctuations (σ_μ^2/σ^2) and on the number of degrees of freedom in the numerator and denominator. This ratio can be estimated, and the chance of making a type II error calculated by a method beyond the scope of this book (see Dixon and Massey, 1969, Chap. 14). Thus, when an investigator accepts a hypothesis H_0, he should be wary, but if he rejects it he is protected because the chance of type I error is set exactly.

The name *analysis of variance* describes the partitioning of the total variance into meaningful pieces. For the fictitious illustrations of areas A and B there are two pieces: one for the variation among sample means and another for the variation of observations within the individual samples. If k samples, each containing n observations, are treated as one large sample, the total sum of squares with $(kn - 1)$ degrees of freedom can be calculated for the large sample. However, the sum of degrees used to estimate the pooled sample variance s_p^2 has $k(n - 1)$ degrees of freedom, and the sum of squares used to estimate the among-sample-means variance $ns_{\bar{w}}^2$ has $(k - 1)$ degrees of freedom. Thus, as illustrated by table 5.4, the total sum of squares is equal to the sum of the among-samples sum of squares and the within-samples sum of squares, and the total number of degrees of freedom is equal to the sum of the numbers of degrees of freedom in the partitioned sum of squares.

According to Cochran's theorem (Wilks, 1962, p. 212), whenever a sum of squares can be partitioned into pieces so that the pieces add up to the total sum of squares and the degrees of freedom of the pieces add up to the total degrees of freedom, the sampling distributions of the individual pieces are independent of one another. Thus in the division into among-sample sum of squares and within-sample sum of squares and the corresponding variances the two statistics $ns_{\bar{w}}^2$ and s_p^2 are independently distributed, and their ratio follows the F-distribution.

Finally, the estimation of the variance of population means σ_μ^2 is of interest, particularly if a hypothesis H_1 is accepted, as in the fictitious illustration of data from area B. The mean value of the variance of sample means $\mu_{s_{\bar{w}}^2}$ may be expressed as the sum of two components (Dixon and Massey, 1969, p. 162). The first, the square of the standard error of the mean σ^2/n, is the common

TABLE 5.8. AVERAGE MEAN-SQUARE VALUES IN ONE-WAY ANALYSIS OF VARIANCE

Source of variation	Degrees of freedom	Mean square	Average mean square
Among-sample means	$k - 1$	$ns_{\bar{w}}^2$	$\sigma^2 + n\sigma_\mu^2$
Within samples	$k(n - 1)$	s_p^2	σ^2

population variance of random fluctuations divided by the sample size. The second is σ_μ^2. Thus

$$\mu_{s_{\bar{w}}^2} = \frac{\sigma^2}{n} + \sigma_\mu^2.$$

In table 5.8 the preceding formula is related to the analysis-of-variance format of table 5.4. Table 5.8 shows the mean-square statistics that estimate the mean-square parameters. As in table 5.4, the variance of sample means $s_{\bar{w}}^2$ is multiplied by the number of sample means n, so that, in table 5.8, the average variance of sample means,

$$\frac{\sigma^2}{n} + \sigma_\mu^2,$$

is multiplied by n to yield

$$\sigma^2 + n\sigma_\mu^2.$$

Hence by subtracting the within-sample mean square from the among-sample mean square and dividing by n the desired estimate of the variance of population means σ_μ^2 is found to be

$$\frac{ns_{\bar{w}}^2 - s_p^2}{n}.$$

If $ns_{\bar{w}}^2$ is less than s_p^2, the result is negative, and it is customary to set this estimate equal to 0. Whenever an average variance is composed of several pieces, the pieces are named *variance components*, and the statistics estimating them are named *variance-component estimates*.

5.4 EXAMPLES OF ONE-WAY ANALYSIS OF VARIANCE

For studying many sets of geological data, the one-way analysis of variance is a powerful tool, even though more complicated analyses of variance, presented subsequently in this chapter, may be required to understand fully all variance components. To emphasize the value of the one-way analysis of variance two example problems are reviewed in this section in some detail.

Comparison of Mean Silver Content of a Vein on Two Sides of a Fault

The first example is a comparison of mean silver content in a vein on two sides of a fault. In the Frisco mine (sec. 1.4) the Don Tomás vein (sec. 4.5, fig. 4.11) is cut by the Frisco fault whose displacement is 35 meters or more. The original data were the assays for five metals of chip samples from 1810 stations in drifts on the vein. In order to reduce and measure local fluctuation, metal contents from successive groups of 10 samples along each drift were averaged, and the resulting means were the example data. Thus there were 181 summary points (1810/10), 63 on the south side and 118 on the north side of the fault.

In the original paper (Koch and Link, 1963) several methods for comparing the mineralization are given. However, in this book only a one-way analysis of variance for the 181 summary-point observations for silver is given as an example. The problem is formulated in the following hypothesis test:

1. *Hypothesis:* The two population means for silver content are equal on the north and south sides of the fault; that is, $H_0 \colon \mu_n = \mu_s$.
2. *Alternative hypothesis:* $H_1 \colon \mu_n \neq \mu_s$, or $\sigma_\mu^2 > 0$.
3. *Assumptions:* Random samples drawn from normal populations with the same variance.
4. *Risk of type I error:* $\alpha = 10$ percent.
5. *Critical region:* $k = 2$, $n_n + n_s = 181$; accordingly, degrees of freedom are 1 and 179. From table A-5, $F_{10\%}$ is 2.74.

Table 5.9 presents the one-way analysis of variance to test this hypothesis. The total variance is partitioned into two pieces: one corresponding to the variance between the north and south sides of the fault, with 1 degree of freedom; and the other corresponding to the variance within vein segments,

TABLE 5.9. ONE-WAY ANALYSIS OF VARIANCE TO TEST THE HYPOTHESIS THAT SILVER CONTENT IS THE SAME IN THE DON TOMÁS VEIN ON BOTH SIDES OF THE FRISCO FAULT

Source of variation	Sum of squares	Degrees of freedom	Mean square	F	$F_{10\%}$
Between north and south sides of Frisco fault	1,726,043	1	1,726,043	34.7	2.74
Within vein segments	8,895,472	179	49,695		

with 179 degrees of freedom. Because it is larger than the tabled $F_{10\%}$-value, the calculated F-value of 34.7 is inside the critical region, and the alternative hypothesis H_1 is accepted, that $\mu_n \neq \mu_s$, or σ_μ^2 is greater than 0. In geological terms the evidence supports a difference in silver mineralization.

Had these silver values been obtained through a designed experiment, this analysis would be one step toward the objective of comparing mineralization on the two sides of the Frisco fault. Some or all of the following variables would have been investigated: the five metals of commercial value, other chemical elements or substances, and physical properties such as rock competency (perhaps reflecting recemented and therefore stronger rock on one side of the fault but not on the other). Target and sampled populations would have been defined, and random statistical samples would have been taken on either side of the fault. If criteria for comparing mineralization were established in enough detail, appropriate sample sizes might have been determined from the estimates of underlying variance derived from preliminary sampling. Such a designed experiment is an ideal to strive for.

Because most of the Don Tomás vein had been mined out, sampling in a designed experiment would have been impossible; therefore, the best available data were the assays obtained in the regular course of mining. In such a situation, which today seems to be more the rule than the exception in geology, the suitability of a particular statistical analysis is always debatable. Even though the question is not further discussed at this point, the issue is raised to warn the reader. With so large a calculated F-value, a difference in mean silver content is clearly established. Had the calculated value been smaller, a more refined statistical analysis might have been worthwhile. The decision would have been made in view of the objectives of the entire study and the total pattern of data.

It should be noted that instead of the one-way analysis of variance, the t-test (sec. 4.7) could have been applied to these data. This test would yield a computed value of $t = \sqrt{F} = \sqrt{34.7}$, and, of course, the same conclusion that the means are unequal would have been drawn.

Comparison of Stope and Development Data from a South African Gold Mine

The second one-way analysis of variance is of gold-assay data from the City Deep mine, Central Witwatersrand, South Africa. The overall problem (Koch and Link, 1966), parts of which are discussed in Chapter 16 of this book, was to determine gold distribution and variability in a typical South African gold mine. The data were assays from a 1000-square-foot block of ground that had been fully developed and stoped out in workings extending throughout most of this section of the City Deep mine. The assays, supplied by City Deep, Ltd., consisted of 503 gold-assay values from all development samples taken in drifts and a raise, and 1536 assays from all stope samples.

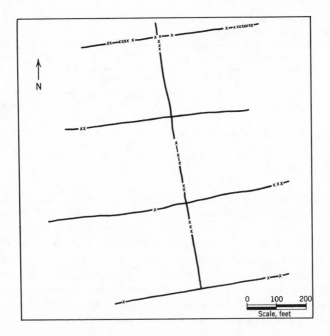

Fig. 5.3. Locations of the 39 of 503 development points accounting for 50 percent of the total in.-dwt gold content, City Deep mine, South Africa.

Figure 5.3 shows that the four parallel drifts and the raise gave good coverage for the development sampling and that the pattern of gold assays is erratic, with 50 percent of the gold content in the development samples concentrated at 39 out of 503 sample points.

We examined the development and stope data separately to learn how well they agreed. Because the purpose of taking the development samples was to predict grade of gold ore to be mined in the stopes, good agreement between the two sets of assay values was desirable for a successful prediction. Table 5.10 lists summary assay data for the stope and development samples. For these data the inch-pennyweight (in.-dwt) variable is the appropriate measure of gold content (sec. 13.2). Table 5.10 shows that the development means far overestimate the grade of gold actually mined in the stopes and also that both development and stope in.-dwt values are highly variable, with coefficients of variation of about 2. In order to learn whether the two sets of data are consistent within limits of expected statistical fluctuation, the one-way analysis of variance of table 5.11 was performed. The hypothesis H_0 that the two population means for the statistical samples of development and stope

Table 5.10. Summary assay data of samples from 1000-foot-square block, City Deep mine, South Africa

	Stope data, 1536 samples			Development data, 503 samples		
Unit	Mean	s	C	Mean	s	C
Dwt	12.00	31.31	2.61	22.58	71.33	3.16
Inches	20.84	9.47	0.45	18.20	10.29	0.57
In.-dwt	192.40	374.70	1.95	273.46	607.80	2.22

Note: Dwt, pennyweights; In.-dwt, inch-pennyweights; s, standard deviation; C, coefficient of variation.

assays are equal was tested against the alternative hypothesis H_1 that the population means are different. This analysis has 1 degree of freedom for variation between stope and development assays and 2037 degrees of freedom $(503 + 1536 - 2)$ for within-group variation. Because the calculated F-value of 12.7 is larger than the tabled $F_{10\%}$-value of 2.71, we accepted the alternative hypothesis H_1 and concluded that the two sets of data are inconsistent. As before, the t-test could have been used, as there are only two sets of means. In section 16.1, in which the discussion of these interesting data is continued, we suggest ways to reconcile the stope and development values.

Table 5.11. One-way analysis of variance to compare development and stope mineralization, City Deep mine

Source of variation	Sum of squares	Degrees of freedom	Mean square	F	$F_{10\%}$
Between stope and development	2,489,747	1	2,489,747	12.7	2.71
Within groups	400,963,342	2037	196,647		

5.5 ANALYSIS OF VARIANCE OF NESTED DATA

In the preceding sections of this chapter we have explained the analysis of variance for the simplest one-way case and have given two examples of the many uses in geology. Now, in this section the analysis of variance of nested data, also named hierarchical analysis of variance, is introduced. The method is explained and in section 5.6 application in geology is discussed. The computing details are omitted; the reader requiring them is referred to Li (1964, I, chap. 18).

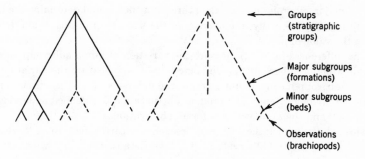

Fig. 5.4. Relation to one another of 60 fictitious observations.

The names *nested* and *hierarchical*, being descriptive, indicate that in this form of the analysis of variance the observations are partitioned into subgroups within larger groups. The method is explained by a fictitious illustration (fig. 5.4) analyzed in the format of table 5.12 for a set of 60 observations classified into five groups, with three major subgroups within each group, two minor subgroups within each major subgroup, and two observations within each minor subgroup. Any desired geological interpretation may be placed on the classification; for instance, the five groups might be stratigraphic groups, the three major subgroups might be formations, the two minor subgroups might be beds, and the observations might be sizes of brachiopods.

For these fictitious data a typical observation w_{ijk} may be written

$$w_{ijk} = \mu + (\mu_i - \mu_j) + (\mu_j - \mu_k) + (\mu_k - \mu) + e_{ijk},$$

where μ is the general mean for all of the data, μ_i is the mean of minor subgroups within major subgroups, μ_j is the mean of major subgroups within groups, μ_k is the mean of groups, and e_{ijk} is the random fluctuation. There would be 30 minor subgroup means μ_i; their averages taken two at a time would determine the 15 major subgroup means μ_k. In turn, the averages of

TABLE 5.12. FICTITIOUS ILLUSTRATION OF NESTED ANALYSIS OF VARIANCE FORMAT

Source of variation	Sum of squares	Degrees of freedom	Mean square
Among groups		4	
Among major subgroups		10	
Between minor subgroups		15	
Within minor subgroups		30	

the 15 major subgroup means taken three at a time would determine the five group means μ_k. Finally, the average of the five group means would determine the value of μ.

The differences $(\mu_i - \mu_j)$ measure the variation of minor subgroup means within a major subgroup; the differences $(\mu_j - \mu_k)$ measure the variation of major subgroup means within a group; the differences $(\mu_k - \mu)$ measure the variation among groups. The random fluctuation e_{ijk} is defined as having a mean of 0 and the same variance for all observations.

Through the model, the degrees of freedom are partitioned. In all there are $5 \times 3 \times 2 \times 2 = 60$ observations (fig. 5.4). Because there are two observations in each minor subgroup, there is one degree of freedom associated with the variation within each minor subgroup. Thus, with 30 minor subgroups, there are altogether 30 degrees of freedom for within-minor-subgroups variation. Because there are two minor subgroups within each of the 15 major subgroups, there are 15 degrees of freedom for the variation of minor subgroups within major subgroups. Because there are three subgroups within each of the five groups, there are $2 \times 5 = 10$ degrees of freedom for the variation of the major subgroups within groups. Finally, because there are five groups, there are 4 degrees of freedom for variation among groups. The sum of degrees of freedom from all levels is 59, the total number of degrees of freedom in 60 observations $(30 + 15 + 10 + 4 = 59)$.

In the fictitious illustration of table 5.12 the appropriate F-value to test whether the major subgroups differ among themselves has 10 and 15 degrees of freedom. Moreover, because both the degrees of freedom and the sum of squares are additive, a hypothesized difference in the groups themselves can be tested by an F-test with 4 and 10 degrees of freedom. A nesting scheme, like that illustrated by the fictitious illustration, can be carried to any number of meaningful levels. The number of degrees of freedom for any level is always the number of degrees of freedom that can be calculated for observations at that level.

5.6 AN EXAMPLE OF NESTED ANALYSIS OF VARIANCE

The statistically designed experiment for sampling ore in the Homestake mine, South Dakota, by diamond-drill-hole core, introduced in section 3.4, provides an excellent example of the analysis of variance of nested data. Our purpose was to investigate several levels of variability in gold distribution: among drill holes, among 1-foot intervals, within 1-foot intervals, and among furnace runs. As shown by figure 3.1, five holes were bored, one pair in one crosscut, and the other three in an adjacent crosscut. The core was sampled

TABLE 5.13. SUMMARY GOLD ASSAY DATA FROM FIVE DIAMOND-DRILL
HOLES, HOMESTAKE MINE, SOUTH DAKOTA

Drill hole	Number of observations	Mean gold assay (oz/ton)	Standard deviation	Coefficient of variation
A	58	0.52	1.45	2.79
B	58	2.19	7.60	3.47
C	24	2.85	8.27	2.90
D	39	0.50	1.16	2.32
E	40	0.69	1.57	2.28
All holes	219	1.25	4.93	3.94

in a random design, and selected geological samples were assayed. With the assay values in hand, the first obvious step was to compute a mean and standard deviation for the individual holes and for all holes (table 5.13). Although all of these statistics are distinctly different, those for holes A, D, and E are rather similar, as are those for holes B and C. The extremely high coefficients of variation reflect the erratic gold mineralization, particularly as observed in these small samples of 1-inch-long cylinders of core.

First of all, taking into account the variability, we wanted to learn if the differences among these five sample means are large enough to indicate a difference in the underlying population means, or if the observed differences merely stem from sample fluctuation. The appropriate analysis of variance, a comparison of among-hole and within-hole variation, is presented in table 5.14A. The measure of within-hole variation is necessarily the among-foot variation rather than the among-observation variation or the among-sample-pair variation that might be expected, because the smallest scale of unrestricted random sampling is at 1-foot intervals. In statistical terminology the sampling is *blocked* within holes at 1-foot intervals, because the *constraint*, or requirement, was made that two pairs of samples be taken for every 1-foot interval. An average assay value was available for each 1-foot interval, the average of the four observations (or fewer, in the few intervals where a potential observation was lost in drilling or assaying). Thus there were 4 degrees of freedom to estimate the variability among the five holes and 51 degrees of freedom to estimate the variability within holes. The hypothesis H_0 that the five population means were equal was tested against the alternative hypothesis H_1 that they were unequal. Because the computed F-value of 0.89 was smaller than the tabled $F_{10\%}$-value of 2.07, it is outside the critical region, and the conclusion is drawn that the population means are the same.

Table 5.14. Three one-way analyses of variance to investigate gold mineralization in diamond-drill core from the Homestake mine

A. Analysis of variance to compare among-hole and within-hole variation (as measured by among-foot variation)

Source of variation	Sum of squares	Degrees of freedom	Mean square	F	$F_{10\%}$
Among-hole	177.883	4	44.471	0.89	2.07
Among-foot	2536.107	51	49.728		

B. Analysis of variance to compare among-foot and within-foot intervals (as measured by between-sample pair variation)

Source of variation	Sum of squares	Degrees of freedom	Mean square	F	$F_{10\%}$
Among-foot intervals	2536.107	51	49.728	1.71	1.45
Between-sample pairs	1511.398	52	29.065		

C. Analysis of variance to compare within-sample-pair and between-sample-pair variation

Source of variation	Sum of squares	Degrees of freedom	Mean square	F	$F_{10\%}$
Between-sample pairs	1511.398	52	29.065	2.90	1.37
Within-sample pairs	1083.141	108	10.029		

We next wanted to investigate the variability on a smaller scale, comparing the variation from foot to foot within holes with the variation within each foot. Again the sampling within each foot is blocked at the scale of sample pairs, so that the appropriate measure is the between-sample-pair variation. The hypothesis H_0 that the variation among 1-foot intervals is no larger than that within 1-foot intervals is tested against the alternative hypothesis H_1 that it is larger. The analysis is performed in table 5.14*B*. As in table 5.14*A*, there are 51 degrees of freedom to measure among-foot variability, and there are 52 degrees of freedom to measure between-sample-pair variability, because there were 52 1-foot intervals with 2 pairs of assays (in the others one or more observations were lost). Because the computed F-value of 1.71 is larger than the tabled $F_{10\%}$-value of 1.45, the decision is to accept the alternative hypothesis H_1 that the variation among 1-foot intervals is larger than that within 1-foot intervals.

TABLE 5.15. NESTED ANALYSIS OF VARIANCE OF GOLD ASSAY DATA FROM THE HOMESTAKE MINE

Line	Source of variation	Sum of squares	Degrees of freedom	Mean square	F	$F_{10\%}$
1	Among hole	177.883	4	44.471	0.89	2.07
2	Among foot intervals	2536.107	51	49.728	1.71	1.45
3	Between sample pairs	1511.398	52	29.065	2.90	1.37
4	Within sample pairs	1083.141	108	10.029		
5	Total	5308.529	215			

Finally, we examined the variation on the smallest scale sampled, within 1-foot intervals, comparing the variation within the paired 1-inch cylinders with that among the paired cylinders. In table 5.14C, the hypothesis H_0 that the variation among sample pairs is no larger than that within sample pairs is tested against the alternative hypothesis H_1 that the variation among sample pairs is larger. As before, the degrees of freedom to estimate the between-sample-pair variation are 52; the degrees of freedom to estimate within-sample-pair variation are 108, because there are 108 pairs of samples (again subtracting lost observations). Because the computed F-value of 2.90 is larger than the tabled $F_{10\%}$-value of 1.37, it lies inside the critical region, and the decision is to accept the alternative hypothesis H_1 that the within-sample-pair variation is smaller than the among-sample-pair variation.

Clearly, the three one-way analyses of variance in table 5.14 are related to one another. The relationship is exhibited by combining the three analyses of variance into a nested analysis of variance in table 5.15. In table 5.15, both the total sum of squares and total degrees of freedom in line 5 have been partitioned into the 4 pieces for each on lines 1 to 4. Certain problems in this analysis are discussed in section 5.11. The Homestake example analysis has been introduced purposely to show that analysis of real data, even that obtained in a designed experiment, is not likely to be clean cut.

5.7   RANDOMIZED BLOCKS

In this section, a statistical procedure named *randomized block* is explained, first by the simplest case of the *t-test of paired observations*, and then by the general procedure for observations that are in groups larger than pairs. Through the randomized-block procedure, the experimental material (perhaps rock units or hand specimens for chemical analysis) is arranged in homogeneous groups named *blocks*, and the experimental treatments (perhaps measurements of grain size by different techniques or chemical analyses by different methods) are applied randomly to the experimental material in each block. Because the randomization is confined within homogeneous groups and because all treatments are applied in each group, most of the among-group variation is eliminated. If the experimental material is not arranged in blocks, any variation in it produces additional random fluctuation.

A randomized block is a simple *experimental design*, a subject developed in sections 8.6 and 12.5. Through an experimental design, as the name implies, a scientific investigation is performed as a statistically designed experiment in order to obtain maximum information from the effort expended.

Application of the *t*-Statistic to Paired Observations

The *t*-test on paired observations is explained by analysis of the fictitious data of table 5.16. Assume that an investigator has nine rock specimens that he wishes to analyze for CaO by two chemical methods. He may prepare two pulverized fractions suitable for chemical analysis from each rock specimen. He may then randomly choose one fraction from each specimen to be analyzed by the first method and the other fraction to be analyzed by the second method. In this way the 18 observations in columns 1 and 2 of table 5.16 might be obtained.

In order to learn whether the two chemical methods furnish consistent results, a confidence interval may be formed for the difference between the population means for the two methods by assuming that the observations are two samples, each of size 9 from populations of infinite numbers of potential analyses. If the confidence interval includes zero, the conclusion is that the two methods are consistent; if it does not, the conclusion is that the two methods are inconsistent. Instead, the hypothesis H_0 that the two population means are equal could be tested against the alternative hypothesis H_1 that they are unequal.

Because, as calculated in table 5.16, the confidence interval for these data extends from -0.833 to 1.455, including zero, the investigator concludes that the two chemical methods are consistent. However, he has made a major blunder in data analysis by ignoring the fact that he had nine pairs of observations from nine rock samples and has treated them as though they were merely 18 unpaired observations. Therefore experimental error (random fluctuation, e) has been increased because the variation in CaO of the rocks from sample to sample has been included in it. As demonstrated next, this uncontrolled variability can readily be eliminated by the proper statistical procedure, with no increase in experimental cost.

TABLE 5.16. FICTITIOUS ILLUSTRATION, COMPARING TWO METHODS OF CHEMICAL ANALYSIS FOR CaO BY THE t-STATISTIC

Data:

	CaO (%)	
	(1)	(2)
	10.4	10.6
	9.9	9.7
	9.1	8.8
	9.6	8.9
	8.5	8.4
	7.4	6.8
	8.1	7.9
	6.6	6.3
	7.2	6.6

Calculations:

$\sum w$	76.8	74.0	
n	9	9	
$\overline{w}$	8.533	8.222	$0.311 = \overline{w}_1 - \overline{w}_2$
$(\sum w)^2$	5898.24	5476.00	
$(\sum w)^2/n$	655.36	608.44	
$\sum w^2$	669.16	625.56	
SS	13.80	17.12	
$SS_1 + SS_2$		30.92	
$n_1 + n_2 - 2$		16	
s_p^2		1.9325	
$1/n_1 + 1/n_2$		0.2222	
$\sqrt{s_p^2(1/n_1 + 1/n_2)}$		0.6553	
$t_{5\%}$ (with 16 d.f.)		1.746	
$t\sqrt{s_p^2(1/n_1 + 1/n_2)}$		1.144	
μ_L		-0.833	
μ_U		1.455	

An incisive way to perform the statistical analysis is outlined in table 5.17. The essential difference from table 5.16 is that now the nine analyses by the two chemical methods are recognized as being *paired*, so that the two analyses of duplicate samples are compared directly with one another. In statistical terminology the two chemical methods are named *treatments*, and the nine

TABLE 5.17. FICTITIOUS ILLUSTRATION, COMPARING TWO METHODS OF CHEMICAL ANALYSIS FOR CaO BY THE t-STATISTIC APPLIED TO PAIRED OBSERVATIONS

Data:

	CaO (%)		Difference, (1) − (2)
	(1)	(2)	(3)
	10.4	10.6	− 0.2
	9.9	9.7	0.2
	9.1	8.8	0.3
	9.6	8.9	0.7
	8.5	8.4	0.1
	7.4	6.8	0.6
	8.1	7.9	0.2
	6.6	6.3	0.3
	7.2	6.6	0.6

Calculations:

$\sum w$	76.8	74.0	2.8
n	9	9	9
$\bar{w}$	8.533	8.222	0.31111
$(\sum w)^2$			7.84
$(\sum w)^2/n$			0.8711
$\sum w^2$	669.16	625.56	1.52
SS			0.64889
s^2			0.081111
s			0.2848
$\sqrt{n}$			3
$s/\sqrt{n}$			0.09493
$t_{5\%}$ (with 8 d.f.)			1.860
$t(s/\sqrt{n})$			0.176
μ_L			0.135
μ_U			0.487

pairs are named *blocks*. As explained in the next subsection, any number of treatments and blocks can be accommodated; the t-statistic applied to paired observations is only the special case for two treatments.

Through application of the t-statistic to the paired observations of table 5.17, a confidence interval is formed for the population mean of the differences between the paired observations. The interpretation of the confidence interval is that, because of experimental error, no one would expect the two chemical methods always to yield exactly the same result. But if the two methods are consistent, one of them should not, on the average, yield higher results than the other. Because the confidence interval for the paired differences extends from 0.135 to 0.487 and does not include zero, the conclusion is that the two methods do not agree. By recognizing that the observations are paired, and by statistically analyzing the experimental data accordingly, one obtains more precise information. The next step would be to find out which chemical method is better or whether one contains a constant bias that can readily be removed by calculation.

Randomized-Block Design with More than Two Treatments

The statistical procedure for reducing unexplained variability by blocking may be readily generalized for more than two treatments. As already explained for paired observations, in any collection of data it is always desirable to organize the observations in several blocks for the purpose of making the amount of unexplained variability as small as possible. The observations are arranged to be as homogeneous as possible within blocks, although the different blocks may be very heterogeneous. Then each treatment is randomly applied within each block. Although the within-block variability is then small, the among-blocks variability may be large, and the randomized-block design allows the effect of the among-block variability to be removed from the statistical analysis.

The general principles of the randomized-block design may best be explained with the aid of an illustration. Suppose that 20 chemical laboratories are to analyze three different rocks for lime. For a comparison of the performance of the laboratories, each laboratory is considered a treatment, and each rock is considered a block. From each rock 20 pulverized fractions are prepared for analysis and are assigned at random to the 20 laboratories. This procedure yields 60 (20×3) chemical analyses or observations.

Even if the 3 rocks have different mean lime contents, the performances of the 20 laboratories may be compared by removing the effect of the different means. The model for an observation w_{LR} is

$$w_{LR} = \mu + (\mu_L - \mu) + (\mu_R - \mu) + e_{LR},$$

where μ is the mean of all observations, μ_L is the mean of the observations for a single laboratory, μ_R is the mean of the observations for a single rock, and e_{LR} is the random fluctuation. The average values of μ_L and μ_R are identical and are equal to μ. The random fluctuation e_{LR} is the deviation of an observation from the average value for laboratory L and rock R. As in other models, the assumption is made that the mean of e_{LR} is 0 and that the variance of e_{LR} is the same for all values of L and R.

In this linear model the term $(\mu_L - \mu)$ is named the *laboratory effect*, and the term $(\mu_R - \mu)$ is named the *rock effect*. In words, the model may be written as

w = general mean + laboratory effect + rock effect + random error.

If the rock effect were not taken into account, the linear model would be

w = general mean + laboratory effect + random error,

and the random error would be larger, if, as generally happens, the lime contents of the rocks were different on the average. Other consequences of this model are that all laboratories are assumed to have about the same random fluctuation and to measure the same deviation from the average of the various rocks.

The calculations for a randomized-block design may be explained by the use of an example from Li (1964, I, p. 244). The data in table 5.18 are observed pH readings from the top, middle, and bottom of six core samples of soil; the problem is to determine if the average reading varies with soil depth. The two bottom rows and the right-hand column of the table list summary quantities needed for the analysis of variance calculations. The next-to-bottom row lists the three totals of the top, middle, and bottom pH readings;

TABLE 5.18. READINGS OF pH FROM THE TOP, MIDDLE, AND BOTTOM OF SIX CORE SAMPLES OF SOIL. (FROM LI, 1964, I, p. 244.)

Core sample no.	Top	Middle	Bottom	Core sample totals, T_r
1	7.5	7.6	7.2	22.3
2	7.2	7.1	6.7	21.0
3	7.3	7.2	7.0	21.5
4	7.5	7.4	7.0	21.9
5	7.7	7.7	7.0	22.4
6	7.6	7.7	6.9	22.2
Location totals, T_t	44.8	44.7	41.8	
Grand total, G				131.3

each total is the sum of the six pH readings from the individual core samples. The right-hand column lists the six totals from the six core samples; each total is the sum of the three pH readings from a single core sample. The entry in the bottom row, the total of all 18 observations, is also the sum of the three totals in the next-to-bottom row, or the sum of the six totals in the right-hand column.

The analysis-of-variance calculations for the randomized-block design in table 5.19 are similar to those for the completely randomized experiment explained in table 5.3; the only difference is the addition to table 5.19 of the lines associated with the replications. The top half of the table contains the preliminary calculations. Column 1 lists the totals of squares for the grand total of all observations squared, G^2; the sum of squares of the replication totals, $\sum T_r^2$; the sum of the squares of the treatment totals, $\sum T_t^2$; and the sum of the squares of all the observations, $\sum w^2$. In column 2 the number of observations per sum is recorded; the entries in column 3 are obtained by dividing the entries in column 1 by those in column 2. The bottom half of table 5.19, the analysis-of-variance format, shows how the sums of squares

TABLE 5.19. COMPUTING METHOD FOR RANDOMIZED BLOCK

	Preliminary calculations		
Type of sum	(1) Total of squares	(2) Number of observations in sum	(3) = (1)/(2) Total of squares per observation
Grand	G^2	kn	G^2/kn (I)
Replication	$\sum T_r^2$	k	$\sum T_r^2/k$ (II)
Treatment	$\sum T_t^2$	n	$\sum T_t^2/n$ (III)
Observation	$\sum w^2$	1	$\sum w^2$ (IV)

	Analysis of variance			
Source of variation	Sum of squares	Degrees of freedom	Mean square	F
Replication	II $-$ I	$n-1$		
Treatment	III $-$ I	$k-1$		
Error	IV $-$ III $-$ II $+$ I	$(k-1)(n-1)$		
Total	IV $-$ I	$kn-1$		

are obtained from the elements labeled I, II, III, and IV in column 3 of the top half of the table.

Table 5.20 gives, in the format of table 5.19, the details of the calculations for the soil sample data in table 5.18. For these data, the hypothesis H_0 is that the pH in the top, middle, and bottom soil samples is the same; or, in other words, that there is no treatment effect. Because the calculated F-value to test this hypothesis is 31.77, which is larger than the tabled value of 2.92 for a 10-percent α risk level, the hypothesis is rejected, and the alternative hypothesis H_1 that there is a treatment effect is accepted. To explore the nature of the treatment effect further would require use of the multiple-comparison techniques discussed in section 5.12.

The analysis of variance in table 5.20 can also be used to investigate whether the six places sampled have, on the average, different soil pH values. The hypothesis H_0 that the soil is the same at the six places, that is, no replication effect exists, is compared with the alternative hypothesis H_1 that the soil is different. Because the calculated F-value to test this hypothesis is 6.43, which is larger than the tabled F-value of 2.52 at the 10-percent α risk level, the alternative hypothesis H_1 that the soil is different is accepted.

TABLE 5.20. COMPUTING METHOD FOR RANDOMIZED BLOCK APPLIED TO DATA OF TABLE 5.18

Type of sum	Preliminary calculations		
	(1) Total of squares	(2) Number of observations in sum	(3) = (1)/(2) Total of squares per observation
Grand	17,239.69	18	957.7606
Replication	2,874.75	3	958.2500
Treatment	5,752.37	6	958.7283
Observation	959.37	1	959.3700

Source of variation	Analysis of variance				
	Sum of squares	Degrees of freedom	Mean square	F	$F_{10\%}$
Replication	0.4894	5	0.09788	6.43	2.52
Treatment	0.9677	2	0.48385	31.77	2.92
Error	0.1523	10	0.01523		
Total	1.6094	17			

5.8 LINEAR COMBINATIONS

A linear combination is a convenient mathematical device for summarizing various observations or statistics. The explanation of linear combinations here is entirely mathematical and is introduced at this point because the material is needed in section 5.9.

A *linear combination* q of n observations w_i is defined as a sum of terms,

$$q = c_1 w_1 + c_2 w_2 + \cdots + c_n w_n,$$

where the quantities c_i are known constants. Alternatively, the above formula may be written in summation notation as

$$q = \sum c_i w_i,$$

where the summation extends from $i = 1$ to $i = n$. A familiar example of linear combination is the sample mean, where the known constants are all equal and are the reciprocal $(1/n)$ of the sample size. Thus, the sample mean may be written

$$q = \frac{1}{n} w_1 + \frac{1}{n} w_2 + \cdots + \frac{1}{n} w_n = \overline{w},$$

or, in summation notation,

$$q = \sum \frac{1}{n} w_i = \overline{w}.$$

If one observation w_i is drawn independently and randomly from each of k populations, the mean and variance of a linear combination q formed from these observations can be calculated by a theorem illustrated by Li (1964, I, p. 246). The mean μ_q of the linear combination q is

$$\mu_q = c_1 \mu_1 + c_2 \mu_2 + \cdots + c_k \mu_k,$$

or

$$\mu_q = \sum c_i \mu_i.$$

The variance σ_q^2 of the same linear combination is

$$\sigma_q^2 = c_1^2 \sigma_1^2 + c_2^2 \sigma_2^2 + \cdots + c_k^2 \sigma_k^2,$$

or

$$\sigma_q^2 = \sum c_i^2 \sigma_i^2.$$

For the linear combination q equal to the sample mean $\overline{w}$, in which all the observations are drawn from a single population, the constant c was shown to be the reciprocal of the sample size $(1/n)$ in the formula

$$q = \sum \frac{1}{n} w_i = \overline{w}.$$

Similarly, if the reciprocal of the sample size is substituted in the formula for μ_q above, then because $k = n$, the result obtained is

$$\mu_q = \sum \frac{1}{n} \mu = \mu = \mu_{\bar{w}},$$

that is, the mean of this linear combination is the mean of sample means. Substituting the reciprocal of the sample size in the formula for σ_q^2 above, we obtain

$$\sigma_q^2 = \sum \left(\frac{1}{n}\right)^2 \sigma^2 = \frac{\sigma^2}{n} = \sigma_{\bar{w}}^2,$$

that is, the variance of this linear combination is the variance of the mean.

Another important linear combination represents the difference between two sample means, $\bar{w}_A - \bar{w}_B$. Again the constant multiplier is the reciprocal of the sample size; for the observations w_A,

$$c_i = \frac{1}{n_A},$$

where

$$i = 1, 2, \ldots, n_A,$$

and, for the observations w_B,

$$c_i = -\frac{1}{n_B},$$

where

$$i = n_A + 1, n_A + 2, \ldots, n_A + n_B.$$

Then the linear combination q is equal to

$$q = \left(\frac{1}{n_A} w_1 + \frac{1}{n_A} w_2 + \cdots + \frac{1}{n_A} w_{n_A}\right)$$

$$- \left(\frac{1}{n_B} w_{n_A+1} + \frac{1}{n_B} w_{n_A+2} + \cdots + \frac{1}{n_B} w_{n_A+n_B}\right),$$

or, in summation notation,

$$q = \sum_{i=1}^{n_A} \frac{1}{n_A} w_i - \sum_{i=n_A+1}^{n_A+n_B} \frac{1}{n_B} w_i = \bar{w}_A - \bar{w}_B.$$

The mean of the linear combination is

$$\mu_q = \left(\frac{1}{n_A} \mu + \frac{1}{n_A} \mu + \cdots + \frac{1}{n_A} \mu\right) - \left(\frac{1}{n_B} \mu + \frac{1}{n_B} \mu + \cdots + \frac{1}{n_B} \mu\right),$$

or, in summation notation,

$$\mu_q = \sum \frac{1}{n_A} \mu - \sum \frac{1}{n_B} \mu = 0,$$

showing that if the two samples are drawn from the same population the mean difference of sample means is 0. The variance of the linear combination is

$$\sigma_q^2 = \left(\frac{1}{n_A^2} \sigma^2 + \frac{1}{n_A^2} \sigma^2 + \cdots + \frac{1}{n_A^2} \sigma^2 \right)$$

$$+ \left(\frac{1}{n_B^2} \sigma^2 + \frac{1}{n_B^2} \sigma^2 + \cdots + \frac{1}{n_B^2} \sigma^2 \right),$$

or, in summation notation,

$$\sigma_q^2 = \sum \frac{1}{n_A^2} \sigma^2 + \sum \frac{1}{n_B^2} \sigma^2 = \sigma^2 \left(\frac{1}{n_A} + \frac{1}{n_B} \right)$$

showing that, if the two samples are drawn from the same population, the variance of the difference of sample means is the sum of the variances of the individual sample means.

A *contrast* is a linear combination for which the sum of constant multipliers $\sum c_i$ is equal to 0. Contrasts, the most important kind of linear combinations, are illustrated by the last example and discussed in detail in section 5.9.

5.9 SINGLE DEGREE OF FREEDOM

In sections 5.1 and 5.3 the one-way analysis of variance is illustrated with fictitious observations of sand content from five outcrops. Now, in this section, contrasts are formed by using the method of the *single degree of freedom* to analyze similar data more incisively. Three different contrasts are used before the method of the single degree of freedom is related to the one-way analysis of variance of section 5.3. The meaning of *single degree of freedom* is explained near the end of this section.

Assume that a geologist is making a geochemical survey for selenium, which is suspected of killing cattle in Wyoming rangeland. Assume further that the geologist wishes to compare two methods of sampling for selenium—method A, taking single hand specimens of outcrops, and method B, taking chip samples across the entire outcrop surface. Finally assume that two formations, numbered 1 and 2, are to be sampled, by taking four groups of observations, each seven in number. Accordingly, the groups of observations

TABLE 5.21. DESIGNATIONS FOR FOUR POPULATIONS
OF FICTITIOUS SELENIUM DETERMINATIONS

| Formation | Method of sampling | |
	A	B
1	μ_1 $(1, A)$	μ_2 $(1, B)$
2	μ_3 $(2, A)$	μ_4 $(2, B)$

come from four populations, with means μ_1 to μ_4, as defined in table 5.21. With this specification of the four statistical samples, three hypotheses about the population means may be tested with contrasts.

The first hypothesis H_0 to be tested is that the two formations have the same mean selenium content. For sampling method A the two formations can be compared by forming the hypothesis that $\mu_1 = \mu_3$, which can be rearranged to the form $\mu_1 - \mu_3 = 0$. Likewise, for sampling method B, the two formations can be compared by the hypothesis that $\mu_2 = \mu_4$, or $\mu_2 - \mu_4 = 0$. These two hypotheses can be added to form the desired hypothesis H_0 that $\mu_1 - \mu_3 + \mu_2 - \mu_4 = 0$, which can be rearranged to the form $\mu_1 + \mu_2 - \mu_3 - \mu_4 = 0$. The alternative hypothesis H_1 is that the sum of the population means is not equal to 0; that is, $\mu_1 + \mu_2 - \mu_3 - \mu_4 \neq 0$. In geological terms the alternative hypothesis H_1 is that the selenium contents of the two formations are different.

Because each of the four population means μ_1 to μ_4 is estimated by a corresponding statistic $\overline{w}_1$ to $\overline{w}_4$, the quantity $\overline{w}_1 + \overline{w}_2 - \overline{w}_3 - \overline{w}_4$ is a statistic that estimates the quantity $\mu_1 + \mu_2 - \mu_3 - \mu_4$. Both quantities are linear combinations whose constant multipliers are 1, 1, -1, and -1. Because for both quantities the sum of constant multipliers is 0, the linear combinations are contrasts (sec. 5.8). To show clearly the role of the constant multipliers, one may write the contrast involving the parameters as follows:

$$\tau = (1)\mu_1 + (1)\mu_2 + (-1)\mu_3 + (-1)\mu_4,$$

where τ is a new symbol defined by the equation, and the contrast involving the statistics may be written

$$q = (1)\overline{w}_1 + (1)\overline{w}_2 + (-1)\overline{w}_3 + (-1)\overline{w}_4.$$

The contrast q formed in the preceding paragraph is essential for testing the hypothesis H_0, which now may be written $\tau = 0$, and also for calculating a confidence interval for τ. Both methods are presented, but the hypothesis test is presented first because it is directly related to the one-way analysis of variance discussed in section 5.3.

Hypothesis Tests with Single Degree of Freedom

Because the appropriate hypothesis test is an F-test for which suitable variances must be found, some preliminary material is presented. From section 5.8, the variance of the contrast q is

$$\sigma_q^2 = \sum c_i^2 \sigma_i^2.$$

The quantities σ_i^2 are the variances of the four sample means, and each is equal to the square of the standard error of the mean; that is,

$$\sigma_i^2 = \sigma_{\bar{w}}^2 = \frac{\sigma^2}{7},$$

where 7 is the number of observations taken in each group. Therefore,

$$\sigma_q^2 = (1)^2 \frac{\sigma^2}{7} + (1)^2 \frac{\sigma^2}{7} + (-1)^2 \frac{\sigma^2}{7} + (-1)^2 \frac{\sigma^2}{7},$$

or, factoring out $\sigma^2/7$,

$$\sigma_q^2 = \left(\sum c_i^2 \right) \frac{\sigma^2}{7} = 4 \frac{\sigma^2}{7}.$$

Each of the $\bar{w}$-values follows a normal distribution; the contrast q also follows a normal distribution with mean equal to τ and variance equal to $4\sigma^2/7$. Because a normally distributed quantity minus its mean divided by its standard deviation is normally distributed with a mean of 0 and a variance of 1 (sec. 3.6), the statistic

$$\frac{q - \tau}{\sqrt{4\sigma^2/7}}$$

follows a normal distribution with a mean of 0 and a variance of 1. If the hypothesis H_0 is true, τ is equal to 0, and the statistic may be rewritten

$$\frac{q - 0}{\sqrt{4\sigma^2/7}}.$$

Since the parameter σ^2 is generally unknown, it is replaced, as usual, with its statistic, the pooled sample variance s_p^2. Then the new quantity,

$$t = \frac{q - \tau}{\sqrt{4s_p^2/7}},$$

Table 5.22. Fictitious selenium contents from four formations

Line	Item	Statistical sample number			
		(1)	(2)	(3)	(4)
1		4.7	5.0	6.7	7.5
2		5.5	7.3	3.3	5.0
3	Observations	3.8	5.4	5.7	4.9
4	selenium, ppm	5.2	5.3	5.4	5.7
5		5.6	5.8	4.5	5.6
6		4.9	6.8	4.2	6.4
7		6.7	8.2	6.4	5.5
8	$\sum w$	36.4	43.8	36.2	40.6
9	n	7	7	7	7
10	$\bar{w}$	5.200	6.257	5.171	5.800
11	$(\sum w)^2$	1324.96	1918.44	1310.49	1648.36
12	$(\sum w)^2/n$	189.28	274.06	187.21	235.48
13	$\sum w^2$	194.08	282.66	196.28	240.32
14	SS	4.80	8.60	9.07	4.84
15	d.f.	6	6	6	6
16	s^2	0.80	1.43	1.51	0.81

$$s_p^2 = (s_1^2 + s_2^2 + s_3^2 + s_4^2)/4 = 1.14$$

follows the t-distribution rather than the standardized normal distribution, as explained in section 4.3. The t-distribution has the 24 degrees of freedom used to calculate s_p^2. Because $t^2 = F$ (sec. 5.13), the squared statistic,

$$F = \frac{(q - \tau)^2}{4s_p^2/7} = \frac{7(q - \tau)^2}{4s_p^2},$$

follows the F-distribution with 1 and 24 degrees of freedom. If the hypothesis H_0 is true, τ is equal to 0, and the statistic becomes

$$F = \frac{7q^2}{4s_p^2}.$$

Now that the method has been explained, the numerical calculations can be performed by using the fictitious data from table 5.22. In table 5.23 the test of the hypothesis that there is no difference in selenium contents of formations 1 and 2 is set up in the same form as that in section 4.7. In lines 1 to 4 are entered the four sample means from table 5.22, the constant multipliers, and the products of the means and the multipliers. In line 5 the value of q is calculated to be 0.486; in line 6 is entered the value 1.14 for the

TABLE 5.23. SINGLE-DEGREE-OF-FREEDOM TEST OF THE HYPOTHESIS
THAT THERE IS NO DIFFERENCE IN THE SELENIUM CONTENTS OF THE
TWO FORMATIONS

Hypothesis:	$H_0: \mu_1 + \mu_2 - \mu_3 - \mu_4 = 0$
Alternative hypothesis:	$H_1: \mu_1 + \mu_2 - \mu_3 - \mu_4 \neq 0$
Statistic:	F
Risk of type I error:	$\alpha = 10\%$
Critical region:	$F > F_{10\%}$ ($F_{10\%} = 2.93$, with 1 and 24 d.f.)

Computation:

Line	Sample number	$\overline{w}$	c	$c\overline{w}$	Remarks
1	1	5.200	1	5.200	Values of $\overline{w}$ from
2	2	6.257	1	6.257	table 5.22
3	3	5.171	-1	-5.171	
4	4	5.800	-1	-5.800	
5	$q = \sum c\overline{w} = 0.486$				
6	$s_p^2 = 1.14$			From table 5.22	
7	$F = \dfrac{7q^2}{4s_p^2} = \dfrac{7(0.486)^2}{4(1.14)} = 0.363$				

Conclusion. Accept hypothesis H_0.

pooled variance, obtained from table 5.22; in line 7, values are substituted in the formula to yield an F-value of 0.363. Because this F-value is outside the critical region, the hypothesis H_0 is accepted.

Two other single-degree-of-freedom hypothesis tests for the same data are now presented. The first is that the selenium content estimated by the two methods of sampling is the same. Through reasoning similar to that employed to form the previous hypothesis, the two hypotheses that $\mu_1 - \mu_2 = 0$ and that $\mu_3 - \mu_4 = 0$ are added to form the desired hypothesis H_0 that $\mu_1 - \mu_2 + \mu_3 - \mu_4 = 0$. In table 5.24 calculations parallel to those in table 5.23 are performed. In the initial data in lines 1 to 4 the only difference is that the constant multipliers in lines 2 and 3 have opposite signs from those in the preceding table. Because the calculated F-value of 4.364 is inside the critical region, the hypothesis H_0 is rejected, and the alternative hypothesis H_1, that the two methods of sampling yield different estimates of selenium content, is accepted.

The last hypothesis to be tested is that the sampling was done consistently in the two formations, rather than sampling method A being applied differently in formation 1 than in formation 2 or sampling method B being applied

TABLE 5.24. SINGLE-DEGREE-OF-FREEDOM TEST OF THE HYPOTHESIS THAT THERE IS NO DIFFERENCE IN SAMPLING METHODS

Hypothesis:	$H_0: \mu_1 - \mu_2 + \mu_3 - \mu_4 = 0$
Alternative hypothesis:	$H_1: \mu_1 - \mu_2 + \mu_3 - \mu_4 \neq 0$
Statistic:	F
Risk of type I error:	$\alpha = 10\%$
Critical region:	$F > F_{10\%}$ ($F_{10\%} = 2.93$, with 1 and 24 d.f.)

Computation:

Line	Sample number	$\overline{w}$	c	$c\overline{w}$
1	1	5.200	1	5.200
2	2	6.257	-1	-6.257
3	3	5.171	1	5.171
4	4	5.800	-1	-5.800

5	$q = \sum c\overline{w} = -1.686$
6	$s_p^2 = 1.14$
7	$F = \dfrac{7q^2}{4s_p^2} = \dfrac{7(-1.686)^2}{4(1.14)} = 4.364$

Conclusion.   Reject hypothesis H_0. Accept alternative hypothesis H_1.

differently in formation 1 than in formation 2. Because the relation, if any, between a sampling method and a formation is investigated, the effect sought is named an *interaction*. If there is no interaction, the result of sampling formation 1 by method A plus the result of sampling formation 2 by method B should be equal, on the average, to the result of sampling formation 2 by method A plus the result of sampling formation 1 by method B (table 5.21). Thus, the hypothesis H_0 is that $\mu_1 + \mu_4 = \mu_2 + \mu_3$ or, rearranging, H_0 is that $\mu_1 - \mu_2 - \mu_3 + \mu_4 = 0$. In table 5.25 the hypothesis H_0 is tested against the alternative hypothesis H_1 that this contrast is not equal to 0. When, in table 5.25, calculations parallel to those in tables 5.23 and 5.24 are performed, the calculated F-value of 0.281 is outside the critical region, and, therefore, the hypothesis H_0 of no interaction is accepted.

In each of the tables 5.23 to 5.25 an F-statistic was calculated from the formula

$$F = \frac{7q^2}{4s_p^2}.$$

TABLE 5.25. SINGLE-DEGREE-OF-FREEDOM TEST OF THE HYPOTHESIS THAT TWO SAMPLING METHODS ARE BEING APPLIED CONSISTENTLY IN TWO FORMATIONS

Hypothesis:	$H_0: \mu_1 - \mu_2 - \mu_3 + \mu_4 = 0$
Alternative hypothesis:	$H_1: \mu_1 - \mu_2 - \mu_3 + \mu_4 \neq 0$
Statistic:	F
Risk of type I error:	$\alpha = 10\%$
Critical region:	$F > F_{10\%}$ ($F_{10\%} = 2.93$, with 1 and 24 d.f.)

Computation:

Line	Sample number	$\overline{w}$	c	$c\overline{w}$
1	1	5.200	1	5.200
2	2	6.257	-1	-6.257
3	3	5.171	-1	-5.171
4	4	5.800	1	5.800

5          $q = \sum c\overline{w} = -0.428$

6          $s_p^2 = 1.14$

7          $F = \dfrac{7q^2}{4s_p^2} = \dfrac{7(-0.428)^2}{4(1.14)} = 0.281$

Conclusion. Accept hypothesis H_0.

If the term $7q^2/4$ is replaced by the new quantity Q^2, the formula becomes

$$F = \frac{Q^2}{s_p^2}.$$

The quantity Q^2 is named the *single-degree-of-freedom statistic*; the name indicates that it has 1 degree of freedom.

With the single-degree-of-freedom statistic Q^2, the relation of the three contrasts can be given to the one-way analysis of variance for these data given in table 5.26. The one-way analysis of variance tests the hypothesis H_0 that $\mu_1 = \mu_2 = \mu_3 = \mu_4$. If this hypothesis is true, the mean selenium content is the same, regardless of sampling method or rock formation. Thus, this hypothesis is much less incisive than those formulated with the single degree of freedom.

In table 5.27 (an analysis of variance incorporating the three contrasts) lines 1, 5, and 6 are copied from table 5.26. The original F-value in line 1, calculated by dividing the mean square in line 1 by the mean square in line 5, is still valid, as is the conclusion that there is no difference among the sample

TABLE 5.26. SIMPLIFIED COMPUTING METHOD FOR ONE-WAY ANALYSIS OF VARIANCE APPLIED TO DATA OF TABLE 5.22

	Preliminary calculations		
Type of sum	(1) Total of squares	(2) Number of observations in sum	(3) = (1)/(2) Total of squares per observation
Grand	24,649.00	28	880.32
Sample	6,202.20	7	886.03
Observation	913.34	1	913.34

	Analysis of variance				
Source of variation	Sum of squares	Degrees of freedom	Mean square	F	$F_{10\%}$
Among-sample means	5.71	3	1.903	1.67	2.33
Within-sample	27.31	24	1.14		
Total	33.02	27			

means if the specification of different formations and methods is not made. In line 2 the numbers are derived from table 5.23. The sum of squares, with 1 degree of freedom, is equal to Q^2, and, because F has already been calculated, Q^2 may be obtained by solving the formula

$$F = \frac{Q^2}{s_p^2}$$

TABLE 5.27. ANALYSIS OF VARIANCE OF FICTITIOUS SELENIUM DATA, INCORPORATING THREE CONTRASTS

Line	Source of variation	Sum of squares	Degrees of freedom	Mean square	F	$F_{10\%}$
1	Among-sample means	5.71	3	1.903	1.67	2.33
2	Between formations	0.41	1	0.410	0.36	2.93
3	Between methods	4.98	1	4.980	4.36	2.93
4	Formation-method interaction	0.32	1	0.320	0.28	2.93
5	Within-sample	27.31	24	1.140		
6	Total	33.02	27			

for Q^2, with the result that

$$Q^2 = Fs_p^2 = 0.363 \times 1.14 = 0.41.$$

In lines 3 and 4 are given the other sums of squares, obtained in the same way from tables 5.24 and 5.25. Notably, the sums of squares for the single-degree-of-freedom comparisons of lines 2, 3, and 4 add up to the sum of squares in line 1; moreover, the degrees of freedom in lines 2 to 4 add up to the 3 degrees of freedom in line 1. By the single-degree-of-freedom method both the sums of squares and the degrees of freedom have been partitioned.

The three single-degree-of-freedom values and Q^2 values add up to the sum of squares with 3 degrees of freedom because the contrasts are *orthogonal*. For equal sample sizes orthogonal contrasts are those in which the sums of the products of the multipliers, taken two at a time in order, add up to 0; that is,

$$\sum c_i c_i' = 0,$$

where the c_i are the coefficients of one of the contrasts, and the c_i' are the coefficients of the other contrast. For instance, in table 5.28 the multipliers from tables 5.23 and 5.24 are listed in two rows. Multiplying the numbers in each column, one column at a time, yields the quantities on the bottom line, which sum to 0.

For the three hypothesis tests with a single degree of freedom the sums of squares add up to the total sum of squares for among-sample means, and the degrees of freedom add up to the degrees of freedom for the among-sample means. Therefore, because of Cochran's theorem (sec. 5.3), these three F-tests are independent insofar as their numerator is concerned. However, some dependence is introduced because of the common denominator, the within-sample mean square. If there is a 10-percent α risk of type I error in each of the three tests, there is at most a 30-percent (3×10 percent) risk of type I error for all three tests combined, and the conclusions of the three tests are essentially independent. Such a joint risk for several tests combined is named the *error rate*.

TABLE 5.28. ILLUSTRATION OF MULTIPLYING
CONSTANTS FOR ORTHOGONAL CONTRASTS

Source	Multiplying constants			
Table 5.23	1	1	-1	-1
Table 5.24	1	-1	1	-1
Product	1	-1	-1	1

The single-degree-of-freedom method should be applied to only a few hypotheses and only to those chosen before the data are collected. First, the technique should be applied to only a few hypotheses because the error rate grows proportionally as the number of single-degree-of-freedom tests increases. If, for example, there are 10 means, 45 single comparisons can be made by contrasts, taking two means at a time. If all of these comparisons are made at a 10-percent risk level, an average of four or five would be rejected merely by chance even if all were true. Second, the single-degree-of-freedom method should be applied only if the hypotheses are chosen before the data are collected in order to ensure honest conclusions. Otherwise, if the data are inspected and a test is then made, the inspection will usually lead to something interesting that a test will "confirm"; for example, if the smallest sample mean in a group of 10 means is compared with the largest, the hypothesis of equality of means is almost sure to be rejected, regardless of the true situation. When, because of one or both of these conditions, the single-degree-of-freedom method is unsuitable, the method of multiple comparisons, explained in section 5.12, should be used.

In the fictitious illustration of sampling selenium there are four means and seven observations per mean. In general, with k means and n observations per mean, the general hypothesis H_0 that $\sum c_i \mu_i$ is equal to 0 is tested against the alternative hypothesis H_1 that $\sum c_i \mu_i$ is not equal to 0. As usual, the i subscripts designate the constant multipliers and means, from the first term $c_1 \mu_1$ to the last, for which i is equal to k. The sum $\sum c_i$ must be equal to 0; that is, it must be a contrast. To test the hypothesis we can calculate the single degree of freedom by the formula

$$Q^2 = \frac{nq^2}{\sum c_i^2},$$

where, as in the fictitious illustration,

$$q = \sum c_i \overline{w}_i.$$

The F-statistic with 1 and $k(n-1)$ degrees of freedom is

$$F = \frac{Q^2}{s_p^2}.$$

The modified method to apply if the numbers of observations in the samples are different is given in section 5.10.

Confidence Intervals with Single Degree of Freedom

In all tests involving contrasts, the hypothesis is that the appropriate linear combination of the population means is equal to 0, that is, that τ is equal to 0. Rather than making a hypothesis test, one can calculate confidence limits for τ by giving reasonable bounds for this value. The principle is like that used to calculate confidence limits for the difference between two means (sec. 4.4); and, as before, if the confidence limits are narrow and include 0, the decision is that the contrast cannot be very different from 0. A detailed explanation for the fictitious illustrations is given after the computations are explained.

Confidence limits for τ may be constructed because the statistic

$$t = \frac{q - \tau}{\sqrt{(\sum c_i^2)s_p^2/n}},$$

introduced earlier in this section, follows the t-distribution. In table 5.29, confidence intervals are calculated for the three cases previously examined by hypothesis tests. In lines 1 to 3, the values of q, s_p^2, and n obtained from tables 5.23 to 5.25 are entered. In line 4 is entered the sum of the squared values of the constant multipliers, equal to 4 in each case. After the calculations are performed, the upper and lower confidence limits for τ, τ_L, and τ_U are obtained (lines 10 and 11).

TABLE 5.29. CONFIDENCE INTERVALS FOR THREE CASES OF FICTITIOUS SELENIUM DATA

Line	Notation	Case I: formation 1 vs. formation 2, table 5.23	Case II: method A vs. method B, table 5.24	Case III: interaction, table 5.25
		Source of data		
1	q	0.486	−1.686	−0.428
2	s_p^2	1.14	1.14	1.14
3	n	7	7	7
4	$\sum c_i^2$	4	4	4
5	$(\sum c_i^2)s_p^2/n$	0.651	0.651	0.651
6	$\sqrt{(\sum c_i^2)s_p^2/n}$	0.807	0.807	0.807
7	d.f.	24	24	24
8	$t_{5\%}$	1.711	1.711	1.711
9	$t\sqrt{(\sum c_i^2)s_p^2/n}$	1.381	1.381	1.381
10	τ_L	−0.90	−3.07	−1.81
11	τ_U	1.87	−0.30	0.95

The confidence intervals in table 5.29 may be interpreted as follows. Because all of the intervals have a 90-percent confidence coefficient, they are correct in 90 percent of the cases for which they are calculated. For case I the interpretation is that reasonable bounds for *twice* the difference in selenium content of the two formations are 1.87 and −0.90. The bounds are for twice the difference rather than for the simple difference because τ is defined by the equation

$$\tau = \mu_1 + \mu_2 - \mu_3 - \mu_4,$$

and therefore the mean associated with formation 1 is

$$\frac{\mu_1 + \mu_2}{2}$$

and that associated with formation 2 is

$$\frac{\mu_3 + \mu_4}{2}.$$

The corresponding interval for the simple difference in selenium content between formations is 0.94 to −0.45, obtained by dividing the values in lines 1 and 9 of the table by 2. Of course, a different contrast could have been formulated initially to yield the desired intervals directly, but this formulation would have obscured the relation to the hypothesis test.

For cases II and III of table 5.29 the same considerations apply as for case I. In case II, after the factor of 2 is removed as before, the reasonable bounds for the difference in selenium content obtained from the two methods of sampling are −0.15 and −1.54, indicating a real but small difference in selenium content. In case III, in which the factor of 2 is also removed, the reasonable bounds for the difference in interaction are 0.48 and −0.90, indicating that if there is an interaction effect it is small, that is, less than 1 percent.

The general formulation of a confidence interval for a contrast is as follows: The contrast τ has been defined by the formula

$$\tau = \sum c_i \mu_i,$$

where $\sum c_i = 0$ and q, the statistic that estimates τ, has been defined by the parallel formula

$$q = \sum c_i \overline{w}_i.$$

Thus, by analogy with section 4.4, the boundaries of the confidence interval for τ are

$$q - t s_p \sqrt{\frac{\sum c_i^2}{n}} \leq \tau \leq q + t s_p \sqrt{\frac{\sum c_i^2}{n}}.$$

Appraisal of the Single-Degree-of-Freedom Method

The single-degree-of-freedom method is one of the most valuable in applied statistics. In section 5.10, example analyses are presented, and many other applications will occur to the reader. To develop the method more fully would expand this book too much. A particularly full account of the method is given by Li (1964, I, p. 252, etc.), who emphasizes its use. The main advantage of the single-degree-of-freedom procedure is that, where it is appropriate, more incisive results can be achieved than with a general F-test. Its disadvantage is that a too liberal use of the method or an application to "interesting results" discovered by an inspection of the data after they are gathered can lead to an error rate that is too large. The multiple-comparison procedures discussed in section 5.12 overcome this objection but lack the incisiveness of the single-degree-of-freedom procedures.

5.10   EXAMPLES OF ANALYSIS WITH SINGLE DEGREE OF FREEDOM

In this section two example analyses with the single degree of freedom are presented with data from the Fresnillo and Homestake mines. The Fresnillo problem is a new one; the Homestake problem is a refinement of the problem of section 5.6, in which the single-degree-of-freedom method leads to a more incisive analysis of the same data.

Single Degree of Freedom Applied to Fresnillo Mine Data

In the Fresnillo mine the 2137 vein (fig. 1.4) splits northwestward into two branches, designated the 2137 vein and the 2137 Footwall vein (fig. 5.5). It was of interest to learn whether in mineralization one of the two branches was more like the 2137 vein before the junction than the other. The purpose of the analysis was twofold: to aid in interpretation of the geologic history of the vein systems by segregating different types of mineralization and to aid mine planning. As usual, we present only part of the analysis from the original paper (Koch and Link, 1967).

In order to make the desired comparison, 200 meters of vein exposed in each of three drifts extending outward from the junction on each level were selected. On each level one of the three drifts followed the 2137 Footwall vein; of the other two, one followed the 2137 vein northwest of the junction (overlapping the 2137 Footwall vein), and the other followed the 2137 vein southwest of the junction. Data were taken from drifts on the 605- to 830-meter levels, which had been sampled at 2-meter intervals.

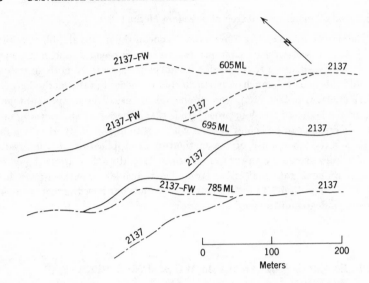

Fig. 5.5. Composite plan of drifts following 2137 and 2137 Footwall veins near their junction, Fresnillo mine.

In order to make an analysis of variance comparing the two segments of the 2137 vein northwestward and southeastward of the vein junction with one another and comparing the combined segments of the 2137 vein with the one segment of the 2137 Footwall vein, the single-degree-of-freedom method was used. In table 5.30 the calculations of the single-degree-of-freedom quantity to compare the two 2137 vein segments with one another are given for the silver-content data. The items in lines 1 to 3 of this table correspond to those in lines 1 to 4 of table 5.23 which is for the fictitious data. There are changes in table 5.30 that make the calculations a little more complex than those in table 5.23. Because the sample sizes are different, the quantity n is no longer a constant multiplier; rather the quantity $\sum (c_i^2/n_i)$ must be evaluated by summing the values in the last column of lines 1 to 3 to yield the result in line 4. For unequal sample sizes

$$Q^2 = \frac{q^2}{\sum (c_i^2/n_i)}.$$

If the n_i quantities are all equal, this formula reduces to the previous formula,

$$Q^2 = \frac{nq^2}{\sum c_i^2}.$$

Table 5.30. Calculation of the single-degree-of-freedom quantity to compare two segments of the 2137 vein, Fresnillo mine, with one another

Line	Vein segment	n	$\overline{w}$	c	$c\overline{w}$	c^2	c^2/n
1	2137 vein southeast of junction	469	161.24286	1	161.24286	1	0.00213220
2	2137 vein northwest of junction	458	167.55109	−1	−167.55109	1	0.00218341
3	2137 Footwall vein	411	128.91070	0	0	0	0

4	$\sum \left(\dfrac{c_i^2}{n_i}\right) = 0.00431561$
5	$q = \sum c_i \overline{w}_i = 6.30823$
6	$q^2 = 39.79377$
7	$Q^2 = 9221$

In lines 5 and 6 q and q^2 are calculated, and finally Q^2 is calculated in line 7.

In table 5.31 parallel calculations to those in table 5.30 are performed to compare the combined segments of the 2137 vein with the one segment of the 2137 Footwall vein.

Finally, in table 5.32 the two Q^2 values from tables 5.30 and 5.31 are entered in the mean-square column. In table 5.32 the F-value comparing the two segments of the 2137 vein is only 0.2; therefore we conclude that silver mineralization in these two segments is the same. On the other hand, because the F-value of 8.6, comparing the 2137 and 2137 Footwall veins, is inside the critical region, we conclude that silver mineralization in the two veins is different.

Repeating this analysis for gold, lead, and zinc data, we (Koch and Link, 1967) found that, except for zinc, mineralization in the 2137 Footwall vein is different from that in the 2137 vein, so that the main vein northwestward from the junction can be distinguished from the branch vein through analysis of the assay data.

In the fictitious illustrations of the single-degree-of-freedom method (sec. 5.9) the contrasts are orthogonal. In the Fresnillo example the contrasts appear to be orthogonal because, in both tables 5.30 and 5.31, $\sum c_i c_i' = 0$

Table 5.31. Calculation of the single-degree-of-freedom quantity to compare the 2137 and 2137 Footwall veins, Fresnillo mine

Line	Vein segment	n	$\bar{w}$	c	$c\bar{w}$	c^2	c^2/n
1	2137 vein southeast of junction	469	161.24286	1	161.24286	1	0.00213220
2	2137 vein northwest of junction	458	167.55109	1	167.55109	1	0.000218341
3	2137 Footwall vein	411	128.91070	-2	-257.82140	4	0.00973236
4	$\sum \left(\dfrac{c_i^2}{n_i}\right) = 0.01404797$						
5	$q = \sum c_i \bar{w}_i = 70.97255$						
6	$q^2 = 5037.10285$						
7	$Q^2 = 358{,}564$						

(table 5.33). However, the two corresponding sums of squares do not add up to the sum of squares with the 2 degrees of freedom because, for unequal sample sizes, the requirement for orthogonality is that

$$\sum \frac{c_i c_i'}{n_i} = 0.$$

Because this requirement is not met, the comparisons for the two single degrees of freedom are not orthogonal; consequently the two tests are not quite statistically independent. Therefore, instead of being exactly 20 percent

Table 5.32. Analysis of variance to assess metal content near the junction of the 2137 and 2137 Footwall veins, Fresnillo mine

Source of variation	Sum of squares	Degrees of freedom	Mean square	F	$F_{10\%}$
Among-vein segments	367,045	2	183,523	4.4	
2137 vein southeast of junction vs. 2137 vein northwest of junction	9,221	1	9,221	0.2	2.71
2137 Footwall vein vs. 2137 vein	358,564	1	358,564	8.6	2.71
Within-vein segments	55,234,376	1335	41,374		

TABLE 5.33. MULTIPLYING CONSTANTS TO COM-
PARE MINERALIZATION IN THE 2137 AND 2137
FOOTWALL VEINS, FRESNILLO MINE

Source	Multiplying constants		
Table 5.30	1	-1	0
Table 5.31	1	1	-2
Product	1	-1	0

(twice 10 percent), the total error rate is between 10 and 20 percent. However, for practical purposes, these two particular single-degree-of-freedom tests may be considered to be independent. A general discussion of to what extent the dependence in such tests can be ignored is beyond the scope of this book; the interested reader is referred to Scheffé (1959, p. 89).

Single Degree of Freedom Applied to Homestake Mine Data

For the Homestake data studied in section 5.6 a more incisive analysis may be made to test whether the mineralization in the two crosscuts (fig. 3.1) is different. To do this the original table 5.15 is expanded by adding several new lines in the new table 5.34. The entries on these two lines partition the among-hole variability into a piece with 1 degree of freedom associated with the between-crosscut variability and a piece with 3 degrees of freedom

TABLE 5.34. NESTED ANALYSIS OF VARIANCE, INCLUDING SINGLE-DEGREE-OF-FREEDOM INTERPRETATION, OF GOLD ASSAY DATA FROM THE HOMESTAKE MINE, SOUTH DAKOTA

Line	Source of variation	Sum of squares	Degrees of freedom	Mean square	F	$F_{10\%}$
1	Total	5308.529	215			
2	Among holes	177.883	4	44.471	1.83	1.97
3	Between crosscuts	75.353	1	75.353	3.10	2.73
4	Within crosscuts	125.433	3	41.811	1.72	2.11
5	Within holes	5130.646	211	24.316		
6	Among-foot intervals	2536.107	51	49.728	3.07	1.32
7	Within-foot intervals	2594.539	160	16.216		
8	Between-sample pairs	1511.398	52	29.065	2.90	1.37
9	Within-sample pairs	1083.141	108	10.029		

associated with the variability among holes within a single crosscut. The table shows that the variability between crosscuts is larger than that within crosscuts, because the calculated F-value of 3.10 is larger than the tabled F-value of 2.73.

The total of the two sums of squares in lines 3 and 4 is larger than the sum of squares on line 2 because, as for the Fresnillo example, the contrasts are not orthogonal. They are not orthogonal because the numbers of observations in the various holes are different.

An alternative way to compare gold in the two crosscuts is to construct confidence limits for the difference between the population means of the assays from the two crosscuts. This difference may be estimated through the contrast

$$q = \tfrac{1}{3}(\overline{w}_A + \overline{w}_B + \overline{w}_C) - \tfrac{1}{2}(\overline{w}_D + \overline{w}_E),$$

where $\overline{w}_A$, $\overline{w}_B$, and $\overline{w}_C$ are mean gold assays from the three drill holes in the first crosscut, and $\overline{w}_D$ and $\overline{w}_E$ are mean gold assays from the two drill holes in the second crosscut. The limits of a 90-percent confidence interval for this difference are -0.46 and 2.98 ounces per ton. Although this estimate includes 0, the interval is quite wide; therefore a substantial difference in gold in the two drifts may exist.

5.11 LINEAR MODELS

In the discussion on soil failure (sec. 1.5) we introduce the concept of models in the example of the simple shear apparatus. In the explanation of the statistical model for shear

$$s = K_1 + (\tan K_2)p + e,$$

several concepts are introduced and discussed. In particular, an observation s depends on the values of variables and parameters explicitly defined in the model and also on implicit variables. Dependence on normal pressure is represented by the variable p; dependence on soil type is represented by the parameter values K_1 and K_2. On the other hand, dependence on temperature, soil structure, and other factors is not explicitly represented in the equation but lumped together in the random fluctuation e.

In Chapters 2 and 4 models are used, although, for simple exposition, they are not discussed explicitly. In section 5.1 models are introduced of the form

$$w_{ij} = \mu + (\mu_i - \mu) + e_{ij},$$

and in sections 5.5 and 5.7 similar but more complicated models are developed.

The model for soil failure in section 1.5 and those introduced previously in this chapter are special cases of the class of models named *linear models*. These models are highly important for several reasons. First, they are fundamental for an understanding of the analyses of variance in this chapter. Also they are fundamental to an understanding of more complicated analyses of variance: trend analysis, in Chapter 9, and multivariate analysis, in Chapter 10. In this section, after general ideas about models are discussed in some detail, the term *linear model* is defined. Then linear models are related to nonlinear models, and some other concepts are introduced.

As discussed in section 1.5, the purpose of a model, whether physical, geological, or mathematical, is to obtain a simple representation of a natural phenomenon useful for gaining insight into it. Thus, through the simple shear apparatus and the accompanying mathematical equation, shear stress is related to three variables: the cohesiveness of the soil, the pressure of the soil, and the coefficient of internal friction. By measuring maximum shear stress and pressure, the coefficients of internal friction can be calculated. Alternatively, shear stress can be calculated if pressure and the two coefficients are known.

Illustration of Mathematical Models

In this subsection functions are defined and discussed; then various kinds of mathematical models, mostly linear, are illustrated.

In the mathematical model for soil failure, the variables are related to one another by a calculating rule or a translation device named a *function*; for instance, the equation

$$s = K_1 + (\tan \theta)p$$

is a linear model in the parameters K_1 and $\tan \theta$ that may be written as

$$s = f(p, K_1, \theta),$$

where f represents the function that encompasses the rules of addition and multiplication and the calculating device for obtaining the tangent of the angle θ. In the general equation,

$$w = f(x_1, x_2, \ldots, x_k, K_1, K_2, \ldots, K_j),$$

w is the dependent variable, the x's are independent variables, the K's are parameters (constants), and f denotes the function comprising the procedures for obtaining the value of w, given the values of the x's and K's.

Functions can be classified by their mathematical properties. If each variable appears to the first degree in a single term, the function is called a *linear function*. The single variable to be calculated, w, is named the *dependent variable*, and the one or more variables, x's, used to calculate it, are named *independent variables.* The graph of a linear function with only one independent variable is a straight line; for example, the graph for the shear test, figure 1.9. Similarly, the graph of a linear function with two independent variables is a plane (sec. 9.4), and that of a linear function with more than two independent variables is a hyperplane (sec. 9.6, fig. 9.15).

In order to illustrate further the meaning of functions, several examples are instructive. For the soil-failure example the relation between shear stress and pressure for a particular soil can be written

$$w = f(x, K_1, K_2),$$

where the dependent variable w is shear stress; the independent variable K_1 is the pressure; and the independent variable K_2 is the coefficient of internal friction. The above equation can be written explicitly as

$$w = K_1 + K_2 x,$$

or, alternatively, because K_2 equals tan θ, as

$$w = K_1 + (\tan \theta)x.$$

Notably, the designation of variables as independent or dependent depends only on how the function is written, for if the above equation is "solved" for x the result obtained is

$$x = \frac{-K_1}{K_2} + \frac{w}{K_2},$$

where x has become the dependent variable and w has become the independent variable.

A second example is the calculation of the specific gravity of a sandy limestone composed entirely of quartz and calcite. Because the specific gravities and proportional weights of the two minerals are known, the specific gravity can be estimated through the functional relationship

$$w = \theta_1 x_1 + \theta_2 x_2,$$

where w is the specific gravity of the rock, θ_1 is the specific gravity of the quartz, x_1 is the proportion of quartz, θ_2 is the specific gravity of calcite, and x_2 is the proportion of calcite. The equation appears to be a linear function with two independent variables and two parameters. It is actually only a function with one independent variable and two parameters, because x_1 and x_2 sum to a constant 100 percent and therefore are not independent variables. The example illustrates the principle that, in determining the number of independent variables, it is necessary to be certain that they are independent and not directly related (chap. 11).

A third example of a functional relationship is afforded by the problem of sampling two beds of phosphate rock, each with a different but uniform phosphate content throughout. Following the approach used in section 5.1, one may write the model as

$$w_i = \mu + (\mu_i - \mu),$$

where w_i is an observation in the ith bed, μ_i is the phosphate content of the ith bed, and μ is the average of the μ_i's. Alternatively, the model may be written in terms of a functional relationship involving other variables explicitly as

$$w = \mu + (\mu_1 - \mu)x_1 + (\mu_2 - \mu)x_2,$$

where w is the observed phosphate content, μ and μ_i have the same meaning as before, and x_1 and x_2 are artificial variables set equal to $(1, 0)$ if the first bed is sampled and to $(0, 1)$ if the second bed is sampled.

The second model appears awkward and the notation needlessly complicated. However, for some purposes, for instance, regression analysis (Chap. 9), the second model may be more suitable. It is introduced here simply to show that more than one model always may be devised for a particular physical situation. The second model can also be written as

$$w = \mu + \beta(x_1 - x_2),$$

where μ is the average phosphate content of the two beds, β is half the difference between the two beds, and x_1 and x_2 have the same meaning as before. The virtue of this approach is that the difference in the average between the two beds is given by a single number, for, if $\beta = 0$, there is no difference between the two beds.

A final example is afforded by the rate of a chemical reaction which may be written

$$w = \alpha e^{-\beta/x},$$

where w is the reaction rate, α and β are parameters, and x is the temperature in degrees Kelvin. Although this equation is a nonlinear model, it may be rewritten

$$w' = \theta - \beta x',$$

where $w' = \ln w$ and $x' = 1/x$.

In terms of the new variables w' and x' and the new parameters θ and β the relationship is linear. The example illustrates that linear models can be used to express many physical relationships provided that the appropriate parameters and variables are chosen, even though the first parameters and variables that come to mind suggest a nonlinear model. However, it is important to choose variables to which a clear physical meaning can be ascribed. This subject is further developed in section 6.3 on transformations.

All the examples of functional relationships are for deterministic models. But each can be converted to a statistical model by adding a random fluctuation e to the right-hand side of the equation, just as the deterministic model for the soil-failure model in section 1.5,

$$s = K_1 + (\tan \theta)p,$$

is converted to the statistical model,

$$s = K_1 + (\tan \theta)p + e.$$

Definition of a Linear Model

A mathematical model that allows an observation w to be expressed as a function of one or more variables and parameters has a functional form that may or may not be linear. However, if the function is linear, a linear model is defined by the general equation

$$w = \alpha + \sum \beta_i x_i + e,$$

where α is a general constant, the x_i terms are independent variables with associated parameters β_i, and e is the random fluctuation. To comprehend this definition, one must recognize that, although the independent variables x_i may themselves be nonlinear functions of other variables, the model is nonetheless a linear model in terms of the parameters α and β_i; for example, the equation

$$w = 3 + 6 \cos 3z_1 + \sin 7z_2 + e$$

is a linear model, because it can be rewritten

$$w = 3 + 6x_1 + x_2 + e,$$

where $x_1 = \cos 3z_1$, and $x_2 = \sin 7z_2$. The ordinary polynomial equation is a linear model; for example, the polynomial equation

$$w = 4 + 2z + 5z^2 + 7z^3 + e$$

can be rewritten

$$w = 4 + 2x_1 + 5x_2 + 7x_3 + e,$$

where $x_1 = z$, $x_2 = z^2$, and $x_3 = z^3$.

Usually numerical values are unknown for the parameters constituting the constant term α and for the coefficients β_i of the independent variables x_i. Then, in order to estimate these parameters, corresponding statistics must be calculated and their frequency distributions determined. The procedure is as follows: Instead of describing an observation w as being drawn at random from a population with mean μ and variance σ^2, it is set equal to the population mean μ plus a random fluctuation e; that is,

$$w = \mu + e.$$

By definition, the random fluctuation e has the following properties:

1. Its mean is equal to 0; that is, $\mu_e = 0$.
2. Its variance is equal to the population variance; that is $\sigma_e^2 = \sigma^2$.
3. It is statistically independent, from observation to observation.
4. It follows a normal frequency distribution.

In order to make this definition of an observation w useful, it must be related to the preceding material by remembering that μ may be written as a linear function of as many independent variables and parameters as is appropriate; that is,

$$\mu = \alpha + \sum \beta_i x_i.$$

For instance, the discussion of the example for the two phosphate beds in the preceding subsection is an illustration of this procedure for two statistical samples.

Interpretation of Analysis of Variance with Linear Models

The fictitious illustration of the geologist sampling two formations for selenium by two different methods (sec. 5.9) serves to demonstrate the relationship between the analysis of variance and linear models. The mathematical structures imposed on the various population means in sections 5.1, 5.5, and 5.7 have all been examples of linear models, although without the detailed exposition of this section. The two sets of notations can be reconciled through the device of artificial variables, which are introduced only to accomplish this purpose and to show that the models discussed in the analysis of variance are indeed linear.

In section 5.9 two formulations of the fictitious illustration of selenium sampling are made, first in a single-degree-of-freedom analysis, and second in the simpler but less incisive one-way analysis of variance. Now, first the one-way analysis of variance formulation is related to artificial variables. In this formulation it is assumed that the geologist was sampling four populations, without connecting them through a further mathematical structure. For a one-way analysis of variance (sec. 5.1) the model for an observation is written

$$w_{ij} = \mu + (\mu_i - \mu) + e_{ij},$$

where i designated the population being sampled and j designated the jth observation from the ith population. This model may be rewritten as follows, where the principal difference in notation is the introduction of the artificial variables x_i, which play essentially the same role as the subscripts i and j in the preceding formulation:

$$w = \mu + \beta_1 x_1 + \beta_2 x_2 + \beta_3 x_3 + \beta_4 x_4 + e,$$

where

1. The general mean μ is the average of μ_1 to μ_4.
2. $\beta_1 = \mu_1 - \mu$, $\beta_2 = \mu_2 - \mu$, $\beta_3 = \mu_3 - \mu$, $\beta_4 = \mu_4 - \mu$.
3. The variables x_i are *artificial variables*, with no intrinsic meaning. They are introduced to designate the population from which an observation comes so that if an observation comes from the first population, $x_1 = 1$, and x_2 to $x_4 = 0$; if an observation comes from the second population, $x_2 = 1$, and x_1, x_3, and $x_4 = 0$; etc.

Several properties of this linear model may be pointed out. If all of the population means μ_i are equal, all of the coefficients β_i are equal to 0. In the analysis-of-variance format of table 5.8 the average within-sample mean square is equal to σ_e^2. Therefore, if the variance of the coefficients β_i is designated σ_μ^2, the average among-sample mean square of that analysis is simply $\sigma^2 + n\sigma_\mu^2$, where n is the number of observations in each sample. The coefficients β_i measure the variation among the means. Because the sum of these coefficients is 0, only three independent parameters are associated with these coefficients, and there are three degrees of freedom for these parameters. The fourth parameter is μ, the general mean of the observations.

For these same data, one can describe another statistical linear model that corresponds to the single-degree-of-freedom interpretation in section 5.9. Three new coefficients β_i and six new artificial variables x_i can be defined as follows:

1. β_1 is the difference associated with the formations.
2. β_2 is the difference associated with the methods.
3. β_3 is the difference associated with the interactions.

Then the linear model

$$w = \mu + \beta_1(x_1 - x_2) + \beta_2(x_3 - x_4) + \beta_3(x_5 - x_6) + e$$

can be written, where the artificial variables x_i have the values in table 5.35. In the previous model there were only four independent parameters. In this new model there are also four independent parameters that appear explicitly; the three β coefficients that are associated with variations among the means account for the three degrees of freedom.

Two explicit linear models using artificial variables have been devised for the two different interpretations of the fictitious selenium data. Many other

TABLE 5.35. VALUES OF THE ARTIFICIAL VARI-
ABLES x_i FOR A LINEAR MODEL CORRESPONDING
TO THE SINGLE-DEGREE-OF-FREEDOM INTERPRE-
TATION OF A FICTITIOUS ILLUSTRATION OF
SAMPLING FOR SELENIUM

	Method	
Formation	A	B
1	$x_1 = 1$	$x_1 = 1$
	$x_2 = 0$	$x_2 = 0$
	$x_3 = 1$	$x_3 = 0$
	$x_4 = 0$	$x_4 = 1$
	$x_5 = 1$	$x_5 = 0$
	$x_6 = 0$	$x_6 = 1$
2	$x_1 = 0$	$x_1 = 0$
	$x_2 = 1$	$x_2 = 1$
	$x_3 = 1$	$x_3 = 0$
	$x_4 = 0$	$x_4 = 1$
	$x_5 = 0$	$x_5 = 1$
	$x_6 = 1$	$x_6 = 0$

models could be associated with this or any other set of data. Often, an
appropriate model is hard to choose; usually the choice is determined by the
investigator's purpose. The same set of data can be therefore interpreted in a
variety of ways simply by using different models. One of the virtues in
designing an experiment before gathering data lies in obtaining insight into
what model might be appropriate. In analysis of data from undesigned
experiments, the choice of an appropriate model and of which variables to
consider can lead to formidable problems that may be difficult if not impossible to resolve.

Variance Components

Through a study of quantities named variance components, the results of
an analysis of variance can be examined in more detail. Variance components
identify the contribution to the observed variability of various terms or
groups of terms in the linear model. The analysis of table 5.26 yields an
analysis-of-variance table whose essential elements are presented in table 5.36.
The last column of the table shows average mean-square values which arise
from the fact that the within-sample variance is the basic fluctuation defined

TABLE 5.36. VARIANCE COMPONENTS FOR ANALYSIS OF VARIANCE OF TABLE 5.26

Source of variation	Degrees of freedom	Mean square	Average mean square
Among-sample means	3	$7s_{\bar{w}}^2$	$\sigma^2 + 7\sigma_\mu^2$
Within sample	24	s_p^2	σ^2

to be σ^2. The among-sample mean square measures $\sigma^2 + 7\sigma_\mu^2$, because the average value of $s_{\bar{w}}^2$ is $\sigma^2/7 + \sigma_\mu^2$, in general, where

$$\sigma_\mu^2 = \frac{\sum \beta^2}{3} = \frac{\sum (\mu_i - \mu)^2}{3}.$$

The β values are those in the first linear model, without the single-degree-of-freedom interpretation associated with the analysis.

Variance components are variances such as σ^2 and σ_μ^2 that identify the sizes of the sources of variability. In the example cited σ^2 represents the basic variability, whereas σ_μ^2 represents the variation introduced by the variance in the β_i values of the linear model, that is, in the variance of the four population means. For any analysis of variance, the variance structure of each line in the analysis-of-variance table can be identified. Although the components are not too difficult to recognize, the calculation of their coefficients is often complicated, too complicated to discuss in this book. The interested reader is referred to Dixon and Massey (1969, p. 153). However, once the variance structure is recognized, the sizes of the different pieces can be estimated and sometimes afford a guide to action (sec. 8.1).

Next, the essential elements of the more complicated analysis of variance (table 5.27) for the fictitious illustration of consistency of selenium sampling can be considered. This analysis is presented in table 5.37. As before, the values for σ_m^2, σ_f^2, and σ_i^2 come from the β_i values in the model. When this

TABLE 5.37. VARIANCE COMPONENTS FOR THE ANALYSIS OF VARIANCE OF TABLE 5.27

Source of variation	Degrees of freedom	Average mean square
Among-sample means	3	$\sigma^2 + 7\sigma_\mu^2$
Between methods	1	$\sigma^2 + 7\sigma_m^2$
Between formations	1	$\sigma^2 + 7\sigma_f^2$
Method-formation interaction	1	$\sigma^2 + 7\sigma_i^2$
Within sample	24	σ^2

more complicated analysis is compared with the previous simple analysis, it is evident that

$$\sigma_\mu^2 = \frac{(\sigma_m^2 + \sigma_f^2 + \sigma_i^2)}{3}.$$

The comparison makes clear the power of the single-degree-of-freedom method used in the more complicated analysis; that is, if only one of the variances on the right side of the equation is relatively large, its chance of showing up is greater in the second analysis, whereas in the first analysis its effect, being diluted by the other smaller variances, is less.

Variance components may be identified in the linear model used for the analysis of Homestake assay data (sec. 5.10). In this analysis, the following sources of variability are identified:

1. Between holes.
2. Among-foot intervals within holes.
3. Between pairs within 1-foot intervals.
4. Within pairs.

With these sources of variability recognized, each observation w may be expressed in the linear model

$$w = \mu + (\mu_i - \mu) + (\mu_j - \mu_i) + (\mu_k - \mu_j) + e,$$

where
μ = general mean,
$(\mu_i - \mu)$ = effects associated with particular holes,
$(\mu_j - \mu_i)$ = effects associated with particular 1-foot intervals,
$(\mu_k - \mu_j)$ = effects associated with particular pairs,
e = random fluctuation.

The purpose of the linear model for the Homestake data is to study the amount of variation among the various distances. The analysis of variance (sec. 5.10, table 5.15) can be used for this study because, for the model chosen, it can be shown that the average mean squares of the analysis-of-variance table (table 5.15) are those given in table 5.38. The numerical values

TABLE 5.38. AVERAGE MEAN-SQUARE VALUES OF VARIANCE COMPONENTS FOR ANALYSIS OF VARIANCE OF HOMESTAKE MINE DATA IN TABLE 5.15

Line	Source of variation	Degrees of freedom	Mean square	Average mean square
1	Among hole	4	44.471	$\sigma^2 + 2\sigma_p^2 + 4\sigma_f^2 + 43\sigma_h^2$
2	Among-foot intervals	51	49.728	$\sigma^2 + 2\sigma_p^2 + 4\sigma_f^2$
3	Between-sample pairs	52	29.065	$\sigma^2 + 2\sigma_p^2$
4	Within-sample pairs	108	10.029	σ^2

Table 5.39. Numerical values of variance components for analysis of variance of Homestake mine data in table 5.15

Type of variance component	Estimate of variance component
Among hole, σ_h^2	0.0 †
Among foot, σ_f^2	5.1
Between-sample pairs, σ_p^2	9.5
Within-sample pairs, σ^2	10.0

† Although the estimated variance component is negative, it is set equal to 0.0 because the parameter σ_h^2 is nonnegative.

calculated for these three quantities are given in table 5.39. Of course, the F-test in the analysis of variance is used to decide whether the positive estimates represent real effects or are merely consequences of statistical fluctuations. Finally, it must be emphasized that the Homestake linear model is only one of many that might have been made for these data but is the most sensible one that we can think of.

Estimation of Parameters

There are several ways to estimate the parameters in a model. In this sub-section we consider two of them, but not in any detail. The reader interested in a full discussion is referred to Wilks (1962).

One way to estimate the parameters is the least-squares method, wherein the statistics are chosen so that the quantity

$$\sum (w - \mu)^2$$

is minimized. The calculation is straightforward, if the model is linear, because the calculations required for a solution involve only simultaneous linear equations. If the model has k parameters, the sum of squares of the observations can be broken into two parts: (a) a sum of squares with k degrees of freedom, which is the variability "explained" by the k parameters, and (b) a sum of squares with $(n - k)$ degrees of freedom, which is the unexplained variability. If the second quantity is divided by $(n - k)$, an estimate of the variance σ^2 of the fluctuations is obtained. These remarks may be summarized in the equation

$$\sum w^2 = \text{SS due to parameters} + \text{residual SS}.$$

This subject is expanded and an illustration is given in section 10.3.

The second way to estimate the parameters in a linear model is the maximum-likelihood method in which the random fluctuations are assumed to follow a particular frequency distribution. Once the assumption is made, the maximum-likelihood method can be used to obtain estimates of the parameters. The theoretical frequency distributions of the n observations and the k parameters are combined to form a joint theoretical frequency distribution, whose equation is named the *likelihood function L*, where

$$L = f(w_1, \ldots, w_n, \beta_1, \ldots, \beta_k).$$

Values of the parameters β_1 to β_k are chosen to maximize this function by setting the k partial derivatives with respect to β_i equal to 0 and solving the k simultaneous equations that result. Details of the procedure are given by Hoel (1962, p. 58).

The two ways of estimating the parameters in a linear model may be compared. The least-squares method requires only the assumption that the fluctuations are independent and have equal variance. If the assumption is made that the observations follow a normal distribution with equal variance, three consequences follow:

1. Statistics calculated by the maximum-likelihood and least-squares methods are identical.

2. The sums of squares associated with the parameters are independently distributed from the sums of squares associated with the fluctuations.

3. The statistics are normally distributed.

Choice of Model

For several reasons a simple model is usually better than a complicated one. One reason is that with complicated models one is seldom certain about how to interpret the results of an analysis, because it is difficult to give a physical meaning to all of the terms of the model.

Another reason is that with a simple model there are fewer parameters to estimate. Since at least one observation is needed to estimate each parameter, with a simple model fewer observations are required, which may be an important consideration if the number of observations is limited because few were collected (as in the analysis of old data) or because few can be collected (as in the analysis of new but expensive data).

Because 1 degree of freedom is required to estimate each parameter in a linear model, the number of degrees of freedom remaining to estimate the *residual variance*, that is, the variance of the fluctuations, is $(n - k)$, where n is the number of observations and k the number of parameters. If a suitable model can be devised, it is desirable to have about 30 degrees of freedom remaining to estimate residual variability. If, for instance, there are fewer than 10 parameters to estimate, having 30 degrees of freedom yields a result only 10 percent less stable than an infinite number of degrees of freedom. With fewer than 5 degrees of freedom, an estimate of residual variability is extremely unreliable. This concept is further explained in section 9.4.

Nonlinear Models

In this section only linear models have been discussed. In principle, the mathematical model can be arbitrary; that is, the measured variable w can depend on independent variables and on parameters through a nonlinear functional relationship. But most mathematical models actually used are linear, even though many physical situations are nonlinear, so that the linear model used to describe them is only an obvious approximation. Linear models are used because the mathematical and statistical techniques available at present do not cope with nonlinear models very well.

Fortunately, any functional relationship can be approximated, at least for some limited range of the variables, by a linear equation; therefore linear models can be used in practice. If nonlinear relations are an essential part of any situation, they must be handled on a case-by-case basis. One must keep in mind that approximations must be used and that predictions, especially extrapolations, beyond the range for which data are available are liable to be very risky.

5.12   MULTIPLE COMPARISONS

In the analysis of variance a general hypothesis test is made, for example, a test of the hypothesis H_0 that the population means of several samples are equal. If hypothesis H_0 is rejected in favor of the alternative hypothesis H_1 that the population means are unequal, it is often interesting to inspect the individual sample means more closely.

The best way to make this inspection is to test, by using the method of the single degree of freedom, one or more predetermined specific hypotheses, perhaps that two sample means from samples having something geologically in common are from the same population. But, as stated in section 5.9, if more than a few comparisons are made with the single-degree-of-freedom method, the error rate, nearly as great as the combined α error, becomes too large. However, it is desirable to inspect the data for unexpected results.

The method of *multiple comparisons*, whose outstanding attribute is that the error rate is set at a constant level regardless of the number of hypotheses tested or confidence limits formed, allows one to carry out such hindsight analyses with the error rate controlled. In this section an illustrative example

TABLE 5.40. FICTITIOUS POTASH OBSERVATIONS IN 10 TRAVERSES

Traverse number	Observations					Sample mean, $\overline{w}$
1	6.4	6.8	6.8	6.0	7.6	6.72
2	2.4	2.5	2.3	3.3	2.5	2.60
3	3.0	1.4	2.4	3.7	1.6	2.42
4	4.3	4.3	4.1	5.7	4.6	4.60
5	6.0	6.6	7.2	6.7	6.2	6.54
6	2.8	3.1	3.7	4.0	2.8	3.28
7	2.2	3.3	3.4	0.4	3.5	2.56
8	5.0	5.7	4.2	3.3	5.1	4.66
9	7.4	7.1	6.4	7.8	5.2	6.78
10	2.7	4.6	6.4	5.6	3.0	4.46

is first given; then the name multiple comparisons is defined, and the calculations are explained. Finally the method is related to the method of the single degree of freedom.

As a fictitious example, suppose that a geologist wishes to investigate potash content in a gneiss exposed in a linear belt of outcrops. He naturally makes traverses at right angles to the foliation. If he makes an arbitrary number of 10 traverses and takes an also arbitrary number of five geological samples from each traverse, he may obtain the 50 fictitious potash observations listed in table 5.40, together with the sample means for the 10 traverses.

With the data in table 5.40 at hand, the first obvious step is to make a one-way analysis of variance to test the hypothesis that the 10 population means associated with the 10 traverses are the same. In table 5.41 this analysis yields an F-value of 17.8, which because it is inside the critical region, leads to acceptance of the alternative hypothesis that the 10 population means are unequal. Inspection of the 10 sample means certainly suggests

TABLE 5.41. ANALYSIS OF VARIANCE FOR FICTITIOUS POTASH DATA FROM 10 TRAVERSES

Source of variation	Analysis of variance				
	Sum of squares	Degrees of freedom	Mean square	F	$F_{10\%}$
Among traverses	137.50	9	15.2775	17.8	1.7929
Within traverses	34.30	40	0.8575		

that some population means are likely to be a good deal larger than the others. In some circumstances, with a predetermined hypothesis, a single-degree-of-freedom test might be reasonable; for example, based on additional geological information, such as changes in rock structure or texture, one could test a hypothesis that the population mean of the first traverse is not equal to the mean of the other traverses, but, if a general overall comparison of all 10 traverses is wanted, the appropriate method is that of multiple comparisons.

The purpose of the following analysis is to segregate the various traverses into groups, so that the traverses within a group will be more or less alike, whereas those in different groups will be somewhat different. With this aim in mind, the means can be ordered according to their sizes and differences can be calculated. Then the multiple-comparison procedure is used to find the size of the largest difference that can be expected between two sample means with a common population mean. If the observed difference is larger than this specified difference, the two sample means can be considered to come from different populations.

The procedure is this. For any contrast (sec. 5.8) a quantity called an *allowance* is calculated, a number that may be interpreted in two ways. The first interpretation is that an allowance is the boundary of the critical region for testing the hypothesis that the population-mean value of the contrast is equal to 0. The second interpretation is that an allowance is the quantity to be added to, and subtracted from, the numerical value of the contrast to form a confidence interval for the population mean of the contrast.

Several uses of allowances are illustrated:

1. To form groups of traverses.
2. To compare any two means to see if they are from the same population.
3. To compare any two *groups* of means to see if they are from the same population.
4. To identify, within a group, means that are extremely large or small.

Two ways to compute allowances have been formulated, and because each is better for some situations, both ways are described.

One way to compute an allowance is described as follows by Tukey (1951): For any desired contrast, the allowance is obtained by multiplying three quantities together. The first is the estimated standard error of the mean, $\sqrt{s_p^2/n}$. The second is the sum of the positive c_i quantities. The third is the critical value of the studentized range from table A-6. For the fictitious illustrative data the calculations are given in table 5.42, in which the contrast is the simple difference between any two means,

$$q = \overline{w}_i - \overline{w}_j.$$

TABLE 5.42. TUKEY MULTIPLE-COMPARISON PROCEDURE APPLIED
TO FICTITIOUS POTASH DATA

Line	Notation	Value	Calculation
1	k	10	From table 5.40
2	n	5	From table 5.40
3	s_p^2	0.8575	From table 5.41
4	s_p^2/n	0.1715	
5	$\sqrt{s_p^2/n}$	0.4140	
6	$\sum (\text{positive } c_i)$	1	
7	$q_{10\%}[k, k(n-1)]$	4.32	From table A-6
8	A_T	1.79	

In lines 2 to 5 the estimated standard error is calculated, recalling that s_p^2 is
the estimate of s^2. For simple differences, the sum of the positive c_i values
must be 1, as recorded in line 6. The value entered in line 7 is from table A-6,
for $k = 10$, and 40 degrees of freedom $[k(n-1) = 10 \times 4]$. The numbers in
lines 5 to 7 are multiplied together to obtain the allowance A_T in line 8.

The results, listed in table 5.43 and plotted in figure 5.6, may be interpreted
as follows. Traverses 9, 1, and 5 form a clearly defined group designated
group 1, whose traverse means are larger than any others. A second group,
group 2, is composed of traverses 8, 4, 10, and 6, which have means of intermediate size. Group 2 lies between group 1 and group 3 and consists of
traverses 6, 2, 7, and 3. Because traverse 6 belongs to both group 2 and
group 3, it may represent an intermediate group between these two groups or
may belong to one of them. More data are needed to resolve this question.

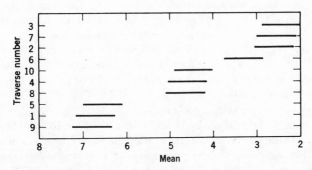

Fig. 5.6. Allowances and traverse groups defined by the Tukey multiple-comparison
method.

TABLE 5.43. GROUPS OF TRAVERSES DEFINED BY THE TUKEY MULTIPLE-COMPARISON PROCEDURE

Traverse number	Mean	Difference between adjacent means	Grouping (numbers are traverse numbers)	Group number
9	6.78		915	
		0.06		
1	6.72		915	1
		0.18		
5	6.54		915	
		1.88		
8	4.66		84106	
		0.06		
4	4.60		84106	2
		0.14		
10	4.46		84106	
		1.18		
6	3.28		84106273	
		0.68		
2	2.60		6273	3
		0.04		
7	2.56		6273	
		0.14		
3	2.42		6273	

Another way to compute an allowance is described by Scheffé (1953). For any desired contrast four quantities are multiplied together. The first is the estimated standard error of the mean, $\sqrt{s_p^2/n}$, as for the Tukey method. The second is the sum of the c_i *squared* quantities, $\sum c_i^2$. The third is the critical value of $F_\alpha[k - 1, k(n - 1)]$. The fourth is $(k - 1)$. For the fictitious illustrative data, the calculations are given in table 5.44 for the same contrast as that in table 5.42. In lines 2 to 5 the standard error is estimated, as before. Line 6 is $\sum c_i^2$, which must be 2 for any simple difference. The F-value in line 7 is obtained from table A-5. Line 8 gives the value of $(k - 1)$. Line 9 gives the product of the numbers in lines 4, 6, 7, and 8. Finally, line 10 gives the square root of the values in line 9, which is A_S.

The results of the analysis by the Scheffé method are listed in table 5.45. Not one of the traverses or their combinations forms a well-defined group. Although there is some evidence of a group of traverses with high means and another with low means, three traverses (8, 4, and 10) belong to both groups, and if one tries to define a group starting with these

TABLE 5.44. SCHEFFÉ MULTIPLE-COMPARISON PROCEDURE APPLIED TO
FICTITIOUS POTASH DATA

Line	Notation	Value	Calculation
1	k	10	
2	n	5	
3	s_p^2	0.8575	
4	s_p^2/n	0.1715	
5	$\sqrt{s_p^2/n}$	0.414	
6	$\sum c_i^2$	2	
7	$F_{10\%}[k-1, k(n-1)]$	1.79	From table A-5
8	$k-1$	9	
9	A_S^2	5.526	Product of lines 4, 6, 7, 8
10	A_S	2.35	

three traverses all the other traverses are included in the single group
thus defined.

What has been done by both methods may be formalized as follows. For
all possible pairs of means the hypothesis H_0 that their population means are
identical has been tested, and those means have been grouped together for
which H_0 is accepted. Alternatively, the interpretation may be made that the
confidence interval for the difference among all possible pairs of means has
been formed, and those means have been grouped together for which the
interval contains 0.

Whenever a hindsight evaluation of data is made, and the method of
multiple comparisons is therefore appropriate, one must decide whether his
primary interest is in simple or in complex contrasts. If primarily in simple
contrasts, the Tukey method is adopted and is used for all contrasts, with
the most incisive results obtained for simple contrasts. If, however, one is
primarily interested in complex contrasts, the Scheffé method is chosen and
again used for all contrasts, with the most incisive results obtained for
complex contrasts. Whichever method an investigator adopts must be used
for all contrasts formed.

Only for illustration are both the Tukey and Scheffé methods used on the
fictitious data. For the particular analysis made the Tukey method is better
because simple contrasts are of interest. With the Tukey method distinct
groups are formed. Had the sample data been used in a complex contrast, for
example, in comparing the first five traverses in a group with the second five
traverses in another group, the Scheffé method would have been preferable.
As stated, with real data only one method can be used for all contrasts.

The allowances calculated by either the Tukey or the Scheffé method have
an interpretation similar to that placed on confidence intervals (sec. 4.4).

Table 5.45. Groups of traverses defined by the Scheffé multiple-comparison procedure

Traverse number	Mean	Difference between adjacent means	Grouping (numbers are traverse numbers)		
9	6.78		9158410	91584106273	
		0.06			
1	6.72		9158410	91584106273	
		0.18			
5	6.54		9158410	91584106273	
		1.88			
8	4.66		9158410	91584106273	84106273
		0.06			
4	4.60		9158410	91584106273	84106273
		0.14			
10	4.46		9158410	91584106273	84106273
		1.18			
6	3.28			91584106273	84106273
		0.68			
2	2.60			91584106273	84106273
		0.04			
7	2.56			91584106273	84106273
		0.14			
3	2.42			91584106273	84106273

Recall that confidence intervals have the property of being correct in a specified percentage of cases, for example, 90 percent of the times that they are tried. Similarly, if an allowance is calculated with a 10-percent error rate, all intervals based on the allowance will be correct for 9 sets of data out of 10. Although the number of contrasts and allowances that can be calculated for each set of data is unlimited, these results still apply. Thus the concept of error rate has been extended from that of single intervals to sets of intervals for a given data set. Therefore, unlike the confidence-interval procedure, no error rate can be associated with a particular interval calculated with an allowance.

5.13 RELATIONSHIPS AMONG BASIC STATISTICAL DISTRIBUTIONS

At this point the basic ways to make statistical inferences by forming confidence intervals and by performing hypothesis tests have been introduced. Most of the remainder of the book explains how to apply these methods to

$d.f._1$ $\diagdown$ $d.f._2$	1	2	3	....	∞
1					
2					
3	t^2				
$\vdots$					
∞	u^2			$\chi^2/d.f._1$	1

Fig. 5.7. Arrangement of F-table.

geological problems. Little new fundamental statistical material is introduced, although special principles and techniques are explained.

The most important sampling distributions in statistics have been introduced: the normal distribution, the t-distribution, the chi-square distribution, and the F-distribution. Although there are innumerable other distributions, some discussed in this book, these four play the central role in statistics, and it is important to realize that they are closely interrelated.

The relationship is best explained in terms of the F-distribution. Because the chi-square distribution is that of SS/σ^2 and $SS = (n - 1)s^2$, dividing the chi-square statistic by its degrees of freedom yields a ratio of s^2/σ^2, which is an F-statistic with $(n - 1)$ and infinity degrees of freedom; that is,

$$\frac{SS}{(n - 1)\sigma^2} = \frac{(n - 1)s^2}{(n - 1)\sigma^2} = \frac{s^2}{\sigma^2}.$$

Thus, if the entries in the chi-square table (table A-3) are divided by their corresponding degrees of freedom, the bottom row of the F-table (table A-5) is obtained as shown diagrammatically in figure 5.7.

If a variable t follows a t-distribution with a particular number of degrees of freedom, the variable t^2 follows an F-distribution with 1 as well as the same particular number of degrees of freedom—because, when squared, the numerator of the t-statistic forms an estimate of the variance with one degree of freedom; whereas, when squared, the denominator forms an estimate of the variance with the certain number of degrees of freedom. Finally, the fact that a t-distribution with infinity degrees of freedom is the normal distribution (sec. 4.3) shows that the entry in the F-table with 1 and infinity degrees of freedom is the square of the corresponding entry in the table for the normal distribution. When squared, the upper 2.5-percent point of the standardized normal distribution becomes the upper 5-percent point of the F-distribution, because the negative lower 2.5-percent point becomes positive when squared.

REFERENCES

Cochran, W. G., and Cox, G. M., 1957, Experimental designs: New York, John Wiley & Sons, 611 p.

Davies, O. L., 1958, Statistical methods in research and production: New York, Hafner Publishing Co., 390 p.

Dixon, W. J., and Massey, F. J., Jr., 1969, Introduction to statistical analysis: New York, McGraw-Hill, 638 p.

Hoel, P. G., 1962, Introduction to mathematical statistics: New York, John Wiley & Sons, 427 p.

Koch, G. S., Jr., and Link, R. F., 1963, Distribution of metals in the Don Tomás vein, Frisco mine, Chihuahua, Mexico: Econ. Geology, v. 58, p. 1061–1070.

———, 1966, Some comments on the distribution of gold in a part of the City Deep mine, Central Witwatersrand, South Africa, *in* Symposium on mathematical statistics and computer applications in ore valuation: South African Inst. of Mining and Metall., Proc., p. 173–189.

———, 1967, Geometry of metal distribution in five veins of the Fresnillo mine, Zacatecas, Mexico: U.S. Bur. Mines Rept. Inv. 6919, 64 p.

Li, J. C. R., 1964, Statistical inference, v. 1: Ann Arbor, Mich., Edwards Bros., 658 p.

Scheffé, Henry, 1953, A method for judging all contrasts in the analysis of variance: Biometrica, v. 40, p. 87–104.

Tukey, J. W., 1951, Quick and dirty methods in statistics, Pt. 2, Simple analyses for standard designs: Am. Soc. Quality Control, 5th Ann. Conf. Proc., p. 189–197.

Wilks, S. S., 1962, Mathematical statistics: New York, John Wiley & Sons, 644 p.

Chapter 6

Distributions and Transformations

Many observed frequency distributions are closely approximated by the theoretical normal distribution; for instance, some distributions of human heights, lives of light bulbs, and weighing errors are more or less normal. Partly because of this correspondence but mainly because the normal distribution is mathematically tractable, most of the statistical theory and methods of inference have been devised for it. But many distributions of geological observations are nonnormal; in fact one of the more common distributions is skewed with a long tail to the right, including distributions of gold and other trace elements (secs. 6.2 and 15.3, Chap. 16). Another common distribution of geological data is a mixed distribution made up of two or more distributions (sec. 6.5).

In this chapter several nonnormal distributions are introduced and methods of working with them explained. The reader is already familiar with one method of working with them: relating the nonnormal distribution to the normal distribution by means of normal approximations. Other methods involve learning a small amount of new statistics and some new tables. We also discuss the use of a *transformation*, a function of an observation that defines a new observation. There are problems involved in working with nonnormal distributions, and the validity of the results depends upon the details of the methods chosen.

Some examples of nonnormal distributions of geological data are introduced in sections 6.1 and 6.2. Then, in section 6.3, some transformations that may be used to normalize or nearly normalize data are explained. Once the data are normalized, hypothesis tests can be made and confidence intervals constructed for the transformed observations. Next follows a discussion of the delicate and complicated questions: Are the results garnered from these transformed observations useful to the geologist's purpose and can they be interpreted in a manner compatible with the original variables? Special graph papers to make transformations visually meaningful and to help the geologist visualize a distribution are discussed in section 6.4. The last section (sec. 6.5) treats mixed-frequency distributions.

6.1 SOME DISCRETE DISTRIBUTIONS

So far the frequency distributions considered have been *continuous distributions* of observations derived, in principle, by measuring; thus they can have any values, not necessarily integers, in a given range. An example of a continuous observed distribution is one of metal assays lying in the range from 0 to 100 percent. An example of a continuous theoretical distribution is the normal distribution. Now introduced in this section are *discrete distributions* of observations derived, in principle, by counting; thus they are integers. Examples of discrete distributions are the number of mines in each township in the western United States, the number of assays more than three standard deviations from the mean on different mine levels, and the number of occurrences of an index fossil in different sedimentary environments.

Actually, in practice the formal distinction between discrete and continuous distributions is seldom important because one kind can be and often is transformed into the other. Thus a list of carbonate percentages can be classified into those above and those below a value defining limestone. By assigning the integer 1 to those above and the integer 0 to those below, one can transform a continuous observation, the carbonate percentage, into a discrete observation. Although information is lost in the process, the loss is not important if one's only interest is in defining limestone. On the other hand, a continuous distribution may be used to represent discrete observations, as in the normal approximation to the discrete Poisson distribution.

The Binomial Distribution

Suppose that the outcome of a trial, such as tossing a coin or classifying a rock as igneous or not igneous, is described as a dichotomy, that is, as one or the other of only two events. One event of the dichotomy is arbitrarily named a *success*, and the other is named a *failure*. Thus a simple "yes, no" classification results. The binomial distribution is the discrete distribution of the number of successes that occur in a fixed number of trials, where the outcomes are statistically independent, and the probability of success is the same for each trial.

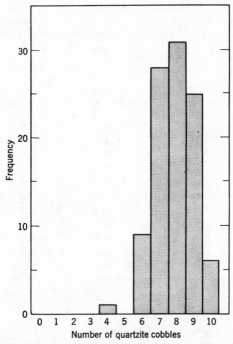

Fig. 6.1. Number of quartzite cobbles in each of 100 samples of 10 cobbles from a gravel deposit on the Gros Ventre river, Wyoming (data from R. Flemal, personal communication, 1967).

The binomial distribution may be introduced by consideration of a small set of data collected by R. Flemal (personal communication). In a study of the geologic environment of gold he counted the number of quartzite cobbles in each of 100 samples of 10 cobbles from a gravel deposit on the Gros Ventre River, Wyoming. The counts of from 0 to 10 quartzite cobbles are given in table 6.1 as a frequency distribution and in figure 6.1 as a histogram. The problem is to estimate the proportion of quartzite cobbles in the gravel. Since it is known intuitively that the number of quartzite cobbles in each sample estimates the proportion of quartzite cobbles in the entire gravel deposit, the only question that remains is how closely.

Under the assumptions that the cobbles can be classified as quartzite and nonquartzite and that the quartzite ones are randomly distributed throughout the gravel, the cobble counts follow the binomial distribution, the formula for which is

$$P(w) = \frac{n!}{w!\,(n-w)!}\,\theta^w(1-\theta)^{n-w},$$

TABLE 6.1. OBSERVED AND THEORETICAL BINOMIAL
FREQUENCY DISTRIBUTION OF QUARTZITE COBBLES IN
EACH OF 100 SAMPLES OF 10 COBBLES FROM A GRAVEL
DEPOSIT ON THE GROS VENTRE RIVER, WYOMING*

| Number of | Frequency | |
quartzite cobbles	Observed	Theoretical
0	0	0.0
1	0	0.0
2	0	0.0
3	0	0.1
4	1	0.8
5	0	3.3
6	9	10.3
7	28	21.7
8	31	30.0
9	25	24.7
10	6	9.1
Item	Observed	Theoretical
Mean	7.87	7.87†
Variance	1.28	1.68

* Data from R. Flemal, personal communication, 1967.
† Assuming true proportion of pebbles is 0.787.

where $P(w)$ is the probability of obtaining w successes in n items, n is the number of items in a sample (10 cobbles in the example), w is the number of successes (number of quartzite cobbles in the example), and θ is the proportion of successes in the population (proportion of cobbles that are quartzite in the area in this example). The mean of this distribution is $n\theta$, and the variance is $n\theta(1 - \theta)$.

Table 6.1 gives the empirical distribution obtained by Flemal and also the mean and variance obtained in the usual way (sec. 3.3). In column 2 the theoretical binomial distribution for the same mean of 7.87 and the theoretical variance calculated by the previous formula are given $[7.87(1 - 0.787) = 1.68]$. Comparison of the two variances shows that the empirical distribution is a little more compressed than the theoretical one because the cobbles were not chosen at random, but the agreement is fairly good. As explained at the end of this section, a chi-square test to compare the two distributions yields a chi-square value of 6.39, which is less than the tabled value of 9.24 for the

10-percent point of the chi-square distribution. Thus the good agreement indicated by the variance is corroborated.

The binomial distribution has only the single parameter θ, from which both the mean and variance are calculated. If both $n\theta$ and $n(1 - \theta)$ are large enough, say 5 or greater, the binomial distribution is satisfactorily approximated by the normal distribution. Because the statistic

$$\frac{\overline{w} - n\theta}{\sqrt{n(\theta)(1 - \theta)}}$$

follows the normal distribution with mean zero and variance 1, it can be used to find confidence limits for the mean and variance (Li, 1964, I, p. 459). For Flemal's data, the estimate of the mean proportion of quartzite cobbles in the gravel is 0.787, and 90-percent confidence limits for the population proportion of quartzite cobbles are 0.766 and 0.808.

In many disciplines, including engineering and the social sciences, there are natural dichotomies; for instance, in quality control of manufacturing certain parts may be grouped into batches of equal size that are tested for good or bad quality. However, geological data tend to be continuous, and there is less occasion to use the binomial distribution. Geological data come mostly from cobble counts in conglomerates, gravels, glacial tills, and the like, and from some paleontological studies.

Consequently, although the binomial distribution is the first subject considered in many statistics books and the mathematics is simple, a complete explanation is not provided in this book because it is only marginally pertinent to geology and because the statistics in this book is not developed from the principles of probability. The reader who needs to work with the binomial distribution can readily follow the explanation in a standard text, for instance, that by Hoel (1962, p. 85, etc.), and will find especially helpful the thorough explanation by Li (1964, I, pp. 443–487), relating the binomial and normal distributions. If there is more than a dichotomy, for instance, if several kinds of cobbles are to be counted rather than only quartzite versus nonquartzite cobbles, the *multinomial distribution*, also treated by Hoel and Li, may be used.

The Poisson Distribution

The number of occurrences of any one item, whether a physical object or an event, may be counted in a group of objects (which may be areas, volumes, time intervals, or something else, usually of equal size or duration). The *Poisson distribution* arises when the occurrences are randomly distributed in the objects being investigated. If, for instance, equal-sized cookies were manufactured from a very large batch of raisin dough, the number of raisins per cookie would follow the Poisson distribution, provided that the raisins were mixed at random in the dough; otherwise the Poisson distribution would not be followed. As a geological example, if oil fields are distributed randomly in the world, the frequency distribution of the number of oil fields per unit area is Poisson.

The equation for the Poisson distribution is

$$f(w) = \frac{\mu^w}{w!} e^{-\mu},$$

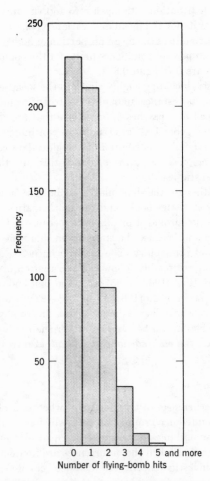

Fig. 6.2. Number of flying-bomb hits in the south of London during World War II, according to areas of 1/4 square kilometer each (after Feller, 1957, p. 150).

Table 6.2. Observed and theoretical Poisson distribution of number of flying-bomb hits in the south of
London during World War II *

Number of flying-bomb hits per $\frac{1}{4}$ sq km	Observed	Theoretical
0	229	226.7
1	211	211.4
2	93	98.5
3	35	30.6
4	7	7.2
5 and more	1	1.6
Summary items:		
Mean	0.932	0.932 †
Variance	0.960	0.932
Total number of cells	576	
Total number of bombs	537	

* After Feller, 1968, p. 161.
† Assuming mean number of bombs per cell is 0.932.

where w is an observation and μ stands for both the mean and variance, which are the same for a Poisson distribution. The observation w may be any nonnegative integer, i.e., 0, 1, 2, 3, ... The sample mean $\overline{w}$ is the estimate of μ. If the sample mean is fairly large, say larger than 5, the Poisson distribution is closely approximated by the normal distribution, and 90-percent confidence limits for μ can be obtained from the relation $\overline{w} \pm 1.645\sqrt{\overline{w}}$. As an example of Poisson-distributed data, Feller (1968) cites the number of flying-bomb hits in the south of London during World War II, counted in areas of $\frac{1}{4}$ square kilometer each. (Table 6.2 and figure 6.2 give the results.) The observed and theoretical distributions are tabulated and a chi-square test (explained at the end of this section) shows that the distribution is Poisson. Two facts should be mentioned: doubtless all or nearly all the flying bombs that fell were noticed, and most people at the time believed that the hits were clustered.

As a geological example of the Poisson distribution, table 6.3 and figure 6.3 list the number of meteorites per square degree in Nebraska, counted from data tabulated by Mason (1962). The table gives both the observed distribution and the theoretical distribution for the same mean of 1.35 meteorites per square degree. We thought that, neglecting the small change in area of a square degree from north to south, the number of meteorites per square degree should be random and therefore Poisson. However, the right-hand columns in the table show that, for the entire United States and for two

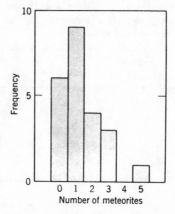

Fig. 6.3. Number of meteorites per square degree in Nebraska (date from Mason, 1962, p. 227).

selected subareas, the distribution is not Poisson because the estimated variance is too much larger than the estimated mean. For these subdivisions and for some others that we investigated, the trouble is that too many cells contain no meteorites.

The point is that these distributions are not of meteorites that fell but rather of meteorites that were found. No doubt many meteorites were not found because they are hard to spot in some areas, such as mountainous or wooded areas, or because no one was looking. The distribution is probably Poisson in Nebraska because the state is reasonably well populated, is flat, and has large plowed areas of fine soil. Not quite as bad as not looking, but also disruptive, is looking too hard; in an area of one square degree in Texas, champion meteorite finders have turned up twelve! The moral of this story is that, to find a Poisson distribution in geology, one probably has to think of a rare event that is observed every time it happens, like a White House wedding. Aside from major earthquakes, we have not been able to think of any.

The Negative Binomial Distribution

The negative binomial distribution is related to the Poisson distribution in its functional form, which is

$$f(x) = \frac{(k + x - 1)!}{x!\,(k - 1)!}\,(1 + \theta)^{-k}\left(\frac{\theta}{1 + \theta}\right)^{x},$$

where x is zero or a positive integer, and k and θ are two parameters. Having two parameters, the negative binomial distribution is somewhat more flexible

Table 6.3. Frequency distributions of meteorites in Nebraska, the entire United States, and two other selected parts of the United States*

Number of meteorites per square degree	Frequency				
	Nebraska			Central U.S.	Central U.S. except N. and S. Dakota
	Observed	Theoretical	U.S.	Central U.S.	
0	6	5.98	818	69	27
1	9	8.05	179	45	34
2	4	5.43	83	30	29
3	3	2.44	23	10	10
4	0	0.82	19	12	12
5	1	0.28	12	8	8
6	0	0	5	2	2
7	0	0	1	1	1
8	0	0	2	1	1
9	0	0	1	1	1
10	0	0	0	0	0
11	0	0	0	0	0
12	0	0	1	1	1
Summary items:					
Mean	1.35	1.35†	0.55	1.51	2.05
Variance	1.60	1.35	1.36	3.63	4.11
Total number of cells	23	23	1144	180	126
Total number of meteorites	31	31	624	271	258

* Data from Mason, 1962, p. 227.
† Assuming mean number of meteorites per square degree is 1.35.

than the Poisson distribution in fitting empirical distributions. Certain mixtures of data, that are Poisson distributed with different means, can produce a negative binomial distribution. Therefore, there is some theoretical justification for using a negative binomial distribution for data that may have arisen from mixtures of Poisson distributions.

The population mean and variance for the negative binomial distributions may be estimated by the formulas

$$\theta = \frac{s^2 - \bar{x}}{\bar{x}},$$

and

$$k = \frac{\overline{x}^2}{s^2 - \overline{x}}.$$

That these estimates are inefficient is unimportant for the geologist because they are used mostly for curve fitting, which is seldom of geological interest.

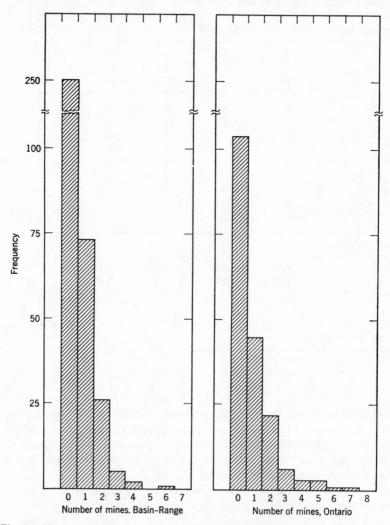

Fig. 6.4. Number of mines in cells each 1000 square kilometers in area in parts of the Basin and Range province, United States and Ontario, Canada (data from Slichter, 1960, p. 574).

Efficient estimates, if required, may be obtained by a more complicated calculation procedure explained by Haldane (1941).

Table 6.4 and figure 6.4 present two sets of data compiled by Slichter (1960). Griffiths (1966) later showed that these data follow a negative binomial distribution. The data are areal distributions of mines in the Basin and Range province in the western United States and Precambrian intrusive and sedimentary areas of Ontario, Canada. In order to obtain the counts in table 6.4, a grid with cells each 1000 square kilometers in area was superimposed on maps of the areas, and the mines in each cell were counted. The table gives the estimated parameters p and k and shows that the agreement between the observed and theoretical distributions is good. One interpretation resulting from the fact that the distributions of mines follow the negative binomial rather than the Poisson distribution is that the mines are not randomly distributed. It then becomes evident that some geological process must have been involved in the distribution. (Section 12.3 describes the situation more fully.)

TABLE 6.4. OBSERVED AND THEORETICAL NEGATIVE
BINOMIAL DISTRIBUTIONS OF MINES IN PARTS OF THE
BASIN AND RANGE PROVINCE, UNITED STATES, AND
ONTARIO, CANADA *

Number of mines per 1000 sq km, n	Frequency, f	
	Observed	Theoretical
A. Basin and Range (154 mines, area 357 units)		
0	250	249.8
1	73	74.7
2	26	22.6
3	5	6.9
4	2	2.1
5	0	.6
6	1	.2
7	0	.1

Estimated parameters for negative binomial distribution:

p	k
0.443	0.975

B. Precambrian intrusive and sedimentary areas of Ontario (147 mines, 185 units)

Number of mines per 1000 sq km, n	Frequency	
	Observed	Theoretical
0	104	104.4
1	45	44.4
2	22	19.8
3	6	8.9
4	3	4.1
5	3	1.9
6	1	.8
7	1	.4
8	0	.2

Estimated parameters for negative binomial distribution:

p	k
0.867	0.917

* From Slichter, 1960, p. 574.

Chi-Square Test

Several theoretical frequency distributions are introduced in this chapter. The question arises of how to compare the agreement of an observed frequency distribution with a theoretical frequency distribution. The *chi-square test*, which is based on the chi-square distribution (sec. 3.7), is the usual way to make this comparison. As always, whenever the hypothesis of agreement is accepted, the risk of type II error, accepting a false hypothesis, must be recognized. Therefore, one can never be certain that a particular set of empirical observations necessarily follows a particular theoretical frequency distribution.

The chi-square test may be explained by applying it to the cobble data introduced at the beginning of this section. First, the observed frequencies O are listed (table 6.1). Then the theoretical frequencies T are obtained from the formula for the binomial distribution,

$$P(w) = \frac{n!}{w! \, (n - w)!} \, \theta^w (1 - \theta)^{n - w},$$

by substituting numerical values and multiplying by the number of samples.

For instance, for $w = 9$ and an assumed value of $\theta = 0.787$, the theoretical frequency of 24.7 is obtained by the formula:

$$T = 100\left(\frac{10!}{9!\,1!}\right)(0.787)^9(0.213) = 24.7.$$

The entries for 0 to 5 cobbles in table 6.1 are combined to obtain a theoretical frequency of 4.2. (Conventionally, each cell entry must have a theoretical frequency of at least 5, but the next entry for size 6 is not included because the resulting value of 14.5 would be too large. For details see Dixon and Massey, 1969, p. 238.)

The next step is to calculate the chi-square statistic, which is

$$\chi^2 = \sum_{i=1}^{k} \frac{(O_i - T_i)^2}{T_i}.$$

The number of degrees of freedom of this statistic is equal to the number of cells minus 1 minus the number of fitted parameters. For the cobble example numerical values are substituted in this formula to obtain the value of chi-square, as follows:

$$\chi^2 = \frac{(1 - 4.2)^2}{4.2} + \frac{(9 - 10.3)^2}{10.3} + \frac{(28 - 21.7)^2}{21.7} + \cdots + \frac{(6 - 9.1)^2}{9.1} = 6.39.$$

The number of degrees of freedom is equal to $6 - 1 - 1 = 4$, because there are six cells (after combining the first five), and one parameter θ is fitted. Because the calculated value of 6.39 is smaller than the tabled value of 9.24 (table A-3) for 4 degrees of freedom the conclusion may be drawn that the cobble data agree with a binomial model.

6.2   THE LOGNORMAL DISTRIBUTION

The lognormal distribution is a continuous distribution characterized by the property that the logarithms of the observations follow a normal distribution. In this section the formula for the distribution is given, estimation of parameters is discussed, calculation of confidence limits by four different methods is explained, and the negatively skewed lognormal distribution is introduced. Chapter 16 is about the interpretation of gold data through use of the lognormal distribution.

The lognormal distribution is discussed in detail in this book for several reasons. First, this distribution is followed more or less closely by many sets of geological data, especially by trace elements, that are characterized by the

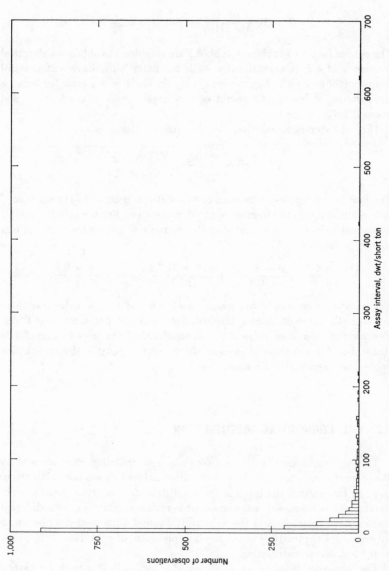

Fig. 6.5. Histogram to show approximately lognormal distribution of 1536 dwt gold assay values from the City Deep mine, South Africa.

fact that, compared with the mean value, most of the observations are small but a few are very large. Such data are encountered in diverse forms, for instance, in the incomes of individuals, magnitudes of earthquakes, gold assay values from a mine, heights of floods in a river, metal analyses from a survey in exploration geochemistry, and heights of buildings in New York City. In every one of these and many more instances most of the values are small relative to the mean, but a few are very large. The large values represent the millionaires, the catastrophic earthquakes (the Richter scale being logarithmic), the bonanza pockets of gold, the disastrous floods, the geochemical anomalies, and the skyscrapers.

Second, the detailed discussion points out that several statistical methods are available for analysis of the same set of data, that pitfalls may be found in using these methods, and that the geologist must choose among methods according to his exact objectives.

Figure 6.5 and table 6.5 present a typical lognormal distribution of gold assays from the City Deep mine, Central Witwatersrand, South Africa. The formula for the lognormal frequency distribution is

$$f(w) = \frac{1}{w\beta\sqrt{2\pi}} \exp\left[-\frac{1}{2\beta^2}(\ln w - \alpha)^2\right],$$

where the two parameters α and β^2 are the mean and variance, respectively, of the natural logarithm of the observation w. The mean of w is given by the formula

$$\mu = e^{\alpha + \frac{1}{2}\beta^2}$$

and the variance of w, by the formula

$$\sigma^2 = \mu^2(e^{\beta^2} - 1).$$

The lognormal distribution has been described in a book by Aitchison and Brown (1957), whose definition of the parameters α, β^2, μ, and σ^2 is the opposite of ours (table 6.6).

Just as there are many normal distributions depending on the values of the parameters μ and σ^2 so are there many lognormal distributions depending on the values of these parameters or on the values of the corresponding parameters α and β^2. The lognormal distribution is always skewed to the right, the amount of skewness depending only on the value of β^2, the variance of the logarithms of the observations. If the value of β^2 is small, so is the skewness, and the frequency distribution is nearly normal. In figure 6.6 are graphs of three lognormal distributions for α equal to 0, with different values of β^2. Although the distribution with β^2 equal to 0.1 is already noticeably skewed, it is more nearly normal than the other distributions with larger values of β^2.

TABLE 6.5. APPROXIMATELY LOGNORMAL FREQUENCY DISTRIBUTION OF 1536 DWT GOLD VALUES FROM THE CITY DEEP MINE, SOUTH AFRICA

Class interval (dwt/short ton)	Frequency	Cumulative frequency	Relative cumulative frequency (%)
0–5	910	910	59.24
5–10	208	1118	72.79
10–15	118	1236	80.47
15–20	80	1316	85.68
20–25	54	1370	89.19
25–30	33	1403	91.34
30–35	24	1427	92.90
35–40	13	1440	93.75
40–45	14	1454	94.66
45–50	8	1462	95.18
50–55	8	1470	95.71
55–60	10	1480	96.36
60–65	4	1484	96.62
65–70	4	1488	96.88
70–75	3	1491	97.07
75–80	1	1492	97.14
80–85	1	1493	97.20
85–90	4	1497	97.46
90–95	1	1498	97.53
95–100	7	1505	97.99
100–105	3	1508	98.18
105–110	2	1510	98.31
110–115	3	1513	98.51
120–125	2	1515	98.63
125–130	1	1516	98.70
130–135	5	1521	99.03
145–150	1	1522	99.09
150–155	1	1523	99.16
155–160	3	1526	99.35
180–185	1	1527	99.42
190–195	2	1529	99.56
205–210	2	1531	99.68
215–220	1	1532	99.72
245–250	1	1533	99.81
305–310	1	1534	99.87
420–425	1	1535	99.93
620–625	1	1536	100.00

Table 6.6. Comparison of notation for lognormal distribution in this book with that of Aitchison and Brown. (Where two symbols are given, the first is the parameter and the second is the statistic.)

	This book	Aitchison and Brown
Observation	w	x
Mean of observation	μ, m	α, a_1
Variance of observation	σ^2, V^2	β^2, b^2
Logarithm of observation	u	y
Mean of logarithms	$\alpha, \overline{u}$	$\mu, \overline{y}$
Variance of logarithms	β^2, s_μ^2	σ^2, v_y^2
Coefficient of variation	γ, C	η
Multiplying factor		
For geometric mean	ψ_n	ψ_n
For variance	Φ_n	χ_n

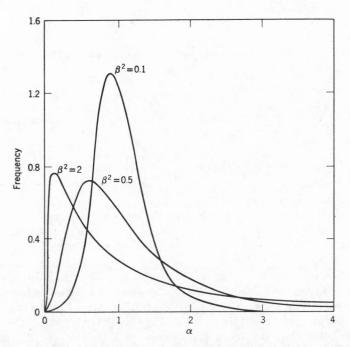

Fig. 6.6. Frequency curves of the lognormal distributions for three values of β^2 (after Aitchison and Brown, 1957, p. 10).

Parameter Estimation

The first problem is to estimate the parameters μ and σ^2. Because, for any distribution, the mean of sample means is equal to the population mean, the sample mean $\bar{w}$ is an unbiased estimate of μ. Similarly, because the mean of sample variances is equal to the population variance, the sample variance s^2 is an unbiased estimate of σ^2. Unfortunately, however, these estimates are not the most efficient ones. The most efficient estimates are obtained by first estimating α and β^2 and then estimating μ and σ^2. The estimate of α is the average value of the natural logarithms of the observations,

$$\hat{\alpha} = \bar{u} = \frac{\sum \ln (w)}{n},$$

which is efficient because the logarithms of the observations are normally distributed, and the sample mean is an efficient estimate of the population mean for observations so distributed. The estimate of β^2 is the variance of the logarithms of the observations,

$$s_u^2 = \frac{\sum (u - \bar{u})^2}{n - 1},$$

where u is the natural logarithm of w. Now, the unbiased efficient estimate m of μ may be obtained from the formula

$$m = e^{\bar{u}} \psi_n(\tfrac{1}{2} s_u^2),$$

and the unbiased efficient estimate V^2 of σ^2 may be obtained from the formula

$$V^2 = e^{2\bar{u}} \left[\psi_n(2s_u^2) - \psi_n\left(\frac{n-2}{n-1} s_u^2 \right) \right] = e^{2\bar{u}} \Phi_n(s_u^2).$$

Values of ψ are given in table A-7.

To illustrate parameter estimation for the lognormal distribution, we make use of 13 gold assay values from boreholes in the Welkom mine, South Africa (Krige, 1961, p. 18). A discussion of the geological significance of the calculations is postponed to Chapter 16. The first illustration is the use of the formulas for m and V^2. In table 6.7 column 1 lists the original 13 values in inch-pennyweights, column 2 lists the natural logarithms of these values, and column 3 lists the common logarithms. The familiar calculations for mean and variance are given below the data list. The note at the bottom of the table verifies that the calculations can be done in either natural or common logarithms because the natural logarithm of a number is 2.302585 times the

Table 6.7. Calculation of means and variances of observations and logarithms of observations, for 13 gold assays from the Welkom mine, South Africa *

	w, in.-dwt		$u = \ln (w)$		$v = \log (w)$
	154		5.03695		2.18752
	525		6.26340		2.72016
	1560		7.35245		3.19312
	1252		7.13247		3.09760
	377		5.93225		2.57634
	70		4.24850		1.84510
	308		5.73010		2.48855
	109		4.69135		2.03743
	1221		7.10743		3.08672
	15		2.70806		1.17609
	48		3.87121		1.68124
	237		5.46806		2.37475
	68		4.21951		1.83251
$\sum w$	5,944	$\sum u$	69.76174	$\sum v$	30.29713
n	13	n	13	n	13
$\overline{w}$	457	$\overline{u}$	5.36629	$\overline{v}$	2.33055
$(\sum w)^2$	35,331,136	$(\sum u)^2$	4866.70036783	$(\sum v)^2$	917.91608624
$(\sum w)^2/n$	2,717,780	$(\sum u)^2/n$	374.361567	$(\sum v)^2/n$	70.608930
$\sum w^2$	6,108,382	$\sum u^2$	398.155209	$\sum v^2$	75.096706
SS_w	3,390,602	SS_u	23.793642	SS_v	4.487776
s_w^2	282,550	s_u^2	1.98280	s_v^2	0.37398
s_w	531	s_u	1.4081	s_v	0.6115

* Data from Krige, 1961, p. 18.

Note: $(2.302585)\overline{v} = \overline{u}$

$(2.302585)^2 s_v^2 = s_u^2$

common logarithm. In table 6.8 $\overline{u}$ and s_u^2 are repeated from table 6.7. Next m is calculated. The value of $e^{\overline{u}}$ is the natural antilogarithm of $\overline{u}$ obtained from a table available in any engineering handbook. The next line gives the value of 13 for n, and the next gives the value for the argument $\frac{1}{2}s_u^2$, used to determine the value of ψ from table A-7, using linear interpolation if necessary. By a similar calculation, the value of V^2 is obtained.

The preceding estimates of the mean and variance of observations that follow a lognormal distribution are quite complicated. Moreover, if the observations do not exactly or closely follow a lognormal distribution, these estimates may lead to biased estimates. Finney (1941) showed that the

TABLE 6.8. CALCULATION BY LOGNORMAL THEORY OF m, A STATISTIC TO ESTIMATE THE MEAN μ, AND V^2, A STATISTIC TO ESTIMATE THE VARIANCE σ^2, FOR THE DATA OF TABLE 6.7

	Calculation of m		Calculation of V^2
$\bar{u}$	5.36629	$\bar{u}$	5.36629
s_u^2	1.98280	s_u^2	1.98280
$e^{\bar{u}}$	214	$e^{2\bar{u}}$	45,825
n	13	n	13
$\frac{1}{2}s_u^2$	0.99	$\Phi_n(s_u^2)$	16.27
$\psi_n(\frac{1}{2}s_u^2)$	2.370	V^2	745,573
m	507	V	863

ordinary sample mean $\bar{w}$ is more than 90-percent efficient as an estimate of μ, provided that the coefficient of variation is less than 1.2, corresponding to a variance of the logarithms β^2 of about 0.9. Figure 6.7, based on Finney's work, gives a plot of the efficiency of the ordinary sample mean $\bar{w}$ as a function of the coefficient of variation. If the coefficient of variation is less than 1.2, we recommend using the ordinary sample mean $\bar{w}$ as the estimate of μ because little efficiency can be lost, considerable arithmetic is saved, and the question of bias does not arise. Because of the importance of the coefficient of variation, several ways of estimating it are now given.

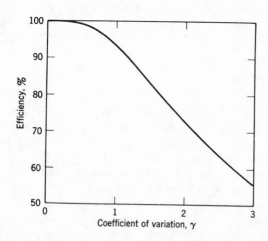

Fig. 6.7. Efficiency of the untransformed mean $\bar{w}$ as a function of the coefficient of variation for lognormally distributed observations (data from Finney, 1941).

TABLE 6.9. METHODS TO CALCULATE C, THE COEFFICIENT OF VARIATION, FOR
LOGNORMALLY DISTRIBUTED OBSERVATIONS

Item	Calculations			
	(1)	(2)	(3)	(4)
Source of data	table 6.7	table 6.7	table 6.8	Krige (1960)
Sample size	13	13	13	33,031
Method of calculation	$\gamma = \sqrt{e^{\beta^2} - 1}$	$\gamma = \sigma/\mu$	$\gamma = \sigma/\mu$	$\gamma = \sqrt{e^{\beta^2} - 1}$
Calculations	$s_u^2 = 1.9828$	$s_w = 531$	$V = 863$	$s_u^2 = 1.05$
	$e^{s_u^2} = 7.26305$	$\overline{w} = 457$	$m = 507$	$e^{s_u^2} = 2.85765$
	$e^{s_u^2} - 1 = 6.26305$			$e^{s_u^2} - 1 = 1.85765$
C	2.50	1.16	1.70	1.36

In order to obtain the estimate C of the coefficient of variation γ, several
methods can be used, as illustrated in table 6.9 for the Welkom data. The first
method, presented in column 1, is to use the estimate of the variance of the
logarithms s_u^2 to find C. The relation between these two quantities is as
follows: the coefficient of variation is found by the formula

$$\gamma = \sigma/\mu,$$

which, if the expression for σ from the first part of this section is substituted,
becomes

$$\gamma = \frac{\sqrt{\mu^2(e^{\beta^2} - 1)}}{\mu},$$

which reduces to

$$\gamma = \sqrt{e^{\beta^2} - 1}.$$

When the calculations in column 1 are performed by using the value of s_u^2
from table 6.7, C is found to equal 2.50. [Alternatively, the value of C may be
obtained from table A-1 of Aitchison and Brown (1957, pp. 154–155).]
Another estimate of s_u^2, based on a large number of assays from the mine as
a whole, may be used in place of the estimate based on only 13 assays. Such
an estimate, based on 33,031 assay values (Krige, 1960, annexures 1 and 2),
is 1.05, which yields a much smaller C of 1.36 when the calculations of column
4 are performed. A second way to estimate C is to divide the estimated
standard deviation of the original observations s_w by the estimated mean of
the original observation $\overline{w}$; the result (column 2) is 1.16. A third way is to
divide m by V (column 3) to yield the result 1.70.

All of the values for C in table 6.9 are different. Because it is based on a large number of mine samples, the value in column 4 should be reliable for the mine as a whole; whether it is applicable to the 13 assays depends on whether they are of typical ore, or, in statistical terms, on whether they are a random sample from the same population. Because they are based on only 13 observations with a large standard deviation, the first three values are different from one another, as well as from the value in column 4. The conclusion is that the coefficient of variation cannot be reliably estimated from so few lognormally distributed observations.

Calculation of Confidence Limits for the Population Mean

The next step is to calculate confidence limits for the population mean by one of several methods. Choice among the methods depends on the number of observations available, background information on consistency and nature of the mineralization, whether both upper and lower confidence limits or only lower confidence limits are required, and the exact purpose. The Welkom data are again utilized to explain the various methods. Although calculation of confidence limits, as summarized in tables 6.10 to 6.17, may appear rather involved, once the calculations are organized, they can be performed without too much trouble.

If a population variance of logarithms β^2 has been established from a large number of measurements and if the sample in question is known to come from this population, confidence limits can be calculated for the logarithms and then transformed into confidence limits for the original observations. As illustrated in table 6.10 for the Welkom assays, the initial data (lines 1 and 2) are a value of β^2 based on some 33,000 observations (table 6.9) and a value of $\bar{u}$ calculated from the 13 observations. In lines 3 to 8 of table 6.10, confidence intervals are calculated for the logarithmic mean α by using the familiar method (sec. 4.2) for normally distributed observations with the population variance known. Next, in lines 9 to 15, these values are transformed to confidence limits for the untransformed mean μ. Finally, in lines 16 to 17, the point estimate m of μ is calculated, based on the β^2 value.

If the population variance of logarithms β^2 is unknown, it must be estimated. Then several methods of calculation can be applied. The first one, the calculation of a confidence region for μ and σ^2 by a method explained by Mood (1950, p. 229), is applied to the Welkom data in table 6.11. Mood's method depends on the fact that, because the distributions of the sample mean and the sum of squares are independent for samples from a normal distribution, a joint confidence region for the parameters α and β^2 can be obtained by finding the values of α and β^2 that satisfy the joint inequalities

$$| \bar{u} - \alpha | < \frac{1.645\beta}{\sqrt{n}},$$

TABLE 6.10. CALCULATION OF CONFIDENCE LIMITS FOR LOGNORMALLY DISTRIBUTED OBSERVATIONS WITH β^2 (THE POPULATION VARIANCE OF LOGARITHMS) ASSUMED TO BE KNOWN

Line	Notation	Value	Calculations	Comments
1	β^2	1.05	From table 6.9	
2	$\bar{u}$	5.36629	From table 6.7	
3	$\beta^2/13$	0.0808		
4	$\sqrt{\beta^2/13}$	0.2842	$\sqrt{0.0808}$	Standard error of mean
5		1.645	From table A-2	5% point of standardized normal distribution
6		0.4675	1.645×0.2842	
7	α_L	4.8988	$5.3663 - 0.4675$	Lower confidence limit for α
8	α_U	5.8338	$5.3663 + 0.4675$	Upper confidence limit for α
9	e^{α_L}	134	From exponential table	
10	e^{α_U}	341	From exponential table	
11	$(n-1)/2n$	0.4615	$(13 - 1)/26$	
12	$[(n-1)/2n]\beta^2$	0.4846		
13	$\exp\left(\dfrac{n-1}{2n}\right)\beta^2$	1.6235	From exponential table	
14	μ_L	218	134×1.6235	
15	μ_U	554	341×1.6235	
16	$e^{\bar{u}}$	214	From table 6.8	
17	m	347	214×1.6235	

and

$$\chi^2_{95\%} < \frac{SS}{\beta^2} < \chi^2_{5\%}.$$

The solution will produce an 81 percent (90 percent squared) confidence region. The first step, illustrated in table 6.11, is to find upper and lower limits for β^2. The second step is to apply these limits to find upper and lower limits for desired values of α between the upper and lower limits for β^2. In table 6.12, four particular values for α have been calculated. Then, in figure 6.8, these four particular α, β^2 values have been plotted in α, β^2 coordinates.

TABLE 6.11. CALCULATION OF PARAMETER ESTIMATES AND THE JOINT CONFIDENCE
REGION FOR LOGNORMALLY DISTRIBUTED OBSERVATIONS BY MOOD'S METHOD:
PART 1, CALCULATION OF UPPER AND LOWER CONFIDENCE LIMITS FOR β^2

Line	Notation	Value	Calculation	Comments
1	$\bar{u}$	5.3663	From table 6.7	
2	SS	23.793642	From table 6.7	
3	n	13	From table 6.7	
4	$\sqrt{n}$	3.6056		
5		1.645	From table A-2	5% point of standardized normal distribution
6		5.23	From table A-3	95% point of χ^2, d.f. = 12
7		21.03	From table A-3	5% point of χ^2, d.f. = 12
8	$SS/\chi^2_{5\%}$	1.1314	23.793642/21.03	β^2_L
9	$SS/\chi^2_{95\%}$	4.5495	23.793642/5.23	β^2_U
10		0.4562	$1.645/\sqrt{13}$	

Additional values to complete the graph were plotted but are not listed in
table 6.12.

Because each α, β^2 point has a corresponding μ, σ point, the region in the
α, β^2 coordinates can be transformed into a region in the μ, σ coordinates.
In table 6.13 the required transformation is calculated for 8 points. In
columns 1 and 2 are given the α, β^2 values at the 8 points to be transformed.
In columns 3, 5, 6, and 7 are given intermediate terms combined according to

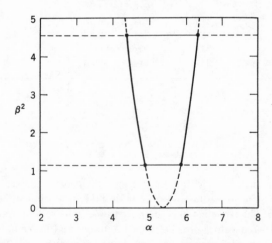

Fig. 6.8. Confidence region in α, β^2 space for lognormally distributed gold observations
from the Welkom mine, South Africa (data from Krige, 1961, p. 18).

TABLE 6.12. CALCULATION OF PARAMETER ESTIMATES AND THE JOINT CONFIDENCE REGION FOR LOGNORMALLY DISTRIBUTED OBSERVATIONS BY MOOD'S METHOD: PART 2, CALCULATION OF UPPER AND LOWER CONFIDENCE LIMITS FOR VARIOUS VALUES OF α

β^2	β	$\dfrac{1.645\beta}{\sqrt{13}}$	$\alpha_L = \overline{u} - \dfrac{1.645\beta}{\sqrt{13}}$	$\alpha_U = \overline{u} + \dfrac{1.645\beta}{\sqrt{13}}$	$\dfrac{\beta^2}{2}$
1.1314	1.064	0.485	4.881	5.851	0.5657
4.5495	2.133	0.973	4.393	6.339	2.2748
2.0000	1.414	0.645	4.721	6.011	1.0000
3.0000	1.732	0.790	4.576	6.156	1.5000

the formulas at the beginning of this section to give the μ, σ values at the 8 points in columns 4 and 8. These points define the 81-percent confidence region graphed in figure 6.9. For comparison the 64-percent confidence region and the points $\overline{w}$, s_w and m, V are also graphed in this figure. The interpretation of the 81-percent confidence region is as follows. At a confidence level of about 90-percent the expected value of μ is between 232 and 5508; these are the μ values plotted as solid circles on the figure. Similarly, with a confidence of about 90 percent, the value of σ is between 336 and 53,824. With a confidence of exactly 81 percent the point μ, σ is expected to lie within the confidence region.

That the point $\overline{w}$, s_w lies outside the confidence *region* is not surprising because this point is calculated only from a sample; however, the individual $\overline{w}$ and s_w values lie within their respective confidence *intervals*. Notably the upper part of the confidence region extends to extremely high values, reflecting the uncertainty about this part of the region.

TABLE 6.13. TRANSFORMATION OF α, β^2 POINTS TO CORRESPONDING μ, σ POINTS

(1)	(2)	(3)	(4)	(5)	(6)	(7)	(8)
α	β^2	$\alpha + \tfrac{1}{2}\beta^2$	μ	μ^2	$e^{\beta^2} - 1$	σ^2	σ
4.881	1.1314	5.447	232	53,852	2.09999	113,100	336
5.851	1.1314	6.417	612	374,741	2.09999	786,900	887
4.393	4.5495	6.668	787	619,369	93.58511	57,963,700	7,613
6.339	4.5495	8.614	5508	30,338,064	93.58511	2,839,191,000	53,284
4.721	2.0000	5.721	305	93,153	6.38906	595,200	771
6.011	2.0000	7.011	1109	1,229,350	6.38906	7,854,000	2,802
4.576	3.0000	6.076	435	189,473	19.08554	3,616,000	1.902
6.156	3.0000	7.656	2113	4,466,000	19.08554	85,240,000	9,233

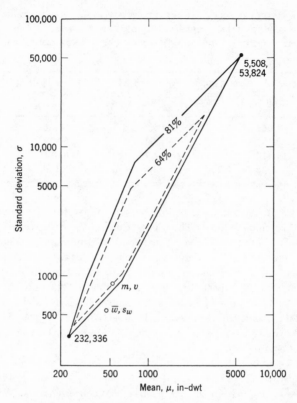

Fig. 6.9. 81-percent confidence region in μ, σ space for lognormally distributed gold observations from the Welkom mine, South Africa.

In both of the next two methods confidence limits are calculated by use of the t-distribution (sec. 4.4), proceeding as though the observations were normally distributed. The value of t for the 5-percent level with 12 degrees of freedom is obtained from table A-4. In the first of these two methods, given in table 6.14, the mean and variance of the untransformed observations from table 6.7 are used, and the calculations follow the familiar form of table 4.2.

In the second of these two methods, given in table 6.15, the mean and variance of the transformed observations from table 6.8 are used. In the first eight lines the 90-percent confidence limits for α are found, following the form in table 6.14. In lines 9 and 10 the powers to the base e of the values in lines 7 and 8 are calculated. Finally, the multiplier for the geometric mean, line 11, is applied to the values in lines 9 and 10 in order to obtain the confidence limits in lines 12 and 13.

TABLE 6.14. CALCULATION OF CONFIDENCE LIMITS FOR LOG-NORMALLY DISTRIBUTED OBSERVATIONS BY USE OF THE t-DISTRIBUTION (METHOD 1 IS BASED ON MEAN AND VARIANCE OF UNTRANSFORMED OBSERVATIONS FROM TABLE 6.7)

Line	Notation	Value	Calculation
1	$\overline{w}$	457	From table 6.7
2	s_w	531	From table 6.7
3	$\sqrt{n}$	3.6056	$\sqrt{13}$
4	$t_{5\%}$	1.782	From table A-4 with 12 d.f.
5	$s_w/\sqrt{n}$	147	531/3.6055
6	$t_{5\%}s_w/\sqrt{n}$	262	1.782×147
7	μ_L	195	$457 - 262$
8	μ_U	719	$457 + 262$

The last method for calculating confidence limits was devised by Sichel (1952, 1966), one of the pioneer workers in the application of statistical methods to the valuation of lognormally distributed gold assays from South African mines. In Sichel's method, calculated in table 6.16, the value of SS_u for the transformed variable u is obtained from table 6.7 and the value of m is obtained from table 6.8. When SS_u is divided by n (line 3 of table 6.16), a

TABLE 6.15. CALCULATION OF CONFIDENCE LIMITS FOR LOGNORMALLY DISTRIBUTED OBSERVATIONS BY USE OF THE t-DISTRIBUTION (METHOD 2, BASED ON MEAN AND VARIANCE OF TRANSFORMED OBSERVATIONS FROM TABLE 6.8)

Line	Notation	Value	Calculation
1	$\overline{u}$	5.36629	From table 6.7
2	s_u	1.4081	From table 6.7
3	$\sqrt{n}$	3.6056	From table 6.11
4	$s_u/\sqrt{n}$	0.3905	
5	$t_{5\%}$	1.782	From table A-4 with 12 d.f.
6	$t_{5\%}(s_u/\sqrt{n})$	0.6959	
7	$\overline{u} - t_{5\%}(s_u/\sqrt{n})$	4.6704	
8	$\overline{u} + t_{5\%}(s_u/\sqrt{n})$	6.0622	
9	$e[\overline{u} - t_{5\%}(s_u/\sqrt{n})]$	106.7	
10	$e[\overline{u} + t_{5\%}(s_u/\sqrt{n})]$	429.3	
11	$\psi n(\frac{1}{2}s_u^2)$	2.370	From table 6.8
12	μ_L	253	
13	μ_U	1017	

Table 6.16. Calculation of confidence limits for lognormally distributed observations by a method devised by Sichel

Line	Notation	Value	Calculation	Comments
1	m	507	From table 6.8	
2	SS_u	23.793642	From table 6.7	
3	SS_u/n	1.83	23.793642/13	V (in Sichel's notation) = SS_u/n
4		0.5124	From Sichel's table B†	Lower multiplying factor
5		4.015	From Sichel's table B†	Upper multiplying factor
6	μ_L	260	507 × 0.5124	
7	μ_U	2036	507 × 4.015	

† Sichel (1966).

value of 1.83 is obtained. The next step is to enter Sichel's table B (1966, p. 14) by using n and V from line 3 to obtain, by interpolation, two multiplying factors (lines 4 and 5), which are used to obtain the confidence limits in lines 6 and 7 of table 6.16.

The estimated means and confidence limits obtained by the five methods are summarized in table 6.17, which shows above all that wide differences are obtained depending upon which method is used. However, the lower bounds agree better than the upper ones, a fortunate circumstance for practical mining application where the main concern is that the minimum expected

Table 6.17. Comparison of confidence limits for lognormally distributed observations calculated by applying five different methods to example data from the Welkom mine, South Africa

Source table	Method	Estimated mean	Confidence limits	
			Lower	Upper
6.10	Apply normal theory; assume β^2 known	347	218	554
6.13	Calculate confidence region	507	232	5508
6.14	Assume $\overline{w}$ normally distributed; use t based on $\overline{w}$ and s_w^2	457	195	719
6.15	Assume m normally distributed; use t based on m and V^2	507	253	1017
6.16	Use Sichel's approximation	507	260	2036

grade of ore be estimated correctly. Of the five methods the best is the application of normal theory under the assumption that β^2 is known (table 6.10), provided that (a) a close estimate of β^2 is available from many observations, and (b) a convincing case can be made that the new observations in question are from the same population. Otherwise, we favor Mood's method of calculating a confidence region (tables 6.11 to 6.13). If the observations follow or seem to follow a lognormal distribution with β^2 larger than 0.9, the two methods that apply the t-distribution, under the assumption that the observations follow a normal distribution (tables 6.14 and 6.15), are inefficient. However, if the observations in fact depart from a lognormal distribution, the use of one of the t-distribution methods will yield conservative results because these methods are robust. Sichel's approximation (table 6.16) is convenient to use, particularly if an electronic computer is not available to implement Mood's method, although problems too involved to describe here can arise (Link, Koch, and Schuenemeyer, 1970).

The Negatively Skewed Lognormal Distribution

The negatively skewed lognormal distribution is one whose frequency curve is the translated mirror image of a lognormal distribution reflected about its natural lower limit, as discussed by Aitchison and Brown (1957, p. 16). Figure 6.10 gives the frequency curve for a negatively skewed lognormal distribution of a variable, $3 - x$, where 3 is the natural upper limit, and x has a lognormal distribution with parameters $\alpha = 0$, $\beta^2 = 0.25$.

Many negatively skewed distributions are found in geology, although not all of these are lognormal. Table 6.18 and figure 6.11 present an example of 40 potassium analyses of granites from South Africa, whose natural upper limit

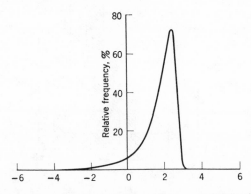

Fig. 6.10. Frequency curve of a negatively skewed lognormal distribution (after Aitchison and Brown, 1957, p. 16).

Table 6.18. Frequency distribution of potassium in granites from South Africa. (After Ahrens, 1963, p. 935.)

Class interval	Frequency	Cumulative frequency	Relative cumulative frequency (%)
1.0–1.6	2	2	5.0
1.6–2.2	1	3	7.5
2.2–2.8	3	6	15.0
2.8–3.4	5	11	27.5
3.4–4.0	9	20	50.0
4.0–4.6	18	38	95.0
4.6–5.2	2	40	100.0

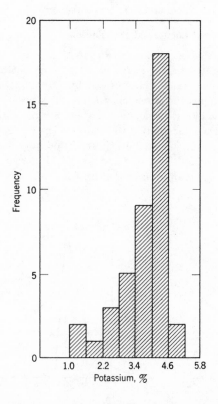

Fig. 6.11. Histogram of potassium in granites from South Africa (after Ahrens, 1963, p. 935).

might be taken as 5.4. In general, negatively skewed distributions are expected if many of the observations are at or near an upper limit: which may correspond to 100 percent, to the composition of a chemical element or other constituents in a pure mineral, or some other upper limit set by some natural process. Thus, quartz analyses in a glass-sand deposit, iron analyses in a hematite deposit, and chromium analyses in chromite may, but need not be, negatively skewed.

6.3   TRANSFORMATIONS

A *transformation* is a function of an observation that defines a new observation, that is,

$$u = f(w),$$

where w is the original observation, f is the transforming function, and u is the new observation. A familiar example is the linear transformation of inches to centimeters by multiplying by the constant 2.54. However, in this section the only kinds of transformation considered are nonlinear ones, that is, those for which a plot of u against w is not a straight line. An example is the logarithmic transformation whereby each observation is changed to its logarithm, according to the formula,

$$u = \ln (w).$$

The basic purpose of a nonlinear transformation is to change the shape of the frequency distribution. A linear transformation does not change the shape of the distribution, but only the scale.

There are three purposes for changing from one frequency distribution to another by a nonlinear transformation. The most important is to stabilize the variance. All transformations discussed in this book are primarily for this purpose, although, fortunately, they also tend to serve the other two purposes. The stabilization of variance may be explained as follows: An observation may be written (sec. 5.1) as its mean value plus a random fluctuation, that is,

$$w = \mu + e.$$

This notation implies that the mean value of e is 0, but its variance may or may not be equal to a constant, which is the requirement for stability. However, if the variance, σ_e^2, of e depends upon μ in some fashion, it may be possible to find a transformation,

$$u = f(w),$$

such that in the transformed equation,

$$u = \mu^* + e^*,$$

where the variance of e^* is a constant independent of the value of μ^*. Constant variance is also named *homoscedasticity*. As an example of nonconstant variance, consider the fluctuations associated with low-grade and with high-grade gold assays. It is known that the variability of gold assays from low-grade ore is much less than the variability of gold assays from high-grade ore. Therefore the average size of the fluctuation e depends upon the grade of the ore. For some gold ores a constant variance may be obtained by taking the logarithms of the observations.

The next most important purpose of a nonlinear transformation is to obtain additivity (sec. 5.11). For instance, if

$$w = e^y e^z e^e,$$

the factors y and z are clearly not additive because the effect of a change in z depends upon the value of y. However, through a transformation the factors can be made additive; in fact, through a logarithmic transformation the result is obtained that

$$u = y + z + e.$$

The third basic purpose is to transform the observations so that they follow a normal distribution. There are two advantages. The first is that, if the observations are normally distributed, the sample mean is an efficient estimate of the population mean (sec. 3.6); but if the observations follow a highly skewed lognormal distribution (sec. 6.2), the sample mean is a rather inefficient estimate of the population mean. The second is that all important statistical procedures for confidence intervals and hypothesis tests assume underlying normality. However, for these procedures departures of data from normality are seldom serious because many of the procedures are robust and work rather well even with observations that are not exactly normally distributed, and the mean is likely to be a quite efficient estimator except for frequency distributions with very long tails. The techniques that are rather robust are t, F as used in the analysis of variance, and any statistic based on sample means. Those statistics discussed so far that are less robust are chi-square and F as used to compare variances.

Transformations should never be made blindly, as a matter of course, because there are disadvantages that may be serious. The most important is that transformations may lead to biased estimates; for example, in estimating the mean of a lognormal distribution, greater efficiency may be obtained by transforming the observations to logarithms, but a bias is introduced because the antilogarithm of $\bar{u}$ is not on the average the population mean μ, as has been explained in detail in section 6.2. For lognormally distributed observations, the bias may be removed because a great deal is known about the lognormal and normal distributions. For observations following other distributions such bias may also be removable in principle, but practical methods to do so may not have been devised. A second disadvantage of transformations is that their use always requires more calculation. Even if an appropriate transformation is known, such additional work may not be economically justified because the efficiency of an estimate based on original untransformed data may be adequate.

In summary, specific advice on when to transform will be offered as the subject is developed. In general, it may be useful to transform observations when the conclusions based on the transformed scale can be understood, when biased estimates are acceptable, or when the amount of bias can be estimated and removed because the details of the distributions are known.

Some Common Transformations

For all of the distributions introduced in this chapter, transformations have been devised to make the distributions normal or nearly so. The only one widely used for geological data is the lognormal transformation whereby skewed distributions are more or less normalized by taking logarithms.

Transformations for observations from discrete distributions are not explained in this book because these data are uncommon in geology. Li (1964, I, pp. 505–526) explains the angular transformation for binomial data and the square-root transformation for Poisson data, but points out that transformations of such data are seldom helpful in practice.

However, one other transformation, the normal-score transformation for ranked data, has sufficient geological application to merit a brief account. Ranked data consist of items ordered according to preference, shades of color, intensity of weathering, or any other attribute not amenable to exact measurement. The procedure may be explained through a fictitious illustration. Suppose that four geologists rank six western states according to favorability for oil exploration and that someone wants to know if their judgments are mutually consistent. Table 6.19 lists the rankings, showing that geologist 1, for example, likes California the most and Arizona the least. Intuitively, one thinks it easier to rank the states at the top and bottom of the list than those in the middle; accordingly the normal-score transformation replaces the ranks by the standardized normal deviates obtained from table A-2. Table 6.20 shows that the differences between items near the center of the list are smaller with the normal-score transformation than those near the extremes.

TABLE 6.19. FICTITIOUS RANKINGS (BY FOUR GEOLOGISTS) OF
SIX STATES ACCORDING TO FAVORABILITY FOR OIL EXPLORATION

State	Geologist			
	1	2	3	4

A. Untransformed data

Washington	3	2	1	1
Oregon	2	3	2	3
California	1	1	3	2
Idaho	4	5	6	6
Nevada	5	6	5	4
Arizona	6	4	4	5

B. Normal-score transformed data

Washington	-0.20	-0.64	-1.27	-1.27
Oregon	-0.64	-0.20	-0.64	-0.20
California	-1.27	-1.27	-0.20	-0.64
Idaho	0.20	0.64	1.27	1.27
Nevada	0.64	1.27	0.64	0.20
Arizona	1.27	0.20	0.20	0.64

TABLE 6.20. COMPARISON OF RANKED AND NORMAL SCORE DATA
FOR FICTITIOUS ILLUSTRATION OF TABLE 6.19

Ranked data		Normal score data	
Rank	Difference between adjacent ranks	Normal score	Difference between adjacent normal scores
1		-1.27	
	1		0.63
2		-0.64	
	1		0.44
3		-0.20	
	1		0.40
4		0.20	
	1		0.44
5		0.64	
	1		0.63
6		1.27	

Table 6.21. Analyses of variance for fictitious data of table 6.19

Source of variation	Sum of squares	Degrees of freedom	Mean square	F	$F_{10\%}$
A. Untransformed data					
States	5.60	5	11.20	12.00	2.27
Residual variability	14.00	15	0.93		
B. Normal-score transformed data					
States	12.50	5	2.50	9.37	2.27
Residual variability	4.00	15	0.27		

Statistical procedures previously discussed can be applied either to the original ranked data or to the transformed data. In table 6.21 the randomized-block design of the analysis of variance (sec. 5.7) is performed for both the original and the transformed fictitious data; the geologists correspond to four treatments and the states to six replications. Because the calculated F-values are larger than the tabled ones, for both kinds of data the judging is deemed inconsistent. However, the calculated F-value is smaller for the transformed than for the original data, a fact which reflects the concept that it is harder to rank the states in the middle of the list.

6.4 GRAPHICAL REPRESENTATION OF DATA

The old saw that a picture is worth a thousand words is as true for understanding geological data as for anything else. Fortunately, geologists like graphs, so all that should be necessary is to point out some special graphs that statisticians have devised for the representation of data.

Graphs are easy to plot and easy to read. If the observations are numerous, computer plots (sec. 10.7 and 17.4) are convenient. Graphs illuminate various facets of data not readily displayed by other means of presentation, in particular extremely large and extremely small values, and, with enough observations, the general shape of a frequency distribution and whether there is one or more than one mode.

This section explains how to plot observations on some special graph papers (available from firms such as Keuffel and Esser, and Dietzgen, or from university bookstores). The techniques are purely for meaningful graphic representation, and they are associated with informal inferences and indications rather than with formal statistical inference.

TABLE 6.22. PLOTTING PERCENTAGES FOR PHOSPHATE DATA OF TABLE 2.1

Class interval ($\%$ P_2O_5)	Relative cumulative frequency ($\%$)	Plotting percentages
14–16	0.45	0.30
16–18	0.90	0.74
18–20	4.47	4.31
20–22	13.84	13.67
22–24	33.48	33.28
24–26	57.60	57.36
26–28	82.60	82.32
28–30	95.99	95.69
30–32	99.10	98.81
32–34	100.00	99.70

Normal Probability Graph Paper

Normal probability graph paper is scaled so that if observations are normally distributed their relative cumulative frequency in percent plots as a straight line. For fewer than 100 observations it is desirable but not essential to replace cumulative frequency (c.f.) by a plotting percentage obtained by the formula

$$\text{plotting percentage} = 100 \times \frac{3(\text{c.f.}) - 1}{3n + 1},$$

where n is the total number of observations. In table 6.22 the plotting percentages are calculated for the phosphate data from table 2.1. For the first class interval the value is

$$100 \times \frac{3(1) - 1}{3(224) + 1} = 0.30.$$

In figure 6.12 the upper bound of the class interval is graphed against the plotting percentage.

If observations are normally distributed, their plot on normal probability graph paper defines a straight line. The straight line fitted by eye to the phosphate data does not exactly fit the points, but it is close enough to them to indicate that they are distributed nearly normally. The mean of the observations is estimated by reading the 50-percent point on the straight line; the value obtained of 25.2 agrees closely with the computed mean of 25.13 (sec. 3.3). The standard deviation of the observations is estimated by sub-

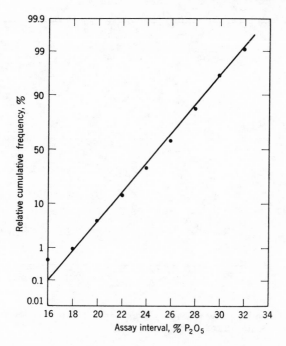

Fig. 6.12. Plot on normal probability graph paper of phosphate data from table 2.1.

tracting the 16-percent point from the 84-percent point and dividing the result by two. For the phosphate data the 16-percent point is 22.2, the 84-percent point is 28.1, the difference is 5.9, and the estimated standard deviation is 2.95, the same as the computed value (sec. 3.3).

Lognormal Probability Graph Paper

Lognormal probability graph paper is like normal probability graph paper except that it is scaled so that, for lognormally distributed observations, their relative cumulative frequency in percent plots as a straight line. In figure 6.13 the 13 Welkom observations from table 6.5 are plotted on this paper. The straight line fitted by eye shows that the points are more or less lognormally distributed, although not all of them are close to the line. The advantage of using the lognormal probability paper is that, by making a quick plot, it is easy to see if the observations are more or less lognormal. The mean and standard deviation cannot be obtained from this graph, however, as they can be from graphs on normal probability paper.

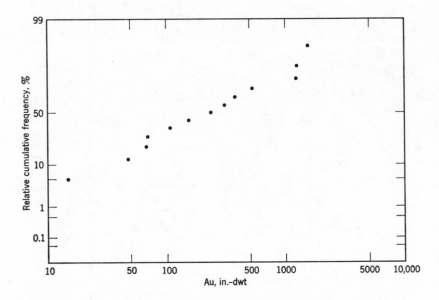

Fig. 6.13. Plot on lognormal probability graph paper of 13 gold observations from the Welkom mine, South Africa.

The Full Normal Plot (FUNOP)

The *FU*ll *NO*rmal *P*lot was devised by Tukey (1962, p. 22) to investigate a set of data for the occurrence of unusually large or small observations to determine which are likely to be outliers that belong in the data and which are probably mistakes. The procedure is similar to making an ordinary plot on normal probability paper, but a new quantity, the Judd, is calculated to make interpretation of the pattern of observations easier. The Judd is equal to the deviation of the observations from the median divided by the standard normal deviate corresponding to the plotting-point percentage. If the observations were distributed perfectly normally, the Judd would be nearly equal to their standard deviation. By inspecting an ordered list of the two-thirds of the Judds associated with the largest one-third and smallest one-third of the original observations one can see if any of the observations seem to be unusually large or small.

The calculations are explained with the 13 Welkom gold observations as an example. In table 6.23 the observations are ordered, and a plotting percentage is calculated for each. For each plotting percentage corresponding to the upper and lower one-third of the observations the standard normal deviate,

Table 6.23. Calculations for full normal plot (fig. 6.14) for 13 gold observations from the Welkom mine, South Africa

(1)	(2)	(3)	(4)	(5)	(6)	(7)	(8)
		Observation					
Rank	Raw, w	Logarithm, $u = \ln(w)$	Plotting percentage	s.n.d.	u − median	Judd = (6)/(5)	Ordered Judds
1	15	2.70806	0.050	−1.6449	−2.76000	1.678	1.146
2	48	3.87121	0.125	−1.1503	−1.59685	1.388	1.330
3	68	4.21951	0.200	−0.8416	−1.24855	1.484	1.388
4	70	4.24850	0.275	−0.5978	−1.21956	2.040	1.447
5	109	4.69135	0.350				
6	154	5.03695	0.425				
7 (median)	237	5.46806	0.500				
8	308	5.73010	0.575				
9	377	5.93225	0.650				
10	525	6.26340	0.725	0.5978	0.79534	1.330	1.484
11	1221	7.10743	0.800	0.8416	1.63937	1.948	1.678
12	1252	7.13647	0.875	1.1503	1.66441	1.447	1.948
13	1560	7.35245	0.950	1.6449	1.88439	1.146	2.040

s.n.d., corresponding to 100 minus the plotting percentage, is obtained from table A-2. In column 6 the deviation of each observation from the median, 5.46806, is calculated, and in column 7 the Judd is calculated by the formula

$$\text{Judd} = \frac{u - \text{median}}{\text{s.n.d.}}.$$

The median, 1.466 (halfway between 1.447 and 1.484), of the Judds is approximately the standard deviation of the observations. If any Judd is more than twice the median, the corresponding observation is suspect, as well as those whose distance from the median is greater than that observation. None of the logarithms of the Welkom observations is suspiciously large or small. In figure 6.14 the Judds are plotted; the plot shows graphically that none of the observations is more than two Judds above the median.

Extreme-Value Graph Paper

Suppose that the daily rainfall at a station is recorded for a year. If the maximum rainfall for the year is taken as an observation and the process is repeated for 50 years, 50 observations, which may reasonably be expected to follow an extreme-value distribution, are collected. An essential part of the concept is that the observations in the distribution are ones *selected* from the

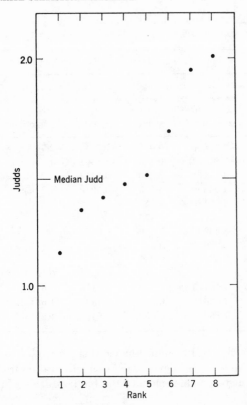

Fig. 6.14. Full Normal Plot (FUNOP) of Judds calculated for the 13 gold observations from the Welkom mine, South Africa.

total number of daily rain-gauge measurements as the largest for each year. Similarly, a radio-tube manufacturer might make life tests of tubes in samples of size 25. If the longest life in each batch of tubes is recorded, this value might also be expected to follow an extreme-value distribution.

Perhaps the extreme-value data of most interest in geology are annual flood records because so many geological processes are associated with unusually high floods. The observations in the extreme-value distribution may be the highest annual discharges or the highest annual floods. Note that these values are *selected* from observations made daily (or at some other period).

A book on extreme-value distributions by Gumbel (1958) explores the subject in great detail; we give only a brief introduction. One practical application is in bridge design, to predict from flood records how often a destructive flood is liable to occur. Another practical application is in rock mechanics,

TABLE 6.24. CALCULATIONS FOR PLOT ON EX-
TREME VALUE GRAPH PAPER OF MAXIMUM DAILY
DISCHARGE YEARLY FOR LAUREL HILL CREEK,
AT URSINA, PENNSYLVANIA *

Year	Maximum daily discharge (ft^3/sec)	Plotting percentages
1923	2070	1.79
1929	2170	4.46
1932	2370	7.14
1931	2780	9.82
1950	2900	12.50
1944	3100	15.18
1919	3110	17.86
1921	3110	20.54
1914	3220	23.21
1921	3330	25.89
1942	3580	28.57
1934	3660	31.25
1925	3780	33.93
1935	4020	36.61
1915	4140	39.29
1929	4270	41.96
1925	4300	44.64
1949	4360	47.32
1920	4410	50.00
1946	4470	52.68
1933	4860	55.36
1937	4860	58.04
1939	5010	60.71
1937	5010	63.39
1948	5150	66.07
1916	5630	68.75
1918	5630	71.43
1927	5630	74.11
1942	5660	76.79
1947	5830	79.36
1917	5970	82.14
1927	5970	84.82
1945	6000	87.50
1940	6510	90.18
1924	8090	92.86
1941	9400	95.54
1936	10300	98.21

* Data from U.S. Geological Survey Circ. 204, p. 21.

to predict from train gauge measurements when rock failure is liable to occur.

For very large samples the distribution of extreme-value observations depends only upon the tails of the frequency distribution from which the observations are drawn. Thus, for the two illustrations, the extreme-value distributions are independent of the distributions of *all* rainfall and radio

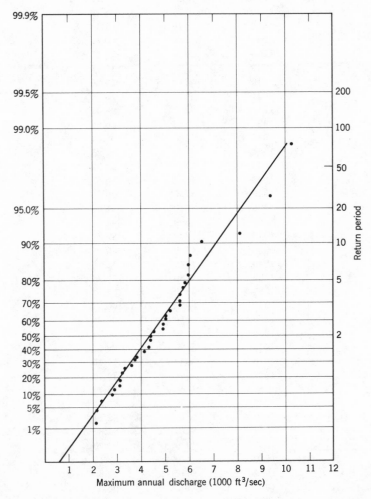

Fig. 6.15. Plot on extreme value graph paper of maximum daily discharge yearly for Laurel Hill Creek, Ursina, Pennsylvania (data from U.S. Geol. Survey Circ. 204, p. 21).

tube measurements. The distribution of extreme-value observations is named the *asymptotic extreme-value distribution* and has the formula

$$F(w) = \exp\left\{-\exp\left[-\alpha(w - a_n)\right]\right\} \qquad \alpha > 0,$$

where $F(a_n) = 0.3678$, $F(a_n - 1/\alpha) = 0.066$, and w is an observation. (Actually there are three similar distributions, but we work only with the one most pertinent for geological data.)

The use of extreme-value graph paper (available from Technical and Engineering Aids for Management, 104 Belrose Avenue, Lowell, Mass.) may be illustrated by plotting maximum daily discharge yearly from 1914 to 1950 for Laurel Hill Creek at Ursina, Pennsylvania (*U.S. Geological Survey Circular* 204, p. 21). In table 6.24 the maximum flood discharges are ordered in size from smallest to largest. For each observation a plotting percentage is calculated and plotted in figure 6.15. Inspection shows that the points lie near to a straight line, and a line was fitted to them by eye.

From the graph the return period can be read—that is, how long one expects it to be before a maximum annual discharge of a certain magnitude reoccurs; for instance, on Laurel Hill Creek, one expects an annual maximum discharge of 8500 cubic feet per second to occur about every 20 years. For bridge design purposes the data could be extrapolated to yield a return period of say, 1000 years; although highly uncertain, the prediction is better than nothing.

Other Graphical Methods

Many other graphical methods, not considered in this book, have been devised for representing data. A good source of information is an article by Mosteller and Tukey (1949).

6.5 MIXED-FREQUENCY DISTRIBUTIONS

Up to this point the geological data reviewed in this book could be related to one of the common theoretical frequency distributions; for instance, the phosphate observations (sec. 2.1) could be referred to the normal distribution and some gold data (sec. 6.2) to the lognormal distribution. However, many sets of geological data follow an empirical distribution that may be represented as a mixture of two or more of the common theoretical distributions. This mixture leads to the complication of having to estimate not only the parameters of the component theoretical distributions but also their proportions.

Mixed-frequency distributions may arise in nature in many ways; for instance, stream-carried sediments, whose particle-size distribution was approximately normal, could be introduced into a basin of distribution. Then, later, volcanic ash may be deposited in the same basin with another particle-size distribution, perhaps also normal but with different parameters. If the two kinds of sediments were mixed by waves and currents, the particle-size distribution of the resulting mixed sediment could be appropriately described by a mixed-frequency distribution, composed of two component distributions, each of them normal.

If the component distributions do not differ very much in mean and variance, if there is no theoretical basis for separating them, and if no other variables or variable to measure are available, a mixed distribution may be difficult to recognize and virtually impossible to decompose into its component distributions. For instance, if an investigator is presented with a set of unidentified measurements, he may not know whether he is dealing with a mixed distribution. However, if he learns that the distribution is one of human heights, he has a basis for expecting it to be a normal distribution or a mixture of normal distributions. If he discovers further that it is a distribution of heights of Caucasians and Pygmies, he will likely assume that the distribution is mixed; and if he is then supplied with data on another variable, for instance, skin color, he may be able to separate the two distributions without too much trouble.

These notions are clarified through a numerical illustration and some real examples. Some concluding remarks assert that most mixed geological distributions are difficult if not impossible to recognize, much less to separate, because the statistical techniques that might be used have such large variances of estimate for the parameters that little reliable information can be obtained about the parameters of the component distributions or about their proportions in the mixture.

An Illustration of Mixed Distributions

As an illustration of mixed distributions, consider five populations, named A to E, each composed of 500 observations with mean μ and standard deviation σ as specified in table 6.25. Each population may be thought of as the silica content of a sedimentary rock measured at various places. From these five populations, five distributions are defined, each consisting of 1000 observations, obtained by mixing these five populations. Distribution 1, which is population A taken twice, has the frequency distribution of table 6.26 and the histogram of figure 6.16-a. Distribution 2 consists of population A plus the 500 observations from population B. The other three distributions are defined similarly, as shown in table 6.26, with the resulting frequency distributions and histograms given in the table and the figure.

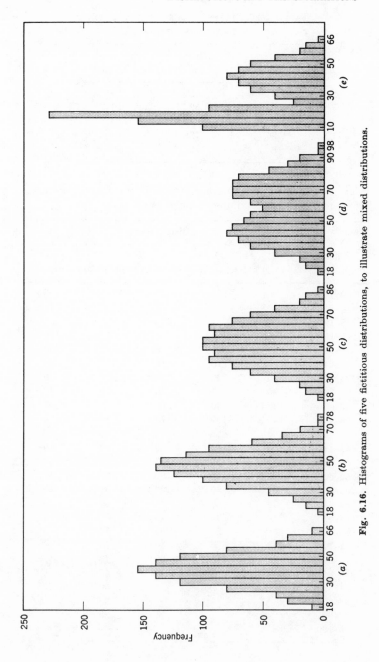

Fig. 6.16. Histograms of five fictitious distributions, to illustrate mixed distributions.

Table 6.25. Means and standard deviations of five fictitious distributions

Population	Mean, μ	Standard deviation, σ
A	42	10
B	52	10
C	62	10
D	72	10
E	42	3.16

Table 6.26. A frequency distribution of five fictitious distributions, to illustrate mixed distributions

Class interval	Distribution				
	1 (2A)	2 (A + B)	3 (A + C)	4 (A + D)	5 (A + E)
0–4					
4–8					
8–12					30
12–16					160
16–20	10	5	5	5	230
20–24	30	15	15	15	95
24–28	40	25	20	20	25
28–32	80	45	40	40	40
32–36	120	80	60	60	60
36–40	140	100	75	70	70
40–44	160	125	95	80	80
44–48	140	140	90	75	70
48–52	120	135	100	65	60
52–56	80	115	100	60	40
56–60	40	90	90	50	20
60–64	30	60	95	60	15
64–68	10	35	75	75	5
68–72		20	60	75	
72–76		5	40	75	
76–80		5	20	70	
80–84			15	45	
84–88			5	30	
88–92				20	
92–96				5	
96–100				5	

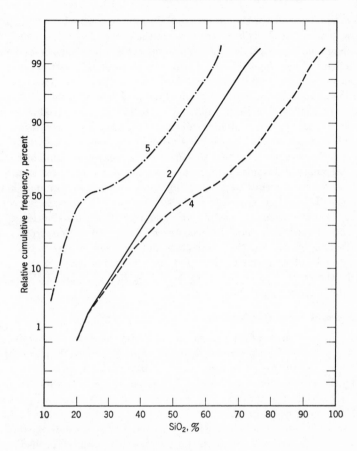

Fig. 6.17. Plot on normal probability paper of fictitious mixed distributions 2, 4, and 5.

The frequency distributions and histograms have the following interesting properties. Distribution 2 is only slightly skewed and is difficult to distinguish from a normal distribution. Distribution 3 is clearly nonnormal and displays three modes with the particular class interval chosen, although if these were real data, one would be hard pressed to decide if there were three modes present or merely one poorly defined mode. Distribution 4 displays two distinct modes, corresponding to the two population means at 42 and 72 and clearly is bimodal. Distribution 5 is skewed. Distributions 2, 4, and 5 are plotted on normal probability paper in figure 6.17. Distribution 2 plots as nearly a straight line, so it would be difficult to determine from the plot that it was not normal. On the other hand, distributions 4 and 5 are definitely nonnormal, a fact that illustrates a use of the probability paper plot.

This simple numerical illustration demonstrates that in some cases it would be difficult if not impossible to distinguish whether the distributions were mixed and to discover the original distributions, provided that there was nothing more to go on than the original data. If, however, the distributions represented silica content in sedimentary rocks, another variable, such as grain size, shape, color, or etching, might be measured as an aid to separation. The purpose for setting up the illustration in this somewhat artificial way, rather than by generating the distributions, frequency tables, and histograms by mathematical formulas implemented by electronic computer, is to point the way for the geologist to perform this work with tables of random normal numbers (Li, 1964, I, appendix table 1) or with punched cards in order to obtain the frequencies by sorting and counting. For a distribution that is evidently mixed, it is helpful to construct mixed distributions from known populations by varying the parameters and numbers of observations. One then grasps unforgettably the fact that many mixtures can yield the same or nearly the same distributions and, therefore, how difficult any statistical procedures for separating them must be.

Examples of Mixed Distributions

The Fresnillo mine affords an example of a mixture of two distinct distributions of silver/lead ratio present through a broad interface (Koch and Link, 1967, pp. 49–60). Although some of each distribution is present in all the veins, the distributions can best be delineated in the 2137 vein. There the area of high silver/lead ratio (greater than 400 grams of silver to 1 percent of lead) extends downward from the 470-meter level, and a transition zone lies in between. These areas were defined by scanning ratios for individual sample points on each level after the basis for selection had been established.

For each of the three areas of different silver/lead ratio we prepared a histogram of the logarithms of the different ratios (fig. 6.18). These fairly symmetrical distributions indicate how silver/lead ratios shift from one zone to the next. They overlap because defining three distinct regions is an oversimplification. The histogram for all three distributions combined is skewed with only one mode.

These high and low ratio types of mineralization could be separated because Stone and McCarthy (1948) have recognized and mapped the two kinds of mineralization designated as high and low sulfide and because there is also a theoretical basis. When the silver/lead ratio is less than 100, the correlation (sec. 9.3) of silver and lead is very high; the slope of the linear regression (sec. 9.2) is typically an increase of about 22 meter-grams of silver for a 1 meter-percent increase in lead; and the intercept is about 96 meter-grams of silver

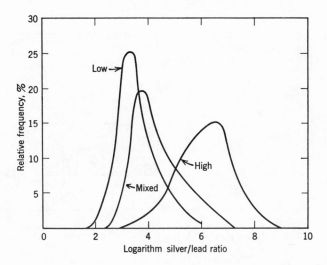

Fig. 6.18. Frequency curves of silver/lead ratios from three areas of the 2137 vein, Fresnillo mine.

per metric ton. These facts are consistent with the geological theory that the silver may be present in solid solution in galena. Elsewhere, where the silver/lead ratio exceeds 400, correlation of silver and lead is low; the slope of the linear regression relating silver to lead is typically about 154; and the intercept is about 631. These facts suggest that most of the silver is independent of lead; and, indeed, galena is sparse, and ruby-silver minerals are plentiful.

One geological field in which mixed distributions have been studied for many years and are reasonably well understood is that of size distributions in sediments. Pettijohn (1957, pp. 45–51) explains mixed-size distributions in sediments and gives references to the literature; for example, some size analyses of California alluvial gravels yield histograms (fig. 6.19) (Conkling and others, 1934, p. 86), which show that sand sizes are unimodal, whereas gravel sizes are bimodal and represent mixed distributions. Conkling writes that "the two maxima of these curves are characteristic of the alluvial gravel samples. . . . Probably the coarsest maximum represents the materials rolled and bounded along the channel bottom, and the finer percentage maximum represents the material deposited in the interstices between the boulders and cobbles from suspension." Again, this example indicates that geologic processes must be invoked and theory is needed to untangle mixed distributions.

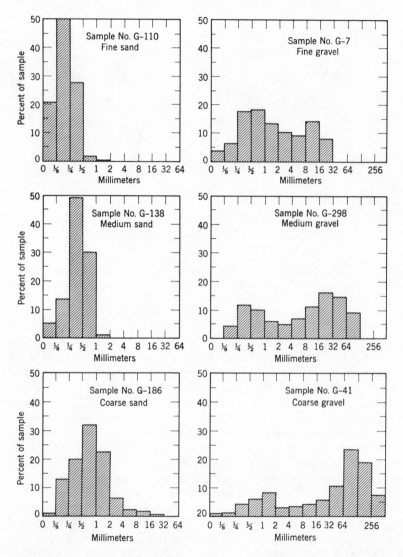

Fig. 6.19. Histograms of size distributions for some California alluvial gravels (after Conkling and others, 1934, p. 86).

Writing about particle-size distributions, Wilks (1963, p. 116) made the following remarks that are also pertinent to mixed distributions in general:

Cumulative particle-size distributions have been determined for many kinds of geological material. . . . The curve obtained in each case is thus characteristic of the sample of particles. As one might expect, these curves vary widely from sample to sample not only from one kind of material to another but from one sample to another of the same material. . . . If the sample consists of a mixture of particles produced by a mixture of two (or more) kinds of geological disintegrating action, each of which would produce . . . normal distributions, the overall particle size distribution . . . would be the sum of the two (or more) normal distributions and would, in general, be non-normal. There is a strong temptation (which occurs not only in geology but in biology and other fields) to make the converse statement, namely, that if a non-normal distribution exists it can be inferred that it is the sum of two or more normal distributions. This is a hazardous inference, since there could be homogeneous grinding, crushing, or sorting actions that do not produce normal particle-size . . . distributions. To make any headway in verifying such a converse statement, one would need reasonably good information or strong grounds for some hypothesis as to how many different major grinding, crushing, or sorting actions were involved in producing the resulting sample of particles, and one would also need to know that each action would by itself produce a normal distribution. . . . One could then decompose the composite sample distribution into its normal components and estimate the fraction of particles involved in each component distribution. This, however, is a rather involved mathematical exercise, even for a mixture of two or three populations, which should be undertaken only when one has good information or a strong hypothesis that a small number of different distributions have been mixed. Such an analysis for a mixture of more than three or four distributions should be made, if at all, only with great caution.

Concluding Remarks

Many frequency distributions encountered in geology are mixed. Sometimes they can be sorted out because the parameters of the component distributions are quite different from one another or because some other variable can be measured. Thus for the California sediments the means and standard deviations, as well as the lithologies, are very different. Other times the mixed distributions are difficult if not impossible to sort out because the parameters of the component distributions are not very different from one another or because some other variable cannot be measured. Thus for the Fresnillo data a large number of ratios had to be classed as mixed mainly because the silver/lead assays were for total silver and lead without differentiating the silver that is in solid solution in galena. If assays were available for distinguishing the two kinds of silver in the mixed-ratio zone, a much cleaner separation would have been possible.

Thus, if mixed distributions are to be separated effectively, one or more of these elements must be present:

1. A theory for the functional forms of the component distributions and preferably also for their parameters.
2. Additional variables, such as lithologies and geographic positions.
3. Rather different parameters for each of the component distributions.
4. Only two, or at most a few, rather than many, component distributions.

REFERENCES

Aitchison, John, and Brown, J. A. C., 1957, The lognormal distribution, with special reference to its uses in economics: Cambridge Univ. Press, 176 p.

Conkling, H., Eckis, R., and Gross, P. L. K., 1934, Ground water storage capacity of valley fill: California Div. Water Resources, Bull. 45, 279 p.

Dixon, W. J., and Massey, F. J., 1969, Introduction to statistical analysis: New York, McGraw-Hill, 638 p.

Feller, William, 1968, An introduction to probability theory and its applications, v. 1: New York, John Wiley & Sons, 509 p.

Finney, D. J., 1941, On the distribution of a variate whose logarithm is normally distributed: Jour. Royal Statistical Soc., Supp. 7, no. 2, p. 155–161.

Griffiths, J. C., 1966, Exploration for natural resources: Jour. Operations Research Soc. Am., v. 14, p. 189–209.

Gumbel, E. J., 1958, Statistics of extremes: New York, Columbia Univ. Press, 375 p.

Haldane, J. B. S., 1941, The fitting of binomial distributions: Annals of Eugenics, v. 11, p. 179–181.

Hoel, P. G., 1962, Introduction to mathematical statistics, 3rd ed.: New York, John Wiley & Sons, 427 p.

Koch, G. S., Jr., and Link, R. F., 1967, Geometry of metal distribution in five veins of the Fresnillo mine, Zacatecas, Mexico: U.S. Bur. Mines Rept. Inv. 6919, 64 p.

Krige, D. G., 1960, On the departure of ore value distributions from the lognormal model in South African gold mines: Jour. South African Inst. Mining Metall., v. 61, p. 231–244.

————, Developments in the valuation of gold mining properties from borehole results: 7th Commonwealth Mining Metall. Cong., v. 1, p. 18.

Li, J. C. R., 1964, Statistical inference, v. 1, Ann Arbor, Mich., Edwards Bros., 658 p.

Link, R. F., Koch, G. S., Jr., and Schuenemeyer, J. H., 1970, The lognormal frequency distribution in relation to gold assay data: U.S. Bur. Mines Rept. Inv., in press.

Mason, Brian, 1962, Meteorites: New York, John Wiley & Sons, 274 p.

Mood, A. M., 1950, Introduction to the theory of statistics: New York, McGraw-Hill, 433 p.

Mosteller, Frederick, and Tukey, J. W., 1949, The uses and usefulness of binomial probability paper: Jour. Am. Statistical Assoc., v. 44, p. 174–212.

Pettijohn, F. J., 1957, Sedimentary rocks: New York, Harper & Bros., 690 p.

Sichel, H. S., and Rowland, R. S., 1961, Recent advances in mine sampling and underground valuation practice in South African gold fields: Trans. 7th Commonwealth Mining Metall. Cong., p. 1–21.

Sichel, H. S., 1966, The estimation of means and associated confidence limits for small samples from lognormal populations, *in* Symposium on mathematical statistics and computer applications in ore valuation: South African Inst. Mining Metall., p. 106–122.

Slichter, L. B., 1960, The need of a new philosophy of prospecting: Mining Eng., v. 12, p. 570–576.

Stone, J. B., and McCarthy, J. C., 1948, Mineral and metal variations in the veins of Fresnillo, Zacatecas, Mexico: Am. Inst. Mining Metall. Engineers Trans., v. 178, p. 91–106.

Tukey, J. W., 1962, The future of data analysis: Annals of Math. Statistics, v. 33, p. p. 1–67.

U.S. Geological Survey, 1952, Floods in Youghiogheny and Kiskminetas river basins, Pennsylvania and Maryland, frequency and magnitude: U.S. Geol. Survey Circ. 204, 22 p.

Wilks, S. S., 1963, Statistical inference in geology, *in* Donnelly, T. W., ed., The earth sciences, problems and progress in current research: Univ. of Chicago Press, p. 105–136.

PART III

SAMPLING AND VARIABILITY
IN GEOLOGY

In Chapters 7 and 8, making up Part III of this book, the univariate statistical methods explained in Part II are applied to geological data.

In Chapter 7 general principles of geological sampling are discussed; and ways of interpreting data derived from several different techniques of sampling, such as diamond-drill-hole and hand-specimen sampling, are examined. The sampling techniques are those most commonly used by geologists; some more specialized methods are discussed in Chapter 15. In the final section of Chapter 7, the range of geological variability of various substances in different rocks is reviewed.

Variability in geological data is the subject of Chapter 8. Different kinds of variability—including that introduced by sampling, sample preparation, and chemical analysis, as well as the inherent natural variability—are appraised.

Chapter 7

Geological Sampling

Having introduced fundamental statistical procedures for data analysis in Chapters 2 to 6, we can now focus our attention on geological problems to a greater extent. The discussion in this chapter is on several methods of geological sampling, starting with collection of hand specimens, and relates them to the statistical concepts, especially that of variability, discussed previously. That the sampling of rocks is not easy is abundantly illustrated by mines that never paid because the grade of ore estimated by sampling was higher than the grade actually mined.

It is convenient to distinguish four kinds of variability in geological data: (a) natural variability, that inherent in the rock body or other geological object being sampled; (b) sampling variability, that introduced by the physical sampling process; (c) preparation variability, that introduced in readying the geological sample for chemical analysis by crushing, splitting, etc.; and (d) analytical variability, that introduced by the chemical or physical determination of substances in the geological samples. This chapter is about the first two kinds of variability, and Chapter 8 is about the other two kinds.

7.1 INTRODUCTION

A geological sample (sec. 3.1) is taken to obtain information about a rock body for a particular purpose. The rock body, whether it be a formation, outcrop, cylinder of rock removed by drill core or cuttings, or whatever else, should first be *classified* (*stratified*) in all the geologically meaningful ways that one can think of. Then, one or more geological samples of the rock body should be obtained to satisfy the following requirements:

1. The geological samples are selected with at least one element of randomness introduced in the sampling process.
2. None, or as little as possible, of any sample is lost during collection.
3. The most inexpensive feasible methods are used to sample.
4. Some way is available to remove the sample from the field to the laboratory or other place in which the sample is to be studied.

Section 7.2 is a discussion of geological factors that affect sampling, and sections 7.3 to 7.7 are about some specific methods of geological sampling. The purpose is not to treat the technology of sampling but rather to discuss statistical analysis of sampling data. Therefore, representative real data are considered, but no attempt is made at an exhaustive examination. Finally, in section 7.8, distributions of coefficients of variation according to types of geology and classes of constituents are discussed. First, some sampling problems that may be even more familiar to the reader than those of rocks are sketched to set the stage for the geological discussion.

A Few Sampling Problems from Disciplines Other Than Geology

In this section we sketch four statistical problems from disciplines other than geology in order to convey some understanding of how statisticians attack real problems. Just as in field mapping a contact between two formations, dating a fossil assemblage, or classifying a rock, the textbook rules are not always followed, so in statistics one must choose, improvise, and be arbitrary. For further reading, a book by Wallis and Roberts (1956), who give several hundred examples, is excellent.

Clearly defined purposes, necessary in order to obtain valid results, are present in these four problems. In the first two, prediction of elections and evaluation of biomedical research, the target populations are unmistakable, and the relations between the target and the sampled populations can be clearly specified. In the other two problems, the relation of smoking to lung cancer and the cause of polio, the target populations are hazy, and their relations to the sampled populations are obscure. The first two problems allow prospective studies to be made because most of the data are gathered after the problems are conceived; the second two problems have retrospective elements because some or all of the data are already at hand. Geological problems parallel to these different combinations are pointed out in the discussion.

Prediction of Elections

For the last few years, the American television networks have competed energetically during nights of the general elections to predict the winning candidates and their shares of the total vote. This frantic activity, live on prime time, is intended to entertain the public and to impress them with the mastery of network men over electronic computers. Their achievements and their mistakes are particularly entertaining to those familiar with statistics and computing. A brief examination of election predicting demonstrates the judicious blending of many elements that is required whenever statisticians work with real data. Ingenuity is required, as well as a knowledge of statistics and the subject matter.

In election prediction both the purpose and the target population are well defined. The purpose is to make an accurate prediction as early as possible, and preferably earlier than another network. The target population consists of the votes cast in that particular election, and the sampled population is those votes, less any that are lost by voting-machine malfunctions, that are stolen by dishonest election judges, etc. Alternatively, the target and sampled populations may be said to coincide if one takes the viewpoint that only the votes that are counted matter. The sample, which consists of early returns, clearly is not a random one, as some precincts consistently provide early returns election after election; and some voters, such as retired people, consistently vote early. But one can pretend that the sample behaves essentially as though it were random, as must be done time and again, more often than not, in geological sampling.

Three kinds of information go into an election prediction: (a) prior information from opinion polls, straws in the wind, conjectures of knowledgeable people, etc.; (b) current information on returns from "key" precincts, which are those chosen because past results indicate that their voting patterns can be more readily interpreted than those of other precincts; and (c) current information derived from the total vote, broken down by states and sometimes into finer subdivisions, usually by county. These three categories of information are combined in a mathematical model, and once the returns begin to supply the two latter kinds of information, decision rules are used either to predict the winner or to conclude that not enough returns are yet in to do so. As the returns come in, these rules are continually applied until all the races being considered are predicted.

The most important element in the mathematical model is the returns from the key precincts, which are chosen judiciously to reflect the electorate. In key precincts such factors are desirable as stable boundaries and populations rather than the changing boundaries and itinerant populations that are found in areas being changed from single-family to multiple-family dwellings.

Once chosen, the statistical properties of the key precincts, obtained from their performance in past elections, are estimated to forecast their behavior in the current election. Two characteristics of the voters must be established, their average behavior and their variability; that is, how they ordinarily vote and how consistently they vote. Once these characteristics are determined by statistical analysis of the historical data, rules can be devised for announcing a winner. The third element, the total vote, enters into the mathematical model mainly in close elections, where the fragmentary results from the early returns in the key precincts are insufficient on which to base a decision.

In summary, election predicting depends upon combining a knowledge of both subject matter and statistics. In order to apply the statistics, many formal assumptions must be ignored, including those about random samples and normal distributions, but experience shows that the applications are successful. The situation is like that in geological sampling where statistical assumptions seldom are fully satisfied. The purposeful selection of key precincts is similar to the geologist's intuitive decisions based on knowledge of the science, perhaps to look for fossils from certain localities where he thinks that they may be found. The continual data analysis is similar to the plan of collecting fossils until enough are found to date a formation. The main difference is that time on election night is crucial, as a prediction is valuable only if made within minutes, but even here one may draw an analogy to the exploration of a prospect by a company that wants to acquire the land as soon as, but no sooner than, it learns that the prospect will make a valuable mine.

Financial Support of Health Programs

In 1964 the National Institutes of Health (NIH) spent about one billion dollars in direct financial support of the health research of the United States. Most of this money was spent on twenty thousand individual projects at more than one thousand universities and medical schools. This system was studied to learn if the American people were getting their money's worth and to see if any changes would increase the program's effectiveness (National Institutes of Health, 1965).

Teams of scientists and administrators visited projects to evaluate the quality of research and the effectiveness of the administrative support. The key to scheduling these visits was an effective sampling scheme to allow the teams to inspect enough projects without spending an exorbitant amount of money, but also to assure that the structure of the NIH was thoroughly evaluated. Because simple random sampling would have required too much time and travel, the statisticians had to devise a sampling scheme to keep time and travel down to a reasonable level.

The statisticians determined that the requirements would be met if teams of six to eight experts spent two or three days each at 25 or 30 institutions and inspected 8 to 10 projects per institution. The institutions were selected as a stratified random sample (Cochran, 1963), stratified first by geography, and second by the amount of NIH dollars contracted to that institution. The number of institutions to be visited in any area was set proportional to NIH dollars for the area, so that those with $2 in support had twice the chance of being visited as those with $1 of support. Because small institutions might have only one or two NIH projects, geographically contiguous small institutions were grouped for sampling into a "single" institution. After an institution was selected the projects to be visited at it were determined in a similar fashion.

Applying this plan led to a sampling scheme with many subjective and arbitrary elements. However, the basic procedure was a random sample taken within constraints to yield an effective sample. The nonrandom elements permitted sampling efficiency, and the random elements ensured the basic integrity of the sample. This solution demonstrates that simple random sampling is seldom the most economical or efficient way to sample, especially if the target population, the objectives, and the resources for sampling are well known.

Smoking and Lung Cancer

One of the controversial topics of our time is whether smoking causes lung cancer. For the last few years both the death rate from lung cancer and cigarette consumption have risen dramatically. Lung cancer has changed from a relatively rare disease to a common one, and the fatality rate from lung cancer is now about equal to that from automobile accidents, which is about 40,000 deaths a year in the United States. There is no doubt that both cigarette smoking and the death rate from lung cancer have increased. The controversy arises from the relationship, if any, between these events. The death rate from automobile accidents has also risen spectacularly since the turn of the century, but no one would blame cigarette smoking for this. Also, other chemical alterations to the environment of our industrial society have risen during the same period.

The studies that bear on this issue have been of two types. Retrospective studies try to match people who have contracted or died from lung cancer with otherwise-alike people with no history of the disease. These studies show that people with lung cancer smoke or smoked more than those without the disease. Prospective studies have been devised to compare the incidence of cancer and the death rates of smokers and nonsmokers. These studies also show higher death rates from lung cancer for smokers than for nonsmokers. Both types of study have been attacked as lacking the proper controls to define properly the populations under study; for example, in one study the health of both the tested and the control groups was much better than that of the general population even though the smokers did have more lung cancer.

Of the many difficulties in studies of smoking, four may be stressed. First, target and sampled populations with which to work are difficult to define. Second, whatever happens takes a long time; lung cancer takes decades to develop, whether or not it is caused by smoking. Third, the effects are confused by other events: a potential cancer victim can be killed accidentally, and most smokers are unhealthier and have higher death rates than other people. Fourth, the cause of lung cancer may be the same factor as whatever induces people to smoke, perhaps a common genetic characteristic that tends to make people smoke and also raises their chance of getting lung cancer.

For this problem and for many others like it an answer is wanted today. Much is at stake socially and economically, but, unfortunately, the unequivocal results will require years to obtain; hence the controversy over the evidence and the conclusions that have so far been reached. The example is an excellent one of how economic interest and rationalizations can sway opinion in both directions. Useful conclusions can already be drawn, but until careful experiments are completed, airtight results cannot be expected.

1954 Polio Vaccine Trials

The 1954 polio vaccine trials were one of the largest medical experiments ever conducted. A brief account may be of interest because many geological experiments are of similar size—comprehensive petrological and paleontological studies on a continent-wide or world scale, analyses of data from one or several oil fields, or sampling of large ore deposits. And these experiments may have the further similarity that a rare occurrence is investigated: polio in the medical experiment, and perhaps oil or ore in the geological experiment. Our discussion and a full account of the vaccine trials by Brownlee (1955) show how complex the application of statistics to a real problem usually is.

Participants in the vaccine trials included 312 state and local health officials, 54 physical therapists, 22 epidemiological intelligence officers, scientists in 28 laboratories, a 17-member advisory committee, countless teachers and physicians, and 1,829,916 children. The purely administrative problems were themselves immense and probably contributed to the imperfection of the scientific work.

Much of the experiment consisted of giving polio vaccine to second-grade children, with first- and third-grade children serving as control groups. This procedure was clearly unsatisfactory because there is no reason to believe that the risk of getting polio was equal for each of the three groups. However, for about 40 percent of the trial, the children of the first three grades were combined. One-half of them, selected at random, received vaccine and the other half, a solution that has no effect on immunity (a placebo). Only the national evaluation center knew which children received the vaccine and which received the placebo.

Some statisticians would say that this part of the trial was the only part with any scientific validity. Out of the total number of 400,000 randomly selected children of mixed grade levels, there were 33 cases of paralytic polio in the group which received the vaccine and 115 cases in the group which did not receive it. Assuming a Poisson distribution (sec. 6.1), it is clear that the incidence of paralytic polio was less in the group which received the vaccine. Unfortunately, there is some evidence that even with the randomized sample the group which received the vaccine had less propensity toward contracting polio (Brownlee, 1955).

Today the effectiveness of polio vaccine is well established, although the type of vaccine now used is much safer and its method of administration more simple than the vaccine used in the trials. It is interesting to note that, despite an intensive effort to conduct scientifically valid tests, only a small part of the work would stand up to criticism. How this state of affairs came about is easy to surmise. The administrative difficulties were so large, the pressure to get on with the trials so great, and the responsibility for the work so diffuse that there cannot have been time enough to think through the experiment before it was begun.

An interesting aftermath to these trials is that once they had concluded and the vaccine was released for public use, it became clear that some of the vaccine was itself causing polio rather than preventing it. Evidently at least some of the vaccine produced commercially was different from that used in the trials. The point is that the trials did not establish that the vaccine could be made commercially in large quantities or establish the safety standards that should be used in its manufacture and in quality control.

The complications that arose in testing polio vaccine were caused primarily by the rarity of polio as a disease—its incidence being about one in a thousand —and the fact that the number of cases varies widely from season to season. Moreover, the vaccine was imperfect and evidently effective only about three-fourths of the time. The combined consequences of the rarity of the disease and the imperfectness of the vaccine meant that an extremely large trial was needed, although the enormousness of the trial actually made was probably not necessary.

For instance—assuming an incidence of only 0.5 per thousand and a 50 percent efficiency of the vaccine—if a total of only 200,000 children had participated in the trial, 100,000 of whom had received the vaccine, the standard deviation between the mean assuming no effect and a 50-percent effect would be about 3. Fifty children on the average would get polio in the untreated group and 25 children would get polio in the treated group. The standard deviation of the difference would be $\sqrt{75} = 8.5$ and $(50 - 25)/8.5 = 2.9$. On the other hand, if it were argued that the vaccine would be worthwhile even if it were only 10 percent effective, a large trial *would* have been necessary.

In summary, this and the other examples show above all that in sampling problems, as in most things, it pays to plan ahead.

7.2   GEOLOGICAL FACTORS THAT AFFECT SAMPLING

The geological factors that affect sampling depend on the goal of the investigator, who must ask first what is the purpose of sampling and then what area should be sampled, what items should be sampled, and finally, what attributes of these items should be sampled. From the answer to these questions can be formulated a *sampling plan*, which consists of the procedures to be followed.

What is the purpose of sampling? It may simply be discovery (Chap. 12), perhaps hunting for a needle in a haystack—the aim being the most efficient possible coverage of an area with the resources available. Another objective is to estimate variability, for instance, comparing rocks to learn if their variability is the same, a circumstance which might suggest that their geology has something in common. This subject is raised by a comparison of coefficients of variation in section 7.8 and is further considered for gold deposits in Chapter 16. However, perhaps the most usual aim of sampling in geology is to estimate the mean value of one or more constituents of a rock.

Unless a geologist carefully thinks out and specifies his goal in sampling, he should expect nothing but trouble. In economic problems, for instance, mine sampling, specifying an objective is much easier than in academic problems. Once a purpose is defined a sampling plan should be devised and followed, at least at first. If it turns out to be unsatisfactory, it can be revised, but each change introduces new subjective elements, difficult to evaluate statistically.

What area should be sampled? Although this question is liable to receive an arbitrary answer, influenced by travel costs, state and national boundaries, or location of an interesting rock body too high in the mountains or too deep in the earth, it is one that deserves careful thought. To promptly find small bodies of mineable chromite, one would be inclined to search in ultramafic rocks; but to play a long shot, one might decide to search for less well-known sources of chromite in the hope of finding the mineral in another environment. On the other hand, for estimating chromite abundance in rocks of all types, enough information on its abundance in ultramafic rocks may be at hand, and all sampling might be concentrated in other lithologies. For other problems similar questions arise; one will suffice to make the point: In a study of granite tectonics, is more learned from a reconnaissance of the Sierra Nevada or from a detailed map of a Maryland quarry? Statistics can aid in thinking through such questions from the stage of sampling to the final data analysis.

What items and what attributes should be sampled? These questions are simpler to answer than the others. Often the entire rock is to be sampled to determine rock chemistry or rock physics. To determine rock chemistry, whether expressed in chemical elements, oxides, minerals, or otherwise, and whether completely or for only one or a few constituents, is the most usual purpose for sampling and the only one discussed in detail in this chapter. In order to determine rock physics; properties such as strength, hardness, rate of transmission of seismic and other waves, and paleomagnetism, may be studied. Besides examining chemical and physical properties of rocks, one samples many other items—fossils and structural properties such as strike and dip and current directions.

We can think of no better way to summarize these introductory remarks than by quoting from Griffiths (1962, p. 604), who writes the following remarks about sedimentary petrology that apply to the whole science of geology:

> Any scientific investigation is no better than its sampling plan; inadequate sampling cannot be subsequently offset by any procedure, experimental or statistical. The problem of sampling arises in the initial stages of an investigation when setting up the most efficient means of achieving the main objective of the experimental program, and it crops up again at various stages throughout the experiment in attaining required levels of precision of estimates from different measuring techniques. Because of its fundamental role in experimentation, the sampling pattern should be decided upon at the same time as the overall strategy of the program, i.e., at the beginning; generally in sedimentary petrography it is resolved as the experimenter becomes aware of it, a certainly inefficient and, possibly, disastrous practice.

Estimation of the Mean

Because estimation of the mean is the principal purpose of most geological sampling, the basic statistical concepts introduced previously are reviewed herewith. Precision of an estimate based on independent observations depends upon the fundamental law about the standard error of the mean, $\sqrt{\sigma^2/n}$, where the variance σ^2 is the *total* variance from all of the sources differentiated in the introduction to this chapter. In order to make the standard error of the mean small, only two ways are possible, as explained in the discussion of confidence intervals (sec. 4.4): either reduce the variance σ^2 or increase the number of observations n.

The largest reduction in the standard error of the mean almost always is afforded by reducing variability through taking advantage of the natural stratification of the rock body sampled, as is discussed throughout this chapter. The importance of deciphering the stratification, if it is not obvious, and then devising an appropriate sampling plan cannot be over-emphasized. The standard error of the mean can also be reduced by better sampling, better sample preparation (sec. 8.4), or better physical or chemical analysis (sec. 8.5).

The second way to reduce the standard error of the mean is to increase n, although the advantage is not commensurate with the increase in effort after a few tens of observations (sec. 4.4). Even though σ^2 may be larger, several observations by an inexpensive method may provide a more accurate estimate than one or a few expensive observations; for instance, several inexpensive rotary-drill holes may yield more information than one expensive diamond-drill hole. This concept, discussed in sections 7.3 to 7.8 and in Chapter 15, demands emphasis because many geologists do not recognize that there may be merit in deliberately accepting an increase in variability (or sloppiness, in plain words) if then more observations can be afforded.

Origin of Natural Variability

The natural variability of substances in rocks plays an important part in planning rock sampling, and this variability depends fundamentally on the geologic processes that formed the rocks. However, although a knowledge of rock genesis may someday enable the variability of a rock to be predicted, this time has not yet come.

One can speculate that, as a general rule, rocks formed at high temperatures should be less variable than those formed at low temperatures, provided that equilibrium has been attained. This generalization is supported by two excellent studies of major and minor elements in basalts by Manson (1967) and Prinz (1967). In rocks formed at lower temperatures, such as sedimentary ones, variability increases, as a rule. However, rocks formed at high temperatures are less likely to have attained equilibrium, as shown by hydrothermal veins and by most igneous extrusive rocks other than basalt flows.

Barth (1962) made the same point when he observed that sedimentary rocks are differentiated into extreme compositional types, thus being, *as a group*, more variable than igneous rocks. However, individual sedimentary rock types, such as mature sandstones, may be extremely uniform in composition. Therefore, for sampling, the most that can be said is that, as a class, igneous rocks formed at high temperatures should have a better understood variability than sedimentary rocks about which information has not yet been obtained. Mason (1962) also discusses this topic.

The variability of rocks also depends on variability in the composition of the constituent minerals. We may cite some examples, but they do not suggest general rules to us. Chemical composition of garnet is highly variable, but in a given rock the limited evidence suggests that composition is rather uniform or varies gradually from place to place (Engel and Engel, 1960). Silver content of native gold in different placer deposits is also highly variable, but in a given placer it is uniform or changes only gradually and regularly (sec. 15.2). Chemical composition of quartz is practically invariant, all quartz being composed of nearly pure SiO_2.

TABLE 7.1. TENTATIVE CLASSIFICATION OF SOME ROCK TYPES
ACCORDING TO THEIR UNIFORMITY FOR SAMPLING

Uniform rocks	Nonuniform rocks
Igneous rocks	
Basalt	Andesite
Diabase	Diorite
Gabbro	Rhyolite
Obsidian	Granodiorite
Anorthosite	Granite
	Most pyroclastic rocks
Metamorphic rocks	
Marble	Hornfels
Quartzite	Mylonites
Eclogite	Slate
	Schist
	Gneiss
Sedimentary rocks	
Limestone	Gravel and conglomerate
Chert	Shale
Diatomite	Marl
Arenite	Wacke
	Dolomite
	Glacial till
	Rocks of intermediate composition, e.g., calcareous sandstone, arenaceous limestone

Thus table 7.1 classifies some rock types according to their uniformity. This table suggests that process of formation is related to sampling, but not in a clear way. A comprehensive petrologic study should generate insights into this subject.

Configuration of the Rock Body to be Sampled

Besides natural variability, the main geologic factor affecting sampling is the configuration of the rock body to be sampled, that is, its shape and position in the ground. It is convenient to distinguish large-scale structure or macrostructure, from small-scale structure, or microstructure, on the scale of a hand specimen; these are discussed in turn with the aid of a few familiar geologic examples. More important for developing a sampling plan than rock history or genesis is rock geometry. Sampling methods explained in this chapter and sampling designs taken up in section 8.6 should be planned to take account of it.

Macrostructures may be tabular or linear in shape and may be concordant or discordant to regional geologic structures and stratigraphy. Tabular and concordant rock bodies include sedimentary beds, pyroclastic beds, lava flows, sills, and the metamorphic equivalents of all of them. The entire rock body may be sampled, as in a petrological study; or only part may be of interest, as in an oil trap or a manto ore deposit.

Most tabular but discordant rock bodies are igneous or clastic dikes and veins. Again, the entire discordant rock body may be of interest; but, especially for veins, sampling may be confined to or concentrated in particular parts of a vein where it intersects another vein or a favorable bed or where it changes in attitude.

Linear or lathlike rock bodies formed in sedimentary rocks at the time of deposition—bodies that are concordant to regional structure and stratigraphy —include shoestring sands and organic reefs, which localize petroleum in many places, and in some places localize ore bodies such as uranium on the Colorado Plateau and lead in southeast Missouri. Formed secondarily, but also essentially concordant, are crests of anticlines, which also localize petroleum and ore bodies such as the gold at Bendigo, Australia; troughs of synclines, which contain zinc ore at Franklin, New Jersey; and sheared-off limbs of folds which localize iron ore at Dover, New Jersey.

A few linear rock bodies are discordant to regional structure—the principal types being volcanic pipes, such as that at Cripple Creek, Colorado, containing gold ore; that at Braden, Chile, containing copper ore; salt domes, which may localize salt and oil; and chimneys of limestone, such as those localizing lead, zinc ores at Santa Eulalia, Mexico.

The position of either tabular or linear bodies relative to the surface of the earth or underground workings affects the physical method of sampling used and also the design of a sampling plan. Tabular bodies may be planar or folded; linear bodies may be straight or curved; and either may be horizontal or inclined. One illustration of the kind of problem encountered is enough. To sample a horizontal coal bed deep below the surface, expensive drilling may be the only feasible way. On the other hand, if the bed is inclined and crops out in the area of interest, sampling the outcrop may be inexpensive. Then a good sampling plan will evaluate the information from the outcrop so that as few holes as possible need be drilled where the inclination takes the bed far below the surface. This scheme depends on the coal bed being uniform laterally. These comments on macrostructures emphasize the importance of making a specific definition of the target population (sec. 3.2).

As used here, microstructure consists of particular size and shape, and changes in these variables from place to place. In at least two instances, microstructure is such that large volumes of rock must be sampled. First, if a rock is coarse-grained, perhaps a conglomerate or a porphyry, large volumes are needed (sec. 7.6), an extreme case being the sampling of a pegmatite (Norton and Page, 1956). Second, if the constituent of interest is sporadically distributed, large volumes of rock must be taken, as in diamond sampling. Thus the variability on the scale of the microstructure may be the dominant source of the natural variability, or it may be small compared with the variability on the scale of the macrostructure.

Sampling Methods to Take Account of Natural Variability

When the geological factors that affect sampling are known, sampling methods to take account of them can be devised. The sampling methods depend on previous factual knowledge about the variability, on factual knowledge gained from preliminary sampling, on the variability expected from geological experience in similar rocks, or on an interpretation of the pattern of variability to be expected from geological theory. Once something is known about the pattern of natural variability, a sampling plan can be devised incorporating a sampling design (sec. 8.6) that specifies the number and pattern of observations and the statistical methods for analyzing the observations.

Two important kinds of sampling are *stratified sampling* and *pattern sampling*. Stratified sampling is always desirable if a meaningful way to do it can be found because the larger the body of rock, the larger the variance, other things being equal. A few examples of stratified sampling follow. Sampling may be stratified based on the bedding of bedded rocks; for example, in sampling gold in the Homestake mine, one takes advantage of the fact that most gold is known to be confined to the Homestake Formation. Therefore, even if one's interest is in the gold in the several rock formations for the purpose of an academic study rather than for mining, sampling should be concentrated in the Homestake Formation because the larger amount of gold there is associated with a higher variability that is more difficult to estimate. Similarly, in sampling for oil, most attention may be concentrated in known reservoir rocks and little in other rocks. In sampling intrusive igneous rocks, stratification might be based on rock type to avoid such a problem as collecting a mixture of zircons from different rock types and thereby producing a multimodal distribution that would be difficult if not impossible to interpret. In sampling metamorphic rocks, stratification should be based on rock types defined by metamorphic zones as well as by original lithologies—for instance, quartz formed by hydrothermal alteration should be distinguished from primary quartz.

The second kind of sampling, pattern sampling, which is discussed in detail in section 8.6, is touched on here because stratified and pattern sampling are almost always combined in practice. As an example, if a well-defined linear structure, such as a shoestring sand, a reef, or a linear outcrop belt, is present, a desirable pattern often would be to traverse across the structure, with sampling within each traverse relatively closely spaced compared with the distances from one traverse to another.

Finally, if natural variability displays a trend recognizable by trend-surface analysis (Chap. 9), the sampling may be stratified to take advantage of the trend.

7.3   HAND-SPECIMEN SAMPLING OF OUTCROPS

Undoubtedly most geological samples, excluding those taken by mining and petroleum geologists, are hand specimens from outcrops. Many fundamental geological theories and conclusions rest on information gained from study of these specimens, and it is therefore surprising that little attention has been paid to the validity of hand-specimen sampling of outcrops. In this section some representative studies of outcrop sampling are first reviewed; then, important work by a group of Pomona College geologists (who compare sampling of outcrops by hand specimens with sampling by the best feasible alternative so far proposed, a hand-carried portable drill) is considered in some detail.

The properties of rock outcrops are seldom like those of the unexposed part of the same rock body. It is well known that, except for accidents such as glaciation, the outcrops of ore bodies are unlike the rocks below the zone of weathering, and a large amount of literature (McKinstry, 1948, pp. 242–276) explains how to determine from the outcrop the mineralogy of the ore body at depth. Studies of rocks other than ore bodies are seldom detailed enough to provide general principles about the relation between outcrops and unexposed rocks near the surface and at greater depth. For rocks formed at surface temperatures and pressures, the changes should not be too severe, at least not as much as those of ore bodies. Because of this unsatisfactory state of knowledge, everything that follows in this section refers to a population of outcrops rather than of rock bodies, as would be preferred if the information were available.

In hand-specimen sampling of outcrops a great amount of bias can be introduced. The human tendency to collect oddities is revealed by the usual rock collection, which contains more strange than common rocks. Figure 7.1 is an example, which Ager (1963, p. 187) believes to be more the rule than the exception, of two histograms picturing proportions of mollusks obtained by bulk sampling and by hand picking from the surface. Ager writes that "the species overrepresented in the hand-picked samples are, in every case, forms

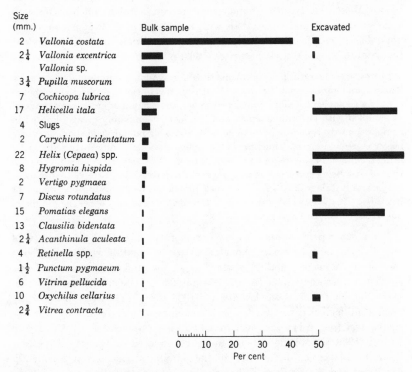

Fig. 7.1. Comparative histograms for quaternary mollusks obtained by bulk sampling (*left*) and by hand picking from the surface (*right*); Arreton Down, Isle of Wight, England (after Ager, 1963, p. 187).

which are conspicuous by reason of size or color . . . paleontologists themselves
add to all their natural difficulties by their subjective and unscientific modes
of collecting, and generally speaking, do not seem to realize the huge subjective errors which come into these normal collecting methods." His remarks
apply all too truly to geology in general as well as to paleontology.

Not only do exposed rocks differ from unexposed ones, but an outcrop
itself is not uniform. Thus, an outcrop may have only a few places where it is
easy to break out a specimen; for instance, most rocks break along joints, so
that the tendency to get specimens with more than the true proportion of
joint surfaces may be disastrous if the specimens are coated with calcite and
the lime content of the rock is under study.

Some Representative Works on Hand-Specimen Sampling

Books on field geology contain instructions on how to take conventional
hand specimens from outcrops; for instance, Compton (1962) gives good
advice. Miesch believes that hand specimens can be taken much better with
a sledge hammer and chisel than with a geologic pick. Relying on as yet
unpublished data from an extensive study of Cambrian outcrops throughout
the western United States, he writes (1967, p. A-3):

In most geochemical field problems the only available sampling devices are a
hammer and chisel or similar tools. (Most rock specimens are collected by means
of only a geologic pick and hammer. However, the portion of an outcrop that can
be sampled with an ordinary pick and hammer is commonly greatly limited in
comparison with that which can be sampled using a 2-pound sledge hammer and a
heavy steel cold chisel. Thus, bias in sampling may be significantly reduced by
using a chisel for sampling many types of rock.) Consequently, depending on the
type of rock being studied, not all parts of the outcrop can be sampled. Specimens
are generally taken only where the rock protrudes, as along bedding planes or
fractures. This is especially true where the rock is hard and dense, as are many
quartzites, dolomites and granites.

If an outcrop is sampled by two or more people, their consistency of
sampling can be appraised, although clearly it will matter whether they visit
the outcrop alone or together, whether they are a professor and his students,
etc. Thus Ager and Guber (Ager, 1963, pp. 240–241) counted species represented by 850 fossils at one locality and 1000 at another in Iowa for a study
of paleontological diversity. Table 7.2 compares their results by the chi-
square test (sec. 6.1) as applied to a two-way classification by Dixon and
Massey (1969, p. 240). Because the tabled chi-squared value at the 10-percent
level with 16 degrees of freedom is 25, the sampling at Rockford was evidently
consistent and that at Bird Hill was mildly inconsistent. Similarly, Griffiths
(1967, Chap. 2) and his students have studied consistency in sampling
sandstones.

TABLE 7.2. COUNTS OF FOSSIL SPECIES AT THE ROCKFORD AND BIRD HILL LOCALITIES, IOWA*

Rockford†			Bird Hill‡		
Name	Number of specimens		Name	Number of specimens	
	Ager	Guber		Ager	Guber
Spirifer hungerfordi Hall	156	83	*Productella walcotti* Fenton and Fenton	146	33
Atrypa devoniana Webster	139	103	*Douvillina arcuata* (Hall)	135	50
Schizophoria iowaensis (Hall)	57	31	*Atrypa rockfordensis* Fenton and Fenton	103	35
Spirifer whitneyi Hall	23	20	*Spirifer whitneyi* Hall	73	23
Spirifer whitneyi subsidus Fenton and Fenton	22	15	*Schizophoria iowaensis* (Hall)	67	29
Atrypa rockfordensis Fenton and Fenton	21	11	*Heliophyllum solidum* (Hall and Whitfield)	43	20
Douvillina arcuata (Hall)	15	14	*Spirifer hungerfordi* Hall	38	10
Floydia gigantea (Hall and Whitfield)	15	3	*Strophonella hybrida* (Hall and Whitfield)	27	4
Heliophyllum solidum (Hall and Whitfield)	13	11	*Spirifer whitneyi subsidus* Fenton and Fenton	25	8
Productella walcotti Fenton and Fenton	12	8	Undetermined stick bryozoans	22	8
Diaphorostoma antiquuum Webster	8	4	*Cranaenella navicalla* (Hall)	13	2
Petalotrypa formosa Fenton and Fenton	5	10	*Charactophyllum nanum* (Hall and Whitfield)	12	2
Strophonella reversa (Hall)	5	2	*Strophonella reversa* (Hall)	10	13
Lioclema occidens (Hall and Whitfield)	4	7	*Atrypa planosulcata* Webster	8	4
Paracyclas sabini White	4	1	Crinoid fragments	6	2
Worm tubes	4	1	*Platyrachella cyrtinaformis* (Hall and Whitfield)	5	0
Miscellaneous	13	10	Miscellaneous	19	15
TOTAL	516	334	TOTAL	752	248

* After Ager, 1963, p. 240–241.
† χ^2 (16 d.f.) = 20.
‡ χ^2 (16 d.f.) = 27.

Grout (1932) studied replicate sampling of granite, banded gneiss, and a large differentiated diabase sill. Some sampling was done by taking hand specimens judged as representative by geologists experienced in the particular rock bodies and some by taking chips from scattered points on the outcrops. Grout's tables are interesting, although not enough data are given for a statistical analysis; they reveal the state of thinking that prevailed among geologists until recently: one of worry and fretfulness—but not enough worry to do anything about it.

Outcrop Sampling by Pomona College Geologists

For several years McIntyre, Baird, Welday, and Richmond, working at Pomona College, have been intensively studying igneous rocks in the San Bernardino Mountains in Southern California, primarily to investigate chemical variation in igneous rocks in much the same way as we study chemical variation in veins of the Frisco and Fresnillo mines. Their work, all of which to date is referenced by Baird and others (1967), provides a fascinating account of the sources of variability in igneous rocks; several points are discussed here.

Richmond (1965) compared outcrop sampling by hand specimens and by diamond drill. Cactus Flat, the area investigated, is part of a larger area studied by the Pomona group in the San Bernardino Mountains and is the type locality for the Cactus quartz monzonite, the rock unit sampled. Outcrops cover about two-thirds of the area.

Although any drill could be used in principle, in practice, when samples are being taken away from roads, a readily portable one is the only one likely to compete with hand sampling. The drill used by Richmond weighs 26 pounds, not counting drill rod, core barrel, water, and tools (manufactured by Packsack Diamond Drills, Ltd., 1385 Hammond St., North Bay, Ontario, Canada). Compared with a conventional diamond drill requiring one or two trucks for transportation, this drill is truly portable; drills of intermediate sizes that can be carried by animals or helicopters are also available.

Richmond (1965, p. 54) describes his sampling plan as follows:

Most of Cactus Flat lies between 5750- and 5900-foot elevation contours, and a tracing with boundaries defined by these conditions was made on graph paper. . . . The tracing included 360 squares, each covering a map area of 200 × 200 feet. Each square was assigned a number, and numbers were drawn at random to determine the order of sampling. The center of each 200-foot square designated for sampling was occupied. Where no outcrops were seen to occur, another 200-foot square was occupied, and it was necessary to examine 46 squares in order to complete the planned sampling of 30 squares.

Sample selection was further randomized within 200-foot squares where outcrops were observed. The location of two 20-foot squares was determined by random numbers drawn from a box of tags numbered 1 to 100, and each small square was then sampled, A, by a 10-inch core (1-inch diameter) obtained by diamond drilling, and B, by a hand specimen collected by hammer and chisel within 5 feet of the core hole. In all, 120 specimens were collected, made up of two samples of 60 specimens each, collected two from each of thirty randomly selected 200-foot squares, and paired by collecting method within a 5-foot radius within squares.

Richmond kindly supplied his original data so that his results could be recomputed in one of the analysis-of-variance formats explained in this book. The results for magnesia, the only one of the six constituents for which a difference was found between the two methods of sampling, are presented here. The analysis of variance, table 7.3, partitions the variability into that associated with the different locations and sampling methods and that which is residual. The single degree of freedom comparing hammer and drill sampling was also broken out. The analysis of variance is a randomized-block design with the 30 replications due to location and sampling methods as treatments. Because the computed F-value of 6.57 comparing the two kinds of sampling with the residual variability is greater than the tabled value, the two methods of sampling are evidently different on the average for magnesium. However, with this being the only constituent for which a significant difference is obtained and with the computed F-value being rather low, one can feel reasonably happy about hammer sampling for this rock.

TABLE 7.3. ANALYSIS OF VARIANCE IN MAGNESIA DATA FROM CACTUS FLAT, SAN BERNARDINO MOUNTAINS, CALIFORNIA *

Source of variation	Sum of squares	Degrees of freedom	Mean square	F	$F_{10\%}$
Sample location	2.5013	29	0.0863	13.96	1.41
Sampling method	0.0544	3	0.0181	2.94	2.15
Hammer versus drill	0.0407	1	0.0407	6.57	2.77
Residual variability	0.5375	87	0.0062		

* Data from Richmond, personal communication, 1967.

Since this section was written studies similar to those of the Pomona College geologists have been started by several investigators, and some results have been published. Before too long much better information on how to sample outcrops should be available. Meanwhile, outcrop sampling by diamond drill rather than by hand tools is the surest way to obtain reliable samples, because of the following advantages:

1. Lack of the bias that is introduced in hammer sampling through collecting specimens that break off easily, for example, along joints.

2. Avoidance of the bias introduced in hammer sampling because the collector sees the rock to be sampled and therefore avoids rocks deemed to be "peculiar" or "atypical," rocks which may actually be part of the population to be sampled. In drill sampling, the first interval of core, say 6 inches or a foot, can be rejected to ensure that the rock type in the second interval of the sample will be initially unknown to the collector.

3. Uniform sample orientation—for example, vertical cores to make consistent the relation of the samples to any layering or lineation in the rock.

4. Uniform size of the geologic samples, which makes it unnecessary to consider sample-volume variance (sec. 7.4), even though evidently this source of variance is seldom large.

5. Uniform sample shape. Although of no known advantage, this factor may have future value. It is at present fanciful, but if someone, someday, learns that the shape of geological specimens is significant, those who collected samples of uniform shape will have data from which information can still be extracted.

7.4 DRILL SAMPLING

The proof of the pudding is in the eating. The best way to investigate the reliability of the sampling of a rock body by drilling was never put better than by I. B. Joralemon some forty years ago, when he wrote (1925, p. 614):

> ... in the development of the New Cornelia Copper Co., at Ajo, Ariz., over 1000 ft. (304 m.) of test pitting was done to check diamond drilling. The average of diamond-drill samples and of large channel samples checked to within 0.005 percent copper. The bulk samples, consisting of every tenth bucketful crushed and quartered mechanically, averaged 0.15 percent higher than the channel and drill samples. The drill and channel samples were accepted as correct. Several million tons of this ore have now been mined and treated. All of this ore to date has averaged 1.51 percent copper, compared with the estimate of 1.54 percent. The grade has been a little higher than the estimate in one part of the orebody and lower in another part. This property has proved that, in a disseminated orebody, it is possible to sample and estimate the ore correctly without reducing the grade by a factor of safety.

To evaluate drill sampling, mining companies generally follow a method like that explained by Joralemon, although the analyses are seldom published. When the actual mining of the rock is impracticable, other methods to investigate consistency within and among drill holes must be followed; it is these methods that are emphasized in this section. Unfortunately, for most of the example data, mining results for empirical evaluation of reliability are unavailable.

Analysis of data is explained in the following subsections. Technology of drill sampling is not discussed because sufficient information and bibliographies are given in standard works, for instance, those by McKinstry (1948), Truscott (1962), and Jackson and Knaebel (1932).

Diamond-Drill Sampling

Because it can bore a hole for a long distance in any direction, a diamond drill is a versatile sampling machine that can yield a reliable sample under the proper conditions. Required are a suitable machine, an experienced operator, appropriate operating rules, and suitable rock conditions. The hole is drilled to a usual diameter of $\frac{3}{4}$ to 3 inches (table 7.4), with an annular bit impregnated with diamonds, while water is being circulated in the hole as a lubricant. Usually a core is obtained, which provides a visual record of the geology as well as a solid-rock sample whose location is certain. Also sludge, the cuttings from the hole, is often collected in settling tanks at the collar of the hole, particularly when core recovery may be impossible or too expensive. Sometimes, composition of a rock body can be more closely determined from diamond-drill cores than from other kinds of samples that are more numerous and larger in volume, as at Naica, Chihuahua, Mexico (G. K. Lowther, personal communication), for lead, zinc ore and at the Giant Yellowknife mine, Northwest Territories, Canada (Dadson and Emery, 1968), for gold ore.

TABLE 7.4. DIAMOND-DRILL HOLE AND CORE DIAMETERS

	Core diameter		Hole diameter	
Name	in.	cm	in.	cm
XRT	0.75	1.905	1.1875	3.01625
EXT	0.9375	2.38125	1.5	3.81
AXT	1.28125	3.254375	1.875	4.7625
BX	1.625	4.1275	2.3125	5.87375
NX	2.125	5.3975	2.9375	7.46125

Besides in the general references cited at the beginning of this section, diamond-drilling technology is fully described in a book by Cumming (1956), who includes references to the periodical literature. Technology is rapidly changing because more and more diamond-drill holes are being bored; demands are made to drill increasingly difficult rock types; and wages, a large part of the total cost, are increasing more rapidly than the prices of equipment and diamond bits. Diamond drilling is increasingly used to sample rocks for purposes other than the directly economic ones such as discovering or estimating grade of ore and evaluating dam sites; for instance, in 1967 the U.S. Geological Survey bored several thousand feet of holes in one area of Nevada alone to investigate regional stratigraphy and structure.

The geologist embarking on diamond-drill sampling must answer a number of questions.

1. Should sludge as well as core be collected?
2. What size core should be taken?
3. At what intervals should the core and sludge be sampled, and, for each interval, how many replicate increments of rock material should be taken for chemical or other analysis?
4. How should the reliability of the sampling results be appraised?

Statistics can aid in making these decisions. Those decisions concerned with relations within a single drill hole of the core and sludge are considered here first; then those that relate the samples to the rock body are discussed.

Sludge and core assays seldom agree, partly because sludge is mixed before coming out of the hole, but mostly because losses of the valuable mineral or minerals are almost certain to be different for the two; for instance, McKinstry (1948, p. 95) records that at the Colquiri tin mine, Bolivia:

> In twelve drill hole intersections, core recovery averaged 84.3 percent, most of the missing core representing sections of friable vein-matter. Core assays averaged 1.44 percent tin; core and sludge assays, combined in proportion to theoretical volume, averaged 1.87 percent tin as compared with channel samples for the same veins, though not in the identical places, which averaged 1.89 percent tin.

Obviously, tin that was lost in the core turned up in the sludge. Because of such biases, sludge should be taken unless (a) the diameter of the hole is large enough so that the amount of sludge proportional to that of core is small, (b) core recovery is excellent, (c) although core recovery is less than excellent, reliable weighting factors have been developed empirically, or (d) cost of obtaining sludge is prohibitive. Systematic core and sludge data from many drilling programs would be highly informative but do not seem to have been published.

An ideal geological sample from a diamond-drill hole would consist of 100 percent of the rock displaced recovered in core and sludge, the chemical analyses of which could then be averaged by weighting them according to the cross-sectional areas of the core and of the annular opening between the core and the wall of the hole. McKinstry (1948) and Cumming (1956) supply full details. However, core recovery is never 100 percent because pieces of rock break up in the hole and are ground up by the drill bit. Sludge recovery is usually worse, since some sludge is not recovered for one or more of many causes: because of insufficient water pressure to wash it out of the hole, because it is washed into cracks in the rock, or because, although washed out of the hole, it does not settle in the tanks provided at the collar of the hole. For these reasons percentages of recovery must be estimated and the weighting factors adjusted.

The weighting factors must be determined empirically because the observations for core and sludge constitute two statistical samples, each with its own mean and variance. For instance, at the Chuquicamata copper mine, Chile, Waterman (1955) published in a careful study the example data of table 7.5.

TABLE 7.5. SLUDGE AND CORE ASSAYS FROM DIAMOND-DRILL HOLES IN THE CHUQUICAMATA COPPER MINE, CHILE *

Item	Core recovered (%)	Core assay (% copper)		Sludge recovered (%)	Sludge assay (% copper)	
		Soluble	Insoluble		Soluble	Insoluble
Observation	44	0.46	0.39	81	0.72	0.13
	50	1.12	0.35	67	0.72	0.10
	58	2.42	0.21	95	1.35	0.10
	57	0.95	0.22	102	1.31	0.10
	61	0.30	0.23	73	0.95	0.11
	80	1.51	0.23	48	0.98	0.12
	64	1.18	0.13	115	1.05	0.12
	83	1.24	0.28	91	1.07	0.10
Mean	62.13	1.14	0.26	84.0	1.02	0.11
Standard deviation	13.52	0.65	0.08	21.3	0.23	0.01
Coefficient of variation	0.22	0.57	0.32	0.25	0.23	0.11

* After Waterman, 1955, p. 60.

Taking for simplicity only the soluble copper, one could theoretically combine these assays according to the formulas

$$\overline{w} = \frac{k_1 \overline{w}_c + k_2 \overline{w}_s}{k_1 + k_2}$$

and

$$s_{\overline{w}}^2 = \left(\frac{k_1}{k_1 + k_2}\right)^2 \frac{s_c^2}{n} + \left(\frac{k_2}{k_1 + k_2}\right)^2 \frac{s_s^2}{n} + 2 \frac{k_1 k_2}{(k_1 + k_2)^2} \frac{s_c s_s}{n} r_{cs},$$

where k_1 and k_2 are percentage recoveries of core and sludge, r is the correlation coefficient (sec. 9.3), and the subscripts c and s stand for core and sludge. As Waterman points out, it is impossible to determine if this estimate has any validity elsewhere in the mine. Therefore Waterman's empirical solution is satisfactory.

Because geological interpretation can be made of the standard deviations and coefficients of variation of table 7.5, a good reason for calculating them can be provided. In the core the coefficient of variation, which is lower for the insoluble than for the soluble copper, presumably reflects a more uniform mineralization. In the sludge the same relation holds; moreover, the coefficients of variation for both soluble and insoluble copper are lower than those for the core, a fact which shows that the sludge must be mixed from one interval to another. More insoluble copper was lost in the sludge than in the core.

Sludge and core observations can be compared by a paired t-test to learn if they are evidently the same or different. Applied to the Chuquicamata data in table 7.5 to compare sludge with core assays, paired t-values are -0.66 for soluble copper and -4.98 for insoluble copper. The first of these values is outside and the second is inside the critical region defined by the tabled t-value of 1.895 for 7 degrees of freedom at the 10-percent level. The conclusion is that core and sludge assays agree for the soluble copper but not for the insoluble copper.

Sample-Volume Variance

For drilling in most geological environments, as core diameter increases, so does cost per foot. Although larger holes are easier and quicker to drill than smaller ones, the higher diamond consumption and cost of the larger equipment outweigh the savings in labor. However, recovery of a large-diameter core is better than that of a small core and should yield a geological sample that is better because its volume is larger. Although the mean of observations based on large volumes of rock should not be expected to be different on the average from those based on small volumes of rock, the corresponding variances might be expected to be smaller. The theory of sample-volume variance that has been advanced states that variance of observations is inversely proportional to sample volume. However, the limited published data suggest that decrease in variance is not related to increase in volume in any simple way, and for some constituents and some rocks the decrease is insignificant within the range of practical sampling.

Some calculations for sample-volume variance are demonstrated for a small set of data published by Krige (1966, p. 74), consisting of gold values from 12 drill holes ($1\frac{3}{8}$ inch diameter) bored through a narrow gold-bearing conglomerate bed. The cores were sawn in half longitudinally, and the two halves assayed. The results are in table 7.6. The in.-dwt values for whole core are the averages of the two halves; the means and variances in the last two lines of the table are nearly the same. If the values from the two halves of the core

TABLE 7.6. TEST OF SAMPLE-VOLUME VARIANCE IN DRILL CORE
FROM A SOUTH AFRICAN GOLD MINE*

Borehole number	In.-dwt value		
	First half core	Second half core	Whole core
1	180	235	207.5
2	131	123	127.0
3	182	173	177.5
4	135	117	126.0
5	91	113	102.0
6	117	112	114.5
7	109	90	99.5
8	179	178	178.5
9	100	58	79.0
10	227	204	215.5
11	304	226	265.0
12	279	285	282.0
$\bar{w}$	169.5	159.5	164.5
s^2	4893.18	4642.45	4508.68

* Data from Krige, 1966, p. 74.

were statistically independent observations from a single population, the
ratio of the variance of the values for the whole core to the average of the
variance of the values for the two halves should be one-half. But, because
the calculated ratio of 0.94 is only slightly less than 1, little is gained by assaying the whole core rather than only one-half of it. Table 7.7, which gives the
results of a randomized-block analysis of variance of these data, shows that
practically all of the variability is among the bore-holes rather than between
the core halves. Thus, the 12 boreholes must be considered to have sampled
different statistical populations with different means.

TABLE 7.7. RANDOMIZED-BLOCK ANALYSIS OF VARIANCE FOR THE DATA OF
TABLE 7.6

Source of variation	Sum of squares	Degrees of freedom	Mean square	F	$F_{10\%}$
Among bore holes	99,103.00	11	9009.36	17.12	2.23
Between core halves	600.00	1	600.00	1.14	3.23
Residual variability	5,789.00	11	526.27		

TABLE 7.8. SAMPLE-VOLUME VARIANCE IN DRILL CORE FROM TWO HOLES IN THE
CLIMAX MOLYBDENUM MINE, COLORADO*

Line	Item	Volumes and variance ratios		
1	Relative volume	1	2	3
2	Volume (cc)	218	436	654
3	Theoretical ratio of variances	1	1/2	1/3
	Observed ratios of variances			
4	Hole 1	1	0.89	0.83
5	Hole 2	1	0.76	0.67
	Observed ratios of variances with location effects removed			
6	Hole 1	1	0.64	0.72
7	Hole 2	1	0.98	0.78

* Data from Hazen, 1962, p. 21.

The results of two other investigations agree with those detailed for Krige's
gold data. In the first, conducted by Hazen (1962), two NX size drill holes
were bored at the Climax molybdenum mine in Colorado. The 101 feet of core
from the first hole and the 52 feet from the second hole were sawn into 2-foot-
long intervals, and then longitudinally at angles of 60, 120, and 180 degrees,
to yield relative volumes of 1, 2, and 3. In lines 4 and 5 of table 7.8 are the
observed variance ratios for the two holes compared with the theoretical
ratios; the observed variance is not inversely proportional to the sample
volume. Because within each hole there is a slight systematic change in ore
grade which might affect the variance ratios, this trend was removed (sec.
9.5), and the data were analyzed again. The results are in lines 6 and 7 of the
table, for which, as before, the variance is not inversely proportional to the
volume, between-location variability being large compared with within-
location variability.

Another study is of gold assays from nine hundred 1-foot intervals of EX
drill core obtained from several drill holes bored in the course of regular
company operations in the Homestake mine. We combined (table 7.9) the
gold assays for the 1-foot intervals to simulate the assays that would have
been obtained, assuming perfect sample preparation and assaying, from 2-,
3-, 4-, and 5-foot intervals. As is true for the previous examples, the observed
variances are not inversely proportional to the sample volume.

For these three examples variance does not decrease linearly as sample
volume increases, although there is a tendency for a small decline. In general,
variance probably declines as volume increases, but we do not know of any
data for which sample-volume variance holds, although doubtless there is
some. The reason is that, for real data, the among-location variability is large

Table 7.9. Sample-volume variance in drill core from the Homestake gold mine, South Dakota

Line	Item	Volumes and variance ratios				
1	Interval length (ft)	1	2	3	4	5
2	Volume (cc)	129	258	387	516	645
3	Observed variances (ppm)	327	218	166	141	125
4	Theoretical variances (ppm)	320	160	107	80	64

compared with the within-location variability; and increasing the sample volume reduces the effect of within-location variability but has no effect on the among-location variability. Because the total variability is a weighted sum of these two kinds of variability, reducing the smaller source does not appreciably reduce the sum.

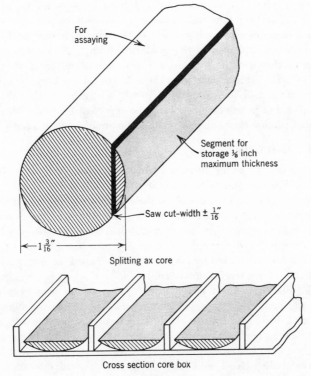

For assaying

Segment for storage ⅛ inch maximum thickness

Saw cut–width ± $\frac{1''}{16}$

$1\frac{3}{16}''$

Splitting ax core

Cross section core box

Fig. 7.2. Core sawn longitudinally with a diamond saw (after Cumming, 1956, p. 238).

If sample-volume variance were important, it would be desirable to assay all or most of the core from a drill hole; and the common industry practice of splitting the core longitudinally, saving one half for a record and assaying the other half, would be bad. A method increasingly used, which is conservative if little is known about the variance relations, is to save a thin segment sawn longitudinally out of the core, as illustrated by figure 7.2. This method is desirable if rock lithology is of particular interest and if the valuable constituent has an erratic distribution, as in sampling precious metal ore from a geologically unknown environment.

Relation of Drill-Hole Samples to the Rock Body

The relation of the drill-hole samples of core and sludge to the rock body sampled can be compared by taking a bulk geological sample of the rock body, by considering the variance of multiple drill holes in a rock body on various scales, and by comparing observations in deflected drill holes. Unfortunately, no generalizations can be made because practically all of the studies extant have been made by individual companies in particular environments, and the information has not been published.

Mining companies commonly check results of drill-hole samples by taking bulk samples, as explained in the quotation from I. B. Joralemon at the beginning of this section. Sometimes the means calculated from the two kinds of sampling agree; sometimes they do not. When they do agree, the variability of the drill-hole samples is generally a little higher than that of the bulk samples, but not as much as would be expected if sample-volume variance applied. Comparisons may be made by an analysis of variance (Chap. 5).

Diamond-Drill-Hole Deflections

The final subject to be discussed about diamond drilling is drill-hole deflections. When diamond-drill holes are bored to obtain information about the grade of an ore body, quite often the reliability of the information obtained from the small-diameter cylinder of rock and/or from the accompanying sludge is in question. Then, the diamond-drill bit may be deflected so that a second hole is bored through the ore body or other geological entity of interest. The bit is deflected by placing a wedge-shaped piece of metal in the original hole to force the diamond-drill bit out of the hole at an angle. Even though one does not expect several samples close to one another to provide as good information as the same number of samples spread more evenly across a rock body, nonetheless the information may be valuable, and the cost of deflecting is much less than that of drilling a new hole, particularly when many feet of barren rock must be penetrated to reach the rock body.

We have studied (Koch and Link, 1969) the value of information from deflected diamond-drill holes by analysis of 795 assays from original intersections and deflections of boreholes in South African gold mines, information supplied us by D. G. Krige (1967, personal communication). When these were arranged in order of size and a cumulative sum tabulated, we found that 21 percent of the value was in only three observations with the extremely high values of 12,528, 15,317, and 23,037 in.-dwt. Although these observations cannot be considered exceptional, they verge on being extreme even for this highly skewed lognormal distribution, with a coefficient of variation of 3.79. The remaining calculations are based on 792 observations—the three highest ones are omitted because their inclusion obscures meaningful comparisons. Had, say 100,000 observations been available, the percentage of extremely high values could have been estimated, and disregarding the three highest ones would have been unnecessary. The problem is one often met in geology where highly skewed distributions are common (secs. 6.2 and 6.3).

Table 7.10, the upper half of the correlation matrix (sec. 9.2) for boreholes with deflections, shows that the correlations are of moderate size. Were they perfect (equal to 1), the deflections would provide no new information about the mean grades of the gold-ore bodies.

The correlations in table 7.10 can be used to calculate the relative variances of the mean for holes with no deflections, one deflection, two deflections, etc. The relative variances for from zero to three deflections are presented in table 7.11. If the observations from the original intersections and the deflections were uncorrelated, the relative variances in the column farthest to the right would be obtained, that is, $1/2 = 0.500$, $1/3 = 0.333$, and $1/4 = 0.250$, because the variance is inversely proportional to the number of observations per hole according to the formula $\sigma_{\bar{w}}^2 = \sigma^2/n$ (standard error of the mean law). However, because the original and deflection observations are correlated, the variances of the means based on combining them are larger than $1/2$, $1/3$, and $1/4$. Thus, an increase in precision is gained by making deflections, but not so large an increase as that obtained by drilling a new borehole, from which to get an independent observation.

TABLE 7.10. CORRELATION MATRIX FOR BOREHOLES WITH DEFLECTIONS IN SOUTH AFRICAN GOLD MINES

Item	P	(0)	(1)	(2)	(3)
Original intersections deflected	(0)	1.0	0.537	0.821	0.592
Deflection 1	(1)		1.0	0.439	0.213
Deflection 2	(2)			1.0	0.589
Deflection 3	(3)				1.0

Table 7.11. Relative variances of from zero to three deflections of observations from boreholes in South African gold mines

Number of deflections per hole	Number of observations per hole	Number of boreholes	Observed relative variance of mean for hole	Relative variance if original intersections and deflections were entirely uncorrelated
0	1	241	1.0	1
1	2	241	0.768	1/2 = 0.500
2	3	101	0.733	1/3 = 0.333
3	4	27	0.649	1/4 = 0.250

Whether these results hold for other geological situations we do not know, but analysis of these data suggests that deflections are well worth the additional cost in these South African mines and presumably in other situations as well.

Drill Sampling with Other Than Diamond Drills

Drill sampling with other than diamond drills is not discussed in detail in this book for two reasons: (a) the general procedures for statistical analysis of diamond-drill-hole sample data should apply, and (b) few detailed recent data have been published although many reside in company files. However, a few differences in the geological characteristics of the samples taken by the various methods of drilling may be mentioned; the references cited at the beginning of this section supply full details on the engineering problems.

Cores are sometimes cut by methods other than diamond drilling. If the cores are of small diameter, obtained by drilling with bits other than diamond bits or by driving a split pipe into sediments, the statistical analysis should be the same as for data from diamond-drill holes. If the cores are of large diameter, as those obtained by calyx drilling, sample preparation must be carefully controlled to avoid introducing variability; in this way, enough core should be obtained so that sludge can be ignored.

Other methods of drilling yield only cuttings, rather than both cuttings and core. In churn drilling a hole usually 12 inches or smaller in diameter, although sometimes larger, is drilled in a wet hole by lifting a cable-hung bit and dropping it to impart a churning motion that breaks up the ground; the mud thus formed can be bailed out at intervals. Holes must, of course, be bored vertically downward. The large volume of sample may compensate for the mixing and for contamination from the walls, contamination that can be controlled by casing the hole. McKinstry (1948, pp. 71–81) gives full details on precautions in churn drilling and in sample handling. Applications to placer sampling are given in section 15.2.

In recent years cuttings have been increasingly obtained by rotary drilling with a noncoring bit, with the advantages over diamond drilling of greater speed and lower cost. Holes are drilled either wet, with the cuttings flushed out of the hole in the return flow of the water or mud, or dry, with the cuttings blown out with air. Studies have been made to compare sampling in rotary drilling with other kinds of sampling; some results, not published with detailed numerical information (Harding, 1956), suggest that mixing of cuttings from one interval to another and incomplete blowing out of particles in dry drilling may be serious problems.

In many underground mines samples are taken from holes bored with the air drills used to drill blast holes (McKinstry, 1948, p. 106). The samples have the advantage of being inexpensive, but the disadvantage is that, in cases of which we are aware, correspondence of results with those from diamond-drill-hole samples or from channel or chip samples (sec. 7.5) is poor. Another problem is that annular corrugations that sometimes form in the holes lead to the riffling out of heavy particles. Again the data necessary to evaluate this kind of sampling quantitatively are not available.

7.5 CHANNEL AND CHIP SAMPLING

In channel sampling a slot or channel is cut in a rock face. The rock fragments broken out of the slot constitute the geological sample. The dimensions of the channel may be maintained by cutting it so that a board of appropriate size, say 2 by 6 inches in cross section, fits into it. In chip sampling, a similar procedure is followed, except that small chips are collected across the face, and no attempt is made to cut a channel. McKinstry (1948, pp. 37–45) and Truscott (1962, pp. 10–12) give full details; Parks (1949, pp. 76–90) explains application in the copper mines at Butte, Montana.

Channel and chip samples are commonly taken in underground metal mines. These samples provide most of the experience and published data on these two methods; but there is no reason why the methods cannot be applied in academic geology to outcrop sampling. Formerly, it was a mark of a competent mining engineer to insist on channel sampling; but today, with increased labor costs, chip sampling or broken-rock sampling (sec. 15.1) is more common for routine control of ore grade. If more accurate results are required, the tendency is to use a much larger bulk sample derived from driving a mine working (sec. 15.4).

In metal mines channel or chip samples are taken systematically at regular intervals in the mine workings. They may be cut from the face, walls, back (roof), or, rarely, floor of tunnel-shaped openings and at various places in the other workings. The length of an individual channel or line of chips traditionally is no longer than 5 feet, but shorter intervals are taken when mineralogy changes obviously, particularly if the minerals differ in hardness or friability. In countries that use the metric system English- and American-trained engineers cut channels that are 2 meters long.

Better results can be expected from samplers who are entirely disinterested in the results of their work rather than from technically trained men who are liable to introduce prejudice. The educated man tends to include too much high-grade material because of optimism or alternatively, too much low-grade material for a "safety factor." In South Africa excellent results are obtained by native samplers, and in Latin America one of us has seen good sampling done by a carefree crew that sampled in the morning, slept in the afternoon, and played guitars all night in a local bar.

Channel or chip sampling of outcrops is done in the same way as of underground workings and chip sampling of outcrops should be inexpensive enough to do routinely in academic geology. Whether most academic geologists sample outcrops by single hand specimens or by chips is a mystery, because this interesting information is seldom revealed in their papers. The best argument for chips is that wider coverage and the advantages of averaging are obtained; the best argument for single hand specimens is that the chemical or modal analysis is related directly to the thin section, the mineralogy of which is studied.

Data Analysis

Common mine practice is to check the assays of channel or chip samples by cutting new samples from the same locations and comparing the assays obtained. Means and standard deviations computed from each set of observations should be close to one another although individual observations seldom agree, the differences between original and check observations naturally being greater for the higher observations. In figure 7.3 are graphed typical data of 267 pairs of check chip samples from an unidentified South African gold mine (Rowland and Sichel, 1960, p. 256).

Therefore, although in channel and chip sampling individual check observations are not expected to be the same, agreement on the average can be sought and tested for by the paired t-test. The procedure is demonstrated in table 7.12, which lists 20 check-sample gold values from another unidentified South African gold mine (Rowland and Sichel, 1960). Because these data were taken by two different samplers, the effect of natural variability is confounded to an unknown extent by the difference in technique between the two men. The African chipper obtains a mean grade that is 1.57 ppm higher than that of the European sampler, but significantly, except for the one

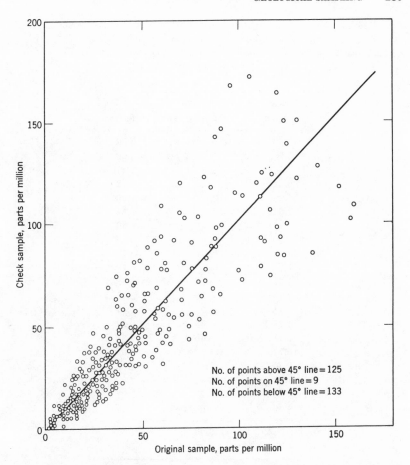

No. of points above 45° line = 125
No. of points on 45° line = 9
No. of points below 45° line = 133

Fig. 7.3. Assays of paired-chip samples from a South African gold mine (after Rowland and Sichel, 1960, p. 256).

extremely high-value location whose paired values are 78.00, 22.29, the African in most places cut lower value samples and most of the discrepancy stems from this one location.

As often happens with real data, the precise statistical analysis to use is obscure. A paired t-test applied to all the observations yields a t-value of 0.55, which is outside the critical region defined by the tabled value of 1.729 with 19 degrees of freedom at the 10-percent significance level. However, arguing that the difference of 55.71 is an outlier (sec. 6.4) that should be removed, one obtains a t-value of -3.92 with 18 degrees of freedom, which is inside the

TABLE 7.12. CHECK-SAMPLE GOLD VALUES FROM
AN UNIDENTIFIED SOUTH AFRICAN GOLD MINE *

| Check-sample gold values (ppm) | | |
African chipper	European sampler	Difference
5.83	8.91	−3.08
3.77	7.37	−3.60
1.54	1.37	0.17
2.91	3.43	−0.52
9.09	10.29	−1.20
4.63	6.17	−1.54
3.77	6.17	−2.40
18.00	21.43	−3.43
78.00	22.29	55.71
1.71	1.54	0.17
1.37	3.26	−1.89
2.23	4.63	−2.40
5.14	3.77	1.37
4.29	4.63	−0.34
4.11	4.46	−0.35
15.77	17.14	−1.37
0.86	1.89	−1.03
1.37	3.94	−2.57
2.74	3.77	−1.03
4.46	3.60	0.86
Sum 171.59	140.06	31.53
Mean 8.57	7.00	1.57

* Data from Rowland and Sichel, 1960.

critical region. Alternatively, the number of positive and negative differences may be compared by a sign test (Dixon and Massey, 1969, p. 335), which has the advantage of being distribution-free and which also leads to the conclusion that the European sampler is somewhat higher on the average than the African chipper. No definite conclusions can be reached from the small amount of published data, although the evidence points to some bias between the two men.

Channel and chip sampling have often been compared with one another, and the conclusion has usually been drawn that the additional expense of channel sampling is unjustified, provided that the chip sampling is carefully done. Likewise channel and chip sampling have been compared with diamond-drill-hole sampling at many mines, with varied results. Few numerical results of these investigations have been published.

The best way to evaluate channel and chip sampling usually is to compare them to bulk sampling from the same mine workings or with mine production. In section 15.4 some comparisons made at the Kilembe mine, Uganda, are given. Another interesting study was made by Bradley (1923) in the Alaska Juneau gold mine, where highly variable ore of a low grade required careful evaluation. Paired t-tests applied to Bradley's data show no inconsistency among channel samples, muck samples, and millheads.

7.6 MODAL ANALYSIS

Modal analysis is a method to estimate the mineralogical composition of rocks, either by superimposing a square grid on a thin section and recording the mineral species present at each grid intersection or, alternatively, by measuring the lengths of mineral intersections on traverses evenly spaced across the grains. Modifications are possible, such as making plane sections other than thin sections. Modal analysis has been discussed in a book by Chayes (1956) and also in other works, of which some of the more significant ones later than Chayes' book are reviewed by Griffiths (1967, Chap. 9).

In this section we review some statistical problems in modal analysis and reinterpret the results of a key experiment performed by Chayes and Fairbairn (1951) to evaluate operator variability. Only point counting is discussed because it yields results equally as good as traversing and is easier to do.

The principles of modal analysis are simple to state. Suppose that a rock is composed of minerals that can be visually distinguished. If n points could be taken at random in the rock and the mineral species present at each point recorded, then the ratio of the number of points for each mineral to the total number of points is an estimate of the volumetric proportion of the particular mineral in the rock. Moreover, if the points were randomly distributed over a plane that randomly cut the rock, the same argument holds.

Although it is impracticable to take points at random throughout a rock and inconvenient to take points at random on a plane (thin section) passing through a rock, taking points on a plane systematically but with a random starting point yields statistics having the same properties. The statistical analyses for the counts are associated with the binomial distribution (sec. 6.1) under the assumptions that no difficulties arise in identifying the mineral at each point and that the thin section is chosen from the rock at random. However, if the grain size of the rock is large, binomial theory will underestimate the counting error unless more than one thin section is taken (Chayes, 1956, p. 93).

The hardest problem in modal analysis is mineral identification. Skill is required to identify the mineral at each point, even if the minerals counted have been selected to minimize the difficulty (as by grouping together all opaque minerals, or both feldspars in a two-feldspar rock). It is even more difficult to measure the length of a traverse across a given mineral because to the mineral identification problem is added the necessity of recognizing accurately the boundaries between minerals.

Aside from the problems of excessive grain size and operator variability, the maximum accuracy to be expected from point counting may be specified. Table 7.13 gives the standard errors of point counts calculated from the binomial distribution; for instance, if a mineral makes up 20 percent of a rock, 100 points must be counted to reduce the standard error of the mean to an absolute accuracy of 4 percent or to 20 percent (4/20) of the content of the mineral in the rock. The table therefore provides a sound basis for calculating how many points should be counted to obtain a desired accuracy in modal analysis of a rock whose mineralogical composition is more or less known.

The crucial problem in modal analysis is operator variability, which is the variability introduced by different persons counting the points. Chayes and Fairbairn (1951) report on two experiments conducted at the Massachusetts Institute of Technology (MIT) in which five graduate students were trained in the point-counting technique. Each operator examined each of five thin sections cut from a single piece of granite from Westerly, Rhode Island, on a schedule devised as a Latin square experimental design (Dixon and Massey, 1957, p. 171). The entire experiment was repeated six months later. The original results are tabulated by Chayes and Fairbairn.

Table 7.14 is an analysis of variance of quartz observations from the second experiment; analyses of variance for the other data gave similar results. The

TABLE 7.13. STANDARD ERRORS OF POINT COUNTS CALCULATED FROM THE BINOMIAL DISTRIBUTION

Number of points counted	Percentage of mineral in rock			
	1	5	20	50
25	1.99	4.36	8.00	10.00
100	1.00	2.18	4.00	5.00
400	0.50	1.09	2.00	2.50
900	0.33	0.73	1.33	1.67
1600	0.25	0.55	1.00	1.25

TABLE 7.14. ANALYSIS OF VARIANCE OF QUARTZ OBSERVATIONS IN AN EXPERIMENT
TO DETERMINE OPERATOR VARIABILITY IN MODAL ANALYSIS *

Source of variation	Sum of squares	Degrees of freedom	Mean square	F	$F_{10\%}$
Replication	0.838	4	0.209	0.19	2.48
Among operators	11.154	4	2.788	2.48	2.48
Among slides	13.438	4	3.359	2.99	2.48
Residual variability	13.481	12	1.123		

* Data from Chayes and Fairbairn, 1951.

four sources of variability are the replications (because there were five tests,
perhaps on different days, or perhaps the operators were more tired for one
than for another), the operators, the slides, and the residual variability. The
table indicates slightly more variability among the analysts and definitely
more variability among slides than would occur by chance. The among-slide
variability is large enough to indicate that the rock was of distinctly different
composition at the five places where the slides were cut. The replication
variability is extremely small. [Chayes and Fairbairn (1951) pooled the
replication and residual mean squares to obtain analytic error, but because
the replication mean square is so small, they probably underestimated the
analytic error in these data.]

Table 7.15 compares the observed and the theoretical mean squares for all
the data. Except for biotite, the residual mean squares are smaller in the
second MIT test, an indication that the operators, after six months' experience, had improved their techniques. Most of the observed mean squares

TABLE 7.15. OBSERVED AND THEORETICAL MEAN SQUARES FROM AN EXPERIMENT
TO DETERMINE OPERATOR VARIABILITY IN MODAL ANALYSIS *

Mineral	Residual mean square		Theoretical binomial mean square
	First MIT test	Second MIT test	
Quartz	1.65	1.12	0.79
Microcline	2.36	1.40	0.91
Plagioclase	1.29	0.689	0.87
Biotite	0.157	0.403	0.12
Muscovite	0.160	0.082	0.047
Opaque minerals	0.045	0.013	0.030
Non-opaque minerals	0.020	0.012	0.016

* Data from Chayes and Fairbairn, 1953, p. 710.

were larger than those predicted by binomial theory. Therefore some variability remained in excess of that expected if the operators were perfect; however, this excess variability was only 50 percent at most. In the first test the results of operator 4 were distinctly different from the others, and in the second test his results were still different, although less so. We have not investigated these data in enough detail to learn whether operator 4 was evidently better or worse than the others. The different results presumably stem from identification problems.

In conclusion, if a competent operator makes the counts, point counting is an effective technique for finding volumetric mineralogy whose accuracy can be reasonably well predicted from binomial theory. The conversion, if required, to chemical analyses is a further step not considered in this book.

7.7   OTHER SAMPLING METHODS

In this section we mention some other methods for geological sampling that do not fit into the previous sections of this chapter. These methods are only a few of many; each specialized field has its particular problems, and we can mention only some.

For sampling in paleoecology and also in paleontology in general, Ager (1963, especially pp. 220 ff.) offers good advice. Krumbein (1965) has also written an article on sampling in paleontology, with a good list of references. Mosimann (1965) has written a long article on statistical methods for pollen analysis, based on binomial and multinomial models (sec. 6.1).

Sampling of uranium ores and statistical analysis of the data present special problems because the composition of the samples changes rather rapidly with time and because chemical analyses must be reconciled with gamma-ray data. These problems have been studied by Grundy and Meehan (1963), by Schottler (1965), and by others; but no very complete investigation has yet been published.

7.8   RANGE OF GEOLOGICAL VARIABILITY

Above all, devising a good plan for geological sampling depends on knowing or estimating the variability in the geological entity to be sampled. Thus in their studies of the Southern California batholith (sec. 7.3) Baird and others (1967, p. 11) write that "evaluation of geochemical distributions over heterogeneous plutonic assemblages . . . requires considerable knowledge of the chemical variability within the individual rock bodies." We have found this true in sampling gold deposits (Chap. 16).

In this section we summarize a few facts and introduce some concepts about the range of geological variability. Variability depends on the quantity of the substance studied in the rock body, on the chemical form of the substance, and on the geological process or processes that formed the rock body and the substance. The dependence is illustrated for the element carbon.

A few data for the element carbon in coal, graphite, and diamond demonstrate the great range of variability and point up the difficulty in obtaining the information needed to evaluate variability. The carbon content of coal ranges from more than 95 percent in anthracite to less than 70 percent in lignite; with such high amounts of carbon present, coefficients of variation are low; a typical one, calculated from 30 observations made on the Pittsburgh coal bed (Pennsylvanian) in 30 counties of Pennsylvania is 0.0087, with a mean of 84.58 percent. For graphite mines, carbon content ranges from a few percent up to 30 percent; a typical coefficient of variation from 188 trench samples in the Benjamin Franklin and Just graphite mines in Pennsylvania is 0.61, with a mean of 1.66 percent.

The variability of carbon content in the form of diamond in diamond deposits must be extreme. Giles (1961, p. 840) writes that typical South African diamond mines that work kimberlite in place have grades of 0.079 ppm at the Premier mine and of 0.018 ppm at the Jagersfontein mine; these grades are only about one-fourth to one-tenth as large as those at the lowest economic limit of gold dredging. Although we have been unable to find size distributions for a complete spectrum of diamond sizes, those for large diamonds are highly skewed. As some mines, such as Jagersfontein, owe their profitability to the larger stones, we may surmise coefficients of variation of the order of magnitude of 1000. In summary, variability in carbon depends mainly on the amount of carbon in the rock body, but also on the geologic processes that formed the carbon-bearing mineral and the rock, which clearly are very different for coal, graphite, and diamond.

A complete study of the range of geological variability would treat all minerals and chemical elements, or at least the common ones, and would be most worthwhile.

In studying geological variability, investigators have used one or another of the methods explained in this book. In a series of papers Ahrens (1963, with references to his earlier papers) confines himself to frequency distributions and histograms, which are not satisfactory for this purpose. A better measure is sample variance, s^2, even if the distribution under study is not normal. However, for most purposes the coefficient of variation, C, is most informative because it gives a relative measure of variability which takes into

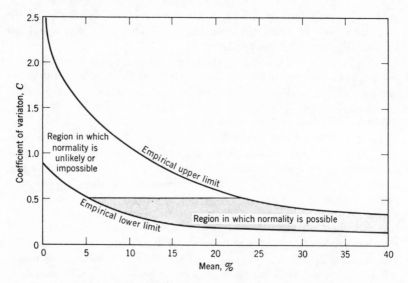

Fig. 7.4. General relation between the mean and the coefficient of variation.

account both the mean and the variance; therefore, the comparisons in this section are made with it.

Figure 7.4 graphs the general relation between the mean and the coefficient of variation. As the mean, expressed in weight percent, increases, the coefficient of variation must tend to decrease until, when the mean reaches 100 percent (a point not shown on the figure), the coefficient of variation must be zero. On the figure are sketched schematically the upper and lower limits for coefficients of variation, based on empirical data discussed in the next subsection; the data are insufficient to define these limits closely. Notably, extremely high coefficients of variation, above say 2 or 2.5, are found only for substances present in minute proportions, such as trace elements or precious metals in ores. On the other hand, extremely low coefficients of variation, below say 0.2, are generally found only for geological substances present in amounts measured in the range of a few tens of percent.

On figure 7.5 the region of expected data, which lies between the empirical upper and lower limits of the coefficient of variation, is divided into two subregions—one in which observations may, but need not, follow a normal frequency distribution and the other in which normality is unlikely or impossible. The dividing line is arbitrarily defined by the fact that, for a distribution of nonnegative observations to be approximately normal, its mean must be a few standard deviation units above zero. If the not very rigorous criterion is adopted that the mean be two standard deviation units above zero, then the

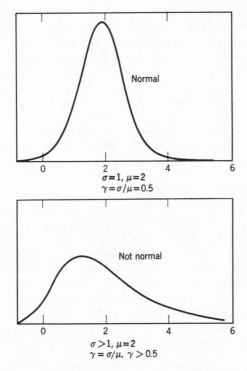

Fig. 7.5. Contrast between normal and non-normal distributions in relation to the coefficient of variation.

coefficient of variation must be less than 0.5 for normality to be possible although not necessary (fig. 7.5). On the graph the plotted horizontal line corresponding to a coefficient of variation of 0.5 shows that substances whose means are between about 5 and 25 percent and whose coefficients of variation are less than 0.5 may be normally distributed; that is, an observed coefficient of variation of less than 0.5 does not guarantee normality, but it does make it a reasonable assumption. As pointed out in section 6.2, the coefficient of variation must be greater than 1.2 before the efficiency of the normal assumption drops enough to make it worthwhile to consider making a transformation.

In summary, figure 7.4 shows that if a substance is present in a certain amount, say 10 percent, the coefficient of variation may be expected to be between about 0.3 and 1.1; and, if it is below 0.5, the observations may be normally distributed. The size of the coefficient of variation must depend fundamentally on the geological variability, provided that the variability introduced by the sampling process is well controlled (Chap. 8). Thus this graph points up once again that postulated "laws" about distribution of elements have little if any meaning. Quantification of the empirical and arbitrary boundaries on this graph would be a fruitful study.

Geological variability depends on the processes that formed the rock. The specific sources of variability include the chemistry and amount of the substance, the geologic processes that formed the rock, and the geologic processes that formed or redistributed the substance in the rock. Thus for gold (Chap. 16), coefficients of variation are, as expected, high for deposits in metamorphic rocks and low for deposits of the Carlin, Nevada, type, where the size of the gold grains is minute. For all substances, data both existing and potential could be organized for the improvement of sampling practice.

Observed Coefficients of Variation

Ideally, observed coefficients of variation for a great variety of geologic substances and environments would be discussed in this subsection, but this ideal cannot be realized because data are lacking. Therefore figure 7.4 had to be based on the data listed in table 7.16 and graphed in figure 7.6—data which are coefficients of variation calculated from 50,057 observations made on various substances in 484 mineral deposits. These data (Hazen and Meyer, 1966, p. 5) are a veritable rag bag of miscellaneous observations derived from samples from churn-drill holes, diamond-drill holes, cut channels, and test pits; the data sources are detailed in the original paper. As these data were taken for many different reasons and were statistically analyzed by Hazen

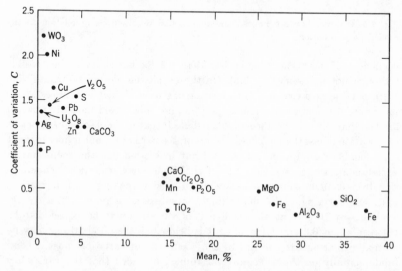

Fig. 7.6. Coefficients of variation from table 7.16 (after Hazen and Meyer, 1966)

TABLE 7.16. SUMMARY STATISTICS FOR 50,057 OBSERVATIONS FROM 484 MINERAL DEPOSITS *

Substance and chemical formula in which analyses are reported	Mean (wt %)	Variance	Coefficient of variation	Number of deposits	Number of observations
Silver, Ag	0.02002	0.000595	1.218	18	971
Alumina, Al_2O_3	29.40	41.16	0.218	39	8888
Lime, CaO	14.53	91.83	0.659	7	1099
Calcite, $CaCO_3$	5.28	39.14	1.185	24	1059
Chromite, Cr_2O_3	15.92	90.66	0.598	40	3592
Copper, Cu	1.66	7.33	1.626	16	843
Magnetite, Fe	26.81	80.58	0.335	30	2810
Iron, miscellaneous deposits, Fe	37.44	105.56	0.274	14	1449
Magnesia, MgO	25.20	146.41	0.480	2	347
Manganese, Mn	14.25	67.63	0.577	63	7711
Nickel, Ni	1.02	4.12	1.981	24	2091
Phosphorus, miscellaneous deposits, P	0.277	0.0668	0.934	18	1307
Phosphate rock, P_2O_5	17.60	83.68	0.520	32	1569
Lead, Pb	2.81	16.05	1.427	8	342
Sulfur, S	4.32	43.84	1.532	5	494
Silica, SiO_2	33.89	143.35	0.353	42	6044
Titanium, TiO_2	14.72	14.74	0.261	4	621
Uranium, U_3O_8	0.402	0.299	1.359	3	519
Vanadium, V_2O_5	1.27	3.36	1.444	25	1286
Tungsten, WO_3	0.587	1.685	2.21	34	4019
Zinc, Zn	4.47	28.17	1.19	36	2976

* From Hazen and Meyer, 1966.

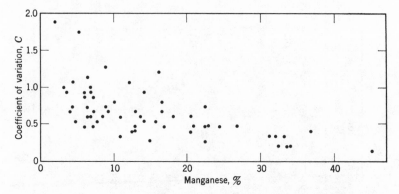

Fig. 7.7. Coefficients of variation for 7711 manganese assays from 63 deposits (after Hazen and Meyer, 1966).

and Meyer for another purpose, they at least are haphazard rather than selected to support our ideas. Other recorded data that we have examined behave similarly.

For individual chemical elements and other substances present in greater than trace amounts a relationship similar to that in figure 7.6 holds; for instance, figure 7.7 graphs the coefficients of variation for 7711 manganese assays from 63 deposits, as tabulated by Hazen and Meyer.

REFERENCES

Ager, D. V., 1963, Principles of paleoecology: New York, McGraw-Hill, 371 p.

Ahrens, L. H., 1963, Element distributions in igneous rocks, pt. 6, Negative skewness of SiO_2 and K: Geochimica et Cosmochimica Acta, v. 27, p. 929–938.

Baird, A. K., McIntyre, D. B., and Welday, E. E., 1967, A test of chemical variability and field sampling methods, Lakeview Mountain tonalite, Lakeview Mountains, Southern California batholith: California Div. of Mines and Geol., Spec. Rept. 92, p. 11–19.

Barth, T. F. W., 1962, Theoretical petrology: New York, John Wiley & Sons, 2nd ed., 416 p.

Bradley, P. R., 1925, Estimation of ore reserves and mining methods in Alaska Juneau Mine: Am. Inst. Mining Metall. Engineers Trans., v. 72, p. 100–120.

Brownlee, K. A., 1955, Statistics of the 1954 polio vaccine trials: Jour. Am. Statistical Assoc., v. 50, no. 272, p. 1005–1013.

———, 1965, A review of "Smoking and Health:" Am. Statistical Assoc. Jour., v. 60, no. 311, p. 722–739.

Chayes, Felix, and Fairbairn, H. W., 1951, A test of the precision of thin-section analysis by point counter: Am. Mineralogist, v. 36, p. 704–712.

Chayes, Felix, 1956, Petrographic modal analysis: an elementary statistical appraisal: New York, John Wiley & Sons, 113 p.

Cochran, W. G., 1963, Sampling techniques: New York, John Wiley & Sons, 413 p.

Compton, R. R., 1962, Manual of field geology: New York, John Wiley & Sons, 362 p.

Cumming, J. D., 1956, Diamond drill handbook: Toronto, Canada, J. K. Smit and Sons, 655 p.

Dadson, A. S., and Emery, D. J., 1968, Ore estimation and grade control at the Giant Yellowknife Mine, *in* Ore reserve estimation and grade control: Canadian Inst. of Mining and Metall., spec. v. 9, p. 215–226.

Dixon, W. J., and Massey, F. J., Jr., 1969, Introduction to statistical analysis: New York, McGraw-Hill, 638 p.

Engel, A. E. J., and Engel, C. G., 1960, Progressive metamorphism and granitization of the major paragneiss, northwest Adirondack mountains, New York: Bull. Geol. Soc. Am., v. 71, p. 1–58.

Giles, G. S., 1961, Diamond mining practice in South Africa: *in* 7th Commonwealth Mining and Metall. Cong., South African Inst. of Mining and Metall., v. 2, p. 839–850.

Griffiths, J. C., 1962, Statistical methods in sedimentary petrography, *in* Sedimentary petrography, v. 1: New York, Macmillan, 609 p., p. 565–617.

————, 1967, Scientific method in analysis of sediments: New York, McGraw-Hill, 508 p.

Grout, F. F., 1932, Rock sampling for chemical analysis: Am. Jour. Sci., v. 24, p. 394–404.

Grundy, W. D., and Meehan, R. J., 1963, Estimation of uranium ore reserves by statistical methods and a digital computer *in* Geology and technology of the Grants uranium region: Soc. Econ. Geologists, Mem. 15, p. 234–246.

Harding, B. W. H., 1956, Prospect sampling by air-flush drill: Inst. Mining Metall. Trans., v. 66, p. 79–87.

Hazen, S. W., Jr., and Berkenkotter, R. D., 1962, An experimental mine-sampling project designed for statistical analysis: U.S. Bur. Mines Rept. Inv. 6019, 111 p.

Hazen, S. W., Jr., and Meyer, W. L., 1966, Using probability models as a basis for making decisions during mineral deposit exploration: U.S. Bur. Mines Rept. Inv. 6778, 83 p.

Jackson, C. F., and Knaebel, J. B., 1932, Sampling and estimation of ore deposits: U.S. Bur. Mines Bull. 356, 155 p.

Joralemon, I. B., 1925, Sampling and estimating disseminated copper deposits: Trans. Am. Inst. Mining Eng., v. 72, p. 607–627.

Koch, G. S., Jr., and Link, R. F., 1969, A statistical analysis of some data from deflected diamond-drill holes: *in* Weiss, Alfred, ed., *A decade of digital computing in the mineral industry*, New York, Am. Inst. Mining Engineers, p. 497–504.

Krige, D. G., 1966, Two-dimensional weighted moving average trend surfaces for ore valuation, *in* Symposium on mathematical statistics and computer applications in ore valuation: South African Inst. of Mining and Metall., p. 13–79.

Krumbein, W. C., 1965, Sampling in paleontology, *in* Handbook of paleontological techniques: San Francisco, W. H. Freeman, p. 137–149.

Link, R. F., Koch, G. S., Jr., and Schuenemeyer, J. H., 1970, The lognormal frequency distribution in relation to gold assay data: U.S. Bur. Mines Rept. Inv., in press.

McKinstry, H. E., 1948, Mining geology: Englewood Cliffs, N.J., Prentice-Hall, 680 p.

Manson, Vincent, 1967, Geochemistry of basaltic rocks: major elements, *in* Hess, H. H. and Poldervaart, Arie, eds., Basalts: New York, John Wiley & Sons, p. 215–270.

Mason, Brian, 1962, Meteorites: New York, John Wiley & Sons, 274 p.

Miesch, A. T., 1967, Theory of error in geochemical data: U. S. Geol. Survey Prof. Paper 574-A. 17 p.

Mosimann, J. E., 1965, Statistical methods for the pollen analyst: multinomial and negative multinomial techniques, *in* Handbook of paleontological techniques: San Francisco, W. H. Freeman, p. 636–673.

Norton, J. J., and Page, L. R., 1956, Methods used to determine grade and reserves of pegmatites: Am. Inst. Mining Metall. Petroleum Eng. Trans., v. 205, p. 401–424.

Parks, R. D., 1949, Examination and valuation of mineral property: Cambridge, Mass., Addison-Wesley Press, 446 p.

Prinz, Martin, 1967, Geochemistry of basaltic rocks: trace elements, *in* Hess, H. H. and Poldervaart, Arie, eds., Basalts: New York, John Wiley & Sons, p. 271–324.

Richmond, J. F., 1965, Chemical variation in quartz monzonite from Cactus Flat, San Bernardino Mountains, California: Am. Jour. Sci., v. 263, p. 53–63.

Rowland, R. S., and Sichel, H. S., 1960, Statistical quality control of routine underground sampling: Jour. South African Inst. Mining Metall., v. 60, p. 251–284.

Schottler, G. R., 1965, Statistical analysis of gamma-ray log sample data from a uranium deposit, Ambrosia Lake area, McKinley County, New Mexico: U.S. Bur. Mines Rept. Inv. 6645, 49 p.

Truscott, S. J., 1962, Mine economics: London, Mining Pubs., Ltd., 471 p.

Wallis, W. A., and Roberts, H. V., 1956, Statistics: a new approach: Glencoe, Illinois, Free Press, 619 p.

Waterman, G. C., 1955, Chuquicamata develops better method to evaluate core drill sludge samples: Mining Eng., January, 1955, p. 54–62.

Chapter 8

Variability in Geological Data

This chapter is closely related to Chapter 7 on geological sampling. Chapter 7 is mainly concerned with natural variability and sampling variability. Now, in Chapter 8, two other kinds of variability, those introduced by preparation and chemical analysis of the samples, are discussed.

For brevity, the term *chemist* is used in this chapter to distinguish workers in several disciplines (including analytical chemistry, spectroscopy, fire assay, and x-ray analysis) from geologists. The word *chemist* is used rather than *analyst*, which would be more appropriate, because the word *analyst* can be confused with statistical analyst or with a statistical analysis. Similarly, the phrase *chemical analysis* is used as a general expression for a variety of chemical and physical operations which are applied to geological materials to yield numerical results.

8.1 PROBLEMS OF GEOLOGICAL VARIABILITY

Geological variability is immense. The first step in an investigation is to decide on the purpose of the investigation, to choose a substance or substances of interest, and to decide in what detail the variability need be known. Total information cannot be extracted from a rock any more than a census enumerator can extract total information from a citizen. Too many purposes or too many substances will make a study fall of its own weight.

The four kinds of variability distinguished in this book may be expressed in a general statistical model, in which any univariate observation is written as

$$w = \mu + \xi_n + \xi_s + \xi_p + \xi_a + e$$

where ξ_n is the natural variability, ξ_s is the sampling variability, ξ_p is the preparation variability, ξ_a is the analytical variability, and e is the random fluctuation not accounted for by the other sources of variability. The coefficients of variation recorded in section 7.8 reflect an unknown mixture of these four variability sources and of the random error. Even in a designed experiment, it is difficult to separate the different sources of variability, but the aim is for the coefficient of variation to reflect natural variability closely. For unevaluated data, one can only hope that the size of the coefficient of variation expresses the natural variability.

In this chapter, natural, sampling, preparation, and analytical variability are taken up in turn in sections 8.2 to 8.5. Natural variability is a property of rock bodies; the other three kinds of variability stem from operations performed on single geological samples or specimens. Then in section 8.6, on experimental designs, variability in rock bodies is related to variability in single geological samples and to the decisions that must be made to determine what precision is required. This subject of decision making, introduced here for the first time, is explored at length in later chapters, especially Chapter 14.

8.2 NATURAL VARIABILITY

Natural variability is the variability inherent in a rock body (sec. 7.2). It is the variability that ideally would be discovered if only it could be separated from the other sources of variability. The concept may be explained by the expository device of considering some kinds of natural variability in the gold-bearing Pinyon Conglomerate of Paleocene age in northwestern Wyoming. Because little has yet been written (Antweiler and Love, 1967), the exposition need not be constrained by the facts.

The Pinyon Conglomerate crops out across an area of more than 100 square miles and has a volume of a few tens of cubic miles, according to Antweiler and Love (1967, fig. 2 and p. 2), who write that

> ... the Pinyon Conglomerate was deposited by rivers that flowed from the quartzite source area eastward and southwestward across Jackson Hole. . . .[In] the area of maximum conglomerate deposition . . . more than 5,000 feet of strata, largely conglomerate, was deposited.

Of the many geologic models that may be postulated for the Pinyon Conglomerate, one could be a model of the composition, and another could be a model of the process of formation. A composition model might be regarded as a tabular body that is bounded by undulating upper and lower surfaces and that thins out laterally. This tabular body is disrupted by stream valleys, folds, and sills; and it is cut by joints, faults, and dikes. Within the defined space are rock fragments ranging in size from grains a few microns across up to boulders. The proportions and arrangement of these rock fragments account for the variability in the composition model. They also provide the basic information for the model of the process of formation; from them the geologist may test hypotheses about geological processes, for instance, about stream channels within the formation which may be marked by changing porosity, permeability, grain size, composition, etc.

Yet at the present state of geological knowledge total variability is far too complicated a concept to be useful. Instead, specific kinds of variability, for instance, that of the gold particles or that of the pebbles, must be surveyed. If one had x-ray eyes, he could see into the Pinyon Conglomerate, learn exactly where the gold is, and measure its variability in terms of grain size, shape, composition, or other variables of interest. One could see the pebbles, identify the composition of each, and thus define the composition model. Still in the realm of make-believe, one could continue backward through time in a time machine to contemplate the process model and determine whence and how these gold grains and pebbles were carried into place. Such speculation is not idle daydreaming but is a fruitful way to contemplate the problem in order to devise a practical sampling scheme.

If the reader keeps in mind the geological model that we have set up for the Pinyon Conglomerate, he will understand why the distinction between target and sampled population concerns us. The reason is that the variability in the sampled population may not be the same as that in the target population. Even if the target population is potentially available for sampling through defining it as the uneroded Pinyon Conglomerate, only a small fraction of the formation is close enough to the surface to actually sample, and there is little hope that this portion is representative. If the target population is defined as the entire formation that was deposited in Paleocene time, no chance at all exists to sample it representatively because some of the formation has been eroded away.

Natural variability is different in the Pinyon Conglomerate for the many attributes that may be studied; for instance, the variability in the proportion of quartzite pebbles is likely to be low (sec. 6.1). On the other hand, the variability of the gold is likely to be high; from one typical locality, Antweiler and Love (1967, personal communication) took 44 geological samples whose mean was 0.2158 ppm and whose coefficient of variation was 3.28.

Natural variability may best be examined by stratifying the rock body. Thus the Pinyon Conglomerate may be stratified in the conventional geologic sense according to time or lithologic breaks, or it may be stratified according to geographic position, if, for instance, the gold increases in a certain direction. There may be a trend, usually a linear one in variability (sec. 9.4), or an area of influence, for instance, a well-sorted conglomerate in which the pebble variability is lower than elsewhere.

8.3 SAMPLING BIAS AND VARIABILITY

Sampling variability, which is that introduced by the process of physical sampling, is discussed for various kinds of sampling in Chapter 7, and in this section only the main conclusions are reviewed and summarized. Because the sampling variability obscures the natural variability, it should be kept small; however, its size should be appropriate to that of the other sources of variability, as there is little advantage to low sampling variability if the preparation and analytical variability are excessive. The main sources of sampling variability stem from losing material that should be in the sample and from contaminating the sample by adding extraneous material to it.

Some of the many sources of contamination are stated in Chapter 7. Sample sacks may contain material left over from a previous job; diamond-drill sludge may contain material from another interval than that being sampled; gravel may cave from the walls of test pits; and steel chips broken off picks and chisels may find their way into the sample. These are accidental sources of contamination; intentional contamination, named *salting*, is discussed entertainingly by McKinstry (1948, pp. 67–69) and by other authors that he references.

Losing material from the sample is serious because one can seldom determine if the loss is biased or not, and usually it is biased. Sample may be lost when rock fragments fly away from the outcrop rather than fall into the sample sack, when diamond-drill core is ground up, when sludge settles in cracks in the rock rather than being washed out of the hole, when dust from a rotary drill blows into the air, etc. These losses almost always introduce bias because the various sized fragments differ in grade (sec. 15.1). Even if bias is absent, variability is surely increased because the desired volume of sample is reduced to an unknown extent that is different from sample to sample.

The scale on which sampling variability is measured must be taken into account by prior information or estimation before sampling begins. For instance, in sampling the Pinyon Conglomerate in Wyoming to determine the proportion of quartz pebbles, geological samples should be larger than one pebble (sec. 6.1); but in sampling to determine the chemical composition of individual pebbles, no geological sample should be larger than one pebble.

8.4 PREPARATION BIAS AND VARIABILITY

Sampling bias and variability, discussed in the preceding section, are clearly the geologist's responsibility. Analytical bias and variability, discussed in the following section, are clearly the chemist's responsibility. Preparation bias and variability, the subject of this section, are not clearly in the province of either of these disciplines. Therefore they are often ignored by both professions, and the task is delegated to technicians whose work may be good, but more often it is bad because of lack of training or interest or both. Probably in most geological sampling more bias and variability are introduced in sample preparation than in any other stage, although we cannot prove this contention because the literature is sparse and lacking in numerical data.

This section is limited to discussing in turn the purposes of sample preparation, then some of the many problems, and finally some of the physical methods. The reader is warned that he should understand the many tiresome details involved because no one else is likely to take the time and care to prepare his samples competently for the chemist.

Purposes of Sample Preparation

Sample preparation has two principal purposes: (a) to reduce the amount of the geological sample and (b) to make it uniform. The first purpose is nearly always necessary because the earth is very large and the amount of material that the chemist can analyze is nearly always very small. The sample may be reduced, either in bulk by removing a certain weight of rock, or in number, perhaps by taking a certain number of sand grains or a certain number of microfossils.

The second purpose of sample preparation is to make a uniform sample by blending (or mixing), which is one of the most difficult processes in sample preparation. After this process a portion usually is removed.

Problems in Sample Preparation

The problems in sample preparation discussed in this book are systematic losses, contamination, blending, concentration, splitting, and alteration. In all of these problems, variability is bound to be introduced, and the goal is to keep it as low as possible. Moreover, serious bias can be introduced extremely easily but usually can be held down to a manageable level by planning. The problems can be tackled with the methods explained in the next subsection.

Systematic losses are those that occur in every sample that is prepared. If, every time a rock is pulverized, a certain weight or proportion of it is lost, trouble results, although it may not be serious unless one constituent of the rock is more liable to be lost than another. One way to determine systematic losses is to weigh the sample at various stages during its preparation, remembering that the sample will gain in weight through contamination as well as decrease in weight through loss. Another way is to prepare samples differently and compare the results. This work should be done on extra material specifically obtained for the test, not on the experimental material itself.

Contamination is the addition to the sample of foreign material; if done intentionally with intent to defraud, contamination amounts to salting (sec. 8.3). Whenever geological samples are crushed or pulverized, contamination is inevitable, as the equipment is abraded as well as the rock. Therefore, one must decide how much and what kind of contamination is acceptable, and then he can take appropriate action. For instance, pulverizers with ceramic rather than metal grinding plates can be used if contamination from iron or other metals is a problem. Because a penalty is paid in greater time and therefore cost, one should determine if the additional expense is justified.

Blending is difficult to do properly, as is well known to industries whose product depends on blending, for instance, the paint industry. It is best avoided in sample preparation. Many ingenious blending devices have been invented, but nearly all of them impart some kind of concentration when blending rock powder, the usual geological material to be blended, especially when the particles vary in shape or specific gravity, as with mica flakes, gold, platinum, and asbestos. Besides the concentration initially present, a concentration is often introduced in blending a sample. For instance, differential settling is well established for such substances as rock powders stored in buildings subject to vibration, molybdenum ore samples carried by airplane after they are pulverized and blended, and samples trucked from a sample preparation plant to a chemical laboratory a few miles away.

The best way to reduce the bulk of samples is by splitting. Although many problems arise, splitting is a better way to reduce the bulk of a sample than blending and scooping some out. Sample splitters of many different designs have been developed, and studies of their performances are discussed in the next subsection. The aim is to ensure that, when a particle goes through a splitter, it has a known chance of going into a particular compartment.

Finally, alteration of a sample during its preparation may be a problem. An example is found in preparing copper sulfide ores, when sulfide minerals can easily be changed to oxides by heat, which is introduced by drying sludges from drill holes or by crushing and grinding. Another example is found in preparing coal samples whose moisture, gas content, etc., are important and can be easily altered. Many other kinds of samples deteriorate with age, and the advantage of speedy analysis must be balanced against the drawbacks of doing chemistry under field conditions.

Methods of Sample Preparation

Of the many methods for preparation of geological samples, some of those most used are listed in table 8.1. The methods have been grouped according to the physical or chemical process used: these may be concentrating, leaching, pulverizing, splitting, or blending. The table also lists some geological materials for which the methods are suitable. Most of the methods are intended for samples weighing a few kilograms or more and are also usually

TABLE 8.1. SOME METHODS OF SAMPLE PREPARATION

Processes	Examples of geological materials
Concentrating processes:	
Washing Gravity concentration Heavy liquid concentration Flotation Elutriation Screening Precipitation Magnetic concentration Electrostatic concentration Hand picking	Unconsolidated or weakly consolidated sediments and sedimentary rocks, especially those with grains varying in size, shape, specific gravity, magnetic properties, or electrostatic properties
Leaching and other chemical processes:	
Solution in water, cyanide solutions, and other reagents, depending on specific nature of the material Amalgamation	Soluble rocks and minerals
Pulverizing:	
Drying Crushing Grinding Pulverizing	Rocks whose grains are relatively uniform in size or specific gravity Other rocks and minerals not amenable to preparation by concentration or leaching and other chemical processes
Splitting:	
Splitting	Any sample composed of discrete grains or fragments
Blending:	
Rolling Shaking	Any sample composed of discrete grains or fragments

used in industry for processing ores, although the machines are scaled down in size for sample preparation. Preparation of smaller samples is explained in specialized works on petrography, for instance, those by Johannsen (1918), Krumbein and Pettijohn (1938), and Milner (1962).

Whole volumes have been written about sample preparation, and to comment in detail on the many methods would expand this book far too much. A few of those most important to a geologist are mentioned, and some of the few published numerical results are cited. Of the works on sample preparation, the following are particularly pertinent. Peele (1941) and Taggert (1945, sec. 19, 208 pp.) explain the preparation of ores and large samples weighing as much as a few tons or even more; Lundell and others (1953) have written on inorganic chemical analysis and rock analysis in particular; and, Milner (1962), German Mueller (1967), and L. D. Muller (1967) discuss sample preparation in sedimentary petrology.

In practice, several of these methods are often combined, and then a flow chart is helpful. Figure 8.1 is a flow chart devised by Schottler (1968, personal communication) for processing gold ore from a saprolite near Kershaw, South Carolina.

We next turn to a discussion of some specific methods of sample preparation. Ideally, each method would be described in terms of experiments performed as follows for all geological materials of interest. Two or more samples would be prepared by each preparation method, with appropriate randomization. When the analytical results were returned, the sources of variability could be partitioned through a nested analysis of variance, as is done for the data from the Kilembe, Uganda, mine (sec. 15.4). However, it is quite impossible to offer any such description because the experiments have not been performed, and the data are not available. Coxon and Sichel's excellent work (1959) discussed below comes as close as any, although they did not quite achieve the overall partitioning of sources of variability.

Concentration

In the first group of methods, those for concentration of samples, quantitative concentration may or may not be required. If it is required, as for preparation of placer samples (sec. 15.2), three classes of material must be distinguished: the *heads*, which constitute the material to be concentrated, the *concentrate*, which is the material enriched in the substance of interest, and the *tails*, which is the remaining material impoverished in the substance of interest. If the concentration were ideal, the concentrate would contain all of the substance of interest and the tails would contain none. However, the tails always contain at least a trace of the valuable substance, often in an amount too small to analyze directly. Therefore, the only check on the efficiency of a quantitative mechanical concentration process is usually to

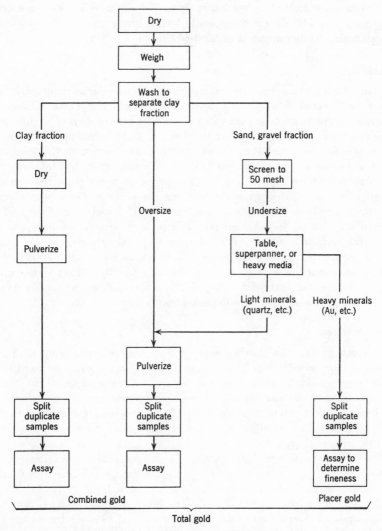

Fig. 8.1. A flow chart for sample preparation of gold in saprolite (after Schottler, 1968).

reconcentrate the tails or a part of them by the same or another method. Most of the recorded information on quantitative concentration is in the placer mining literature (sec. 15.2).

If quantitative concentration of samples is not required, the problem of sample preparation is simplified. Usually the aim is to obtain fairly pure mineral concentrates for chemical analysis. Methods used include concentration by magnetic forces; electrostatic forces; heavy liquids; elutriation in water, air, or other media; and hand picking.

Leaching

In table 8.1, the second group of methods listed includes leaching and other chemical methods for dissolving samples. Solution of water-soluble substances is seldom a problem. In exploration geochemistry various ways of leaching are commonly used, as explained by Hawkes and Webb (1962, pp. 35 ff.). Once the samples are in solution, mixing and then sampling from the solutions present a few problems, but they have been thoroughly studied by chemists.

Another interesting problem, whose explanation might point the way for sampling other substances, is the leaching of gold. In gold ores, where large volume samples are desirable because the size and distribution of the gold is extremely varied, it has been proposed to leach samples of from one to several kilograms with cyanide solutions for total extraction of the gold. Because nearly complete recoveries of gold can be made with cyanide solutions in production-scale mills, the problems, evidently not yet surmounted (Green, 1968, personal communication), lie in scaling down the operation and in devising laboratory-scale equipment that is easy to clean.

Pulverizing

Another group of methods comprises those that pulverize samples. If the samples are wet, the first step is drying, which should cause no particular problems provided that the temperatures are kept low enough so that nothing is lost through sublimation or is altered through oxidation or other chemical change. Also, no contamination from such sources as air currents blowing material about can be allowed. The amount of water evaporated is measured if its amount is important.

After drying, pulverization requires several (occasionally only one) steps to reduce the particle size of the sample. Although the names crushing, grinding, and pulverizing are conventionally given to three stages of reducing size gradationally, the physical actions involved may overlap, and the fundamental distinctions if any are controversial. Sample fragments are broken by both hitting and rubbing; in both of these processes, contamination by the apparatus is inescapable; also dust losses of the geological sample may be unavoidable. Pulverizing is probably the worst necessary part of sample preparation; the other bad part, blending, is a step that nearly always can and should be avoided.

While setting up a laboratory for pulverizing ore samples a few years ago we investigated some of the many machines for crushing and pulverizing rocks on a laboratory scale. Most of these machines are scaled-down models of ones used to crush ore for production. All that we have seen have some or all of these faults: excessive dust losses, difficult cleanup, excessive contamination, or shoddy construction (misaligned parts, defective bearings, broken castings, etc.). The crushers and grinders can quite readily be modified to reduce losses in receiving and discharging material; once this modification is done, these machines, which reduce material down to approximately 8 mesh per inch, introduce less contamination and have fewer problems than do the pulverizers. The modifications needed are to catch any entering material that is thrown back out and to collect without loss the material after it has been reduced in size. Pulverizers are of two general types. In the first the sample is fed between two plates, one or both of which rotate, and is collected below them; dust losses tend to be high; the plates frequently have to be refaced, and they become misaligned. In the second the sample is contained in a sealed chamber in which it is pulverized by the action of hard pucks or balls; there is little dust loss in this type, but much excessively fine powder is produced, as all of the rock stays in the chamber until the pulverizing action is completed.

Better machines for crushing and pulverizing may be made; we have not seen them all. For none have we seen any authentic records evaluating quality of the sample preparation, an evaluation which is very important as indicated by an interesting experiment that Coxon and Sichel (1959, p. 496) performed on high- and low-grade gold ore. Figure 8.2 shows that, for both high- and low-grade ore, reproducibility increases rapidly as the sample is crushed finer. In fact, one pulverized sample has about 100 times the precision (square of the ratio of the coefficients of variation) of one coarse-crushed sample. Moreover, Coxon and Sichel point out that "from the histograms it is also evident that ideally the total sample as received should be pulverized before assay . . . economically this is not possible, and hence sample splitting will be with us for some time to come." But an experiment like theirs provides a basis for establishing the necessary steps to obtain acceptable preparation variability and cost.

A sample preparation laboratory always has many samples coming into it; through Parkinson's law they increase to clog the capacity of the plant, and the aim will be to get rid of them as fast as possible. To do this, the tendency is to coarse-crush, split, and throw out the material rejected in the splitting, which is an acceptable procedure unless it leads to the trouble pictured in figure 8.2. Accordingly, tables of "safe" sizes at which to split have been constructed. Those by Taggert (1945) are the most widely used, but as Davis (1963, p. 259) points out in a study of sample preparation at the Kilembe, Uganda, mine, the generalities are of little value, and each case should be evaluated on its own.

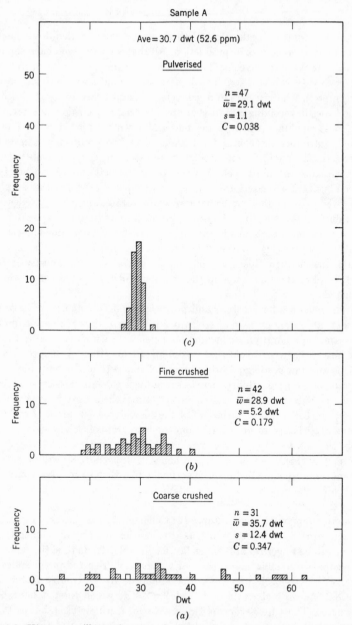

Fig. 8.2. Histograms illustrating reproducibility of mine assay values when identical sample is split after (*a*) coarse crushing, (*b*) fine crushing, and (*c*) pulverizing (after Coxon and Sichel, 1959, p. 496).

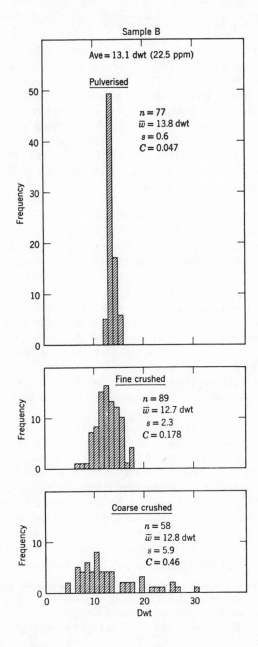

If contamination is a problem, several precautions can be taken. Lundell and others (1953, p. 809) state that breaking rocks by hand methods that emphasize hitting rather than rubbing reduces iron contamination. Also ceramic plates in machine pulverizers, or ceramic or agate mortars, can be used to substitute silica contamination for metal contamination if silica contamination is less undesirable. Klugman (1966, personal communication) reports that most pulverizer plates are cast from scrap metal that may contain many kinds of metal impurities different from one lot to the next, and presumably crusher plates are made in the same way. One could have these plates custom cast to control this problem.

Splitting

Splitting, the fourth method in table 8.1, is used to divide fragmented material—either originally disaggregated material such as gravel or soil, or pulverized rock—into two or more parts. One or more of several devices may be used; the most common is the Jones splitter, in which the material to be split is dumped onto a knife edge so that one-half falls on one side and the one-half on the other. Splitting is almost always needed to prepare samples for chemical analysis and is often needed to reduce samples for grain-count analysis. The requirements differ, depending on the precise purpose of the splitting and on the amount of material to be handled. For grain-count analyses a specified fraction of the total material, often one-half, must go to a given compartment of the splitter; whereas in order to get a standard amount of material for chemical analysis, a specified volume or weight may be collected.

The fairly extensive literature on sample splitting, which is more complete than for most other kinds of sample preparation, reflects the various requirements. The *Journal of Sedimentary Petrology* has published several papers over the years, among the best of which are those by Wentworth and others (1934), Otto (1937), Kellagher and Flanagan (1956), and Flanagan and others (1959). Milner (1962) and Griffiths (1967) discuss sampling for sedimentary petrology, and Boericke (1939) and Coxon and Sichel (1959) discuss splitting of mine samples. Manufacturers of splitters also have information.

Flanagan and others (1959, p. 108) clearly explain desirable qualities for a laboratory splitter that should also apply to larger ones:

1. The splitter should be able to reduce both small and large amounts of sample.
2. The splitter should be capable of sampling a wide variety of particle sizes.
3. The time required for the operation should be short.
4. All particles in the lot to be sampled should have an equal chance of being sampled.
5. The sampler should be easily cleaned and require a minimum of maintenance.
6. The operation of the sampler should be simple.

7. The materials of which the sampler is made should not contaminate the sample.

A review of the relative merits of different types of splitters would take too much space in this book. However, the reader is warned that bias and unnecessary variability can easily be introduced by the Jones splitter, which has been more or less the standard one. For an extensive research project, one is well advised to test a splitter on representative material of the type to be split. Experiments made with other materials, particularly artificial ones, are liable not to apply to the problem at hand; for instance, a splitter may satisfactorily split lead shot in sand but may not do at all well on flakes and flattened nuggets of gold in fibrous gangue minerals.

The most-used splitters are the Jones splitter and cone splitters, which divide the material outward from the apex of a cone. Another splitter that has been recommended for laboratory use was devised for the U.S. Atomic Energy Commission office in Grand Junction, Colorado. This splitter feeds powder down a chute by a vibrating action (as in a Frantz magnetic separator) to discharge onto a rotating wheel about the size of a roulette wheel, with compartments rather like those that catch the roulette ball. These compartments catch the powder; and by varying their sizes, one can collect a sample of a certain fixed volume or divide an initial volume into fractions of different proportions.

Blending

Finally, blending, the last of the methods in table 8.1, should be avoided in favor of splitting because it is difficult to do properly. We have found no experimental data of particular interest, although presumably something is known in industries like paint manufacturing that do a lot of blending. Materials may be blended wet or dry; dry blending is more often done in geology. It seems probable that blending usually segregates rather than blends particles differing from the average in specific gravity, size, or shape. If blending cannot be avoided, one should take samples for analysis from several parts of the blended pile—or from top and bottom of the blending container—to establish the blending variability. The literature shows that problems that arise are extremely specific for the substances that are blended.

8.5   ANALYTICAL BIAS AND VARIABILITY

If a geologist could have his wish, he would send replicate increments of perfectly prepared geological samples to several chemists, none of whom could communicate with each other, and each chemist would make determinations by several chemical methods. After the resulting observations were statistically analyzed, the geologist would know within definite limits the composition of his rock.

This ideal state of affairs does not seem likely to be realized in everyday analytical practice in the forseeable future, although Britten (1961, p. 1008) writes that in South Africa "under good conditions of application two assayers work independently using separate reagents and separate equipment including assay balances. At the Rand Refinery the obviously beneficial provisions of entirely separate and independent accommodation and equipment has been an unqualified success for many years." Until it becomes economically feasible to perform all chemical analyses of geological samples under such controlled conditions, geologists must realize that mistakes in chemical analysis can always happen and probably will.

As already mentioned in this chapter, the problems of analytical bias and variability lie primarily in the discipline of analytical chemistry, thought of in its broadest sense as including other methods of analysis, like spectroscopic or x-ray analysis. In this book, space permits only a brief treatment of a few of the many ways to evaluate analytical data statistically. Dalrymple and Lanphere (1969, Chap. 7) give an excellent example of the application of statistics to the evaluation of the reliability of data for potassium-argon dating.

That the accuracy of chemical analyses made in well-run laboratories has been increasing in recent years is illustrated by Tatlock (1966) in an excellent review of 55 older and 110 modern chemical analyses of Australasian tektites. Tatlock suggests that "more than half of these older alkali and titania determinations are decidedly inaccurate and misleading." For instance, in figure 8.3a K_2O is plotted against Na_2O for 110 modern tektite analyses made by five different chemists to define a joint range-of-variability region. Then, in figure 8.3b, 55 older analyses are plotted, one-half of which fall outside the joint region.

However, in some laboratories, the accuracy of routine chemical analyses of geological materials may have actually deteriorated over the years, probably because fewer geologists and mining engineers do their own analyses and are familiar with the analytical methods. Therefore the geologist must rely in part at least on consistency of the observations. Some ways to investigate consistency are explained in this section. One of the simplest is to determine if the analyst favors certain digits in reporting analyses, a subject fully developed by Hazen (1967, pp. 80–83). For instance, for 90 assays from a drill-sampling project, Hazen states that the frequency of occurrence of digits in the tenths position, from 0 to 9, was as follows: 8, 10, 7, 7, 10, 6, 10, 17, 2, 13, instead of the expected average frequency of 9 for each digit. The

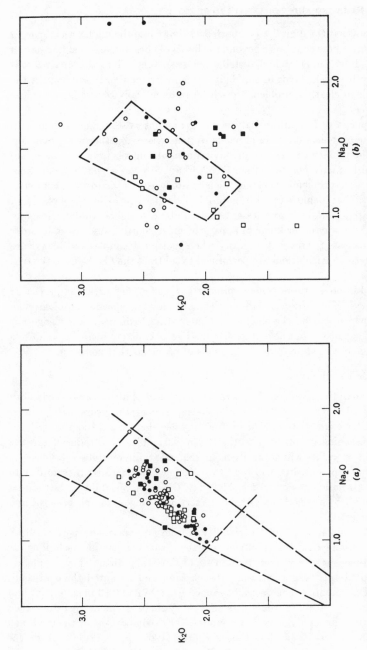

Fig. 8.3. Variation of K_2O with respect to Na_2O in analyses of Australian tektites (after Tatlock, 1966, p. 124). In (*a*) are plotted 110 precisely monitored analyses; in (*b*) are plotted 55 older analyses. Symbols identify analysts cited in the original paper.

list shows that the digit 7 was reported 17 times, and the digit 8 was reported only twice; these counts are beyond the limits of normal statistical fluctuation predicted by the binomial distribution (sec. 6.1). Thus the analyst was strongly biased in favor of the digit 7 (his lucky number?) and against the digit 8, and clearly a problem arose if he systematically reduced most values of 8 to 7.

The purpose of chemical analysis is to get an observation, and in doing this all of the problems stated in previous sections arise—together with some new ones. The problems in setting up environmental controls for a laboratory are detailed in many books; Wilson (1952, Chap. 5) gives a good introduction. One of these problems is getting the laboratory apparatus to work correctly, a process that includes calibration problems, which also arise in sampling and preparation, although to a lesser extent. Balance weights, volumetric glassware, and many other kinds of apparatus must be calibrated. Also, in order to compensate for reagent impurity when geological samples and blanks are run together, calculations are performed to subtract the blanks from the true samples.

In addition to these general problems, specific ones arise for particular substances, for instance, certain elements. Solving specific problems like these, which can be challenging and interesting, requires a knowledge not only of chemistry but also of geology and geochemistry. Fleischer and Chao (1960) give a good account of the kinds of problems that come up.

Operator Variability

The analytical problem of operator variability, which is that variability introduced by the people who make the experimental measurements, should always be evaluated. The operators may be chemists, geologists, petrographers, engineers, or other workers. The variability is introduced because different operators will naturally make somewhat different observations, and even the same operator will tend to be somewhat inconsistent from day to day and from hour to hour on a single day. Operator variability is introduced in section 7.6 on modal analysis; in this section, a single example is considered in more detail.

The example of operator variability is from an experiment performed by Rosenfeld and Griffiths (1953) to test the visual-comparison technique for estimating sphericity and roundness of quartz grains. They randomly selected 21 quartz grains from a beach sand and mounted them in semipermanent mounts. The 21 grains ranged from 0.71 to 0.93 in sphericity and from 0.10 to 0.80 in roundness. Seven graduate-student operators then estimated the sphericity and roundness by comparison with published standard charts (reproduced by Griffiths, 1967, pp. 112–113, from the original papers). In this book we discuss only the analysis for sphericity

Table 8.2. Average mean squares in an experiment to test visual-comparison techniques for estimating sphericity *

Source of variation	Average mean squares
Operators	$\sigma^2 + 3\sigma^2_{OXG} + 21\sigma^2_{D(O)} + 63\sigma^2_O$
Days	$\sigma^2 + 3\sigma^2_{OXG} + 21\sigma^2_{D(O)} + 147\sigma^2_D$
Grains	$\sigma^2 + 3\sigma^2_{OXG} + 36\sigma^2_G$
Day (operator)	$\sigma_2 + 3\sigma^2_{OXG} + 21\sigma^2_{D(O)}$
Day–grain interaction	$\sigma^2 + 7\sigma^2_{DXG}$
Operator–grain interaction	$\sigma^2 + 3\sigma^2_{OXG}$
Day (operator)–grain interaction	σ^2

* Data from Rosenfeld and Griffiths, 1953.

measurements of the raw data; Rosenfeld and Griffiths present further interesting analyses of transformed sphericity data and of the roundness data.

An analysis of the raw sphericity data is presented in tables 8.2 and 8.3. The set of average mean squares (table 8.2) is similar to but not exactly like that postulated by Rosenfeld and Griffiths, a difference that we deliberately introduce to show that different people can interpret the same data differently. Thus, unlike Rosenfeld and Griffiths, we believe that although the day-to-day variation of a single operator can be estimated, this variation is not an operator–day interaction because days are nested with respect to operators. (An interaction would be like the operator–grain interaction in this experiment, or the method-formation interaction in section 5.9.) Moreover, it is unlikely that the sphericity of the quartz grains could change from day to day, so we presume that the day–grain interaction is 0.

Table 8.3. Analysis of variance in an experiment to test visual-comparison techniques for estimating sphericity *

Source of variation	Sum of squares	Degrees of freedom	Mean square	F	$F_{10\%}$	Degrees of freedom used for test
Operators	865.14	6	144.19	6.94	2.33	(6, 12)
Days	42.13	2	21.07	1.01	2.81	(2, 12)
Grains	4879.55	20	243.98	24.72	1.48	(20, 120)
Day (operator)	249.51	12	20.79	2.11	1.60	(12, 120)
Day–grain interaction	203.01	40	5.08	0.91	1.28	(40, 240)
Operator–grain interaction	1184.19	120	9.87	1.76	1.21	(120, 240)
Day (operator)–grain interaction	1344.02	240	5.60			

* Data from Rosenfeld and Griffiths, 1953.

The analysis of variance (table 8.3) for our postulated model leads to the following conclusions:

1. There is a small day-to-day difference in the estimates of roundness, which stems mostly from the day-to-day variability of the individual operators. The additional day-to-day variability found by averaging among operators is negligible.

2. The next largest source of variability is the variation among operators; this variability is larger than the variability of a single operator from day to day.

3. The largest source of variability is that in the grains, because their sphericities are very different. Notably, had the operator variability been confused with the sphericity variability through some operators measuring only certain grains and other operators measuring only certain other grains, the operator and sphericity variabilities could not have been separated. Then it might have been difficult to identify the sphericity differences because of the larger operator variability.

Some Problems of Analytical Sensitivity

Chemical and other analytical methods are generally, if not always, suitable for certain concentration ranges of the substance being analyzed. Then observations too large or too small must be grouped as being either above or below the detection limits. When the value "zero" or "not detected" is reported for some of a group of determinations from an area where the substance is usually present, as in a set of gold assays from a mine, one believes that, if most of the geological samples contain detectable gold, the others also contain a little. Other sorts of grouped determinations are those reported as "trace" or as "1 ppm," etc. Although the problem usually arises at the low end of the frequency distribution, trouble also may occur at the high end of the range, as with spectrographic analyses. Miesch (1967b, p. B3) explains clearly the four kinds of problems, which come from distributions that are *left censored* (concentrations too small to be evaluated), *right censored* (concentrations too large to be evaluated), *left truncated* (an unknown number of values below the detection limit), and *right truncated* (an unknown number of values above the detection limit). He writes that

Every analytical method has limits of sensitivity beyond which it is ineffective for determination of concentration values. These limits may occur at both the lower and the upper bounds for a concentration range. . . . [A] spectrographic method . . . for example, may be used to determine silicon within the range from 0.002 to 10 percent. Concentrations judged to be lower or higher than this range are reported as < 0.002 percent or > 10 percent, respectively. Thus, the spectrographic data may be either left or right censored. The term "censored" is applicable here because the number of values beyond each sensitivity limit is known for any set of analyzed rock samples. In other types of problems, where the number of values beyond certain limits is unknown, the data are said to be truncated. . . .

An example of a left-truncated distribution is a size distribution of beach sand grains from which those grains below a certain size have been blown away by the wind. The geologist measuring this distribution might be certain that some grains had been blown away but not know how many. Another example of a left-truncated distribution is the weights of nuggets found in Victoria, South Australia; complete records establish the numbers and weights of large nuggets from this famous placer area, but unfortunately information on the small nuggets was never recorded. An example of a right-truncated distribution is assay values from mines where the assayer discards high values without recording their number and assays additional sample splits, on the grounds that high values are erratic.

Of the four problems left-censored distributions arise most often and therefore cause the most trouble, particularly if a large proportion of a set of observations is reported as trace or 0. The difficulty in forming ratios, because division by 0 is not allowed, is discussed in section 11.2. Another problem arises in computing standard deviations because, if many observations are the same, a "spike" is introduced into the distribution, and the standard deviation is reduced below its correct size. This problem exists even if observations of 0 or trace are replaced by a single arbitrary small value, perhaps the lowest limit of analytical sensitivity or halfway between this lowest limit and 0, as is often done, particularly if a logarithmic transformation is to be made.

Miesch (1967b) clearly explains several special statistical methods that have been devised to extract information from censored and truncated data. For three reasons these methods would seem to have limited use for most geological problems. First, if only relative sizes of observations are needed, as in exploration geochemistry and many problems of theoretical geology, censored and truncated distributions cause little trouble because the statistics, although they may be biased, will reflect relative sizes. Many of the kinds of data that are censored and truncated, such as spectroscopic analyses, are prevalent in these fields. Second, if observations are below the detection limit, even the most complicated statistical technique cannot help much because little information exists. Third, if observations are above the detection limit, a different method of chemical analysis should be used if the information is important to the investigator; money is better spent in this way than on statistical analysis.

Evaluation of Analytical Data

The reliability of analytical data can be evaluated by the statistical methods already introduced in this book, particularly by the analysis of variance. In this subsection, two representative examples are discussed.

TABLE 8.4. ANALYSIS OF VARIANCE IN GOLD DETERMINATIONS MADE BY THREE DIFFERENT METHODS*

Source of variation	Sum of squares	Degrees of freedom	Mean square	F	$F_{10\%}$
Samples	229.3	21	10.92	42.9	1.60
Analytical methods	0.456	2	0.228	0.9	2.44
Residual variability	10.69	42	0.255		

* Data from Erickson and others, 1966, p. 3.

In a study of gold mineralization in the Cortez district, Lander County, Nevada, Erickson and others (1966, p. 3) analyzed 22 gold samples by three different analytical methods—fire assay, colorimetric, and atomic absorption. In table 8.4 we compare the determinations by an analysis of variance to learn if, on the average, the three methods give the same answer. Because of the belief that gold tends to be lognormally distributed, the analysis of variance presented is the one performed on the logarithms. Since the calculated F-value of 0.9 is lower than the tabled F-value of 2.44 at the 10-percent significance level, evidently the three methods give the same results. However, if an analysis is carried out on the untransformed values, the multiple-comparison procedure (sec. 5.12) shows that the colorimetric analysis gives higher results than the other two analytical methods, which nearly agree with each other. Thus, the final conclusion must be that insufficient data were taken to draw a clear-cut conclusion about analytical variability.

As a part of the extensive study of analytical problems in analysis of the G-1 and B-1 rocks, Flanagan (1951) reported on a cooperative study of the lead content of the granite, G-1. In an interlaboratory comparison, each of six laboratories made six duplicate determinations (12 chemical analyses in all) on six samples of G-1 replicated six times. Thus there were 36 pairs of

TABLE 8.5. ANALYSIS OF VARIANCE FOR COOPERATIVE STUDY OF THE LEAD CONTENT OF GRANITE, G-1*

Source of variation	Sum of squares	Degrees of freedom	Mean square	F	$F_{10\%}$
Samples	41.2	5	8.2	0.9	2.16
Laboratories	705.9	5	141.2	15.9	2.16
Replications	153.6	5	30.7	3.4	2.16
Residual variability	177.4	20	8.9		

* Data from Flanagan, 1951.

Table 8.6. Multiple-comparison of means for lead content estimated by six laboratories[*]

Laboratory	Type of analysis	Mean of Pb (ppm)
A	Spectrographic	55.7
B	Spectrographic	49.2
D	Spectrographic	47.8
E	Spectrographic	47.2
F	Chemical	47.1
C	Chemical	46.5

[*] Data from Flanagan, 1951.

analyses or 72 in all. The six samples of rock distributed to the six laboratories were randomly selected, and the experimental arrangement was that of a Latin square (Dixon and Massey, 1969, p. 310).

Table 8.5 is an analysis of variance for the results of this experiment; the numbers are those calculated by Flanagan (1951, p. 119), but their labels are changed slightly to correspond to the notation of this book. There appear to be no differences among the six samples, but there are replication differences within the laboratories, and also the laboratories find, on the average, different means. The replication variation shows that the individual laboratories cannot repeat determinations from day to day, or from week to week, as well as they can duplicate determinations at one particular time. This situation is common in chemical analysis.

In table 8.6 the means for lead estimated by the six laboratories are compared by the multiple-comparison procedure (sec. 5.12). The results show that the results from laboratory A appear to be discrepant from the rest. If the results from this laboratory are excluded, the mean value of the determinations from the other laboratories is 48 ppm, compared with the nominal value of 27 ppm for lead content of G-1 accepted before Flanagan's study. Moreover, the discrepancy between the spectrographic and chemical analyses stems from the determinations made by laboratory A. When the determinations of this laboratory are excluded, no difference between the spectrographic and colorimetric determinations is apparent.

8.6 EXPERIMENTAL DESIGNS

An experimental design is a procedure for gathering data systematically. One example is the randomized block design introduced in section 5.7. The purpose of an experimental design is to evaluate different kinds of variability.

There would be no point in defining kinds of variability if there were no way to sort them out. Experimental designs may be used to investigate the types of variability discussed in this chapter and are also pertinent for geologists who perform laboratory experiments, such as in experimental geochemistry, geophysics, and sedimentology.

In order to use an experimental design, one must have something clearly in mind that he wants to study. The variables (named *factors* in the language of experimental design) and their ranges that are to be considered must be identified. Also, the observations (often named *responses*) to be measured, whether univariate or multivariate, must be defined. The size of the experiment must be decided upon, including both the range of the factors and their number; these decisions specify the total number of observations.

Once these decisions have been made, an experimental design for the number and pattern of observations may be devised. The number of observations may be fixed initially or may be determined sequentially; the pattern may be haphazard, random, or one of many systematic types. Patterns of variables may exist with respect to space, time, or matter. Although for geology most patterns are spatial, problems may be patterned on time or on the experimental material, which often has a spatial or time pattern associated with it. Usable patterns do not include all those that might be invented; they include only those for which a statistical analysis has been devised.

A workable experimental design is achieved only through carefully specifying aims and through prescribing methods of collecting observations, which often must be done in a rigidly ordered manner. The key is planning. The reward, measured in information obtained for a certain amount of experimental effort, can be enormous compared with hit-or-miss experimentation.

These ideas may be illustrated by considering a simple fictitious experiment of sampling gold ore. Suppose that the natural variability on a scale of 100 feet is to be investigated. A decision is then made to sample at two locations 100 feet apart and to take at each location two geological samples of 5 kilograms each, as outlined in the diagram of figure 8.4. Because of local variability and sampling losses, the four samples will differ in composition. If each one is crushed, ground, pulverized, and split, eight fractions result. If

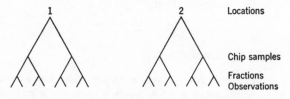

Fig. 8.4. Sampling design for a fictitious experiment of sampling gold ore.

two random assay tons (29.167 grams) are taken from each of the eight fractions and assayed, sixteen observations (assays) are obtained. Care should be taken to assay the individual assay tons in a random order and in different furnace runs (perhaps on different days and maybe even in different weeks).

From a statistical analysis of the observations, something can be learned about the variability due to assaying, to preparation, to combined local and sampling variability, and to natural variability on a scale of 100 feet. The design is an application of the nested analysis of variance (sec. 5.5), and the analysis-of-variance table has the following format:

Source of variation	Sum of squares	Degrees of freedom	Mean square	F	$F_{10\%}$
Location		1			
Geological samples		2			
Fractions		4			
Assay		8			

The previous analysis of variance sorts out and compares the four different kinds of variability.

The chain that has been outlined is perhaps the simplest that might be used to obtain an observation. In this chain the sample locations and their spacing 100 feet apart were chosen by geological judgment. The number of degrees of freedom to be obtained at each level would also require a choice based on geological and statistical judgment about the expected variability. The randomization would enter only at the sample preparation and assaying stages. The number of components of the chain to be examined separately depends on judgment about how much detail concerning variability is needed and about how many observations must be taken to learn about this variability. The randomization is a small but essential element of the whole design.

All experimental designs are basically like the one outlined. They contain arbitrary elements of judgment; they have a pattern of data collection for which a meaningful statistical analysis can be performed; and they include some places where randomization can and must be applied. The rest of this section takes up some salient points of experimental design; details about this important topic are given in an excellent book by Cochran and Cox (1957).

Pattern of Observations

The pattern of observations defines the experimental design. The best way to collect data unfortunately depends upon the functional form of the mathematical model and even upon the relationship of the coordinate system to this function. The problems can be extremely complicated, and answers are seldom available except for trivial cases. More often than not, once the data are collected, it becomes clear they they should have been collected in a different manner, one that can be specified. For instance, if one wanted to collect observations in order to investigate how a response changes with changing location, no trouble would arise if the nature of the surface to be fitted were known. Thus, if one knew that the surface was a plane, maximum information about its strike, dip, and elevation could be obtained by taking all of the observations in each of the four corners of the area of interest. In practice, one seldom if ever knows that a plane rather than some other surface is appropriate; and if a plane is inappropriate, sampling at the corners would be among the worst possible ways to gather observations.

This dilemma implies that some sort of sequential collecting process is often useful, particularly if data are expensive to collect and if only a small penalty is incurred through the delay needed to collect and analyze the data in sequence rather than all at once. The size of the penalty should be reduced as much as possible by advance planning, high-speed data processing by computer, etc. Appropriate elements of randomization must be introduced if possible into sequential sampling; otherwise, experimental effects can be confused with time trends, which may be introduced by such causes as increasing skill of the operator (sec. 7.6), changes in reagents, or seasonal changes (sec. 15.3).

Systematic Patterns

Sampling is usually done on systematic patterns because the most thorough coverage of the area of interest is obtained. In a systematic pattern, points are arranged on a grid (fig. 8.5), which is usually square; but it may be rectangular if a linear trend is suspected in the data (sec. 7.2), or it may be on polar coordinates for a radial pattern as in pyroclastic rocks dispersed outward from a volcano. Systematic patterns have been used for years in drill-hole and soil sampling. Their main drawback is that unexpected periodicity in the data may wreak havoc with conclusions. If one systematically sampled traffic flow on alternate streets in midtown New York City, he might wrongly conclude that all traffic is westbound if he picked odd-numbered streets and that those few streets with eastbound traffic represented experimental error. In truth, almost all odd-numbered streets are one-way westbound, almost all even-numbered streets are one-way eastbound, and a few streets are two-way. In geology periodic effects such as folding of the Appalachian type, joint patterns, and wave patterns are well known; and other periodic patterns are doubtless still unrecognized.

If a systematic pattern is adopted, the experimental design depends on the rock body and on the physical method of sampling. If stratification is

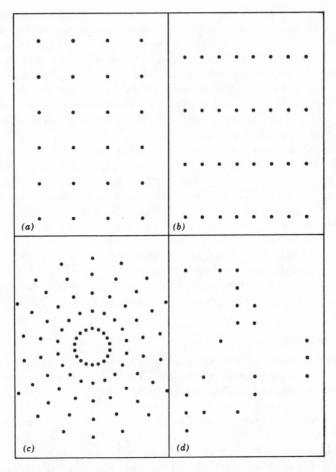

Fig. 8.5. Some arrangements of sample points. Patterns (a) to (c) are systematic; pattern (d) is random.

present, a systematic pattern should be used for the physical field sampling, which is concerned with natural and sampling variability. Then a random pattern should be adopted for the sample preparation and the chemical analysis.

Random Patterns

As first stated in section 3.4, and repeatedly emphasized since, a random pattern is not the same as a haphazard one. An example of a haphazard pattern that can cause trouble is the preference of chemists for certain digits (sec. 8.5). Figure 8.5 illustrates a random sampling pattern on rectangular coordinates, and random patterns of points in stereographic projection are shown in section 10.7. For a geologist, the main drawback to a random sampling pattern is that evenly spaced areal coverage is not provided. Readers who travel in Nevada can readily grasp this concept by playing the game of Keno. In this game, the player chooses one or more of 80 numbers, after which the banker selects 20 at random. Rarely do the 20 numbers chosen, which appear on an illuminated 10 by 8 grid, afford what a geologist would regard as suitable coverage for the grid area.

Thus random patterns are generally unsuitable for setting up the experimental design for the major factors of an investigation, but they become essential at one or more of the detailed working stages of data gathering, as in the sample preparation or chemical analysis. However, a random sampling pattern should be used whenever areal coverage is not required or if no stratification is present in the data, as generally occurs in sample preparation and chemical analysis.

Sometimes in geological sampling, random and systematic patterns may be used together. Thus for sampling granitic rocks in the San Bernardino Mountains (sec. 7.3) Richmond (1965) laid out a coarse grid. Then a sampler went to the grid points and at each one randomly chose the exact spot to sample.

Cluster Sampling

Cluster sampling refers to samples taken in local groups or clusters. In order to sample a large area, say, the state of California, it may be feasible to visit only 50 locations; but once one has taken the trouble to reach them, it may be desirable to take several geological samples at each location. If the local and regional variability are of about the same size, cluster sampling is a statistically efficient sampling technique as measured by information gained per dollar spent. In the extreme case one could visit a single location and obtain all the information available about California. Although this extreme seldom happens, it may be found where the stratification is vertical, as in sampling an oil field, a coal bed, a phosphate deposit, or sea water. Sampling vertically downward from a single point through all strata may be sufficient, and, indeed, samples from other locations may yield essentially identical results.

Cluster sampling has been extensively used in social science surveys, as explained in books by Stephan and McCarthy (1958) and Cochran (1963, Chaps. 9 and 10). Because of the potential gains, one should always try to apportion sampling so that an appropriate amount is done at individual locations as opposed to occupying other locations. In this book this subject is discussed with respect to drill-hole deflections (sec. 7.4), the contrast of local and regional variability in the Fresnillo mine (sec. 9.4), and geochemical sampling (sec. 15.3).

A General Rule for Choosing a Pattern of Observations

A general rule for choosing a pattern of observations is to be as systematic as possible, down to the place in the list of sources of variability where no advantage is gained by being systematic. However, it is essential to introduce some randomness at the very end of the sampling process, if not before.

NUMBER OF OBSERVATIONS. If observations are regarded as the group of values obtained at the end of a sampling process, which includes the acts of geological sampling, preparation, and chemical analysis, the appropriate number of observations to obtain may be considered. From this point of view the number of observations depends only on the following:

1. The variability associated with each of the several sources—natural, sampling, preparation, and analytical.

2. The statistic or statistics to be computed—mean, standard deviation, coefficient of variation, or other.

3. The distributional forms of the statistics—normal, lognormal, or other—and how well these forms are known.

4. The required precision.

5. The required accuracy.

6. The cost of obtaining observations, including planning, field sampling, preparation, and chemical analysis.

TYPES OF EXPERIMENTAL DESIGNS. The purposes of experimental designs may be summarized in three categories:

1. To estimate the population mean for one or more variables of interest.

2. To investigate how changes in different variables affect the population mean of a variable of interest. (The changes may be studied by varying quantitative or qualitative variables.)

3. To investigate how the population mean changes with changes in geographic location through trend-surface analysis (Chap. 9). (Although this category is a special case of the second one, it is separated because of its importance in geology.)

Several types of variable are recognized. In section 5.11 variables are classed as dependent or independent in a mathematical sense. They may be further classified as quantitative or qualitative. Quantitative variables have definite scales such as units of length and weight associated with them, whereas qualitative variables have only classifications such as species and type of basalt associated with them. Although quantitative variables may be used in a qualitative sense, as when rocks are classified into several types according to their iron content, the original ordering remains and may be important for the statistical analysis. Generally, the variables defined as dependent ones are quantitative, and the variables that affect their behavior may be either quantitative or qualitative. The variable or variables being studied are usually named the *response variable* or *variables*, and the variables that may influence their average values are named *factors*.

Although the ideal experimental design that would eliminate all sources of extraneous variability is never realized, it should be striven for, because often extraneous variability can be greatly reduced. Of the many ways to reduce experimental fluctuation, only two outstanding ones are mentioned here. First, *blocking* is the grouping together of similar material to take advantage of stratification (sec. 5.7). The term *block* has an abstract meaning; a laboratory, day, furnace run, or rock type may be a block. *Pairing* is the special name for the blocking of two items. Second, trends, which in geology are usually geographic, can be removed through trend-surface analysis to reduce the variability, a subject discussed in Chapter 9 (see in particular the Mi Vida example, sec. 9.8). Both methods take advantage of prior information about how the mean value of the response variable may be affected by some of the factors to be considered in an experiment.

Of the many types of experimental design, those for which a suitable statistical analysis exists or can be devised are of practical interest. Some experimental designs are introduced in this volume and others are discussed in the second volume. The simplest experimental design, the *completely randomized design*, is associated with the one-way analysis of variance (sec. 5.3). The *randomized-block design* takes advantage of blocking to reduce extraneous variability or experimental error and is also associated with an analysis of variance explained in detail (sec. 5.7). In order to consider more than one factor, *factorial designs* have been devised. Being relatively complex, they are not discussed in this book; the reader is referred to Cochran and Cox (1957) and Scheffé (1959). The example of operator variability in measuring quartz grains (sec. 8.5) is analyzed through a factorial design although the complexities of analysis are not explained. *Response surface designs* take advantage of the fact that, if several factors in an experiment are quantitative, other patterns than factorial designs may offer more efficient ways to gather data. One response surface design is explained in section 12.5. Experimental designs for interpretation of nonlinear models are being developed, but the work is still in its infancy, as explained by Draper and Smith (1966).

In any experimental design the replicating of observations is desirable in order to assess local variability, which by definition is independent of the trend-surface or other factors. Learning whether an observation has been replicated or merely duplicated may not be easy. If duplicate geological samples are sent to two chemists for analysis, their determinations clearly furnish replications, provided only that they do not communicate with one another. On the other hand, if duplicate geological samples are sent to the same chemist, his determinations furnish replications only if the samples are randomly numbered by the sender and if the chemist has no way, such as color, to match the duplicates. Also, a consideration of scale arises. If two diamond-drill holes are bored 10 feet apart, replicated observations undoubtedly are obtained if the scale of investigation is measured in miles. Whether a replication has been achieved or merely a duplication has resulted must always be considered. Duplicity, whether willful or not, must always be expected and guarded against because it can lead to underestimating the real variability.

REFERENCES

Antweiler, J. C., and Love, J. D., 1967, Gold-bearing sedimentary rocks in northwest Wyoming—a preliminary report: U.S. Geol. Survey Circ. 541, 12 p.

Boericke, W. F., 1939, The Jones riffle in cutting down samples: Eng. Mining Jour., v. 140, no. 6, p. 55–56.

Britten, H., 1961, Laboratory control, part 1: gold assaying: Trans. 7th Commonwealth Metall. Cong., Johannesburg, p. 1007–1021.

Cochran, W. G., and Cox, G. M., 1957, Experimental designs: New York, John Wiley & Sons, 611 p.

Cochran, W. G., 1963, Sampling techniques: New York, John Wiley & Sons, 413 p.

Coxon, C. H., and Sichel, H. S., 1959, Quality control of routine mine assaying and its influence on underground valuation: Jour. South African Inst. Mining and Metall., v. 59, no. 10, p. 489–517.

Dalrymple, G. B. and Lanphere, M. A., 1969, Potassium-argon dating: San Francisco, W. H. Freeman, 258 p.

Davis, G. R., 1963, Observations on sample preparation: Trans. Inst. Mining and Metall., v. 73, p. 255–267.

Dixon, W. J., and Massey, F. J., Jr., 1969, Introduction to statistical analysis: New York, McGraw-Hill, 638 p.

Draper, N. R., and Smith, Harry, 1966, Applied regression analysis: New York, John Wiley & Sons, 407 p.

Erickson, R. L., van Sickle, G. H., Nakagama, H. M., McCarthy, J. H., Jr., and Leong, K. W., 1966, Gold geochemical anomaly in the Cortez district, Nevada: U.S. Geol. Survey Circ. 534, 9 p.

Flanagan, F. J., 1960, The lead content of G-1, *in* Stevens, R. E., and others, Second report on a cooperative investigation of the composition of two silicate rocks: U.S. Geol. Survey Bull. 1113, p. 113–121.

Flanagan, F. J., Kellagher, R. C., and Smith, W. L., 1959, The slotted cone splitter: Jour. Sed. Petrology, v. 29, no. 1, p. 108–115.

Fleischer, Michael, and Chao, E. C. T., 1960, Some problems in the estimation of abundances of elements in the earth's crust: Internat. Geol. Cong., 21st, Copenhagen, Rept., pt. 1, p. 141–148.

Griffiths, J. C., 1967, Scientific method in analysis of sediments: New York, McGraw-Hill, 508 p.

Hawkes, H. E., and Webb, J. S., 1962, Geochemistry in mineral exploration: New York, Harper & Row, 401 p.

Hazen, S. W., Jr., 1967, Some statistical techniques for analyzing mine and mineral-deposit sample and assay data: U.S. Bur. Mines Bull. 621, 223 p.

Johannsen, Albert, 1914, Manual of petrographic methods: New York, McGraw-Hill, 649 p.

Kellagher, R. C., and Flanagan, F. J., 1956, The multiple-cone sample splitter: Jour. Sed. Petrology, v. 26, no. 3, p. 213–221.

Krumbein, W. C., and Pettijohn, F. J., 1938, Manual of sedimentary petrography: New York, Appleton-Century-Crofts, 531 p.

Lundel, G. E. F., Bright, H. A., and Hoffman, J. I., 1953, Applied inorganic analysis: New York, John Wiley & Sons, 1034 p.

McKinstry, H. E., 1948, Mining geology: Englewood Cliffs, N.J., Prentice-Hall, 680 p.

Miesch, A. T., 1967, Methods of computation for estimating geochemical abundance: U.S. Geol. Survey Prof. Paper 574-B, 14 p.

Milner, H. B., 1962, Laboratory technique: pt. 1, preparatory methods, *in* Sedimentary petrography, v. 1: New York, Macmillan, p. 80–128.

Muller, L. D., 1967, Mineral separation, *in* Zussman, J., ed., Physical methods in determinative mineralogy: New York, Academic Press, p. 1–30.

Müller, German, 1967, Methods in sedimentary petrology: New York, Hafner Publishing 283 p.

Otto, G. H., 1937, The use of statistical methods in effecting improvements on a Jones sample splitter: Jour. Sed. Petrology, v. 7, no. 3, p. 110–132.

Peele, Robert, ed., 1941, Mining engineers' handbook: New York, John Wiley & Sons, 2 vols., 2442 p.

Richmond, J. F., 1965, Chemical variation in quartz monzonite from Cactus Flat, San Bernardino Mountains, California: Am. Jour. Science, v. 263, p. 53–63.

Rosenfeld, M. A., and Griffiths, J. C., 1953, An experimental test of visual comparison technique in estimating two dimensional sphericity and roundness of quartz grains: Am. Jour. Science, v. 251, p. 553–585.

Scheffé, Henry, 1959, Analysis of variance: New York, John Wiley & Sons, 476 p.

Stephan, F. F., and McCarthy, P. J., 1958, Sampling opinions, an analysis of survey procedure: New York, John Wiley & Sons, 451 p.

Taggert, A. F., ed., 1945, Handbook of mineral dressing: New York, John Wiley & Sons. 1905 p.

Tatlock, D. B., 1966, Some alkali and titania analyses of tektites before and after G-1 precision monitoring: Geochimica et Cosmochimica Acta, v. 80, p. 123–128.

Wentworth, C. K., Wilgus, W. L., and Koch, H. L., 1934, A rotary type of sample splitter: Jour. Sed. Petrology, v. 4, no. 3, p. 127.

Wilson, E. B., Jr., 1952, An introduction to scientific research: New York, McGraw-Hill, 375 p.

APPENDIX

TABLE A.1. RANDOM DIGITS*

10000	11164	36318	75061	37674	26320	75100	10431	20418	19228	91792
10001	21215	91791	76831	58678	87054	31687	93205	43685	19732	08468
10002	10438	44482	66558	37649	08882	90870	12462	41810	01806	02977
10003	36792	26236	33266	66583	60881	97395	20461	36742	02852	50564
10004	73944	04773	12032	51414	82384	38370	00249	80709	72605	67497
10005	49563	12872	14063	93104	78483	72717	68714	18048	25005	04151
10006	64208	48237	41701	73117	33242	42314	83049	21933	92813	04763
10007	51486	72875	38605	29341	80749	80151	33835	52602	79147	08868
10008	99756	26360	64516	17971	48478	09610	04638	17141	09227	10606
10009	71325	55217	13015	72907	00431	45117	33827	92873	02953	85474
10010	65285	97198	12138	53010	94601	15838	16805	61004	43516	17020
10011	17264	57327	38224	29301	31381	38109	34976	65692	98566	29550
10012	95639	99754	31199	92558	68368	04985	51092	37780	40261	14479
10013	61555	76404	86210	11808	12841	45147	97438	60022	12645	62000
10014	78137	98768	04689	87130	79225	08153	84967	64539	79493	74917
10015	62490	99215	84987	28759	19177	14733	24550	28067	68894	38490
10016	24216	63444	21283	07044	92729	37284	13211	37485	10415	36457
10017	16975	95428	33226	55903	31605	43817	22250	03918	46999	98501
10018	59138	39542	71168	57609	91510	77904	74244	50940	31553	62562
10019	29478	59652	50414	31966	87912	87154	12944	49862	96566	48825
10020	96155	95009	27429	72918	08457	78134	48407	26061	58754	05326
10021	29621	66583	62966	12468	20245	14015	04014	35713	03980	03024
10022	12639	75291	71020	17265	41598	64074	64629	63293	53307	48766
10023	14544	37134	54714	02401	63228	26831	19386	15457	17999	18306
10024	83403	88827	09834	11333	68431	31706	26652	04711	34593	22561
10025	67642	05204	30697	44806	96989	68403	85621	45556	35434	09532
10026	64041	99011	14610	40273	09482	62864	01573	82274	81446	32477
10027	17048	94523	97444	59904	16936	39384	97551	09620	63932	03091
10028	93039	89416	52795	10631	09728	68202	20963	02477	55494	39563
10029	82244	34392	96607	17220	51984	10753	76272	50985	97593	34320
10030	96990	55244	70693	25255	40029	23289	48819	07159	60172	81697
10031	09119	74803	97303	88701	51380	73143	98251	78635	27556	20712
10032	57666	41204	47589	78364	38266	94393	70713	53388	79865	92069
10033	46492	61594	26729	58272	81754	14648	77210	12923	53712	87771
10034	08433	19172	08320	20839	13715	10597	17234	39355	74816	03363
10035	10011	75004	86054	41190	10061	19660	03500	68412	57812	57929
10036	92420	65431	16530	05547	10683	88102	30176	84750	10115	69220
10037	35542	55865	07304	47010	43233	57022	52161	82976	47981	46588
10038	86595	26247	18552	29491	33712	32285	64844	69395	41387	87195
10039	72115	34985	58036	99137	47482	06204	24138	24272	16196	04393
10040	07428	58863	96023	88936	51343	70958	96768	74317	27176	29600
10041	35379	27922	28906	55013	26937	48174	04197	36074	65315	12537
10042	10982	22807	10920	26299	23593	64629	57801	10437	43965	15344
10043	90127	33341	77806	12446	15444	49244	47277	11346	15884	28131
10044	63002	12990	23510	68774	48983	20481	59815	67248	17076	78910
10045	40779	86382	48454	65269	91239	45989	45389	54847	77919	41105
10046	43216	12608	18167	84631	94058	82458	15139	76856	86019	47928
10047	96167	64375	74108	93643	09204	98855	59051	56492	11933	64958
10048	70975	62693	35684	72607	23026	37004	32989	24843	01128	74658
10049	85812	61875	23570	75754	29090	40264	80399	47254	40135	69911

TABLE A.1. (CONTINUED)

10050	40603	16152	83235	37361	98783	24838	39793	80954	76865	32713
10051	40941	53585	69958	60916	71018	90561	84505	53980	64735	85140
10052	73505	83472	55953	17957	11446	22618	34771	25777	27064	13526
10053	39412	16013	11442	89320	11307	49396	39805	12249	57656	88686
10054	57994	76748	54627	48511	78646	33287	35524	54522	08795	56273
10055	61834	59199	15469	82285	84164	91333	90954	87186	31598	25942
10056	91402	77227	79516	21007	58602	81418	87838	18443	76162	51146
10057	58299	83880	20125	10794	37780	61705	18276	99041	78135	99661
10058	40684	99948	33880	76413	63839	71371	32392	51812	48248	96419
10059	75978	64298	08074	62055	73864	01926	78374	15741	74452	49954
10060	34556	39861	88267	76068	62445	64361	78685	24246	27027	48239
10061	65990	57048	25067	77571	77974	37634	81564	98608	37224	49848
10062	16381	15069	25416	87875	90374	86203	29677	82543	37554	89179
10063	52458	88880	78352	67913	09245	47773	51272	06976	99571	33365
10064	33007	85607	92008	44897	24964	50559	79549	85658	96865	24186
10065	38712	31512	08588	61490	72294	42862	87334	05866	66269	43158
10066	58722	03678	19186	69602	34625	75958	56869	17907	81867	11535
10067	26188	69497	51351	47799	20477	71786	52560	66827	79419	70886
10068	12893	54048	07255	86149	99090	70958	50775	31768	52903	27645
10069	33186	81346	85095	37282	85536	72661	32180	40229	19209	74939
10070	79893	29448	88392	54211	61708	83452	61227	81690	42265	20310
10071	48449	15102	44126	19438	23382	14985	37538	30120	82443	11152
10072	94205	04259	68983	50561	06902	10269	22216	70210	60736	58772
10073	38648	09278	81313	77400	41126	52614	93613	27263	99381	49500
10074	04292	46028	75666	26954	34979	68381	45154	09314	81009	05114
10075	17026	49737	85875	12139	59391	81830	30185	83095	78752	40899
10076	48070	76848	02531	97737	10151	18169	31709	74842	85522	74092
10077	30159	95450	83778	46115	99178	97718	98440	15076	21199	20492
10078	12148	92231	31361	60650	54695	30035	22765	91386	70399	79270
10079	73838	77067	24863	97576	01139	54219	02959	45696	98103	78867
10080	73547	43759	95632	39555	74391	07579	69491	02647	17050	49869
10081	07277	93217	79421	21769	83572	48019	17327	99638	87035	89300
10082	65128	48334	07493	28098	52087	55519	83718	60904	48721	17522
10083	38716	61380	60212	05099	21210	22052	01780	36813	19528	07727
10084	31921	76458	73720	08657	74922	61335	41690	41967	50691	30508
10085	57238	27464	61487	52329	26150	79991	64398	91273	26824	94827
10086	24219	41090	08531	61578	08236	41140	76335	91189	66312	44000
10087	31309	49387	02330	02476	96074	33256	48554	95401	02642	29119
10088	20750	97024	72619	66628	66509	31206	55293	24249	02266	39010
10089	28537	84395	26654	37851	80590	53446	34385	86893	87713	26842
10090	97929	41220	86431	94485	28778	44997	38802	56594	61363	04206
10091	40568	33222	40486	91122	43294	94541	40988	02929	83190	74247
10092	41483	92935	17061	78252	40498	43164	68646	33023	64333	64083
10093	93040	66476	24990	41099	65135	37641	97613	87282	63693	55299
10094	76869	39300	84978	07504	36835	72748	47644	48542	25076	68626
10095	02982	57991	50765	91930	21375	35604	29963	13738	03155	59914
10096	94479	76500	39170	06629	10031	48724	49822	44021	44335	26474
10097	52291	75822	95966	90947	65031	75913	52654	63377	70664	60082
10098	03684	03600	52831	55381	97013	19993	41295	29118	18710	64851
10099	58939	28366	86765	67465	45421	74228	01095	50987	83833	37276

TABLE A.1 (CONTINUED)

10100	37100	62492	63642	47638	13925	80113	88067	42575	44078	62703
10101	53406	13855	38519	29500	62479	01036	87964	44498	07793	21599
10102	55172	81556	18856	59043	64315	38270	25677	01965	21310	28115
10103	40353	84807	47767	46890	16053	32415	60259	99788	55924	22077
10104	18899	09612	77541	57675	70153	41179	97535	82889	27214	03482
10105	68141	25340	92551	11326	60939	79355	41544	88926	09111	86431
10106	51559	91159	81310	63251	91799	41215	87412	35317	74271	11603
10107	92214	33386	73459	79359	65867	39269	57527	69551	17495	91456
10108	15089	50557	33166	87094	52425	21211	41876	42525	36625	63964
10109	96461	00604	11120	22254	16763	19206	67790	88362	01880	37911
10110	28177	44111	15705	73835	69399	33602	13660	84342	97667	80847
10111	66953	44737	81127	07493	07861	12666	85077	95972	96556	80108
10112	19712	27263	84575	49820	19837	69985	34931	67935	71903	82560
10113	68756	64757	19987	92222	11691	42502	00952	47981	97579	93408
10114	75022	65332	98606	29451	57349	39219	08585	31502	96936	96356
10115	11323	70069	90269	89266	46413	61615	66447	49751	15836	97343
10116	55208	63470	18158	25283	19335	53893	87746	72531	16826	52605
10117	11474	08786	05594	67045	13231	51186	71500	50498	59487	48677
10118	81422	86842	60997	79669	43804	78690	58358	87639	24427	66799
10119	21771	75963	23151	90274	08275	50677	99384	94022	84888	80139
10120	42278	12160	32576	14278	34231	20724	27908	02657	19023	07190
10121	17697	60114	63247	32096	32503	04923	17570	73243	76181	99343
10122	05686	30243	34124	02936	71749	03031	72259	26351	77511	00850
10123	52992	46650	89910	57395	39502	49738	87854	71066	84596	33115
10124	94518	93984	81478	67750	89354	01080	25988	84359	31088	13655
10125	00184	72186	78906	75480	71140	15199	69002	08374	22126	23555
10126	87462	63165	79816	61630	50140	95319	79205	79202	67414	60805
10127	88692	58716	12273	48176	86038	78474	76730	82931	51595	20747
10128	20094	42962	41382	16768	13261	13510	04822	96354	72001	68642
10129	60935	81504	50520	82153	27892	18029	79663	44146	72876	67843
10130	51392	85936	43898	50596	81121	98122	69196	54271	12059	62539
10131	54239	41918	79526	46274	24853	67165	12011	04923	20273	89405
10132	57892	73394	07160	90262	48731	46648	70977	58262	78359	50436
10133	02330	74736	53274	44468	53616	35794	54838	39114	68302	26855
10134	76115	29247	55342	51299	79908	36613	68361	18864	13419	34950
10135	63312	81886	29085	20101	38037	34742	78364	39356	40006	49800
10136	27632	21570	34274	56426	00330	07117	86673	46455	66866	76374
10137	06335	62111	44014	52567	79480	45886	92585	87828	17376	35254
10138	64142	87676	21358	88773	10604	62834	63971	03989	21421	76086
10139	28436	25468	75235	75370	63543	76266	27745	31714	04219	00699
10140	09522	83855	85973	15888	29554	17995	37443	11461	42909	32634
10141	93714	15414	93712	02742	34395	21929	38928	31205	01838	60000
10142	15681	53599	58185	73840	88758	10618	98725	23146	13521	47905
10143	77712	23914	08907	43768	10304	61405	53986	61116	76164	54958
10144	78453	54844	61509	01245	91199	07482	02534	08189	62978	55516
10145	24860	68284	19367	29073	93464	06714	45268	60678	58506	23700
10146	37284	06844	78887	57276	42695	03682	83240	09744	63025	60997
10147	35488	52473	37634	32569	39590	27379	23520	29714	03743	08444
10148	51595	59909	35223	44991	29830	56614	59661	83397	38421	17503
10149	90660	35171	30021	91120	78793	16827	89320	08260	09181	53622

Table A.1 (continued)

10150	54723	56527	53076	38235	42780	22716	36400	48028	78196	92985
10151	84828	81248	25548	34075	43459	44628	21866	90350	82264	20478
10152	65799	01914	81363	05173	23674	41774	25154	73003	87031	94368
10153	87917	38549	48213	71708	92035	92527	55484	32274	87918	22455
10154	26907	88173	71189	28377	13785	87469	35647	19695	33401	51998
10155	68052	65422	88460	06352	42379	55499	60469	76931	83430	24560
10156	42587	68149	88147	99700	56124	53239	38726	63652	36644	50876
10157	97176	55416	67642	05051	89931	19482	80720	48977	70004	03664
10158	53295	87133	38264	94708	00703	35991	76404	82249	22942	49659
10159	23011	94108	29196	65187	69974	01970	31667	54307	40032	30031
10160	75768	49549	24543	63285	32803	18301	80851	89301	02398	99891
10161	86668	70341	66460	75648	78678	27770	30245	44775	56120	44235
10162	56727	72036	50347	33521	05068	47248	67832	30960	95465	32217
10163	27936	78010	09617	04408	18954	61862	64547	52453	83213	47833
10164	31994	69072	37354	93025	38934	90219	91148	62757	51703	84040
10165	02985	95303	15182	50166	11755	56256	89546	31170	87221	63267
10166	89965	10206	95830	95406	33845	87588	70237	84360	19629	72568
10167	45587	29611	98579	42481	05359	36578	56047	68114	58583	16313
10168	01071	08530	74305	77509	16270	20889	99753	88035	55643	18291
10169	90209	68521	14293	39194	68803	32052	39413	26883	83119	69623
10170	04982	68470	27875	15480	13206	44784	83601	03172	07817	01520
10171	19740	24637	97377	32112	74283	69384	49768	64141	02024	85380
10172	50197	79869	86497	68709	42073	28498	82750	43571	77075	07123
10173	46954	67536	28968	81936	95999	04319	09932	66223	45491	69503
10174	82549	62676	31123	49899	70512	95288	15517	85352	21987	08669
10175	61798	81600	80018	84742	06103	60786	01408	75967	29948	21454
10176	57666	29055	46518	01487	30136	14349	56159	47408	78311	25896
10177	29805	64994	66872	62230	41385	58066	96600	99301	85976	84194
10178	06711	34939	19599	76247	87879	97114	74314	39599	43544	36255
10179	13934	46885	58315	88366	06138	37923	11192	90757	10831	01580
10180	28549	98327	99943	25377	17628	65468	07875	16728	22602	33892
10181	40871	61803	25767	55484	90997	86941	64027	01020	39518	34693
10182	47704	38355	71708	80117	11361	88875	22315	38048	42891	87885
10183	62611	19698	09304	29265	07636	08508	23773	56545	08015	28891
10184	03047	83981	11916	09267	67316	87952	27045	62536	32180	60936
10185	26460	50501	31731	18938	11025	18515	31747	96828	58258	97107
10186	01764	25959	69293	89875	72710	49659	66632	25314	95260	22146
10187	11762	54806	02651	52912	32770	64507	59090	01275	47624	16124
10188	31736	31695	11523	64213	91190	10145	34231	36405	65860	48771
10189	97155	48706	52239	21831	49043	18650	72246	43729	63368	53822
10190	31181	49672	17237	04024	65324	32460	01566	67342	94986	36106
10191	32115	82683	67182	89030	41370	50266	19505	57724	93358	49445
10192	07068	75947	71743	69285	30395	81818	36125	52055	20289	16911
10193	26622	74184	75166	96748	34729	61289	36908	73686	84641	45130
10194	02805	52676	22519	47848	68210	23954	63085	87729	14176	45410
10195	32301	58701	04193	30142	99779	21697	05059	26684	63516	75925
10196	26339	56909	39331	42101	01031	01947	02257	47236	19913	90371
10197	95274	09508	81012	42413	11278	19354	68661	04192	36878	84366
10198	24275	39632	09777	98800	48027	96908	08177	15364	02317	89548
10199	36116	42128	65401	94199	51058	10759	47244	99830	64255	40550

* From "A Million Random Digits" published by the Rand Corporation, Free Press Glencoe, Ill., 1955.

TABLE A-2. PERCENTAGE POINTS OF
THE STANDARDIZED NORMAL DISTRIBU-
TION *

Percentage point	s.n.d.
99.9	− 3.0902
99.5	− 2.5758
99.0	− 2.3263
97.5	− 1.9600
95.0	− 1.6449
90.0	− 1.2816
80.0	− 0.8416
75.0	− 0.6745
70.0	− 0.5244
60.0	− 0.2533
50.0	0.0000
40.0	0.2533
30.0	0.5244
25.0	0.6745
20.0	0.8416
10.0	1.2816
5.0	1.6449
2.5	1.9600
1.0	2.3263
0.5	2.5758
0.1	3.0902

* From Table 3, Biometrika Tables for Statisticians, Volume 1, Second Edition, Cambridge University Press, 1962.

TABLE A.3. PERCENTAGE POINTS OF THE χ^2-DISTRIBUTION *

α df	99.5%	99%	97.5%	95%	90%
1	392704.10⁻¹⁰	157088.10⁻⁹	982069.10⁻⁹	393214.10⁻⁸	0·0157908
2	0·0100251	0·0201007	0·0506356	0·102587	0·210720
3	0·0717212	0·114832	0·215795	0·351846	0·584375
4	0·206990	0·297110	0·484419	0·710721	1·063623
5	0·411740	0·554300	0·831211	1·145476	1·61031
6	0·675727	0·872085	1·237347	1·63539	2·20413
7	0·989265	1·239043	1·68987	2·16735	2·83311
8	1·344419	1·646482	2·17973	2·73264	3·48954
9	1·734926	2·087912	2·70039	3·32511	4·16816
10	2·15585	2·55821	3·24697	3·94030	4·86518
11	2·60321	3·05347	3·81575	4·57481	5·57779
12	3·07382	3·57056	4·40379	5·22603	6·30380
13	3·56503	4·10691	5·00874	5·89186	7·04150
14	4·07468	4·66043	5·62872	6·57063	7·78953
15	4·60004	5·22935	6·26214	7·26094	8·54675
16	5·14224	5·81221	6·90766	7·96164	9·31223
17	5·69724	6·40776	7·56418	8·67176	10·0852
18	6·26481	7·01491	8·23075	9·39046	10·8649
19	6·84398	7·63273	8·90655	10·1170	11·6509
20	7·43386	8·26040	9·59083	10·8508	12·4426
21	8·03366	8·89720	10·28293	11·5913	13·2396
22	8·64272	9·54249	10·9823	12·3380	14·0415
23	9·26042	10·19567	11·6885	13·0905	14·8479
24	9·88623	10·8564	12·4011	13·8484	15·6587
25	10·5197	11·5240	13·1197	14·6114	16·4734
26	11·1603	12·1981	13·8439	15·3791	17·2919
27	11·8076	12·8786	14·5733	16·1513	18·1138
28	12·4613	13·5648	15·3079	16·9279	18·9392
29	13·1211	14·2565	16·0471	17·7083	19·7677
30	13·7867	14·9535	16·7908	18·4926	20·5992
40	20·7065	22·1643	24·4331	26·5093	29·0505
50	27·9907	29·7067	32·3574	34·7642	37·6886
60	35·5346	37·4848	40·4817	43·1879	46·4589
70	43·2752	45·4418	48·7576	51·7393	55·3290
80	51·1720	53·5400	57·1532	60·3915	64·2778
90	59·1963	61·7541	65·6466	69·1260	73·2912
100	67·3276	70·0648	74·2219	77·9295	82·3581

Table A.3 (continued)

α df	10%	5%	2.5%	1%	0.5%
1	2·70554	3·84146	5·02389	6·63490	7·87944
2	4·60517	5·99147	7·37776	9·21034	10·5966
3	6·25139	7·81473	9·34840	11·3449	12·8381
4	7·77944	9·48773	11·1433	13·2767	14·8602
5	9·23635	11·0705	12·8325	15·0863	16·7496
6	10·6446	12·5916	14·4494	16·8119	18·5476
7	12·0170	14·0671	16·0128	18·4753	20·2777
8	13·3616	15·5073	17·5346	20·0902	21·9550
9	14·6837	16·9190	19·0228	21·6660	23·5893
10	15·9871	18·3070	20·4831	23·2093	25·1882
11	17·2750	19·6751	21·9200	24·7250	26·7569
12	18·5494	21·0261	23·3367	26·2170	28·2995
13	19·8119	22·3621	24·7356	27·6883	29·8194
14	21·0642	23·6848	26·1190	29·1413	31·3193
15	22·3072	24·9958	27·4884	30·5779	32·8013
16	23·5418	26·2962	28·8454	31·9999	34·2672
17	24·7690	27·5871	30·1910	33·4087	35·7185
18	25·9894	28·8693	31·5264	34·8053	37·1564
19	27·2036	30·1435	32·8523	36·1908	38·5822
20	28·4120	31·4104	34·1696	37·5662	39·9968
21	29·6151	32·6705	35·4789	38·9321	41·4010
22	30·8133	33·9244	36·7807	40·2894	42·7956
23	32·0069	35·1725	38·0757	41·6384	44·1813
24	33·1963	36·4151	39·3641	42·9798	45·5585
25	34·3816	37·6525	40·6465	44·3141	46·9278
26	35·5631	38·8852	41·9232	45·6417	48·2899
27	36·7412	40·1133	43·1944	46·9630	49·6449
28	37·9159	41·3372	44·4607	48·2782	50·9933
29	39·0875	42·5569	45·7222	49·5879	52·3356
30	40·2560	43·7729	46·9792	50·8922	53·6720
40	51·8050	55·7585	59·3417	63·6907	66·7659
50	63·1671	67·5048	71·4202	76·1539	79·4900
60	74·3970	79·0819	83·2976	88·3794	91·9517
70	85·5271	90·5312	95·0231	100·425	104·215
80	96·5782	101·879	106·629	112·329	116·321
90	107·565	113·145	118·136	124·116	128·299
100	118·498	124·342	129·561	135·807	140·169

* From Table 8, Biometrika Tables for Statisticians, Volume 1, Second Edition, Cambridge University Press, 1962.

TABLE A.4. PERCENTAGE POINTS OF THE t_m DISTRIBUTION *

df \ α	10%	5%	2.5%	1%
1	3.078	6.314	12.706	31.821
2	1.886	2.920	4.303	6.965
3	1.638	2.353	3.182	4.541
4	1.533	2.132	2.776	3.747
5	1.476	2.015	2.571	3.365
6	1.440	1.943	2.447	3.143
7	1.415	1.895	2.365	2.998
8	1.397	1.860	2.306	2.896
9	1.383	1.833	2.262	2.821
10	1.372	1.812	2.228	2.764
11	1.363	1.796	2.201	2.718
12	1.356	1.782	2.179	2.681
13	1.350	1.771	2.160	2.650
14	1.345	1.761	2.145	2.624
15	1.341	1.753	2.131	2.602
16	1.337	1.746	2.120	2.583
17	1.333	1.740	2.110	2.567
18	1.330	1.734	2.101	2.552
19	1.328	1.729	2.093	2.539
20	1.325	1.725	2.086	2.528
21	1.323	1.721	2.080	2.518
22	1.321	1.717	2.074	2.508
23	1.319	1.714	2.069	2.500
24	1.318	1.711	2.064	2.492
25	1.316	1.708	2.060	2.485
26	1.315	1.706	2.056	2.479
27	1.314	1.703	2.052	2.473
28	1.313	1.701	2.048	2.467
29	1.311	1.699	2.045	2.462
30	1.310	1.697	2.042	2.457
40	1.303	1.684	2.021	2.423
60	1.296	1.671	2.000	2.390
120	1.289	1.658	1.980	2.358
∞	1.282	1.645	1.960	2.326

* From Table 12, Biometrika Tables for Statisticians, Volume 1, Second Edition, Cambridge University Press, 1962.

TABLE A.5. PERCENTAGE POINTS OF THE F-DISTRIBUTION *

$\alpha = 10\%$

df_2 \ df_1	1	2	3	4	5	6	7	8	9
1	39.864	49.500	53.593	55.833	57.241	58.204	58.906	59.439	59.858
2	8.5263	9.0000	9.1618	9.2434	9.2926	9.3255	9.3491	9.3668	9.3805
3	5.5383	5.4624	5.3908	5.3427	5.3092	5.2847	5.2662	5.2517	5.2400
4	4.5448	4.3246	4.1908	4.1073	4.0506	4.0098	3.9790	3.9549	3.9357
5	4.0604	3.7797	3.6195	3.5202	3.4530	3.4045	3.3679	3.3393	3.3163
6	3.7760	3.4633	3.2888	3.1808	3.1075	3.0546	3.0145	2.9830	2.9577
7	3.5894	3.2574	3.0741	2.9605	2.8833	2.8274	2.7849	2.7516	2.7247
8	3.4579	3.1131	2.9238	2.8064	2.7265	2.6683	2.6241	2.5893	2.5612
9	3.3603	3.0065	2.8129	2.6927	2.6106	2.5509	2.5053	2.4694	2.4403
10	3.2850	2.9245	2.7277	2.6053	2.5216	2.4606	2.4140	2.3772	2.3473
11	3.2252	2.8595	2.6602	2.5362	2.4512	2.3891	2.3416	2.3040	2.2735
12	3.1765	2.8068	2.6055	2.4801	2.3940	2.3310	2.2828	2.2446	2.2135
13	3.1362	2.7632	2.5603	2.4337	2.3467	2.2830	2.2341	2.1953	2.1638
14	3.1022	2.7265	2.5222	2.3947	2.3069	2.2426	2.1931	2.1539	2.1220
15	3.0732	2.6952	2.4898	2.3614	2.2730	2.2081	2.1582	2.1185	2.0862
16	3.0481	2.6682	2.4618	2.3327	2.2438	2.1783	2.1280	2.0880	2.0553
17	3.0262	2.6446	2.4374	2.3077	2.2183	2.1524	2.1017	2.0613	2.0284
18	3.0070	2.6239	2.4160	2.2858	2.1958	2.1296	2.0785	2.0379	2.0047
19	2.9899	2.6056	2.3970	2.2663	2.1760	2.1094	2.0580	2.0171	1.9836
20	2.9747	2.5893	2.3801	2.2489	2.1582	2.0913	2.0397	1.9985	1.9649
21	2.9609	2.5746	2.3649	2.2333	2.1423	2.0751	2.0232	1.9819	1.9480
22	2.9486	2.5613	2.3512	2.2193	2.1279	2.0605	2.0084	1.9668	1.9327
23	2.9374	2.5493	2.3387	2.2065	2.1149	2.0472	1.9949	1.9531	1.9189
24	2.9271	2.5383	2.3274	2.1949	2.1030	2.0351	1.9826	1.9407	1.9063
25	2.9177	2.5283	2.3170	2.1843	2.0922	2.0241	1.9714	1.9292	1.8947
26	2.9091	2.5191	2.3075	2.1745	2.0822	2.0139	1.9610	1.9188	1.8841
27	2.9012	2.5106	2.2987	2.1655	2.0730	2.0045	1.9515	1.9091	1.8743
28	2.8939	2.5028	2.2906	2.1571	2.0645	1.9959	1.9427	1.9001	1.8652
29	2.8871	2.4955	2.2831	2.1494	2.0566	1.9878	1.9345	1.8918	1.8560
30	2.8807	2.4887	2.2761	2.1422	2.0492	1.9803	1.9269	1.8841	1.8498
40	2.8354	2.4404	2.2261	2.0909	1.9968	1.9269	1.8725	1.8289	1.7929
60	2.7914	2.3932	2.1774	2.0410	1.9457	1.8747	1.8194	1.7748	1.7380
120	2.7478	2.3473	2.1300	1.9923	1.8959	1.8238	1.7675	1.7220	1.6843
∞	2.7055	2.3026	2.0838	1.9449	1.8473	1.7741	1.7167	1.6702	1.6315

$\alpha = 10\%$

10	12	15	20	24	30	40	60	120	∞
60.195	60.705	61.220	61.740	62.002	62.265	62.529	62.794	63.061	63.328
9.3916	9.4081	9.4247	9.4413	9.4496	9.4579	9.4663	9.4746	9.4829	9.4913
5.2304	5.2156	5.2003	5.1845	5.1764	5.1681	5.1597	5.1512	5.1425	5.1337
3.9199	3.8955	3.8689	3.8443	3.8310	3.8174	3.8036	3.7896	3.7753	3.7607
3.2974	3.2682	3.2380	3.2067	3.1905	3.1741	3.1573	3.1402	3.1228	3.1050
2.9369	2.9047	2.8712	2.8363	2.8183	2.8000	2.7812	2.7620	2.7423	2.7222
2.7025	2.6681	2.6322	2.5947	2.5753	2.5555	2.5351	2.5142	2.4928	2.4708
2.5380	2.5020	2.4642	2.4246	2.4041	2.3830	2.3614	2.3391	2.3162	2.2926
2.4163	2.3789	2.3396	2.2983	2.2768	2.2547	2.2320	2.2085	2.1843	2.1592
2.3226	2.2841	2.2435	2.2007	2.1784	2.1554	2.1317	2.1072	2.0818	2.0554
2.2482	2.2087	2.1671	2.1230	2.1000	2.0762	2.0516	2.0261	1.9997	1.9721
2.1878	2.1474	2.1049	2.0597	2.0360	2.0115	1.9861	1.9597	1.9323	1.9036
2.1376	2.0966	2.0532	2.0070	1.9827	1.9576	1.9315	1.9043	1.8759	1.8462
2.0954	2.0537	2.0095	1.9625	1.9377	1.9119	1.8852	1.8572	1.8280	1.7973
2.0593	2.0171	1.9722	1.9243	1.8990	1.8728	1.8454	1.8168	1.7867	1.7551
2.0281	1.9854	1.9399	1.8913	1.8656	1.8388	1.8108	1.7816	1.7507	1.7182
2.0009	1.9577	1.9117	1.8624	1.8362	1.8090	1.7805	1.7506	1.7191	1.6856
1.9770	1.9333	1.8868	1.8368	1.8103	1.7827	1.7537	1.7232	1.6910	1.6567
1.9557	1.9117	1.8647	1.8142	1.7873	1.7592	1.7298	1.6988	1.6659	1.6308
1.9367	1.8924	1.8449	1.7938	1.7667	1.7382	1.7083	1.6768	1.6433	1.6074
1.9197	1.8750	1.8272	1.7756	1.7481	1.7193	1.6890	1.6569	1.6228	1.5862
1.9043	1.8593	1.8111	1.7590	1.7312	1.7021	1.6714	1.6389	1.6042	1.5668
1.8903	1.8450	1.7964	1.7439	1.7159	1.6864	1.6554	1.6224	1.5871	1.5490
1.8775	1.8319	1.7831	1.7302	1.7019	1.6721	1.6407	1.6073	1.5715	1.5327
1.8658	1.8200	1.7708	1.7175	1.6890	1.6589	1.6272	1.5934	1.5570	1.5176
1.8550	1.8090	1.7596	1.7059	1.6771	1.6468	1.6147	1.5805	1.5437	1.5036
1.8451	1.7989	1.7492	1.6951	1.6662	1.6356	1.6032	1.5686	1.5313	1.4906
1.8359	1.7895	1.7395	1.6852	1.6560	1.6252	1.5925	1.5575	1.5198	1.4784
1.8274	1.7808	1.7306	1.6759	1.6465	1.6155	1.5825	1.5472	1.5090	1.4670
1.8195	1.7727	1.7223	1.6673	1.6377	1.6065	1.5732	1.5376	1.4989	1.4564
1.7627	1.7146	1.6624	1.6052	1.5741	1.5411	1.5056	1.4672	1.4248	1.3769
1.7070	1.6574	1.6034	1.5435	1.5107	1.4755	1.4373	1.3952	1.3476	1.2915
1.6524	1.6012	1.5450	1.4821	1.4472	1.4094	1.3676	1.3203	1.2646	1.1926
1.5987	1.5458	1.4871	1.4206	1.3832	1.3419	1.2951	1.2400	1.1686	1.0000

TABLE A.5 (CONTINUED)

$\alpha = 5\%$

df_1 df_2	1	2	3	4	5	6	7	8	9
1	161.15	199.50	215.71	224.58	230.16	233.99	236.77	238.88	240.54
2	18.513	19.000	19.164	19.247	19.296	19.330	19.353	19.371	19.385
3	10.128	9.5521	9.2766	9.1172	9.0135	8.9406	8.8868	8.8452	8.8123
4	7.7086	6.9443	6.5914	6.3883	6.2560	6.1631	6.0942	6.0410	5.9988
5	6.6079	5.7861	5.4095	5.1922	5.0503	4.9503	4.8759	4.8183	4.7725
6	5.9874	5.1433	4.7571	4.5337	4.3874	4.2839	4.2066	4.1468	4.0990
7	5.5914	4.7374	4.3468	4.1203	3.9715	3.8660	3.7870	3.7257	3.6767
8	5.3177	4.4590	4.0662	3.8378	3.6875	3.5806	3.5005	3.4381	3.3881
9	5.1174	4.2565	3.8626	3.6331	3.4817	3.3738	3.2927	3.2296	3.1789
10	4.9646	4.1028	3.7083	3.4780	3.3258	3.2172	3.1355	3.0717	3.0204
11	4.8443	3.9823	3.5874	3.3567	3.2039	3.0946	3.0123	2.9480	2.8962
12	4.7472	3.8853	3.4903	3.2592	3.1059	2.9961	2.9134	2.8486	2.7964
13	4.6672	3.8056	3.4105	3.1791	3.0254	2.9153	2.8321	2.7669	2.7144
14	4.6001	3.7389	3.3439	3.1122	2.9582	2.8477	2.7642	2.6987	2.6458
15	4.5431	3.6823	3.2874	3.0556	2.9013	2.7905	2.7066	2.6408	2.5876
16	4.4940	3.6337	3.2389	3.0069	2.8524	2.7413	2.6572	2.5911	2.5377
17	4.4513	3.5915	3.1968	2.9647	2.8100	2.6987	2.6143	2.5480	2.4943
18	4.4139	3.5546	3.1599	2.9277	2.7729	2.6613	2.5767	2.5102	2.4563
19	4.3808	3.5219	3.1274	2.8951	2.7401	2.6283	2.5435	2.4768	2.4227
20	4.3513	3.4928	3.0984	2.8661	2.7109	2.5990	2.5140	2.4471	2.3928
21	4.3248	3.4668	3.0725	2.8401	2.6848	2.5727	2.4876	2.4205	2.3661
22	4.3009	3.4434	3.0491	2.8167	2.6613	2.5491	2.4638	2.3965	2.3419
23	4.2793	3.4221	3.0280	2.7955	2.6400	2.5277	2.4422	2.3748	2.3201
24	4.2597	3.4028	3.0088	2.7763	2.6207	2.5082	2.4226	2.3551	2.3002
25	4.2417	3.3852	2.9912	2.7587	2.6030	2.4904	2.4047	2.3371	2.2821
26	4.2252	3.3690	2.9751	2.7426	2.5868	2.4741	2.3883	2.3205	2.2655
27	4.2100	3.3541	2.9604	2.7278	2.5719	2.4591	2.3732	2.3053	2.2501
28	4.1960	3.3404	2.9467	2.7141	2.5581	2.4453	2.3593	2.2913	2.2360
29	4.1830	3.3277	2.9340	2.7014	2.5454	2.4324	2.3463	2.2782	2.2229
30	4.1709	3.3158	2.9223	2.6896	2.5336	2.4205	2.3343	2.2662	2.2107
40	4.0848	3.2317	2.8387	2.6060	2.4495	2.3359	2.2490	2.1802	2.1240
60	4.0012	3.1504	2.7581	2.5252	2.3683	2.2540	2.1665	2.0970	2.0401
120	3.9201	3.0718	2.6802	2.4472	2.2900	2.1750	2.0867	2.0164	1.9588
∞	3.8415	2.9957	2.6049	2.3719	2.2141	2.0986	2.0096	1.9384	1.8799

$\alpha = 5\%$

10	12	15	20	24	30	40	60	120	∞
241.88	243.91	245.95	248.01	249.05	250.09	251.14	252.20	253.25	254.32
19.396	19.413	19.429	19.446	19.454	19.462	19.471	19.479	19.487	19.496
8.7855	8.7446	8.7029	8.6602	8.6385	8.6166	8.5944	8.5720	8.5494	8.5265
5.9644	5.9117	5.8578	5.8025	5.7744	5.7459	5.7170	5.6878	5.6581	5.6281
4.7351	4.6777	4.6188	4.5581	4.5272	4.4957	4.4638	4.4314	4.3984	4.3650
4.0600	3.9999	3.9381	3.8742	3.8415	3.8082	3.7743	3.7398	3.7047	3.6688
3.6365	3.5747	3.5108	3.4445	3.4105	3.3758	3.3404	3.3043	3.2674	3.2298
3.3472	3.2840	3.2184	3.1503	3.1152	3.0794	3.0428	3.0053	2.9669	2.9276
3.1373	3.0729	3.0061	2.9365	2.9005	2.8637	2.8259	2.7872	2.7475	2.7067
2.9782	2.9130	2.8450	2.7740	2.7372	2.6996	2.6609	2.6211	2.5801	2.5379
2.8536	2.7876	2.7186	2.6464	2.6090	2.5705	2.5309	2.4901	2.4480	2.4045
2.7534	2.6866	2.6169	2.5436	2.5055	2.4663	2.4259	2.3842	2.3410	2.2962
2.6710	2.6037	2.5331	2.4589	2.4202	2.3803	2.3392	2.2966	2.2524	2.2064
2.6021	2.5342	2.4630	2.3879	2.3487	2.3082	2.2664	2.2230	2.1778	2.1307
2.5437	2.4753	2.4035	2.3275	2.2878	2.2468	2.2043	2.1601	2.1141	2.0658
2.4935	2.4247	2.3522	2.2756	2.2354	2.1938	2.1507	2.1058	2.0589	2.0096
2.4499	2.3807	2.3077	2.2304	2.1898	2.1477	2.1040	2.0584	2.0107	1.9604
2.4117	2.3421	2.2686	2.1906	2.1497	2.1071	2.0629	2.0166	1.9681	1.9168
2.3779	2.3080	2.2341	2.1555	2.1141	2.0712	2.0264	1.9796	1.9302	1.8780
2.3479	2.2776	2.2033	2.1242	2.0825	2.0391	1.9938	1.9464	1.8963	1.8432
2.3210	2.2504	2.1757	2.0960	2.0540	2.0102	1.9645	1.9165	1.8657	1.8117
2.2967	2.2258	2.1508	2.0707	2.0283	1.9842	1.9380	1.8895	1.8380	1.7831
2.2747	2.2036	2.1282	2.0476	2.0050	1.9605	1.9139	1.8649	1.8128	1.7570
2.2547	2.1834	2.1077	2.0267	1.9838	1.9390	1.8920	1.8424	1.7897	1.7331
2.2365	2.1649	2.0889	2.0075	1.9643	1.9192	1.8718	1.8217	1.7684	1.7110
2.2197	2.1479	2.0716	1.9898	1.9464	1.9010	1.8533	1.8027	1.7488	1.6906
2.2043	2.1323	2.0558	1.9736	1.9299	1.8842	1.8361	1.7851	1.7307	1.6717
2.1900	2.1179	2.0411	1.9586	1.9147	1.8687	1.8203	1.7689	1.7138	1.6541
2.1768	2.1045	2.0275	1.9446	1.9005	1.8543	1.8055	1.7537	1.6981	1.6377
2.1646	2.0921	2.0148	1.9317	1.8874	1.8409	1.7918	1.7396	1.6835	1.6223
2.0772	2.0035	1.9245	1.8389	1.7929	1.7444	1.6928	1.6373	1.5766	1.5089
1.9926	1.9174	1.8364	1.7480	1.7001	1.6491	1.5943	1.5343	1.4673	1.3893
1.9105	1.8337	1.7505	1.6587	1.6084	1.5543	1.4952	1.4290	1.3519	1.2539
1.8307	1.7522	1.6664	1.5705	1.5173	1.4591	1.3940	1.3180	1.2214	1.0000

TABLE A.5 (CONTINUED)

α = 1%

df_1 / df_2	1	2	3	4	5	6	7	8	9
1	4052.2	4999.5	5403.3	5624.6	5763.7	5859.0	5928.3	5981.6	6022.5
2	98.503	99.000	99.166	99.249	99.299	99.332	99.356	99.374	99.388
3	34.116	30.817	29.457	28.710	28.237	27.911	27.672	27.489	27.345
4	21.198	18.000	16.694	15.977	15.522	15.207	14.976	14.799	14.659
5	16.258	13.274	12.060	11.392	10.967	10.672	10.456	10.289	10.158
6	13.745	10.925	9.7795	9.1483	8.7459	8.4661	8.2600	8.1016	7.9761
7	12.246	9.5466	8.4513	7.8467	7.4604	7.1914	6.9928	6.8401	6.7188
8	11.259	8.6491	7.5910	7.0060	6.6318	6.3707	6.1776	6.0289	5.9106
9	10.561	8.0215	6.9919	6.4221	6.0569	5.8018	5.6129	5.4671	5.3511
10	10.044	7.5594	6.5523	5.9943	5.6363	5.3858	5.2001	5.0567	4.9424
11	9.6460	7.2057	6.2167	5.6683	5.3160	5.0692	4.8861	4.7445	4.6315
12	9.3302	6.9266	5.9526	5.4119	5.0643	4.8206	4.6395	4.4994	4.3875
13	9.0738	6.7010	5.7394	5.2053	4.8616	4.6204	4.4410	4.3021	4.1911
14	8.8616	6.5149	5.5639	5.0354	4.6950	4.4558	4.2779	4.1399	4.0297
15	8.6831	6.3589	5.4170	4.8932	4.5556	4.3183	4.1415	4.0045	3.8948
16	8.5310	6.2262	5.2922	4.7726	4.4374	4.2016	4.0259	3.8896	3.7804
17	8.3997	6.1121	5.1850	4.6690	4.3359	4.1015	3.9267	3.7910	3.6822
18	8.2854	6.0129	5.0919	4.5790	4.2479	4.0146	3.8406	3.7054	3.5971
19	8.1850	5.9259	5.0103	4.5003	4.1708	3.9386	3.7653	3.6305	3.5225
20	8.0960	5.8489	4.9382	4.4307	4.1027	3.8714	3.6987	3.5644	3.4567
21	8.0166	5.7804	4.8740	4.3688	4.0421	3.8117	3.6396	3.5056	3.3981
22	7.9454	5.7190	4.8166	4.3134	3.9880	3.7583	3.5867	3.4530	3.3458
23	7.8811	5.6637	4.7649	4.2635	3.9392	3.7102	3.5390	3.4057	3.2986
24	7.8229	5.6136	4.7181	4.2184	3.8951	3.6667	3.4959	3.3629	3.2560
25	7.7698	5.5680	4.6755	4.1774	3.8550	3.6272	3.4568	3.3239	3.2172
26	7.7213	5.5263	4.6366	4.1400	3.8183	3.5911	3.4210	3.2884	3.1818
27	7.6767	5.4881	4.6009	4.1056	3.7848	3.5580	3.3882	3.2558	3.1494
28	7.6356	5.4529	4.5681	4.0740	3.7539	3.5276	3.3581	3.2259	3.1195
29	7.5976	5.4205	4.5378	4.0449	3.7254	3.4995	3.3302	3.1982	3.0920
30	7.5625	5.3904	4.5097	4.0179	3.6990	3.4735	3.3045	3.1726	3.0665
40	7.3141	5.1785	4.3126	3.8283	3.5138	3.2910	3.1238	2.9930	2.8876
60	7.0771	4.9774	4.1259	3.6491	3.3389	3.1187	2.9530	2.8233	2.7185
120	6.8510	4.7865	3.9493	3.4796	3.1735	2.9559	2.7918	2.6629	2.5586
∞	6.6349	4.6052	3.7816	3.3192	3.0173	2.8020	2.6393	2.5113	2.4073

$$\alpha = 1\%$$

10	12	15	20	24	30	40	60	120	∞
6055.8	6106.3	6157.3	6208.7	6234.6	6260.7	6286.8	6313.0	6339.4	6366.0
99.399	99.416	99.432	99.449	99.458	99.466	99.474	99.483	99.491	99.501
27.229	27.052	26.872	26.690	26.598	26.505	26.411	26.316	26.221	26.125
14.546	14.374	14.198	14.020	13.929	13.838	13.745	13.652	13.558	13.463
10.051	9.8883	9.7222	9.5527	9.4665	9.3793	9.2912	9.2020	9.1118	9.0204
7.8741	7.7183	7.5590	7.3958	7.3127	7.2285	7.1432	7.0568	6.9690	6.8801
6.6201	6.4691	6.3143	6.1554	6.0743	5.9921	5.9084	5.8236	5.7372	5.6495
5.8143	5.6668	5.5151	5.3591	5.2793	5.1981	5.1156	5.0316	4.9460	4.8588
5.2565	5.1114	4.9621	4.8080	4.7290	4.6486	4.5667	4.4831	4.3978	4.3105
4.8492	4.7059	4.5582	4.4054	4.3269	4.2469	4.1653	4.0819	3.9965	3.9090
4.5393	4.3974	4.2509	4.0990	4.0209	3.9411	3.8596	3.7761	3.6904	3.6025
4.2961	4.1553	4.0096	3.8584	3.7805	3.7008	3.6192	3.5355	3.4494	3.3608
4.1003	3.9603	3.8154	3.6646	3.5868	3.5070	3.4253	3.3413	3.2548	3.1654
3.9394	3.8001	3.6557	3.5052	3.4274	3.3476	3.2656	3.1813	3.0942	3.0040
3.8049	3.6662	3.5222	3.3719	3.2940	3.2141	3.1319	3.0471	2.9595	2.8684
3.6909	3.5527	3.4089	3.2588	3.1808	3.1007	3.0182	2.9330	2.8447	2.7528
3.5931	3.4552	3.3117	3.1615	3.0835	3.0032	2.9205	2.8348	2.7459	2.6530
3.5082	3.3706	3.2273	3.0771	2.9990	2.9185	2.8354	2.7493	2.6597	2.5660
3.4338	3.2965	3.1533	3.0031	2.9249	2.8442	2.7608	2.6742	2.5839	2.4893
3.3682	3.2311	3.0880	2.9377	2.8594	2.7785	2.6947	2.6077	2.5168	2.4212
3.3098	3.1729	3.0299	2 8796	2.8011	2.7200	2.6359	2.5484	2.4568	2.3603
3.2576	3.1209	2.9780	2.8274	2.7488	2.6675	2.5831	2.4951	2.4029	2.3055
3.2106	3.0740	2.9311	2.7805	2.7017	2.6202	2.5355	2.4471	2.3542	2.2559
3.1681	3.0316	2.8887	2.7380	2.6591	2.5773	2.4923	2.4035	2.3099	2.2107
3.1294	2.9931	2.8502	2.6993	2.6203	2.5383	2.4530	2.3637	2.2695	2.1694
3.0941	2.9579	2.8150	2.6640	2.5848	2.5026	2.4170	2.3273	2.2325	2.1315
3.0618	2.9256	2.7827	2.6316	2.5522	2.4699	2.3840	2.2938	2.1984	2.0965
3.0320	2.8959	2.7530	2.6017	2.5223	2.4397	2.3535	2.2629	2.1670	2.0642
3.0045	2.8685	2.7256	2.5742	2.4946	2.4118	2.3253	2.2344	2.1378	2.0342
2.9791	2.8431	2.7002	2.5487	2.4689	2.3860	2.2992	2.2079	2.1107	2.0062
2.8005	2.6648	2.5216	2.3689	2.2880	2.2034	2.1142	2.0194	1.9172	1.8047
2.6318	2.4961	2.3523	2.1978	2.1154	2.0285	1.9360	1.8363	1.7263	1.6006
2.4721	2.3363	2.1915	2.0346	1.9500	1.8600	1.7628	1.6557	1.5330	1.3805
2.3209	2.1848	2.0385	1.8783	1.7908	1.6964	1.5923	1.4730	1.3246	1.0000

* From Table 18, Biometrika Tables for Statisticians, Volume 1, Second Edition, Cambridge University Press, 1962.

TABLE A.6. PERCENTILES OF THE STUDENTIZED RANGE *

$\alpha = 10\%$

df \ k	2	3	4	5	6	7	8	9	10
1	8.93	13.44	16.36	18.49	20.15	21.51	22.64	23.62	24.48
2	4.13	5.73	6.77	7.54	8.14	8.63	9.05	9.41	9.72
3	3.33	4.47	5.20	5.74	6.16	6.51	6.81	7.06	7.29
4	3.01	3.98	4.59	5.03	5.39	5.68	5.93	6.14	6.33
5	2.85	3.72	4.26	4.66	4.98	5.24	5.46	5.65	5.82
6	2.75	3.56	4.07	4.44	4.73	4.97	5.17	5.34	5.50
7	2.68	3.45	3.93	4.28	4.55	4.78	4.97	5.14	5.28
8	2.63	3.37	3.83	4.17	4.43	4.65	4.83	4.99	5.13
9	2.59	3.32	3.76	4.08	4.34	4.54	4.72	4.87	5.01
10	2.56	3.27	3.70	4.02	4.26	4.47	4.64	4.78	4.91
11	2.54	3.23	3.66	3.96	4.20	4.40	4.57	4.71	4.84
12	2.52	3.20	3.62	3.92	4.16	4.35	4.51	4.65	4.78
13	2.50	3.18	3.59	3.88	4.12	4.30	4.46	4.60	4.72
14	2.49	3.16	3.56	3.85	4.08	4.27	4.42	4.56	4.68
15	2.48	3.14	3.54	3.83	4.05	4.23	4.39	4.52	4.64
16	2.47	3.12	3.52	3.80	4.03	4.21	4.36	4.49	4.61
17	2.46	3.11	3.50	3.78	4.00	4.18	4.33	4.46	4.58
18	2.45	3.10	3.49	3.77	3.98	4.16	4.31	4.44	4.55
19	2.45	3.09	3.47	3.75	3.97	4.14	4.29	4.42	4.53
20	2.44	3.08	3.46	3.74	3.95	4.12	4.27	4.40	4.51
24	2.42	3.05	3.42	3.69	3.90	4.07	4.21	4.34	4.44
30	2.40	3.02	3.39	3.65	3.85	4.02	4.16	4.28	4.38
40	2.38	2.99	3.35	3.60	3.80	3.96	4.10	4.21	4.32
60	2.36	2.96	3.31	3.56	3.75	3.91	4.04	4.16	4.25
120	2.34	2.93	3.28	3.52	3.71	3.86	3.99	4.10	4.19
∞	2.33	2.90	3.24	3.48	3.66	3.81	3.93	4.04	4.13

TABLE A.6 (CONTINUED)

df \ k	11	12	13	14	15	16	17	18	19	20
1	25.24	25.92	26.54	27.10	27.62	28.10	28.54	28.96	29.35	29.71
2	10.01	10.26	10.49	10.70	10.89	11.07	11.24	11.39	11.54	11.68
3	7.49	7.67	7.83	7.98	8.12	8.25	8.37	8.48	8.58	8.68
4	6.49	6.65	6.78	6.91	7.02	7.13	7.23	7.33	7.41	7.50
5	5.97	6.10	6.22	6.34	6.44	6.54	6.63	6.71	6.79	6.86
6	5.64	5.76	5.87	5.98	6.07	6.16	6.25	6.32	6.40	6.47
7	5.41	5.53	5.64	5.74	5.83	5.91	5.99	6.06	6.13	6.19
8	5.25	5.36	5.46	5.56	5.64	5.72	5.80	5.87	5.93	6.00
9	5.13	5.23	5.33	5.42	5.51	5.58	5.66	5.72	5.79	5.85
10	5.03	5.13	5.23	5.32	5.40	5.47	5.54	5.61	5.67	5.73
11	4.95	5.05	5.15	5.23	5.31	5.38	5.45	5.51	5.57	5.63
12	4.89	4.99	5.08	5.16	5.24	5.31	5.37	5.44	5.49	5.55
13	4.83	4.93	5.02	5.10	5.18	5.25	5.31	5.37	5.43	5.48
14	4.79	4.88	4.97	5.05	5.12	5.19	5.26	5.32	5.37	5.43
15	4.75	4.84	4.93	5.01	5.08	5.15	5.21	5.27	5.32	5.38
16	4.71	4.81	4.89	4.97	5.04	5.11	5.17	5.23	5.28	5.33
17	4.68	4.77	4.86	4.93	5.01	5.07	5.13	5.19	5.24	5.30
18	4.65	4.75	4.83	4.90	4.98	5.04	5.10	5.16	5.21	5.26
19	4.63	4.72	4.80	4.88	4.95	5.01	5.07	5.13	5.18	5.23
20	4.61	4.70	4.78	4.85	4.92	4.99	5.05	5.10	5.16	5.20
24	4.54	4.63	4.71	4.78	4.85	4.91	4.97	5.02	5.07	5.12
30	4.47	4.56	4.64	4.71	4.77	4.83	4.89	4.94	4.99	5.03
40	4.41	4.49	4.56	4.63	4.69	4.75	4.81	4.86	4.90	4.95
60	4.34	4.42	4.49	4.56	4.62	4.67	4.73	4.78	4.82	4.86
120	4.28	4.35	4.42	4.48	4.54	4.60	4.65	4.69	4.74	4.78
∞	4.21	4.28	4.35	4.41	4.47	4.52	4.57	4.61	4.65	4.69

TABLE A.6 (CONTINUED)

$\alpha = 5\%$

df \ k	2	3	4	5	6	7	8	9	10
1	17.97	26.98	32.82	37.08	40.41	43.12	45.40	47.36	49.07
2	6.08	8.33	9.80	10.88	11.74	12.44	13.03	13.54	13.99
3	4.50	5.91	6.82	7.50	8.04	8.48	8.85	9.18	9.46
4	3.93	5.04	5.76	6.29	6.71	7.05	7.35	7.60	7.83
5	3.64	4.60	5.22	5.67	6.03	6.33	6.58	6.80	6.99
6	3.46	4.34	4.90	5.30	5.63	5.90	6.12	6.32	6.49
7	3.34	4.16	4.68	5.06	5.36	5.61	5.82	6.00	6.16
8	3.26	4.04	4.53	4.89	5.17	5.40	5.60	5.77	5.92
9	3.20	3.95	4.41	4.76	5.02	5.24	5.43	5.59	5.74
10	3.15	3.88	4.33	4.65	4.91	5.12	5.30	5.46	5.60
11	3.11	3.82	4.26	4.57	4.82	5.03	5.20	5.35	5.49
12	3.08	3.77	4.20	4.51	4.75	4.95	5.12	5.27	5.39
13	3.06	3.73	4.15	4.45	4.69	4.88	5.05	5.19	5.32
14	3.03	3.70	4.11	4.41	4.64	4.83	4.99	5.13	5.25
15	3.01	3.67	4.08	4.37	4.59	4.78	4.94	5.08	5.20
16	3.00	3.65	4.05	4.33	4.56	4.74	4.90	5.03	5.15
17	2.98	3.63	4.02	4.30	4.52	4.70	4.86	4.99	5.11
18	2.97	3.61	4.00	4.28	4.49	4.67	4.82	4.96	5.07
19	2.96	3.59	3.98	4.25	4.47	4.65	4.79	4.92	5.04
20	2.95	3.58	3.96	4.23	4.45	4.62	4.77	4.90	5.01
24	2.92	3.53	3.90	4.17	4.37	4.54	4.68	4.81	4.92
30	2.89	3.49	3.85	4.10	4.30	4.46	4.60	4.72	4.82
40	2.86	3.44	3.79	4.04	4.23	4.39	4.52	4.63	4.73
60	2.83	3.40	3.74	3.98	4.16	4.31	4.44	4.55	4.65
120	2.80	3.36	3.68	3.92	4.10	4.24	4.36	4.47	4.56
∞	2.77	3.31	3.63	3.86	4.03	4.17	4.29	4.39	4.47

TABLE A.6 (CONTINUED)

df \ k	11	12	13	14	15	16	17	18	19	20
1	50.59	51.96	53.20	54.33	55.36	56.32	57.22	58.04	58.83	59.56
2	14.39	14.75	15.08	15.38	15.65	15.91	16.14	16.37	16.57	16.77
3	9.72	9.95	10.15	10.35	10.52	10.69	10.84	10.98	11.11	11.24
4	8.03	8.21	8.37	8.52	8.66	8.79	8.91	9.03	9.13	9.23
5	7.17	7.32	7.47	7.60	7.72	7.83	7.93	8.03	8.12	8.21
6	6.65	6.79	6.92	7.03	7.14	7.24	7.34	7.43	7.51	7.59
7	6.30	6.43	6.55	6.66	6.76	6.85	6.94	7.02	7.10	7.17
8	6.05	6.18	6.29	6.39	6.48	6.57	6.65	6.73	6.80	6.87
9	5.87	5.98	6.09	6.19	6.28	6.36	6.44	6.51	6.58	6.64
10	5.72	5.83	5.93	6.03	6.11	6.19	6.27	6.34	6.40	6.47
11	5.61	5.71	5.81	5.90	5.98	6.06	6.13	6.20	6.27	6.33
12	5.51	5.61	5.71	5.80	5.88	5.95	6.02	6.09	6.15	6.21
13	5.43	5.53	5.63	5.71	5.79	5.86	5.93	5.99	6.05	6.11
14	5.36	5.46	5.55	5.64	5.71	5.79	5.85	5.91	5.97	6.03
15	5.31	5.40	5.49	5.57	5.65	5.72	5.78	5.85	5.90	5.96
16	5.26	5.35	5.44	5.52	5.59	5.66	5.73	5.79	5.84	5.90
17	5.21	5.31	5.39	5.47	5.54	5.61	5.67	5.73	5.79	5.84
18	5.17	5.27	5.35	5.43	5.50	5.57	5.63	5.69	5.74	5.79
19	5.14	5.23	5.31	5.39	5.46	5.53	5.59	5.65	5.70	5.75
20	5.11	5.20	5.28	5.36	5.43	5.49	5.55	5.61	5.66	5.71
24	5.01	5.10	5.18	5.25	5.32	5.38	5.44	5.49	5.55	5.59
30	4.92	5.00	5.08	5.15	5.21	5.27	5.33	5.38	5.43	5.47
40	4.82	4.90	4.98	5.04	5.11	5.16	5.22	5.27	5.31	5.36
60	4.73	4.81	4.88	4.94	5.00	5.06	5.11	5.15	5.20	5.24
120	4.64	4.71	4.78	4.84	4.90	4.95	5.00	5.04	5.09	5.13
∞	4.55	4.62	4.68	4.74	4.80	4.85	4.89	4.93	4.97	5.01

TABLE A.6 (CONTINUED)

$$\alpha = 1\%$$

df \ k	2	3	4	5	6	7	8	9	10
1	90.03	135.0	164.3	185.6	202.2	215.8	227.2	237.0	245.6
2	14.04	19.02	22.29	24.72	26.63	28.20	29.53	30.68	31.69
3	8.26	10.62	12.17	13.33	14.24	15.00	15.64	16.20	16.69
4	6.51	8.12	9.17	9.96	10.58	11.10	11.55	11.93	12.27
5	5.70	6.98	7.80	8.42	8.91	9.32	9.67	9.97	10.24
6	5.24	6.33	7.03	7.56	7.97	8.32	8.61	8.87	9.10
7	4.95	5.92	6.54	7.01	7.37	7.68	7.94	8.17	8.37
8	4.75	5.64	6.20	6.62	6.96	7.24	7.47	7.68	7.86
9	4.60	5.43	5.96	6.35	6.66	6.91	7.13	7.33	7.49
10	4.48	5.27	5.77	6.14	6.43	6.67	6.87	7.05	7.21
11	4.39	5.15	5.62	5.97	6.25	6.48	6.67	6.84	6.99
12	4.32	5.05	5.50	5.84	6.10	6.32	6.51	6.67	6.81
13	4.26	4.96	5.40	5.73	5.98	6.19	6.37	6.53	6.67
14	4.21	4.89	5.32	5.63	5.88	6.08	6.26	6.41	6.54
15	4.17	4.84	5.25	5.56	5.80	5.99	6.16	6.31	6.44
16	4.13	4.79	5.19	5.49	5.72	5.92	6.08	6.22	6.35
17	4.10	4.74	5.14	5.43	5.66	5.85	6.01	6.15	6.27
18	4.07	4.70	5.09	5.38	5.60	5.79	5.94	6.08	6.20
19	4.05	4.67	5.05	5.33	5.55	5.73	5.89	6.02	6.14
20	4.02	4.64	5.02	5.29	5.51	5.69	5.84	5.97	6.09
24	3.96	4.55	4.91	5.17	5.37	5.54	5.69	5.81	5.92
30	3.89	4.45	4.80	5.05	5.24	5.40	5.54	5.65	5.76
40	3.82	4.37	4.70	4.93	5.11	5.26	5.39	5.50	5.60
60	3.76	4.28	4.59	4.82	4.99	5.13	5.25	5.36	5.45
120	3.70	4.20	4.50	4.71	4.87	5.01	5.12	5.21	5.30
∞	3.64	4.12	4.40	4.60	4.76	4.88	4.99	5.08	5.16

TABLE A.6 (CONTINUED)

df \ k	11	12	13	14	15	16	17	18	19	20
1	253.2	260.0	266.2	271.8	277.0	281.8	286.3	290.4	294.3	298.0
2	32.59	33.40	34.13	34.81	35.43	36.00	36.53	37.03	37.50	37.95
3	17.13	17.53	17.89	18.22	18.52	18.81	19.07	19.32	19.55	19.77
4	12.57	12.84	13.09	13.32	13.53	13.73	13.91	14.08	14.24	14.40
5	10.48	10.70	10.89	11.08	11.24	11.40	11.55	11.68	11.81	11.93
6	9.30	9.48	9.65	9.81	9.95	10.08	10.21	10.32	10.43	10.54
7	8.55	8.71	8.86	9.00	9.12	9.24	9.35	9.46	9.55	9.65
8	8.03	8.18	8.31	8.44	8.55	8.66	8.76	8.85	8.94	9.03
9	7.65	7.78	7.91	8.03	8.13	8.23	8.33	8.41	8.49	8.57
10	7.36	7.49	7.60	7.71	7.81	7.91	7.99	8.08	8.15	8.23
11	7.13	7.25	7.36	7.46	7.56	7.65	7.73	7.81	7.88	7.95
12	6.94	7.06	7.17	7.26	7.36	7.44	7.52	7.59	7.66	7.73
13	6.79	6.90	7.01	7.10	7.19	7.27	7.35	7.42	7.48	7.55
14	6.66	6.77	6.87	6.96	7.05	7.13	7.20	7.27	7.33	7.39
15	6.55	6.66	6.76	6.84	6.93	7.00	7.07	7.14	7.20	7.26
16	6.46	6.56	6.66	6.74	6.82	6.90	6.97	7.03	7.09	7.15
17	6.38	6.48	6.57	6.66	6.73	6.81	6.87	6.94	7.00	7.05
18	6.31	6.41	6.50	6.58	6.65	6.73	6.79	6.85	6.91	6.97
19	6.25	6.34	6.43	6.51	6.58	6.65	6.72	6.78	6.84	6.89
20	6.19	6.28	6.37	6.45	6.52	6.59	6.65	6.71	6.77	6.82
24	6.02	6.11	6.19	6.26	6.33	6.39	6.45	6.51	6.56	6.61
30	5.85	5.93	6.01	6.08	6.14	6.20	6.26	6.31	6.36	6.41
40	5.69	5.76	5.83	5.90	5.96	6.02	6.07	6.12	6.16	6.21
60	5.53	5.60	5.67	5.73	5.78	5.84	5.89	5.93	5.97	6.01
120	5.37	5.44	5.50	5.56	5.61	5.66	5.71	5.75	5.79	5.83
∞	5.23	5.29	5.35	5.40	5.45	5.49	5.54	5.57	5.61	5.65

* From Table of Upper Percentage Points of the Studentized Range, James Pachares, Biometrika, Vol 46, parts 3–4, 1959.

TABLE A.7. MULTIPLYING FACTOR FOR THE GEOMETRIC MEAN *

	Sample Size							
T	2	5	8	10	13	15	25	50
0.05	1.025	1.041	1.045	1.046	1.047	1.048	1.049	1.050
0.10	1.050	1.082	1.091	1.093	1.096	1.097	1.100	1.103
0.15	1.076	1.125	1.138	1.143	1.147	1.149	1.154	1.158
0.20	1.102	1.169	1.187	1.194	1.200	1.203	1.210	1.216
0.25	1.128	1.214	1.238	1.247	1.255	1.259	1.268	1.276
0.30	1.154	1.260	1.291	1.302	1.312	1.317	1.330	1.340
0.35	1.180	1.307	1.345	1.359	1.372	1.378	1.393	1.406
0.40	1.207	1.356	1.401	1.418	1.433	1.441	1.460	1.476
0.45	1.234	1.406	1.459	1.479	1.498	1.506	1.530	1.548
0.50	1.261	1.457	1.519	1.542	1.564	1.574	1.602	1.625
0.55	1.288	1.509	1.581	1.608	1.633	1.645	1.678	1.705
0.60	1.315	1.563	1.645	1.675	1.705	1.719	1.757	1.789
0.65	1.343	1.618	1.711	1.746	1.780	1.796	1.840	1.876
0.70	1.371	1.675	1.779	1.818	1.857	1.876	1.926	1.968
0.75	1.399	1.733	1.849	1.894	1.938	1.958	2.016	2.064
0.80	1.427	1.792	1.922	1.971	2.021	2.045	2.110	2.165
0.85	1.456	1.853	1.996	2.052	2.108	2.134	2.208	2.270
0.90	1.485	1.915	2.074	2.135	2.197	2.227	2.310	2.381
0.95	1.514	1.979	2.153	2.221	2.291	2.323	2.417	2.496
1.00	1.543	2.044	2.235	2.310	2.387	2.424	2.528	2.617
1.05	1.573	2.111	2.320	2.403	2.487	2.528	2.644	2.744
1.10	1.602	2.180	2.407	2.498	2.591	2.636	2.765	2.876
1.15	1.632	2.250	2.497	2.596	2.698	2.748	2.891	3.014
1.20	1.662	2.321	2.589	2.698	2.810	2.864	3.022	3.159
1.25	1.693	2.395	2.685	2.803	2.926	2.985	3.159	3.311
1.30	1.724	2.470	2.783	2.911	3.045	3.111	3.301	3.470
1.35	1.754	2.547	2.884	3.023	3.169	3.241	3.450	3.636
1.40	1.786	2.626	2.988	3.139	3.298	3.376	3.604	3.809
1.45	1.817	2.706	3.096	3.259	3.431	3.515	3.766	3.991
1.50	1.849	2.788	3.206	3.382	3.569	3.661	3.933	4.181
1.55	1.880	2.873	3.320	3.510	3.711	3.811	4.108	4.379
1.60	1.913	2.959	3.437	3.642	3.859	3.967	4.291	4.587
1.65	1.945	3.047	3.558	3.777	4.012	4.129	4.480	4.804
1.70	1.977	3.137	3.682	3.918	4.171	4.297	4.678	5.031
1.75	2.010	3.229	3.810	4.062	4.334	4.471	4.883	5.269
1.80	2.043	3.323	3.942	4.212	4.504	4.651	5.097	5.517
1.85	2.077	3.420	4.077	4.366	4.680	4.838	5.320	5.776
1.90	2.110	3.518	4.216	4.525	4.861	5.031	5.552	6.048
1.95	2.144	3.619	4.359	4.688	5.049	5.232	5.794	6.331
2.00	2.178	3.721	4.506	4.857	5.243	5.439	6.045	6.628

* From The Lognormal Frequency Distribution in Relation to Gold Assay Data, U.S. Bureau of Mines Report of Investigations, in press

TABLE A.8. PERCENTILES OF THE DISTRIBUTION OF r WHEN $\rho = 0$

N	$r_{.95}$	$r_{.975}$	$r_{.99}$	$r_{.995}$	$r_{.9995}$	N	$r_{.95}$	$r_{.975}$	$r_{.99}$	$r_{.995}$	$r_{.9995}$
5	.805	.878	.934	.959	.991	20	.378	.444	.516	.561	.679
6	.729	.811	.882	.917	.974	22	.360	.423	.492	.537	.652
7	.669	.754	.833	.875	.951	24	.344	.404	.472	.515	.629
8	.621	.707	.789	.834	.925	26	.330	.388	.453	.496	.607
9	.582	.666	.750	.798	.898	28	.317	.374	.437	.479	.588
10	.549	.632	.715	.765	.872	30	.306	.361	.423	.463	.570
11	.521	.602	.685	.735	.847	40	.264	.312	.366	.402	.501
12	.497	.576	.658	.708	.823	50	.235	.279	.328	.361	.451
13	.476	.553	.634	.684	.801	60	.214	.254	.300	.330	.414
14	.457	.532	.612	.661	.780	80	.185	.220	.260	.286	.361
15	.441	.514	.592	.641	.760	100	.165	.196	.232	.256	.324
16	.426	.497	.574	.623	.742	250	.104	.124	.147	.163	.207
17	.412	.482	.558	.606	.725	500	.074	.088	.104	.115	.147
18	.400	.468	.543	.590	.708	1000	.052	.062	.074	.081	.104
19	.389	.456	.529	.575	.693	∞	0	0	0	0	0
N	$-r_{.05}$	$-r_{.025}$	$-r_{.01}$	$-r_{.005}$	$-r_{.0005}$	N	$-r_{.05}$	$-r_{.025}$	$-r_{.01}$	$-r_{.005}$	$-r_{.0005}$

Greek Alphabet of Capital and Lower-Case Letters

alpha	A α a	nu	N ν
beta	B β	xi	Ξ ξ
gamma	Γ γ	omicron	O o
delta	Δ δ ∂	pi	Π π
epsilon	E ϵ ε	rho	P ρ
zeta	Z ζ	sigma	Σ σ s
eta	H η	tau	T τ
theta	Θ θ ϑ	upsilon	Υ υ
iota	I ι	phi	Φ ϕ φ
kappa	K κ	chi	X χ
lambda	Λ λ	psi	Ψ ψ
mu	M μ	omega	Ω ω

Partial List of Symbols

363

Greek Letters

θ	proportion of successes in a binomial population	204
θ	mean of the negative binomial distribution	208
μ	population mean	34
μ_L	lower confidence limit for the population mean	89
μ_q	mean of a linear combination q	161
$\mu s_{\bar{w}}^2$	mean of the variance of sample means	143
μ_U	upper confidence limit for the population mean	89
$\mu_{\bar{w}}$	mean of sample means	67
ξ_a	analytical variability	304
ξ_n	natural variability	304
ξ_p	preparation variability	304
ξ_s	sampling variability	304
σ	population standard deviation	34
σ^2	population variance	50
σ_μ^2	variance of population means	134
σ_q^2	variance of a linear combination q	161
$\sigma_{\bar{w}}^2$	variance of a sample mean	67
τ	linear combination	164
ϕ_n	multiplying factor for the variance of lognormally distributed observations	217
χ^2	chi-square statistic	73
ψ_n	multiplying factor for the geometric mean of lognormally distributed observations	217
Ω	cutting point in a hypothesis test	109

Index

STATISTICAL ANALYSIS
OF GEOLOGICAL DATA

Volume II

Preface

This book is the second of two volumes. The purpose is to present some methods of statistical analysis of geological data that are more advanced than those in the first volume. While this volume stands alone, it also is a logical continuation of the earlier one.

This volume is written for the person with a knowledge of basic statistics equivalent to that taught in a one-semester college course or obtained through study of the previous volume. The reader is assumed to understand statistical methods for the analysis of univariate observations, the relation between a statistical sample and a population, and the basic principles of geological sampling.

From the preface to Volume 1, we repeat that our aim is to explain statistical procedures effective for the analysis of geological data and to discuss methods of obtaining reliable data that are worth analyzing. We write for the person with numerical data from which he wants to draw conclusions or who has a problem whose solution will require obtaining and interpreting numerical data. We stress basic statistical methods and emphasize that thoughtful application of these methods will yield valid results. Because geology is a complicated and diverse science, we purposely include some involved geological arguments. The mathematics is relatively simple, however, and requires for its understanding only elementary algebra and geometry, and simple operations of matrix algebra.

This volume is for readers with some geological training who may or may not have professional experience. Although most readers may be geologists, the book will also interest mining and petroleum engineers, geochemists and geophysicists, mineral economists, and others.

While most statistical analyses in this volume are of real data, some are of fictitious data chosen for numerical properties that illustrate specific statistical methods. Real data come from several fields of geology, especially economic and structural geology, and petrology.

This volume contains Parts IV, V, and VI of the complete book. In Part IV several methods for the statistical analysis of multivariate geological observations are explained and are applied to various geological problems. In Part V some problems in applied geology are reviewed, including several applications of operations research in geology and in the mineral industries, and some specialized methods of geological sampling. Because the statistical methods in this volume usually require electronic computers for implementation in practice, Part VI is a chapter that outlines the role of electronic computing in geological data analysis.

We acknowledge again with thanks the financial support of the institutions and the cooperation of the mining companies detailed in the preface to Volume 1. We have profited from the advice and criticism of the following men, who read all or parts of this volume: L. R. Drew, R. C. Flemal, S. W. Hazen, Jr., David Hoaglin, A. T. Miesch, D. B. Morris, J. H. Schuenemeyer, G. S. Watson, and Alfred Weiss. The errors and misjudgments that remain are ours alone.

We acknowledge again the help of the librarians cited in the preface to Volume 1. G. W. Johnson, Martha Koch, and Sally Konnak improved the English expression by careful editing.

George S. Koch, Jr.
Richard F. Link

Denver, Colorado
New York, New York
June 1970

Contents

Chapter 11

Ratios and Variables of Constant Sum

PART V

SOME PROBLEMS IN APPLIED GEOLOGY

Chapter 12

Exploration

Chapter 13

Delineation of Ore Deposits

MULTIVARIATE STATISTICAL METHODS

In Chapters 9 to 11 selected statistical methods for analysis of multivariate geological data are introduced. Multivariate data are those for which a single observation has more than one variable, as distinguished from univariate data, for which a single observation has only one variable. Examples of multivariate observations are rock analyses for more than one constituent, several size and shape measurements on a single fossil specimen, several sedimentary petrological measurements on a single sand sample, and multi-element analyses in exploration geochemistry.

Chapter 9 is about geological trend analysis, including representation of observations in two or three dimensions with mathematical models, interpretation of residual values from trend surfaces, and calculation of confidence bands and volumes. Chapter 10 is on the interpretation of multivariate observations by methods that include multiple regression, the generalized analysis of variance, generalized discriminant analysis and the discriminant function, and factor analysis. Analysis of directional data, a special kind of multivariate observations, is also considered. Chapter 11 discusses three linked topics in multivariate analysis: ratios, variables of constant sum, and interpretation of triangular diagrams.

Chapter 9

Geological Trends

9.1 INTRODUCTION

A *geological trend* is a systematic change, usually in a geographical direction, in the value of a geological variable or variables, such as the feldspar content of a granite or the porosity of a sandstone. The term was introduced into geology by Grant (1957). The first extensive work on analysis of geological trends was done at Northwestern University by Krumbein and Whitten; this work has been summarized by Krumbein and Graybill (1965, Chap. 13), who provide a full list of references.

Geological trend analysis always involves a graph, which may have one, two, three, or (rarely) even more than three dimensions. Krumbein (1966a) points out that a two-dimensional graph, or map, is the "basic tool of communication in geology." This statement is surely not exaggerated. Maps facilitate the presentation of summary information in particular, and are often used by geologists for this purpose. Indeed, it is this familiarity with maps that has made trend analysis popular among geologists.

In geological trend analysis the coefficients of a statistical model are estimated and the equation is presented as a contour map. The similarity of this map to a topographic contour map suggests that both may be made in the same way, but important differences must be emphasized. Consider how a geologist makes a topographic contour map with a plane table and telescopic alidade. The late Kirk Bryan, who thought about details of plane-table mapping as carefully as he thought about fundamental geologic principles, gave directions like these: Locate control points where slopes change and on the noses of slopes. Survey other control points only where there is a large hole in the net of points. Do not attempt to walk out contours. Draw contour lines in the field, where details of topography can be seen while more control points can be surveyed if necessary.

By following directions like these, the geologist draws a contour map representing the land surface to any required degree of accuracy. Each control or data point is obtained purposefully, and the integrating power of the human brain establishes the continuity among points. Because the map is made in the field by a man trained in land forms and geomorphology, it is better than one made by a surveyor who locates points in the field with a transit and then draws contours mechanically in the office.

In trend-surface analysis, on the other hand, one is seldom if ever concerned with topographic contouring, except perhaps for an illustrative exercise. Instead, the data to be contoured may be well or borehole data, or perhaps mineralogical data or chemical analyses from surface outcrops. The observations to be contoured are obtained at points, among which there is generally no observed evidence of natural continuity. Although the data points may be gridded, more commonly they are scattered haphazardly or are clustered. Additional observations, especially those from the subsurface, can be obtained, if at all, only at substantial expense. By contrast, the plane-table topographer can obtain additional points exactly where he needs them with small effort. Therefore a model that is suitable for contouring data in trend-surface analysis is different from one suitable for contouring topographic data. Although the geologist recognizes that the details, if they were all known, might require an intricate surface, he is aware that the best feasible representation must contain relatively smooth contour curves, because the data are never as numerous as for topographic mapping. The resulting map may be an isopach map, porosity map, structure contour map, ore-grade map, mineralogical-variation map, or still another type.

Think of how a geologist hand-contours nontopographic data, say, subsurface elevations on a key bed. Every geologist will draw a different contour map, as anyone who has assigned contouring as a student exercise will attest. The maps will differ because each geologist has a different geological model in mind; the best map will correspond to the best model. For instance, the geologist familiar with the Los Angeles basin has a pattern of faults in mind and can contour subsurface data from this area better through his model than can the man thinking of an inappropriate model.

Next consider how the same data may be contoured by a statistical method. The geologist begins with a statistical model in mind, just as the geologist contouring by hand is guided by a geological model. The difference is that the statistical model must be precisely formulated, while the geological model need not be. This point is emphasized because some investigators have suggested wrongly that trend-surface analysis is "objective" while hand-contouring is "subjective." On the contrary, both methods have subjective and objective elements.

The usual statistical model for trend-surface analysis is some form of *regression analysis*. In regression analysis a dependent variable w is expressed as a function of one or more independent variables x_i and constants β_i, plus a random fluctuation e. In mathematical notation

$$w = f(x_1, x_2, ..., x_k; \beta_1, \beta_2, ..., \beta_k) + e ,$$

where e follows a frequency distribution, such as the normal distribution. Today the name "regression" is arbitrary rather than descriptive, although historically it had a descriptive meaning.

In any problem of regression analysis, explicit assumptions are made about the mathematical function and about the frequency distribution of e. For a single set of data alternative assumptions can always be made, and one set must be subjectively chosen as most appropriate.

A Simple but Real Example

To make the introductory remarks more meaningful, we now give a simple but real example of how data can be contoured. The discussion is related first to a geological and then to a statistical model. The example is a set of subsurface observations in the Rangely oil-and-gas field of western Colorado (Thomas et al., 1945). The data are elevations, measured in 26 wells (fig. 9.1), of the top of the Weber sandstone (Pennsylvanian or Permian age) above a datum plane.

One way to summarize these data is to draw structural contours on the elevations in the 26 holes. The geological model to guide the contouring is evident from viewing a surface geological map, or from flying over Rangely on one of the commercial air routes from Denver to Salt Lake City, since the structure is readily seen from the air. At the surface the geologic structure, expressed by the outcropping Mancos and Mesaverde Formations of Cretaceous age, is a doubly plunging anticline. Because the underlying formations conform essentially to those that outcrop, the geologic model for the subsurface is also a doubly plunging anticline.

For a statistical model corresponding to the geological model, a smooth continuous concave or convex function is appropriate. From the many statistical models meeting this specification, we choose a quadratic equation of the form

$$w = \alpha + \beta_1 x + \beta_2 y + \beta_3 x^2 + \beta_4 y^2 + \beta_5 xy + e,$$

where w is the elevation of the top of the Weber sandstone at a given geographic point, x is the east coordinate, y is the north coordinate, α and β_1 to β_5

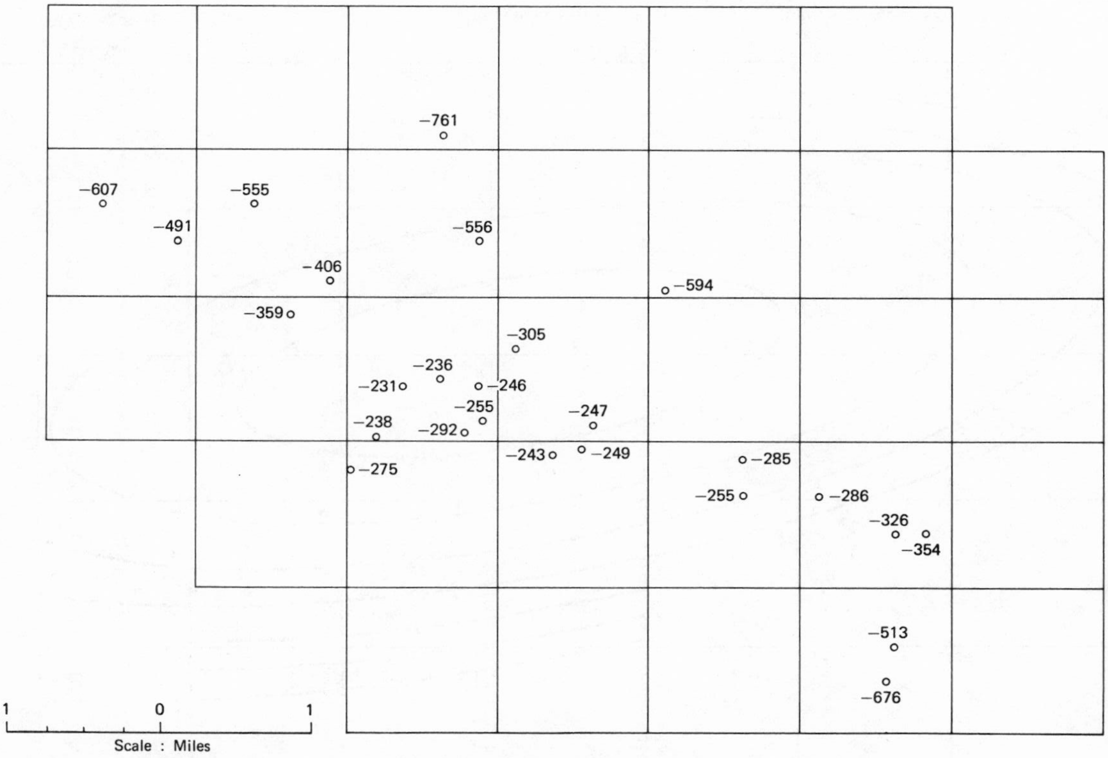

Fig. 9.1. Elevations of the Weber sandstone in 26 wells in the Rangely oil field, Colorado.

Scale : Miles

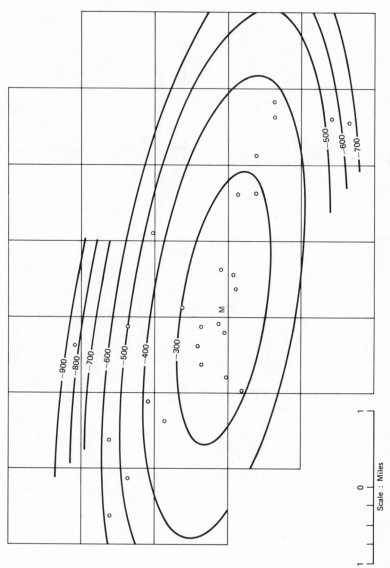

Fig. 9.2. Structure contour map of the top of the Weber sandstone prepared by quadratic regression analysis.

are parameters, and e is the assumed random fluctuation. When the parameters are estimated in a way described below (sec. 9.3), the equation becomes

$$\hat{w} = 2629 - 393.0x - 968.8y + 20.7x^2 + 153.9y^2 + 48.2xy,$$

where $\hat{w}$ is the statistic that estimates the smoothed function. In order to obtain the structure-contour map graphed in figure 9.2, it is only necessary to evaluate this equation for closely spaced (x, y) grid points corresponding to selected values of w, which for the chosen contour interval are in even hundreds of feet. Complications and qualifications of this analysis are postponed to section 9.5.

Purposes of Trend-Surface Analysis

Trend-surface analyses have been made of different kinds of geological data for many purposes. Data have come from igneous petrology (Whitten, 1962), sedimentary petrology (Krumbein, 1966; Griffiths, 1967), mining geology (Koch and Link, 1967), oil geology (Harbaugh, 1964), and other fields.

Most of all, trend-surface analysis has been used to summarize data that are too numerous to be grasped readily by eye. Trend-surface analysis also allows values within and outside the net of points to be predicted, along with a quantitative appraisal of the reliability of the prediction. Another purpose is to estimate a volume between the trend surface and a datum plane, or between two trend surfaces. Other uses are to narrow the confidence limits in estimating means, which may be the average grade of ore or the average feldspar content of a rock, and to obtain residual values, which are the differences between the observed and predicted values at data points. These and other purposes of trend-surface analysis are examined later in this chapter.

There are many things that trend-surface analysis is not. It is not a method for automatic contouring of data by computer when the aim is to fit a surface exactly to all of the data points. Nor is it a method to be used blindly, as some authors, in their enthusiasm for new methods, suggest. Although a topographic contour map can be made with various types of interchangeable transits or alidades, a trend-surface map cannot be made by various interchangeable computer programs and computers, given the present state of the computer art. Rather, the exact purposes of a trend-surface analysis and the behavior of the computer programs need to be evaluated individually.

9.2 LINEAR REGRESSION ANALYSIS OF DATA IN ONE DIMENSION .

Having introduced geological trends in the previous section, we now turn to a formal analysis. In this section the simplest kind of regression, that of linear analysis of data in one dimension, is explained. The explanation is divided into two parts: a descriptive analysis to provide intuitive understanding, and a statistical analysis suitable for actual computing.

In the rest of the chapter the principles introduced in this section are extended to more complex situations in two and three dimensions.

Descriptive Analysis

An example serves to introduce the descriptive analysis of fitting a straight line to points. Consider a straight drift, following a vein, with assays (w-values) for some metal beginning with high values at one end and gradually diminishing in grade in a regular linear fashion toward the other end. The general equation for this straight line or linear grade trend is

$$w = \alpha + \beta x + e,$$

where x-values are distances along the drift, w is the assay value at any x point, α is a regression coefficient termed *intercept*, β is a regression coefficient termed *slope*, and e is the random error.

The terms α and β are parameters. (In this book the usual convention is followed of designating parameters by Greek letters and statistics by Roman letters.) In the example drift, an essentially unlimited number of paired location and assay values are potentially available from repeated mine sampling; these observations comprise the population. Yet only a few mine samples are actually taken in the drift to determine the values that comprise the statistical sample. Therefore a new equation,

$$\hat{w} = a + bx,$$

is required—one that corresponds to the original equation

$$w = \alpha + \beta x + e,$$

with the parameters α and β replaced by the corresponding statistics a and b.

The meaning of the previous equations may be demonstrated by fitting a line to 22 points, representing values of silver content from successive mine samples taken in a drift following a vein in the Fresnillo mine (fig. 9.3 and table 9.1). Distances along the drift are plotted on the horizontal x-axis; the numbers correspond to the sample locations spaced 2 meters apart. Values of silver content for each sample location are plotted on the vertical w-axis. The straight line fitted to these sample points has the equation

$$\hat{w} = 480.1 - 11.74x,$$

where $\hat{w}$ is estimated mean silver content and x is distance. The intercept a, 480.1, is the estimated silver content at zero distance; the slope b, -11.74, is the estimated decrease in silver content per 1-meter increase in distance.

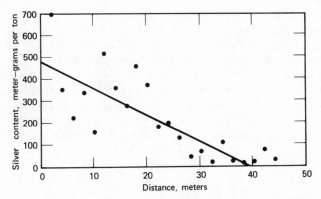

Fig. 9.3. Silver content at 22 contiguous sample points in a drift in the Fresnillo mine, with fitted line of linear regression.

TABLE 9.1. OBSERVATIONS OF SILVER CONTENT AT 22 CONTIGUOUS SAMPLE POINTS IN A DRIFT IN THE FRESNILLO MINE

x-value distance from starting point (m)	w-value silver content (m-g/ton)
2	698
4	365
6	223
8	335
10	156
12	512
14	357
16	274
18	454
20	369
22	179
24	194
26	137
28	40
30	65
32	16
34	100
36	22
38	13
40	19
42	72
44	23

The utility of a fitted equation generally depends on the proportion of the variability in the w-values that is "explained." If all of the variability is accounted for, because all of the sample points fall on the fitted straight line, the equation for the line summarizes the data concisely and accurately. If, on the other hand, almost none of the variability is accounted for, because none of the sample points fall on or even close to the straight line, then the fitted equation may provide a useless summary. For real data, a complete range in degree of "explanation" is found, and there is no simple answer to the obvious question: How much variation must a fitted equation explain to be useful?

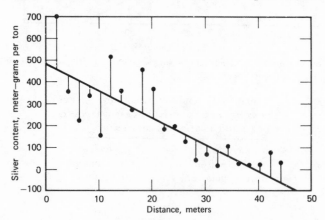

Fig. 9.4. Vertical deviations of the fitted line of figure 9.3 from the 22 data points.

The coefficients a and b for the fitted line are calculated by a procedure described in the next subsection, a procedure which fits a line so that the average of the squared vertical deviations of the data points from the line is as small as possible. In figure 9.4, which is figure 9.3 redrawn, the vertical deviations are the vertical line segments. The sum of the squared values of these segments is a minimum for the particular straight line fitted to the data.

Statistical Analysis

For statistical analysis, a random fluctuation e (sec. 5.11) is allowed for by writing the regression equation as

$$w = \alpha + \beta x + e.$$

This linear equation is a statistical model, and the fluctuation e is assumed to have the following properties: the mean of the fluctuations is equal to zero, and the variance of the fluctuations is equal to the mean value of the squared distances from the regression line. Moreover, the fluctuations from point to point are assumed to be statistically independent.

In summary, the problem is: Given a set of data, that is, a set of paired values for x and y, how may the best estimate be made of the coefficients α and β, and of the variance σ_e^2 of e, which is assumed to be constant for all values of x?

The least-squares solution to this problem requires finding the values of a and b that minimize the expression

$$\sum (w - [a + bx])^2 = f(a, b).$$

These values are

$$a = \overline{w} - b\overline{x}$$

and

$$b = \frac{\text{SS}_{xw}}{\text{SS}_{xx}},$$

where a is the estimate of α, b is the estimate of β,

$$\overline{w} = \frac{\sum w}{n},$$

$$\overline{x} = \frac{\sum x}{n},$$

$$\text{SS}_{xw} = \sum (x - \overline{x})(w - \overline{w}),$$

and

$$\text{SS}_{xx} = \sum (x - \overline{x})^2.$$

[The values of a and b are obtained by calculating partial derivatives, setting them equal to zero, and solving the resulting equations (Link, Koch, and Gladfelter, 1964).]

The estimate s_e^2 of the variance σ_e^2 is obtained by putting the values of a and b from the above solution into the expression for $f(a, b)$ and then dividing the result by $(n - 2)$, the number of observations minus the number of estimated parameters, as discussed in section 5.11. Thus the estimate of σ_e^2 is

$$s_e^2 = \frac{\sum (w - [a + bx])^2}{n - 2}.$$

Once a and b have been obtained, they may be used to partition the sum of squares of the w-values into two components by using the sum of squares as the measure of variability. The first component is the amount of the sum of squares "explained" by the fitted line, named the *regression sum of squares* (regression SS). The other component is the amount of the sum of squares unexplained by the fitted line, named the *residual sum of squares* (residual SS). Geometrically, the residual sum of squares is the sum of the squared vertical deviations illustrated by figure 9.4. It may be shown (Li, 1964, I, p. 293) that

$$\text{regression SS} = \frac{(\text{SS}_{xw})^2}{\text{SS}_{xx}},$$

hence

$$\text{residual SS} = \text{SS}_{ww} - \text{regression SS},$$

where

$$\text{SS}_{ww} = \sum (w - \overline{w})^2.$$

The amount of variation "explained" by the fitted line may be compared with the residual variation through the one-way analysis of variance format (sec. 5.3) as follows:

Source of variation	Sum of squares	Degrees of freedom	Mean square	F
Regression		1		
Residual variability		$n-2$		
Total		$n-1$		

Table 9.2 is a linear regression analysis of the data in table 9.1 and figure 9.3. In the preliminary calculations, lines 1 to 7, the sums of squares of the two variables x and w and the cross-product sum of squares SS_{xw} are calculated by

Table 9.2. Linear regression analysis of data from table 9.1

Preliminary calculations

Line						
1	n	22				
2	$\sum x$	506			$\sum w$	4,623
3	$\overline{x}$	23			$\overline{w}$	210.1364
4	$(\sum x)^2$	256,036	$(\sum x)(\sum w)$	2,339,238	$(\sum w)^2$	21,372,129
5	$(\sum x)^2/n$	11,638	$(\sum x)(\sum w)/n$	106,329	$(\sum w)^2/n$	971,460
6	$\sum x^2$	15,180	$\sum xw$	64,732	$\sum w^2$	1,724,919
7	SS_{xx}	3,542	SS_{xw}	$-41,597$	SS_{ww}	753,459

Analysis

8	Slope:	$b = \text{SS}_{xw}/\text{SS}_{xx} = -11.74$
9	Regression equation:	$\hat{w} = a + bx = 480.1 - 11.74x$
10	Regression SS:	$(\text{SS}_{xw})^2/\text{SS}_{xx} = (-41,597)^2/3,542 = 488,512$
11	Residual SS:	$\text{SS}_{ww} - \text{regression SS} = 753,459 - 488,512 = 264,947$
12	Residual MS:	$\text{residual SS}/(n-2) = 264,947/20 = 13,247$

the short-cut formula (sec. 3.3). In lines 8 to 12 the linear regression analysis is performed. The corresponding one-way analysis of variance of the results from lines 10 to 12 is given in table 9.3.

TABLE 9.3. ANALYSIS OF VARIANCE FOR LINEAR REGRESSION ANALYSIS OF TABLE 9.2

Source of variation	Sum of squares	Degrees of freedom	Mean square	F	$F_{10\%}$
Regression terms	488,512	1	488,512	36.88	2.97
Residual terms	264,947	20	13,247		
Total	753,459	21			

Other statistical techniques could have been chosen to fit a straight line to these data. We chose the least-squares approach for two reasons. The first and more important one is that if the e-values are normally distributed, the least-squares solution will produce an unbiased solution with minimum variance. The second reason is that the least-squares solution is conventional.

In the example the regression of silver content on distance was considered, that is, the variation of silver content with increasing distance. For these data it would be difficult to choose the wrong model and study the change of distance with increasing silver content. But for other data the situation may not be so clear, as has been discussed by Tukey (1951) and Berkson (1950). In fact, in the geological literature, the wrong model has been used on occasion without the author even being aware of it, because he made his interpretation as if he had used the right model. In order to emphasize this point, the roles of w and x are interchanged for the silver observations in the example.

If the equation

$$w = a + bx$$

is rewritten as

$$x = d + cw,$$

it is true that $c = 1/b$ and $d = a/b$. However, if the statistical model for linear regression is changed in an apparently analogous way, the results are not interchangable. Thus, while the statistical model might be written

$$x = \Sigma + \gamma w + e,$$

if γ and Σ are estimated as before, the result obtained is

$$\hat{x} = g + fw,$$

where $f = \mathrm{SS}_{xw}/\mathrm{SS}_{ww}$. In particular, the value -0.552 of f is different from the value -0.085 of $1/b$. However, the calculated F-value is the same.

In terms of the analysis for silver, these two models may be interpreted as follows. In the correct one the deviations from the linear model were ascribed to random fluctuations in the silver content at evenly spaced distances along the drift. In the incorrect model the silver content was wrongly considered to vary smoothly, and the distances along the drift were thought to fluctuate randomly. In table 9.4, the calculations for the data of table 9.1 are repeated

TABLE 9.4. LINEAR REGRESSION ANALYSIS WITH WRONG MODEL
OF DATA FROM TABLE 9.1

Line	Item
8	Slope: $f = \mathrm{SS}_{xw}/\mathrm{SS}_{ww} = -0.552$
9	Regression equation: $\hat{x} = g + fw = 139 - 0.552w$
10	Regression SS: $(\mathrm{SS}_{xw})^2/\mathrm{SS}_{ww} = (-41{,}597)^2/753{,}459 = 2{,}296$
11	Residual SS: $\mathrm{SS}_{xx} -$ regression SS $= 3{,}542 - 2{,}296 = 1{,}246$
12	Residual MS: residual SS$/(n-2) = 62.3$

with the incorrect model. For this problem the first model is clearly the right one, and it is only necessary to keep the calculations straight. For other problems that are more difficult and confusing, because both variables contain some fluctuation, a more complicated statistical procedure, functional analysis (Lindley, 1953), must be used.

Confidence Band for a Regression Line

For a statistic such as the mean or variance, a confidence interval is defined by two numbers specifying upper and lower limits. For a regression line the limits are curved lines that delimit a *confidence band* centered on the regression line. For a two-dimensional surface the confidence band is a volume defined by two surfaces, and for a three-dimensional surface the confidence band is a hypervolume defined by two hypersurfaces. In this book the band for the one-dimensional regression line is introduced, for which Li (1964, I, p. 321) explains the statistical method. The two- and three-dimensional cases are discussed by Krumbein (1963), by Link, Koch, and Gladfelter (1964, pp. 13, 16), and by Link, Yabe, and Koch (1966, p. 5).

The confidence band for a regression line is obtained by calculating quantities

$$\hat{w} \pm \sqrt{F_{10\%}(2, n-2)} \sqrt{s_e^2 \left(\frac{1}{n} + \frac{(x-\bar{x})^2}{\mathrm{SS}_x} \right)}$$

for values of x over the range of x. The quantity $s_e^2 (1/n + (x-\bar{x})^2/\mathrm{SS}_x)$ is $s_{\hat{w}}^2$ (Li, 1964, I, p. 326). If several values of w are available for a given x-value, an estimate of μ_w can be found in two ways: (1) from the fitted line, or (2) from averaging the w-values at the x point. The preferred method is the one that gives the smaller standard error of estimate. The standard error of (1) is $s_{\hat{w}}$, while that of (2) is $s/\sqrt{n}$, where n is the number of x-values and s is the standard deviation of the corresponding w-values.

For the 22 silver values of figure 9.3, the confidence band lies between the two curved lines plotted in figure 9.5. As in this example, the confidence band

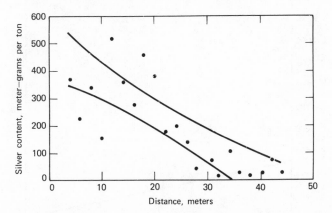

Fig. 9.5. Confidence band for linear regression line fitted to the data of figure 9.3.

is always wider at the ends of the fitted line than at its center, because the slope line passes through the mean of the observations. Therefore the error in estimating the direction of the slope is magnified as distance from the mean increases, just as the error in estimating a compass direction is magnified as one travels away from the origin.

9.3   THE CORRELATION COEFFICIENT

When, as in section 9.2, a straight line is fitted to data in one dimension, all degrees of adequacy of fit are found, from the ideal fit (where all of the plotted sample points lie exactly on a line) to a poor fit (where the sample points are haphazardly scattered about the line). A usual measure of adequacy of fit is the *correlation coefficient*, a number ranging from $+1$ to -1, with values of exactly ± 1 corresponding to a perfect fit and those of zero corresponding to no linear fit. For the example of figure 9.3, the correlation coefficient -0.81 indicates that 66 percent $(100[-0.81]^2)$ of the silver variability is "explained" by the fitted straight line. In contrast, in the example of figure 9.6 for data from a segment of drift adjacent to that of figure 9.3, the correlation coefficient -0.27 indicates that only 7 percent $(100[-0.27]^2)$ of the variability is "explained" by the fitted straight line. A comparison of figures 9.3 and 9.6 shows that the plotted sample points lie closer to the fitted straight line in figure 9.3.

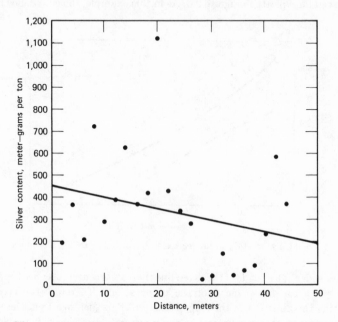

Fig. 9.6. Silver content at 22 contiguous sample points in a drift in the Fresnillo mine, to show unsatisfactory fitted line of linear regression.

The correlation coefficient is defined by the equation

$$\rho_{xw} = \frac{\mu(x-\mu_x)(w-\mu_w)}{(\sigma_x^2 \sigma_w^2)^{1/2}}.$$

The statistic r_{xw}, the estimate of ρ_{xw}, may be obtained from the equation

$$r_{xw} = \frac{\dfrac{\Sigma(x-\bar{x})(w-\bar{w})}{n-1}}{\left[\dfrac{\Sigma(x-\bar{x})^2}{n-1} \cdot \dfrac{\Sigma(w-\bar{w})^2}{n-1}\right]^{1/2}} = \frac{\mathrm{SS}_{xw}}{[\mathrm{SS}_{xx}\,\mathrm{SS}_{ww}]^{1/2}} = \left[\frac{\text{regression SS}}{\mathrm{SS}_{ww}}\right]^{1/2}.$$

TABLE 9.5. CALCULATION OF CORRELATION COEFFICIENT FOR DATA FROM TABLE 9.1

Line	Notation	Numerical value	Remarks
1	SS_{xw}	$-41,597$	From table 9.2
2	SS_{xx}	$3,542$	,,
3	SS_{ww}	$753,459$	,,
4	$\sqrt{SS_{xx}\,SS_{ww}}$	$51,660$	
5	r_{xw}	-0.8052	Line 1/line 4

Because the correlation coefficient (table 9.5) is obtained from the same sums-of-squares terms calculated for the analysis of variance of table 9.4, the two methods are redundant. This brief discussion is given because the correlation coefficient is often found in the geological literature. In a new investigation one usually learns more by calculating the regression line and making an analysis of variance than by working with the correlation coefficient. Although the generalized or multiple correlation coefficient may be calculated to measure the correspondence of observations to any fitted equation, only the one-dimensional case is explained in this book; the generalized case is discussed by Wilks (1962, p. 91).

Like any other parameter, the correlation coefficient can be investigated by hypothesis tests and confidence intervals. Dixon and Massey (1969) give clear expositions with the necessary tables. These methods are omitted from this book, except for the test of the hypothesis that ρ is equal to zero.

TABLE 9.6. TEST OF THE HYPOTHESIS H_0 THAT $\rho_{xw} = 0$ FOR THE DATA OF FIGURES 9.3 AND 9.6

Hypothesis:	$H_0: \rho_{xw} = 0$
Alternative hypothesis:	$H_1: \rho_{xw} > 0$, or $\rho_{xw} < 0$
Statistic:	r_{xw}
Risk of type I error:	$\alpha = 10\%$
Critical region:	$r_{xw} > 0.360$, or $r_{xw} < -0.360$
Figure 9.3 data:	$r_{xw} = -0.81$
Conclusion:	Reject H_0, accept alternative hypothesis H_1 that $\rho_{xw} < 0$
Figure 9.6 data:	$r_{xw} = -0.27$
Conclusion:	Accept H_0 that $\rho_{xw} = 0$

Application of the test for $\rho_{xw} = 0$ reveals many cases in the geological literature where ρ_{xw} is not evidently different from zero, even though the r_{xw} statistic is quite far from zero. The method is explained for the example data of figures 9.3 and 9.6 by means of the hypothesis test in table 9.6. For a 10 percent α risk level, because the alternative hypothesis H_1 is that ρ is either positive or negative, the 95 percent points in table A-1 are used to yield for 22 observations the critical region stated in table 9.6. The meaning of the critical region is that, if ρ_{xw} is zero, 90 percent of the values of r_{xw} calculated from random samples of normally distributed observations will lie between 0.360 and -0.360. For the data of figure 9.3 H_0 is rejected, and for those of figure 9.6 it is accepted.

9.4 LINEAR AND QUADRATIC REGRESSION ANALYSIS OF DATA IN TWO DIMENSIONS

In section 9.2, the fitting of a straight line to data in one dimension was explained. Now, in section 9.4, the fitting of data in two dimensions is discussed, with the Rangely oil-field data of section 9.1 as an example. Two cases are taken up in subsections: first, the fitting of a plane, which is the two-dimensional equivalent of a straight line, and second, the fitting of a quadratic surface, which is the two-dimensional equivalent of a quadratic curved line. In a concluding subsection the representation of fitted equations by contour lines is explained.

Fitting a Plane

The plane is the simplest surface in two dimensions. The general equation for a plane is

$$w = \alpha + \beta_1 x + \beta_2 y + e,$$

where x and y are map coordinate points on a rectangular grid; w is the elevation of the surface related to the datum plane at any particular (x, y) grid point; α, β_1, and β_2 are coefficients whose values depend on the strike, dip, and elevation of the plane; and e is the random fluctuation. Accordingly, when the coefficients are estimated, the $\hat{w}$-values of the height of the fitted surface at any (x, y) grid point or points can be calculated. The assumptions are made that the random fluctuations are independent, their mean is zero, and their variance is a constant.

The statistics a, b_1, and b_2 estimate the coefficients α, β_1, and β_2, and the statistic s_e^2 estimates the variance σ_e^2. The following procedure (Fraser, 1958, p. 295) may be used to obtain the least-squares estimates of α, β_1, and β_2. The estimate of α is

$$a = \overline{w} - b_1 \bar{x} - b_2 \bar{y}.$$

The estimates of β_1 and β_2 are obtained by solving the two simultaneous linear equations

$$b_1 \, SS_{xx} + b_2 \, SS_{xy} = SS_{xw}$$

and

$$b_1 \, SS_{xy} + b_2 \, SS_{yy} = SS_{yw},$$

where

$$SS_{xx} = \sum (x - \bar{x})^2,$$
$$SS_{yy} = \sum (y - \bar{y})^2,$$
$$SS_{xy} = \sum (x - \bar{x})(y - \bar{y}),$$
$$SS_{xw} = \sum (x - \bar{x})(w - \bar{w}),$$

and

$$SS_{yw} = \sum (y - \bar{y})(w - \bar{w}).$$

It is understood that the summation extends over all the data.

Although the solution of these two equations by ordinary algebra is easy, it will be useful as the subject develops and for subjects treated in later chapters to use matrix notation, which accordingly is introduced at this point. Matrix notation allows concepts to be presented more compactly and also is useful in programming a computer that uses an algebraic compiler such as FORTRAN or ALGOL (sec. 17.2). Only the most elementary properties of matrices are used.

In matrix notation the two simultaneous linear equations may be represented by

$$\begin{bmatrix} SS_{xx} & SS_{xy} \\ SS_{xy} & SS_{yy} \end{bmatrix} \cdot \begin{bmatrix} b_1 \\ b_2 \end{bmatrix} = \begin{bmatrix} SS_{xw} \\ SS_{yw} \end{bmatrix}$$

or, more compactly, by

$$\mathbf{S} \cdot \mathbf{b} = \mathbf{P},$$

where

$$\mathbf{S} = \begin{bmatrix} SS_{xx} & SS_{xy} \\ SS_{xy} & SS_{yy} \end{bmatrix},$$

$$\mathbf{b} = \begin{bmatrix} b_1 \\ b_2 \end{bmatrix},$$

and

$$\mathbf{P} = \begin{bmatrix} SS_{xw} \\ SS_{yw} \end{bmatrix}.$$

For these two simultaneous equations, $\mathbf{S}$ is a 2×2 matrix, $\mathbf{b}$ is a 2×1 matrix, and $\mathbf{P}$ is also a 2×1 matrix.

The total sum of squares of the w-values, SS_{ww}, may be partitioned into two components: the regression sum of squares and the residual sum of squares.

The regression sum of squares is the amount of variability of w "explained" by the fitted plane and is calculated by the formula

$$\text{regression SS} = b_1 \, \text{SS}_{xw} + b_2 \, \text{SS}_{yw},$$

or, in matrix notation,

$$\text{regression SS} = \mathbf{P'} \cdot \mathbf{b},$$

where $\mathbf{P'}$ is the transpose of $\mathbf{P}$. The residual sum of squares is the amount of variability remaining unexplained by the regression sum of squares and is calculated by the formula

$$\text{residual SS} = \text{SS}_{ww} - \text{regression SS}.$$

The amount of variability "explained" by the plane may be compared with the residual variability through the one-way analysis of variance format (sec. 5.3), as follows:

Source of variation	Sum of squares	Degrees of freedom	Mean square	F
Regression		2		
Residual variability		$n-3$		
Total		$n-1$		

The quantity s_e^2 is the basic measure of variability, and the reliability of the analysis must be assessed by using this quantity as a yardstick.

The variances of b_1 and b_2 may be obtained from the following expressions:

$$\sigma_{b_1}^2 = \frac{\sigma_e^2 \, \text{SS}_{yy}}{\text{SS}_{xx} \, \text{SS}_{yy} - (\text{SS}_{xy})^2}$$

and

$$\sigma_{b_2}^2 = \frac{\sigma_e^2 \, \text{SS}_{xx}}{\text{SS}_{xx} \, \text{SS}_{yy} - (\text{SS}_{xy})^2},$$

where n is the number of data points. These quantities may be estimated by substituting s_e^2 for σ_e^2 in the expressions.

Confidence Limits for the Strike

The strike of the plane and confidence limits for the strike are estimated from the regression coefficients b_1 and b_2. The strike is estimated by the formula

$$A = \tan^{-1}\left(\frac{b_1}{b_2}\right),$$

where A is the angle between the strike and the y-direction of the rectangular coordinate system. If b_1/b_2 is positive, the angle is measured clockwise from y; if b_1/b_2 is negative, the angle is measured counterclockwise from y. If the y-direction of the coordinate system does not correspond to north, an appropriate correction must be applied.

Confidence limits for the strike are obtained by calculating confidence limits for β_1/β_2 and then finding the arc tangents of the upper and lower limits. Exact confidence limits for β_1/β_2 cannot be calculated, but approximate ones may be obtained from the expression

$$\frac{b_1}{b_2} \pm t_{5\%}(n-3)\left(\frac{b_1}{b_2}\right)\left\{\frac{s_{b_1}^2}{b_1^2} + \frac{s_{b_2}^2}{b_2^2} - \frac{2r_{b_1 b_2} s_{b_1} s_{b_2}}{b_1 b_2}\right\}^{1/2},$$

which is most nearly exact when $s_{b_1}^2$ and $s_{b_2}^2$ are small relative to b_1^2 and b_2^2, respectively. Limits are calculated with $t_{5\%}(n-3)$, which specifies the upper 5 percent point of the t-distribution with $(n-3)$ degrees of freedom.

Fitting a Quadratic Surface

The logical procedure to consider after fitting a plane is to fit the next more complicated figure, a quadratic surface. The general equation for a quadratic surface is

$$w = \alpha + \beta_1 x + \beta_2 y + \beta_3 x^2 + \beta_4 y^2 + \beta_5 xy + e.$$

As for a plane, the equation is still a linear model in the regression coefficients, because the squared terms x^2 and y^2 and the cross-product term xy can be replaced by terms such as X_1, X_2, and X_3. We do not make this replacement, because the meaning of the equation with substituted variables is more obscure and its meaning may be forgotten (sec. 10.3).

In fitting a quadratic equation, as with a plane, least-squares regression analysis yields statistics to estimate the coefficients α and β_1 to β_5 and the variance σ_e^2. As before, s_e^2 is the statistic that estimates σ_e^2, and a, the statistic that estimates α, is found from the formula

$$a = \bar{w} - b_1 \bar{x} - b_2 \bar{y} - b_3 \overline{x^2} - b_4 \overline{y^2} - b_5 \overline{xy}.$$

The statistics b_1 to b_5, which estimate β_1 to β_5, are obtained by solving five linear equations, which, as for the plane, can be represented in matrix notation as

$$\mathbf{S \cdot b} = \mathbf{P}.$$

For the quadratic surface $\mathbf{S}$ is a 5×5 matrix, $\mathbf{b}$ is a 5×1 matrix, and $\mathbf{P}$ is a 5×1 matrix. The elements of these matrices are not explicitly exhibited in this book; details with instructions for computer programming are published elsewhere (Link, Koch, and Gladfelter, 1964).

Next, as for the plane, the sum of squares of the w-value is partitioned into a part "explained" by the fitted quadratic surface and a residual part. The regression sum of squares (that part of the sum of squares "explained" by the fitted surface) is further divided into a linear regression sum of squares (that part of the sum of squares "explained" by fitting a plane) and a quadratic regression sum of squares (that part of the sum of squares "explained" by the quadratic terms x^2, y^2, and xy). The resulting one-way analysis of variance (sec. 5.3) is as follows:

Source of variation	Sum of squares	Degrees of freedom	Mean square	F
Linear regression		2		
Quadratic regression		3		
Residual variability		$n-6$		
Total		$n-1$		

Location of a Local Extremum

An extremum of a quadratic surface may be a maximum, minimum, or saddle point. The location of such an extreme point may be found first by differentiating the fitted function with respect to the variables x and y, then by setting the partial derivatives equal to zero, and finally by solving the two resulting simultaneous linear equations. In principle the method is like that used to locate inflection points for curves in a plane. The following equations are obtained:

$$\frac{\partial w}{\partial x} = b_1 + 2b_3 x + b_5 y$$

and

$$\frac{\partial w}{\partial y} = b_2 + b_5 x + 2b_4 y.$$

When the partial derivatives are set equal to zero,

$$\frac{\partial w}{\partial x} = \frac{\partial w}{\partial y} = 0;$$

the solution is

$$x_m = \frac{(b_2 b_5 - 2b_1 b_4)}{(4b_3 b_4 - b_5^2)}$$

and

$$y_m = \frac{(b_1 b_5 - 2b_2 b_3)}{(4b_3 b_4 - b_5^2)},$$

where (x_m, y_m) is the location of the extremum.

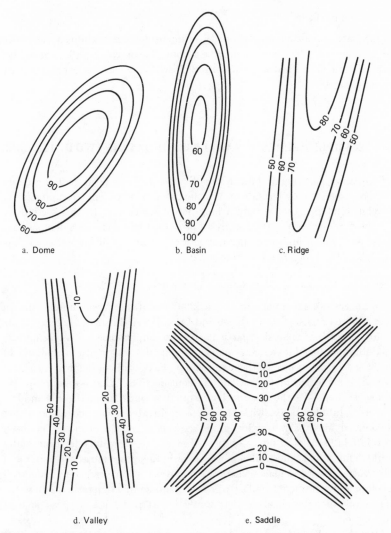

Fig. 9.7. Typical forms of quadratic regression equations fitted to data in two dimensions.

Depending on the analysis, many interpretations may be made of an extremum. It may be a structural high, low, or inflection point; the central point of a halo of hydrothermal mineralization; a volcanic vent; etc. While easy to find in two dimensions, the location of an extremum in three dimensions may be difficult without a computer program.

Contouring Surfaces

Quadratic equations for data in two dimensions may exhibit a variety of forms of surfaces. The surface may be a dome, basin, ridge, valley, or saddle. Typical forms are illustrated in figure 9.7, with the surfaces represented by contour maps.

9.5   EXAMPLE REGRESSION ANALYSES OF DATA IN TWO DIMENSIONS

Regression analyses of data in two dimensions have been made by many workers, especially by Krumbein, Whitten, and their students. Krumbein and Graybill (1965, p. 348) supply a long list of references. In this section we provide three examples: two analyses of mine assay data that demonstrate particular points, and a further inspection of some problems raised by the analysis of the Rangely data of section 9.1.

Regression Analysis of Silver Content in a Vein

An analysis of silver content in the 2200 vein of the Fresnillo mine, drawn from a more extensive study (Koch and Link, 1967), illustrates several points about regression analysis, particularly the importance of assessing the reliability of a fitted regression equation. Figure 9.8 is a longitudinal section of the 2200 vein, one typical of the many that we have studied. Before a statistical analysis is made, the first question to be considered is: What is an appropriate geological model for the distribution of ore in a vein? If one regards an ideal vein as a tabular body, with the ore concentrated in a central ore shoot surrounded by a barren or low-grade gangue, a geological model has been postulated. Translated into a statistical model, assay values for each metal would define a mathematical dome, with best mineralization at the center and a decrease outward to places where the vein was too low-grade for commercial development. The model applies to gross vein mineralization rather than to the fine details of mineralization that would be studied for a purpose such as blocking ore.

After the geological model and the corresponding statistical model have been chosen, data may be obtained and then processed to get the summary points suitable for the statistical model chosen. Data from 22 mine-sample points were averaged to obtain a summary point. The number 22 was chosen to correspond to 44 meters of drift sampled. This interval yielded summary points on an essentially orthogonal grid because, with few exceptions, the mine levels are 45 meters apart. Where the number of original points were not equal to 22 times an integer, the extra points were disregarded.

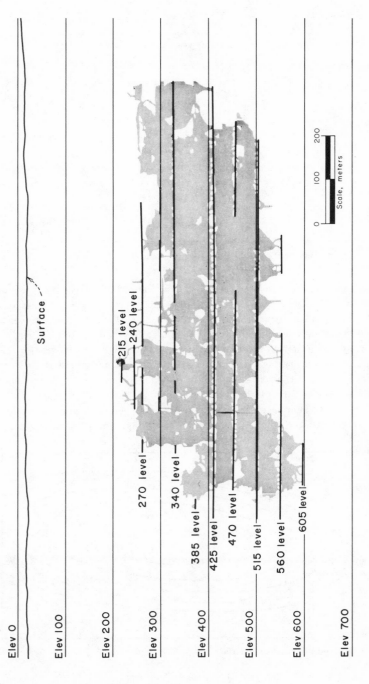

Fig. 9.8. Vertical longitudinal section, 2200 vein, Fresnillo mine.

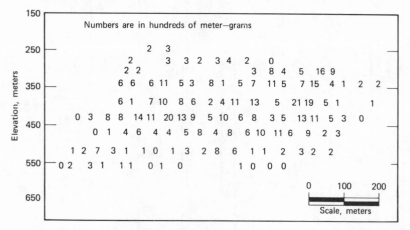

Fig. 9.9. Summary assay data for silver content, 2200 vein, Fresnillo mine.

The 121 summary points are plotted in figure 9.9. Already a readable summary of the data at the 2662 original sample points has been obtained by simple averaging to reduce the original observations, too numerous for over-all comprehension, to a number that can be integrated by eye. For instance, it is now easy to see that all the summary points with silver content greater than 1000 meter-grams (1 meter-kilogram), plotted as two-digit numbers, are more or less centered in the vein and can be grouped into four high-grade ore shoots by the hand-contoured dashed lines of figure 9.9.

However, to summarize the observations more generally and to test their correspondence to the domical model, we fitted a quadratic regression surface

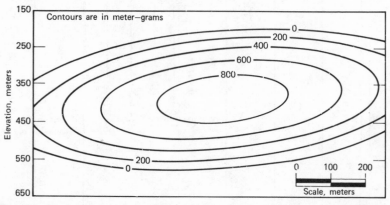

Fig. 9.10. Quadratic regression surface representing silver content, 2200 vein, Fresnillo mine.

to the data. The quadratic surface is contoured in figure 9.10. One can see that the surface corresponds well to the simple geological model of a dome. Notably, the quadratic surface, because of its simplicity, cannot distinguish the four high-grade ore shoots of silver greater than 1000 meter-grams; the highest point of the surface is less than 900 meter-grams. Nonetheless, if the quadratic model is appropriate for these data, the purpose of summarizing gross vein mineralization has been achieved.

TABLE 9.7. ANALYSIS OF VARIANCE FOR QUADRATIC REGRESSION ANALYSIS OF SILVER DATA FROM THE 2200 VEIN OF THE FRESNILLO MINE

Source of variation	Sum of squares	Degrees of freedom	Mean square	F	$F_{10\%}$
Linear terms	1,840,621	2	920,310	6.8	2.35
Quadratic terms	6,384,475	3	2,128,158	15.7	2.13
Residual variability	15,624,883	115	135,868		

We used two analyses of variance to investigate the statistical validity of the quadratic model for these data. The first, table 9.7, compares the amount of variation "explained" by the linear and quadratic terms with the residual unexplained variation in a one-way analysis of variance. Because about one-third of the total sum of squares is "explained," and because calculated F-values are inside the critical region, a correspondence to the model is indicated.

The second analysis of variance, table 9.8, compares two kinds of variability that can be separated for these data: (1) the local variability calculated when

TABLE 9.8. ANALYSIS OF VARIANCE TO COMPARE LOCAL AND RESIDUAL VARIABILITY FROM QUADRATIC FIT FOR SILVER DATA FROM THE 2200 VEIN OF THE FRESNILLO MINE

Source of variation	Sum of squares	Degrees of freedom	Mean square	F	$F_{10\%}$
Among-summary-points variation	529,699,562	120	4,372,496		
Linear regression	40,451,279	2	20,225,640		
Quadratic regression	140,503,469	3	46,839,490		
Deviation from quadratic model	343,744,814	115	2,989,085	11.82	1.18
Within-summary-points variation	692,724,772	2,541	252,942		
Total	1,167,424,334	2,661			

the original assay data are summarized to obtain the summary data and (2) the residual variability remaining after the quadratic model is fitted to the summary assay data. Local variability can be estimated for each of the 121 summary assay points, because each one is obtained from 22 original points. In table 9.8, which is a form of the single-degree-of-freedom analysis of variance (sec. 5.9), the among-summary-points variation (as measured by the among-summary-points sum of squares) has 120 (121 − 1) degrees of freedom. These 120 degrees of freedom are partitioned into 2 degrees of freedom for linear regression, 3 degrees of freedom for quadratic regression, and 115 degrees of freedom for deviation from the quadratic model. The within-summary-points variation (as measured by the within-summary-points sum of squares) has 2541 (121 × [22 − 1]) degrees of freedom. Finally, the total variation (as measured by the total sum of squares) has 2661 (2662 − 1) degrees of freedom.

By the calculated F-ratio, equal to 11.82, the mean-square value with 115 degrees of freedom, corresponding to deviation from the quadratic model, is compared with the within-summary-points mean-square value with 2541 degrees of freedom, corresponding to the within-summary-points variation. Because the calculated F-value is larger than the tabled $F_{10\%}$ value of 1.18, the quadratic model is too simple to reveal all of the "explainable" details, for the within-summary-points variation is smaller than that associated with deviation from the quadratic model.

Thus, for the silver data from the 2200 vein, the quadratic model is too simple. Nonetheless, even the "best fit" would leave over half the total variability unexplained, because the within-summary-points sum of squares is more than one-half the total sum of squares.

In conclusion, we offer a few reflections about these models. Both the geological and statistical models are extreme simplifications of the actual geological situation. If one is familiar with ore veins or reads descriptions of ideal veins such as the excellent accounts by McKinstry (1948) or Wisser (1941), he recognizes that many other factors could enter into a geological model and then be translated into a corresponding mathematical model. But even though data are insufficient to account for all geological factors in a particular situation, study of a simple model may be informative. A simple statistical model is often better than none at all, because, as in the silver example, a quantitative summary of geological data is made that provides geological insight.

An Analysis of More Complicated Data

As a second example of trend-surface analysis of data in two dimensions, we summarize an investigation published elsewhere (Link, Koch, and Gladfelter, 1964). The study is of another Fresnillo mine vein, the 2137 vein

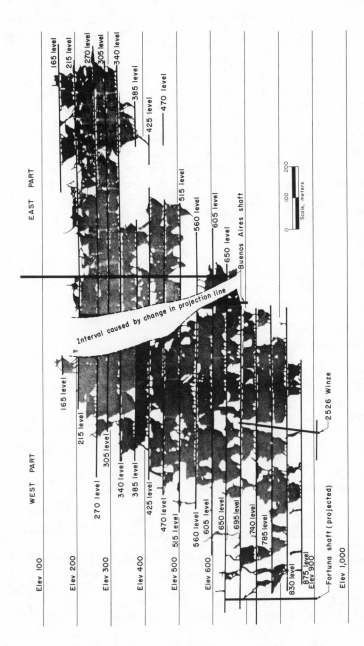

Fig. 9.11. Longitudinal section, 2137 vein, Fresnillo mine.

(fig. 9.11), which is much larger than the 2200 vein. Because the 2137 vein changes strike and steepens in dip to the east, it has been divided into an eastern and a western part, separated on the longitudinal section by an interval caused by change in projection line. The data from the western part of the vein, where there are 4510 original mine-sample points and 205 summary-sample points, were studied. Only two metals, silver and zinc, are discussed, in order to demonstrate one surface that is mainly linear and another that is mainly quadratic.

In tables 9.9 and 9.10 are given the results of the regression analyses, and in figures 9.12 and 9.13 are plotted the fitted surfaces, as well as the summary data. That the two fitted equations are quite different is shown graphically by the contoured surfaces and numerically by the values of the estimated coefficients b_1 to b_5. For silver content the fitted surface is strongly quadratic, because the reduction in the total sum of squares is about as large for the quadratic terms as for the linear terms. In contrast, for zinc content the fitted

TABLE 9.9. QUADRATIC REGRESSION ANALYSIS OF SILVER DATA FROM THE 2137 VEIN OF THE FRESNILLO MINE

Linear equation: $\hat{w} = 1072.3 - 0.21788x - 0.83499y$
Estimated strike: $105°$ (N 75° W)
Quadratic equation: $\hat{w} = 3830.4 - 2.1537x - 7.0362y -$
$$- 0.00041368x^2 - 0.0022997y^2 + 0.0039234xy$$

Linear analysis of variance

Source of variation	Sum of squares	Degrees of freedom	Mean square	F	$F_{10\%}$
x	1,547,291	1	1,547,291	21.4	2.73
y	4,404,526	1	4,404,526	60.8	2.73
Residual variability	14,623,835	202	72,395		

Quadratic analysis of variance

Source of variation	Sum of squares	Degrees of freedom	Mean square	F	$F_{10\%}$
Linear terms	5,951,816	2	2,975,908	58.7	2.33
Quadratic terms	4,527,656	3	1,509,219	29.7	2.11
Residual variability	10,096,179	199	50,735		
Total	20,575,651				

surface is essentially linear; that is, the quadratic terms do not appreciably reduce further the total sum of squares.

As for the 2200 vein, the quadratic regression analyses do not fit the silver and zinc data of the 2137 vein exactly, as appropriate analyses of variance, not printed in this book, indicate. Therefore the model is too simple, but it is nonetheless useful. Comparison of the plotted data with the fitted surfaces (figs. 9.12 and 9.13) shows the success of the fit.

Comparison of the quadratic surfaces for silver and zinc content (figs. 9.12 and 9.13) shows that across most of the area the contours for one variable are nearly at right angles to those for the other; these facts imply a lack of correlation between silver and zinc content. Even though only 50 percent of the silver variability and 35 percent of the zinc variability are "explained" by the fitted surfaces, the implied lack of correlation is corroborated by a calculated correlation coefficient r of only 0.014, a value that leads to acceptance of the hypothesis that ρ equals zero, at an α risk level of 10 percent (table A-1).

TABLE 9.10. QUADRATIC REGRESSION ANALYSIS OF ZINC DATA FROM THE 2137 VEIN OF THE FRESNILLO MINE

Linear equation: $\hat{w} = 5.4608 - 0.0071334x + 0.013830y$
Estimated strike: 63° (N 63° E)
Quadratic equation: $\hat{w} = 14.14050 - 0.21275x + 0.011523y -$
$$- 0.0000013704x^2 - 0.000016347y^2 + 0.000023896xy$$

Linear analysis of variance

Source of variation	Sum of squares	Degrees of freedom	Mean square	F	$F_{10\%}$
x	136.61	1	136.61	10.10	2.73
y	1208.28	1	1208.28	89.32	2.73
Residual variability	2737.64	202	13.53		

Quadratic analysis of variance

Source of variation	Sum of squares	Degrees of freedom	Mean square	F	$F_{10\%}$
Linear terms	1344.89	2	672.45	50.52	2.33
Quadratic terms	83.83	3	27.94	2.10	2.11
Residual variability	2648.81	199	13.31		
Total	4077.53				

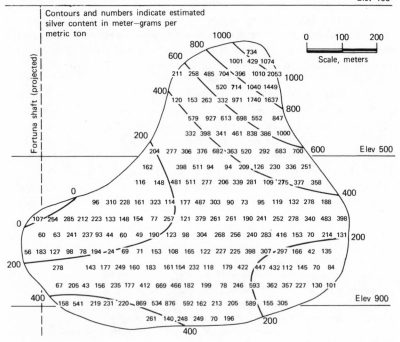

Fig. 9.12. Comparison of quadratic surface representing silver data with fitted data, 2137 vein, Fresnillo mine.

If the contours for silver and zinc were nearly parallel, a high correlation between the two metals would be implied.

A Further Examination of the Rangely Example

A further examination of the Rangely example, introduced in section 9.1, points up a contrast to the 2200 vein example. In table 9.11, the 26 elevations of the key bed are listed to provide a convenient small set of practice data for the reader. The well locations are measured in an arbitrary northeast-quadrant coordinate system.

Table 9.12, an analysis of variance for these data, shows that about 93 percent of the total sum of squares is "explained" by the fitted equation, mostly by the quadratic rather than by the linear terms, and that calculated F-values are large. The fit is among the best that we have obtained for real data with a quadratic equation.

Elev 100

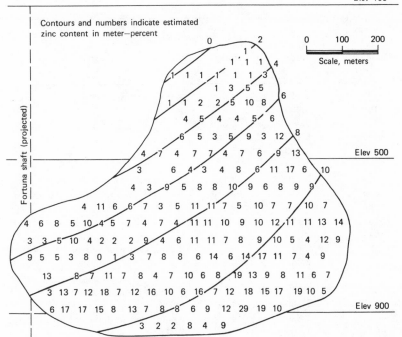

Fig. 9.13. Comparison of quadratic surface representing zinc content with fitted data, 2137 vein, Fresnillo mine.

In figure 9.14, the fitted quadratic surface is compared with Thomas' hand-drawn surface; the dashed lines are Thomas' contours. The two contoured domes are similar, the main difference being in the strikes of the major axes of the domes. Two other differences are that the north and south limbs of the hand-drawn dome dip at unequal angles, whereas the quadratic equation is constrained to the same angles; the noses of Thomas' dome are somewhat sharper.

Which of the two domes best corresponds to the real world? Disregarding new information since these data were obtained, and assuming that the only information is that in the 26 wells, we would choose Thomas' contours rather than the statistical ones, primarily because Thomas was able to evaluate subjectively the reliability of each point. Thomas classified the points into several categories (table 9.11), and doubtless had even finer distinctions in his mind. Moreover, he was able to weight the points according to their spacing,

Table 9.11. Key-bed elevations from 26 wells, Rangely oil field, Colorado

Elevation, feet below sea level, w	North coordinate, miles, y	East coordinate, miles, x	Notes
607	3.640	4.350	Weber datum elevation
491	3.380	4.870	,,
555	3.640	5.380	,,
406	3.100	5.880	,,
556	3.360	6.860	,,
761	4.100	6.620	,,
594	3.020	8.100	Abandoned well, datum interpolated
359	2.860	5.610	Weber datum elevation
231	2.380	6.360	,,
236	2.430	6.610	,,
246	2.380	6.870	,,
238	2.010	6.200	Abandoned well, datum interpolated
292	2.050	6.800	,,
255	2.130	6.900	,,
305	2.620	7.160	Weber datum elevation
247	2.100	7.640	,,
275	1.810	6.020	Abandoned well, datum interpolated
243	1.900	7.380	Weber datum elevation
249	1.940	7.560	,,
255	1.630	8.630	,,
285	1.880	8.630	,,
286	1.620	9.130	Drilling well, datum interpolated
326	1.370	9.630	,,
354	1.370	9.840	Producing well, datum interpolated
676	0.350	9.590	Drilling well, datum interpolated
513	0.600	9.620	Drilled to Weber sand, not yet producing, datum interpolated

Table 9.12. Analysis of variance for quadratic regression analysis of Rangely data

Source of variation	Sum of squares	Degrees of freedom	Mean square	F	$F_{10\%}$
Linear terms	153,658.9	2	76,829.4	33.9	2.59
Quadratic terms	435,574.6	3	145,191.5	64.1	2.38
Residual variability	45,311.1	20	2,265.6		
Total	654,544.5				

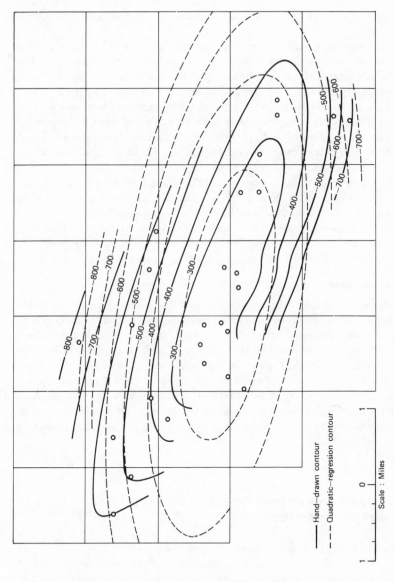

Fig. 9.14. Comparison of structure contour maps of the Weber sandstone, Rangely oil field, Colorado, prepared by quadratic regression analysis and by hand.

Hand—drawn contour
Quadratic—regression contour

Scale : Miles

to give more or less weight to those far from neighbors. Thus in his mind he could integrate these various kinds of information and, postulating the simple dome model, draw the structure contours.

In the regression analysis we used only the elevation and location of each data point and ignored the subjective information that Thomas had. The major advantage of the quadratic equation is the analysis of variance; unfortunately, the analysis of local variability, so informative for the 2200 vein example, could not be repeated because the data points were too few. Of course, some of the 26 points could have been omitted from the regression analysis, or the points could have been weighted according to their reliability, but these procedures did not seem appropriate for these data because the published information was not detailed.

In summary, the Rangely example provides a small data set for the reader's practice, introduces the concept of trend-surface analysis, and shows the difference between a contour map drawn by an experienced geologist familiar with local geology and one made by fitting a mathematical equation that gives a nearly perfect statistical fit. The difference suggests that the statistical contours disagree because of sampling errors and because the model is crude.

Concluding Remarks

For the small set of Rangely data, with the geologic model clear, the mathematical model provides little if any illumination. For the 2200 vein at Fresnillo, the mathematical model provides a valuable comparison of local and regional variability. For the 2137 vein, with 4510 data points, the mathematical model provides an indispensable summary of data that otherwise would be far too complicated to comprehend by eye. In general, trend-surface analysis may be very informative if data are numerous, if comparisons among variability of two or more scales are desired, and if there is no well-defined geological model or if there are several geological models that may be compared statistically.

9.6 LINEAR AND QUADRATIC REGRESSION ANALYSIS OF DATA IN THREE DIMENSIONS

The methods of linear and quadratic regression analysis in section 9.4 may readily be extended for data in three dimensions, where the three independent variables usually are elevation, north coordinate, and east coordinate. The main conceptual difficulty is that, because the data points are located in three dimensions, the w-values must be thought of as lying in a fourth dimension. While the situation may be difficult to visualize geometrically, an algebraic description is straightforward. Therefore the following explanation is purely algebraic, and no attempt is made to give a geometric description in terms of hyperplanes and hypersurfaces.

Fitting a Linear Surface

A linear hypersurface, which is the simplest surface to fit, exhibits a direction of increase of the dependent variable along a straight line (fig. 9.15). The general equation of a linear surface is

$$w = \alpha + \beta_1 x + \beta_2 y + \beta_3 z + e,$$

where x, y, and z are the independent variables corresponding to the coordinates of a point in a three-dimensional system of rectangular coordinates; w is a value associated with each (x, y, z) point; α, β_1, β_2, and β_3 are coefficients whose values are a function of the strike, dip, and plunge of the linear surface; and e is the random fluctuation.

As for the two-dimensional case, the estimate of α is

$$a = \overline{w} - b_1 \overline{x} - b_2 \overline{y} - b_3 \overline{z}.$$

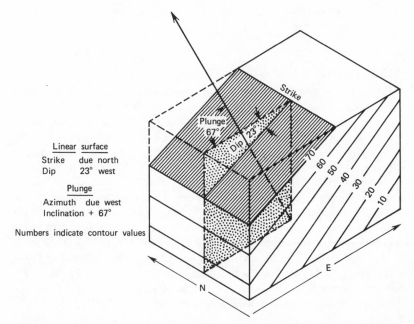

Linear surface

Strike   due north
Dip   23° west

Plunge

Azimuth   due west
Inclination + 67°

Numbers indicate contour values

Fig. 9.15. Typical linear surface fitted to data in three dimensions, illustrating bearing, strike, and plunge.

The statistics that estimate β_1 to β_3 are obtained from the solution of three linear equations, which can be represented in matrix notation as

$$\mathbf{S} \cdot \mathbf{b} = \mathbf{P}.$$

For this case $\mathbf{S}$ is a 3×3 matrix, $\mathbf{b}$ is a 3×1 matrix, and $\mathbf{P}$ is a 3×1 matrix. The elements of the matrices, not given here, are published elsewhere (Link, Yabe, and Koch, 1966).

Now the total sum of squares of the w-values may be partitioned into two components, the regression sum of squares and the residual sum of squares. The regression sum of squares is the amount of variability of w "explained" by the fitted linear surface. The residual sum of squares is the amount of variability remaining unexplained by the regression sum of squares.

The amount of variability "explained" by the linear surface may be compared with the residual variability through the following one-way analysis of variance format:

Source of variation	Sum of squares	Degrees of freedom	Mean square	F
Regression		3		
Residual variation		$n - 4$		
Total		$n - 1$		

Plunge of a Fitted Linear Surface

For regression analysis of data in three dimensions, the plunge is defined as the direction vector perpendicular to the linear surface corresponding to the increase of the dependent variable along a straight line (fig. 9.15). The plunge is described by two angles, designated the inclination and the azimuth. The inclination is the vertical angle between the xy-plane and the plunge, and the azimuth is the horizontal angle formed by the positive y-axis and the projected line segment of the plunge onto the xy-plane. The inclination angle is restricted to values between plus and minus 90 degrees. The azimuth angle is measured clockwise from the positive y-axis, with a range from 0 to 360 degrees. If the positive y-axis is chosen as the north direction and the positive z-axis as the vertical upward direction, the notation is consistent with that conventionally used in paleomagnetism for directional data. (In this convention (sec. 10.7), the vertical angle $20°$ denotes the plunge of a line inclined $20°$ *below* the horizon; the vertical angle $-20°$ denotes an inclination *above* the horizon.)

The inclination ϕ is calculated by the formula

$$\phi = \pm \tan^{-1} \frac{b_3}{\sqrt{b_1^2 + b_2^2}};$$

the plus sign is used when b_3 is less than zero and the minus sign when b_3 is greater than zero. The azimuth θ is calculated by one of the following formulas:

$$\theta = \tan^{-1}\left|\frac{b_1}{b_2}\right| \quad (b_1, b_2 \text{ positive}),$$

$$= 180° - \tan^{-1}\left|\frac{b_1}{b_2}\right| \quad (b_1 \text{ positive}, b_2 \text{ negative}),$$

$$= 180° + \tan^{-1}\left|\frac{b_1}{b_2}\right| \quad (b_1, b_2 \text{ negative}),$$

$$= 360° - \tan^{-1}\left|\frac{b_1}{b_2}\right| \quad (b_1 \text{ negative}, b_2 \text{ positive}).$$

Fitting a Quadratic Surface

After a linear surface is fitted, the next logical step is to fit the next more complicated figure, a quadratic surface, the general equation for which is

$$w = \alpha + \beta_1 x + \beta_2 y + \beta_3 z + \beta_4 x^2 + \beta_5 y^2 + \beta_6 z^2 + \beta_7 xy + \beta_8 xz + \beta_9 yz + e.$$

As for a linear equation, least-squares regression analysis is done to obtain statistics that estimate the parameters α and β_1 to β_9 and to obtain an estimate of the variance σ_e^2. The procedure is more complicated for the quadratic surface, because there are ten coefficients rather than the four for a linear surface. The estimates of the nine coefficients β_1 to β_9 are obtained by solving nine simultaneous linear equations, which are not exhibited in this book.

Next, the sum of squares of the w-values is partitioned into a part "explained" by the fitted quadratic surface and into a residual part. The regression sum of squares is further partitioned into a linear regression sum of squares and a quadratic regression sum of squares. The resulting one-way analysis of variance is as follows:

Source of variation	Sum of squares	Degrees of freedom	Mean square	F
Linear regression		3		
Quadratic regression		6		
Residual variation		$n - 10$		
Total		$n - 1$		

Location of a Local Extremum

An extremum of a quadratic surface with three independent variables, x, y, and z, may be a maximum or minimum at the center of an ellipsoid or may be a point of maximum constriction or enlargement of a hyperboloid. The location of the extremum is found as for the two-dimensional case by differentiating the fitted quadratic function to yield the following equations:

$$x = \frac{2b_3 b_5 b_8 + 2b_2 b_6 b_7 + b_1 b_9^2 - 4b_1 b_5 b_6 - b_3 b_7 b_9 - b_2 b_8 b_9}{8b_4 b_5 b_6 + 2b_7 b_8 b_9 - 2b_4 b_9^2 - 2b_5 b_8^2 - 2b_6 b_7^2},$$

$$y = \frac{2b_3 b_4 b_9 + 2b_1 b_6 b_7 + b_2 b_8^2 - 4b_2 b_4 b_6 - b_1 b_8 b_9 - b_3 b_7 b_8}{8b_4 b_5 b_6 + 2b_7 b_8 b_9 - 2b_4 b_9^2 - 2b_5 b_8^2 - 2b_6 b_7^2},$$

and

$$z = \frac{2b_1 b_5 b_8 + 2b_2 b_4 b_9 + b_3 b_7^2 - 4b_3 b_4 b_5 - b_2 b_7 b_8 - b_1 b_7 b_9}{8b_4 b_5 b_6 + 2b_7 b_8 b_9 - 2b_4 b_9^2 - 2b_5 b_8^2 - 2b_6 b_7^2}.$$

Contouring Surfaces

It is difficult to draw contour maps of surfaces for four variables, three independent and one dependent, because a four-dimensional plot is required.

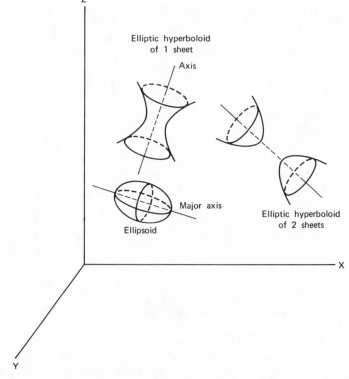

Fig. 9.16. Typical forms of quadratic surfaces fitted to data in three dimensions.

The forms that these quadratic surfaces may exhibit are shown in figure 9.16. The drawings represent x-, y-, and z-values corresponding to a single fixed w-value. For each form the full contour surface consists of a series of such drawings (forming onion-like layers) corresponding to the selected w-values. In figure 9.16, the drawings are ellipsoids and elliptic hyperboloids; the elliptic paraboloid and hyperbolic paraboloid are omitted because they are special cases that seldom if ever correspond to actual data.

For a discussion of the forms that a three-dimensional quadratic surface may exhibit, the interested reader may consult a book on solid analytical geometry. For instance, Albert (1949) gives a clear account of the forms, their standardized equations, and (pp. 110–111) the relation of the general quadratic equation to the standardized ones.

Plane sections are easy to contour for detailed study or for rapid visualization of the surfaces. By setting one independent variable at a time equal to a constant and by solving the equation, two-dimensional equations are obtained that can be plotted by the method described in section 9.4; a computer program is published elsewhere (Link, Yabe, and Koch, 1966). In principle, the plane sections can be oriented obliquely to the x-, y-, or z-axes; in practice, however, it is more convenient to rotate the axes before the regression analysis is made than to program the additional calculations.

9.7 EXAMPLE REGRESSION ANALYSIS OF DATA IN THREE DIMENSIONS

Because many geologists work with numerical data in only two dimensions, relatively few regression analyses of data in three dimensions have been published. Examples are Harbaugh's (1964) study of oil-gravity data, Peikert's (1962) study of specific gravity measurements, and Davis's (1969) study on the distribution of hydrocarbons.

In mining geology, extensive three-dimensional data are common. Many three-dimensional analyses have been made by the Mine Systems Engineering Group of the U.S. Bureau of Mines, of which only a few have been published. As an example, we abstract a study of assay data from the Frisco mine (Koch and Link, 1967).

We made a three-dimensional quadratic regression analysis of Frisco mine data to learn the over-all pattern of metal distribution in the veins. The Frisco veins (sec. 1.4) form an interlacing pattern and are mined through a vertical extent of some 800 meters, on levels 1 to 15, which are spaced either 50 or 60 meters apart. The assay data are from levels 7, 10, 13, 14, and 15. These data can be visualized as branching networks of data points spaced 2 meters

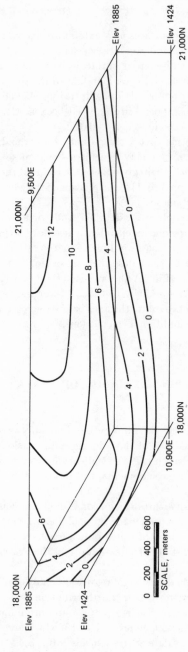

Fig. 9.17. Isometric projection of quadratic regression surface, in three dimensions, fitted to lead data from the Frisco mine.

apart along the interlacing drifts on the five levels. Because mineralization is essentially confined to the veins, there are many gaps in the data net; our over-all summary is intended to show the pattern of mineralization if there is a vein present, not to suggest that mineralization is to be expected in the absence of a vein fracture.

Only the analysis for lead is reviewed in this book. By summarizing original data as for the Fresnillo data (sec. 9.5), we reduce the original Frisco sample points to 1150, a 15-fold decrease. Applying the three-dimensional regression analysis to these summary data yields the regression analysis of table 9.13

TABLE 9.13. QUADRATIC REGRESSION ANALYSIS IN THREE DIMENSIONS OF LEAD DATA FROM THE FRISCO MINE

Linear equation: $\hat{w} = -21.888 - 0.0022178x + 0.0013811y + 0.014607z$
Plunge: Azimuth: 302°   Inclination: 80°
Quadratic equation:
$$\hat{w} = -1684.2 + 0.20398x + 0.063328y + 0.056092z - 0.0000063877x^2 -$$
$$- 0.00000060181y^2 - 0.000026131z^2 - 0.00000416484xy +$$
$$+ 0.0000013931xz + 0.0000016471yz$$

	Quadratic analysis of variance				
Source of variation	Sum of squares	Degrees of freedom	Mean square	F	$F_{10\%}$
Linear terms	5,892.9	3	1,964.3	50.8	2.09
Quadratic terms	1,103.9	6	184.0	4.8	1.78
Residual variation	44,120.2	1,140	38.7		
Total	51,117.0				

and the quadratic surface plotted in figure 9.17. The analysis of variance shows that about 14 percent of the total sum of squares is "explained," mostly by a strong linear trend. The nearly vertical plunge of the linear surface ($+80$ degrees) indicates a decrease in lead downward. Accordingly, in the linear equation, the coefficient with the largest absolute value (other than the constant) is positive and is the multiplier for the z-variable. While the plotted surface also indicates some tendency for lead to increase northwestward, the analysis is dominated by the nearly vertical linear decrease downward.

9.8 RESIDUALS FROM FITTED TREND SURFACES

When a quadratic or other statistical equation is fitted to data, the resulting surface seldom if ever corresponds exactly to the actual observations at the sample points, because the fit is not perfect. Rather, between the surface and the observations there is a residual variation, measured by the vertical distance at each point between the elevation of the point and the elevation of the fitted surface, both with reference to the data plane. This residual variation, named simply the *residual*, is conventionally designated positive if the observation is above the fitted surface and negative if the observation is below. For the Rangely example, the residuals are listed in table 9.14.

Residuals may represent random noise or they may contain geologically significant information. They represent random noise if the chosen mathematical model truly represents the observations. If the residuals contain

TABLE 9.14. RESIDUALS FROM QUADRATIC
REGRESSION ANALYSIS OF RANGELY DATA

Observed elevation, feet below sea level	Predicted elevation	Residual
607	587	20
491	483	8
555	571	− 16
406	389	17
556	501	55
761	858	− 97
594	462	132
359	338	21
231	263	− 32
236	265	− 29
246	261	− 15
238	264	− 26
292	248	44
255	247	8
305	300	5
247	254	− 7
275	290	− 15
243	248	− 5
249	249	0
255	289	− 34
285	286	− 1
286	316	− 30
326	364	− 38
354	380	− 26
676	608	68
513	519	− 6

geologically significant information, as is more common, the basic underlying behavior of the dependent variable may be easier to recognize once a generalized trend, say a quadratic trend, is removed. Geologists are accustomed to generalized linear trends. In structural petrology, data may be plotted on a stereographic net and then rotated to make some element, such as a hypothetical "*b*"-axis or fold axis, horizontal. Or regional dip may be removed for a problem in petroleum geology. Or details of vein structures may be plotted according to the Conolly (1936) method. In each of these examples linear trend is removed. The same method can be applied mathematically, with the advantage that a quadratic or higher-order trend can also be removed. Whitten (1961) has discussed the relation between the degree of equation fitted and the pattern of residuals.

One searches a plot of the residuals for patterns, which may or may not be geographic. In the following subsections some of the patterns that may be found are examined.

Frequency Distributions of Residuals

By performing a linear or quadratic regression, a trend in data can often be removed so that the residuals follow a more or less normal frequency distribution, even though the original distributions are highly skewed.

We illustrate the effect of removing geographic trend through quadratic regression by considering uranium, vanadium, and lime assays from the Mi Vida mine near Moab, Utah (Koch, Link, and Hazen, 1964). For each metal (assays weighted by widths), residuals were calculated separately for the northern and southern areas of the mine. The frequency distributions of the residuals (fig. 9.18) are reasonably symmetric and resemble normal distributions, although their tails are somewhat too long. They certainly are nearly enough normal so that hypothesis tests based on the normal distribution are valid. The symmetry of the distributions was established by chi-square tests (sec. 6.1), applied singly to each of the three sets of 79 values and jointly to the combined 237 (3×79) values. Each test led to accepting the hypothesis that there are as many negative as positive residuals, under the assumption that the residuals are randomly distributed.

The skewness of the frequency distributions of observations is removed to the extent indicated by comparing figure 9.18-a, a frequency distribution of uranium residuals, with figure 9.18-d, a frequency distribution of original uranium observations. Figure 9.18-a is symmetric in contrast to the highly skewed figure 9.18-d and is more informative because the trend variation of the assays with respect to location is taken into account. If there were no discernible trend, the average value for uranium would be the same throughout the sampled area, and the outlined procedure would be unnecessary; however, if trend is present, it must be removed to examine satisfactorily the basic

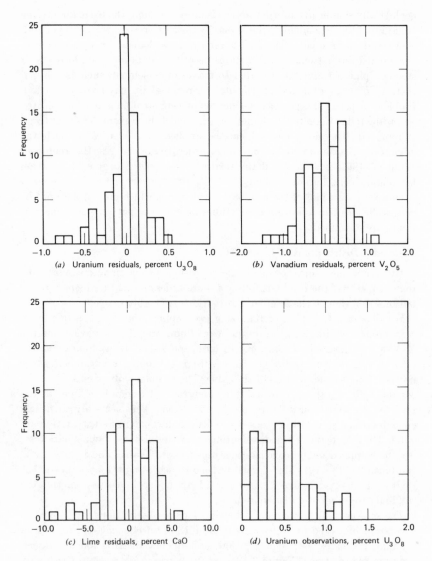

Fig. 9.18. Histograms of uranium, vanadium, and lime residuals, and of original uranium data, from the Mi Vida mine.

underlying frequency distribution. Trend, although usually geographic, may also be a trend with regard to time, petrographic differentiation, or other geologic process.

Homoscedasticity of Residuals

One reason for studying the geographic distribution of residuals is that their pattern may reveal more about the geology than a map of the original data. One interesting pattern is of homoscedasticity, or constant variance, of residuals. If the variance of the residuals is more or less constant across the fitted map area, the reliability of the fitted equation can appropriately be investigated by F-tests (sec. 5.3). On the other hand, if the variance changes radically from place to place, this change must be noted because F-tests concerning the fitted equation will be unreliable, and it may be better to fit the surface in different pieces. Or a study of the original fitted surface may reveal something of geological interest associated with the changing variance.

An example is furnished by a quadratic regression analysis of silver in the 2137 vein of the Fresnillo mine. Using the assay data, we were able (Koch and Link, 1967) to separate the vein into three parts with different ratios of silver to lead. These three zones of high, low, and mixed ratio of silver to lead— plotted in figure 9.19—correspond to ore types previously identified (Stone and McCarthy, 1948) by mine mapping as "light" sulfide zones (high ratio of silver to lead), "mixed," and "heavy" sulfide zones (low ratio of silver to lead). In figure 9.19 are also plotted residuals of silver content from the fitted quadratic surface.

TABLE 9.15. STANDARD DEVIATIONS DEMON-
STRATING LACK OF HOMOSCEDASTICITY OF
RESIDUALS FROM QUADRATIC REGRESSION
ANALYSIS OF SILVER CONTENT IN THE 2137
VEIN OF THE FRESNILLO MINE

Area	Number of observations	Standard deviation
High-ratio	104	505.1
Mixed-ratio	33	337.3
Low-ratio	174	157.6

The behavior of the residuals is summarized in histograms from the three parts of the vein (fig. 9.20) and by the means and standard deviations of residuals (table 9.15). The fact that in the high-ratio area the residuals have a high standard deviation whereas in the low-ratio area they have a low standard deviation demonstrates the lack of homoscedasticity.

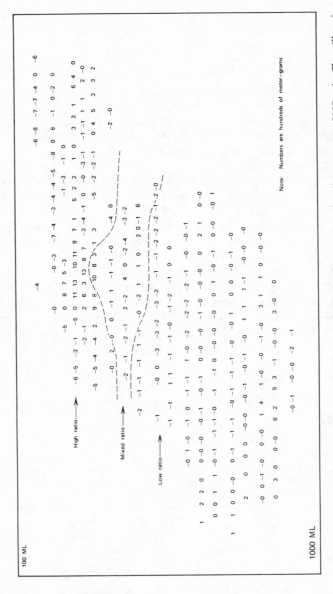

Fig. 9.19. Residuals from quadratic regression surface fitted to summary data for silver content, 2137 vein, Fresnillo mine.

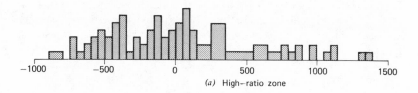

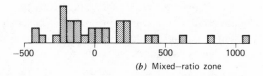

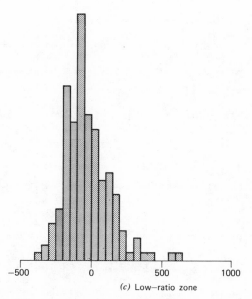

Fig. 9.20. Histograms of residuals of silver content from high—(a), low—(b), and mixed—(c) Ag/Pb ratio zones, Fresnillo mine.

Lack of homoscedasticity was not a serious problem because of our purpose in studying the 2137 vein (sec. 9.5). One way to attain homoscedasticity would have been to take logarithms of the observations, but taking such logarithms would lead to new problems for these data (sec. 11.2).

Ghost Stratigraphy

A familiar geological problem in deciphering the history of metamorphic rocks is to identify the nature of the rock before metamorphism. The same problem exists for rocks changed by processes not always considered to be metamorphic, such as diagenesis, weathering, and hydrothermal alteration. When such rocks are studied, it is sometimes possible through statistics to identify numerical information that has not been obliterated by the metamorphism, just as a mineral pseudomorph can be recognized even though the

28.6	34.0	30.2	29.7	24.7	16.1	24.3	30.2	28.6	28.4	27.4	32.9	36.1	35.5	42.5	36.0
27.6	33.0	31.2	32.6	19.3	23.1	24.1	24.3	28.4	29.0	30.4	26.6	27.9	29.7	39.9	33.7
26.5	25.6	28.2	31.2	17.0	18.9	24.7	19.7	33.3	27.2	36.2	34.8	31.6	34.3	40.1	37.1
25.9	29.3	22.8	29.8	27.2	15.0	28.1	25.1	29.1	22.1	30.4	24.2	37.8	32.0	37.9	32.1
30.1	24.7	28.5	25.2	20.2	22.8	19.8	20.9	23.7	24.7	22.9	22.0	26.8	43.0	33.6	36.7
21.1	31.3	27.9	28.1	16.0	26.6	18.8	19.8	21.5	32.5	22.3	32.1	38.6	34.8	31.8	40.4
36.8	21.9	29.3	25.1	19.7	17.6	23.4	20.8	21.2	24.3	32.1	24.7	33.5	33.0	33.6	39.0
23.4	21.6	22.3	24.5	17.5	16.9	16.8	20.2	21.8	26.0	28.7	32.1	33.3	28.3	30.2	34.8
23.6	19.4	25.2	24.7	22.1	15.5	20.1	33.2	22.8	29.8	25.2	24.3	35.9	34.1	29.5	33.0
27.4	23.2	21.0	26.8	16.7	12.1	19.9	26.9	25.0	25.6	23.4	25.6	32.1	26.3	32.1	33.9
22.7	23.4	22.0	29.4	12.0	18.3	18.1	10.7	22.3	24.2	23.2	20.6	26.2	36.1	33.1	30.5
26.5	27.9	30.6	22.8	19.0	16.0	13.1	20.9	24.5	21.5	26.2	23.6	26.4	34.2	33.7	26.3
25.9	21.7	17.5	26.6	6.0	9.8	14.0	13.5	23.5	20.5	24.7	21.8	26.2	21.6	31.8	30.1
24.5	27.1	18.5	22.7	12.6	14.0	16.2	22.0	24.5	22.3	19.7	33.5	32.4	43.4	32.4	35.8
21.8	21.3	15.5	22.5	17.5	12.6	14.8	13.8	18.2	15.7	19.9	22.1	28.1	32.0	25.0	34.0
18.4	17.4	22.1	21.9	12.1	17.9	18.2	6.4	14.8	27.4	23.7	20.7	21.9	32.5	28.8	27.8

N ↑

Bed 1: μ_1 = 20   Bed 2: μ_2 = 10   Bed 3: μ_3 = 15   Bed 4: μ_4 = 20

Fig. 9.21. Simulated observations and means of four beds, for study of ghost stratigraphy.

mineralogy has been changed. Thus Whitten (1959) has recognized "ghost stratigraphy," stating that, in the Donegal area, Ireland, "the deviations [residuals] have geological significance and may be closely correlated with the metasediments extant prior to the emplacement of the granitoid rocks."

In order to demonstrate ghost stratigraphy, we present a fictitious illustration with data obtained by simulation. Simulation, the method of obtaining numerical data by random processes to test a statistical model, is a sometimes useful method usually implemented with electronic computers (sec. 17.1).

For the simulated illustration, we assume a square surface area, as outlined in figure 9.21. The area, of arbitrary size, is divided into four equal parts, which may be imagined to represent outcropping sedimentary beds striking due north, which are to be sampled at 256 points on a 16 × 16 grid. The mean value of an imaginary constituent is different in the four beds, as shown in the figure (20 in bed 1, for example). However, the observations in each bed are not equal to the mean because two additional elements have been introduced. First, to simulate natural fluctuation, a random number from a normal distribution with a mean of zero and a standard deviation of 2 has been added to each of the original mean values. Second, a northeast linear trend, to simulate a regional metamorphic front, has been added, numerically equal to zero at the southwest corner of the map area, increasing linearly to 10 on the northwest-southeast diagonal and to 20 at the northeast corner. The resulting observations are plotted in figure 9.21.

If we now pretend that the construction of the model is unknown and that only the observations at the sample points are available, we can test whether the 256 observations provide evidence of northward-trending ghost stratigraphy in the simulated metamorphic area. When a quadratic equation is fitted to the data, the surface of figure 9.22 and the analysis of variance of table 9.16 are obtained. As the element of linear northeast gradient is strong, most of the regression is explained by the linear terms, and a strike of 141 degrees (N 39° W) is estimated. The plotted surface shows the general increase in gradient to the northeast, complicated by a quadratic trend introduced because of the initially different means in the four beds.

TABLE 9.16. ANALYSIS OF VARIANCE FOR QUADRATIC REGRESSION ANALYSIS OF SIMULATED GHOST-STRATIGRAPHY OBSERVATIONS

Source of variation	Sum of squares	Degrees of freedom	Mean square	F	$F_{10\%}$
Linear terms	4,698.23	2	2,349.11	119.0	2.32
Quadratic terms	2,533.90	3	844.63	42.8	2.10
Residual variation	4,935.71	250	19.74		
Total	12,167.84				

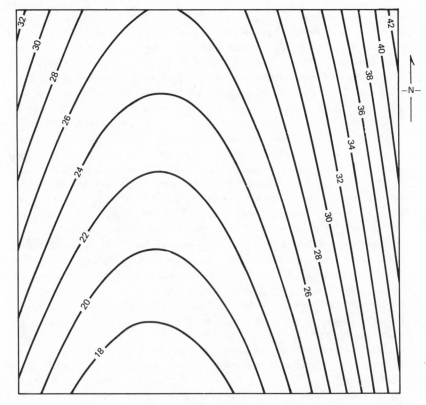

Fig. 9.22. Quadratic regression surface fitted to simulated ghost-stratigraphy observations.

The next step is to determine the pattern of residuals from the quadratic trend. When the residuals are calculated and plotted, the pattern shown in figure 9.23 and summarized in table 9.17 is obtained. For simplicity, on the

TABLE 9.17. DISTRIBUTION OF POSITIVE, ZERO, AND NEGATIVE RESIDUALS FROM QUADRATIC REGRESSION ANALYSIS OF GHOST-STRATIGRAPHY OBSERVATIONS

Bed	Number +	Number 0	Number −
1	37	6	20
2	11	8	45
3	28	17	19
4	27	10	27

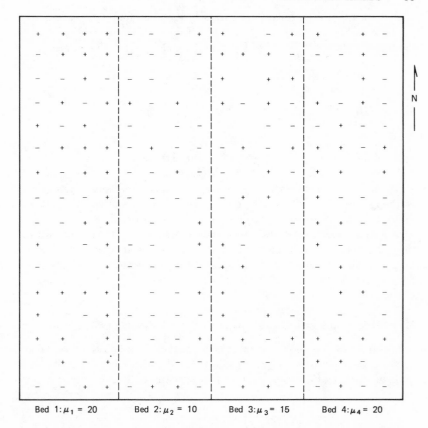

Fig. 9.23. Positive and negative residuals from quadratic regression surface.

figure the residuals are plotted as plus or minus signs, with blanks indicating those from $+1$ to -1. The residuals indeed define a pattern; for example, in bed 1 there are 37 positive residuals and only 20 negative ones. To carry the analysis further, we could have compared the residuals among beds by one of several F- or t-tests, depending on the details of the hypothesis chosen; and we could then have confirmed the pattern, already evident from the counting of positive and negative residuals.

In summary, this simulation of ghost stratigraphy shows the kind of assumptions that can be made and the complications of a specific model. The presentation also gives an idea of the value of simulation from which insight can conveniently be gained about how real data might behave under specified model conditions. While preparing this illustration, we simulated several sets of ghost stratigraphy with various parameters, and we found it instructive to see how, as the trend, the initial bed differentiation, and the amount of randomness changed, the "ghost" got fainter and fainter until it vanished.

9.9   OTHER TYPES OF MAP ANALYSIS

The previous sections of this chapter explain how to find trends in geological data through regression analysis with linear and quadratic equations. However, many mathematical methods more complicated than these low-order polynomial equations are also used for map analysis (Miesch and Connor, 1968; Harbaugh and Merriam, 1968; Agterberg, 1969). In this section three of these methods are summarized; they depend on Fourier series, higher-order polynomial equations, and arbitrary functions. These methods are compared with the contouring of data by linear interpolation. Statistical designs for purposeful gathering of map data are a major factor in exploration and are therefore discussed in Chapter 12 (particularly secs. 12.4 and 12.5), even though they are closely related to the topics of this chapter as well.

Models Based on Fourier Series and High-Order Polynomial Equations

In this subsection one of several models based on Fourier series is explained and compared with polynomial models. The Fourier model depends on a periodic mathematical function that may be expressed in one dimension as

$$w = \frac{a_0}{2} + \sum_{n=1}^{N} \left(a_n \cos \frac{n\pi x}{L} + b_n \sin \frac{n\pi x}{L} \right),$$

where L is the half range of the x-values, and where the constants to be estimated, $2N+1$ in number, are $a_0, a_1, a_2, ..., a_N$ and $b_1, b_2, ..., b_N$. Typical Fourier functions are graphed in figure 9.24.

This outline of the Fourier model covers only the one-dimensional case. The two-dimensional case as applied to geological data is explained by Preston and Harbaugh (1965) and by James (1966); the three-dimensional case can be derived by analogy. Wylie (1960) gives a good general account of Fourier series.

The principal advantage of the Fourier model is that, being a periodic function, it matches a geological model for periodic data. For instance, the model should be suitable for mapping a key stratigraphic horizon in rocks with Appalachian type folding. To demonstrate the model, James (1966) made a Fourier analysis of topographic elevations at 90 gridded data points in the

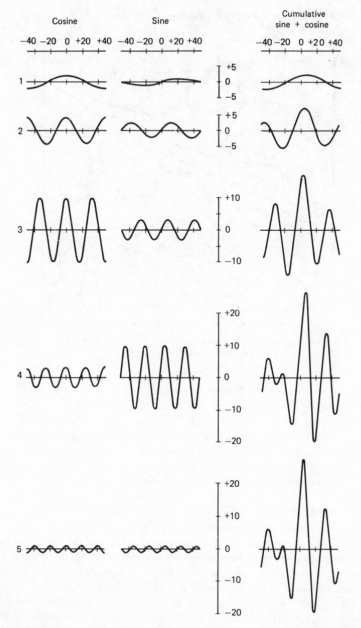

Fig. 9.24. Synthetic single Fourier series showing (1) waveforms of individual terms and (2) waveforms of series representing the summed individual terms (after Preston and Harbaugh, 1965, p. 5).

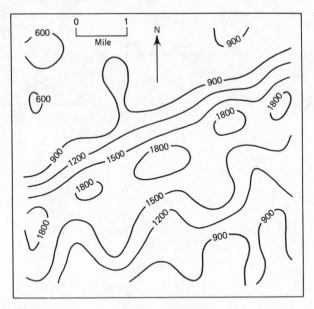

Fig. 9.25. Fourier map fitted to topographic elevations at 90 data points in the McClure, Pennsylvania, quadrangle (after James, 1966, p. 24).

folded Appalachians and obtained a map (fig. 9.25) based on all 90 Fourier coefficients that can be associated with the 90 observations.

By including a few terms with high harmonics, the Fourier model yields a fitted surface with many undulations and, for data that are nearly periodic, should provide a fit with fewer coefficients than are needed for polynomial models. Fourier models may become confused, however, if there are linear trends in the data, because the sine and cosine functions have no linear components, and the fit is therefore distorted. For extrapolation Fourier models are usually worse than polynomial models, because they must repeat the pattern established over the region of data. In short, Fourier models are most useful for geologic data that are periodic but have no linear or quadratic trend, or for geologic data from which these trends have been removed. For periodic data, time-series analysis (Grenander and Rosenblatt, 1957), which is beyond the scope of this book, should also be considered.

In higher-order polynomial models, terms higher than quadratic ones are added to the model. The linear polynomial model has two independent variables x and y, besides the constant term, to which three more variables, x^2, y^2, and xy, are added in the quadratic model. Similarly, in the cubic model, four more terms, x^3, y^3, x^2y, and xy^2, are added. In higher-order polynomials the number of terms increases rapidly, as shown by table 9.18, which compares the polynomial and Fourier models. The advantage of this increase in the number of independent variables is that it allows for the possibility of a rapidly increasing number of extrema, as shown in the table, instead of the one extremum of the quadratic model. The disadvantages are that computing problems become formidable, and the predictive value of the fitted equation becomes small—points that are reviewed extensively in the literature and discussed in section 9.11.

TABLE 9.18. COMPARISON OF FOURIER AND POLYNOMIAL MAP MODELS

Polynomial model				Fourier model		
Order		Number of independent variables	Maximum number of extrema	Number of harmonics	Number of independent variables	Maximum number of extrema
Number	Name					
1	Linear	2	0			
2	Quadratic	5	1			
3	Cubic	9	4	1	8	4
4	Quartic	14	9			
5	Quintic	20	16			
6	Sextic	27	25	2	24	16
7	Septic	35	36			
8	Octic	44	49	3	48	36

Figure 9.26, from a paper by Krumbein (1966b, p. 25), compares an eighth-order polynomial surface based on 45 coefficients with its nearest Fourier counterpart (table 9.18) based on 49 coefficients. Although the two surfaces are superficially similar, Krumbein (1966b, p. 24) believes that the Fourier surface is a much better representation.

Mapping with a Function Containing Arbitrary Terms

In the Fourier, the polynomial and, in some other statistical models not treated explicitly in this book, the mathematical terms are defined by a function that corresponds to a geological and statistical model specified by the investigator. Another approach to mapping is to use a function that is developed by the investigator and that contains arbitrary terms based on computation rather than on the physical situation. Inverse logarithms, high-order polynomials, and hyperbolic sines are some examples of arbitrary terms. Hewlett (1964) gives a good account of this technique as applied to mining data.

A stepwise regression analysis (sec. 10.3) is used to implement this approach.

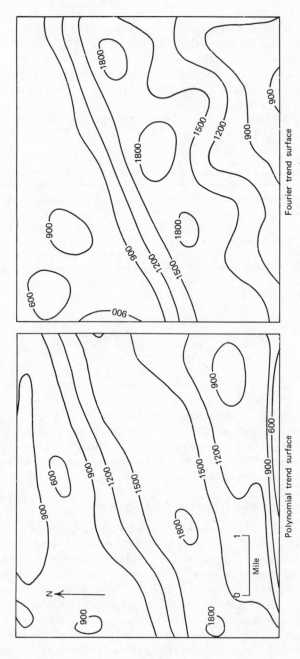

Fig. 9.26. Eighth-order polynomial and Fourier trend surfaces fitted to topographic data from the McClure quadrangle, Pennsylvania (after Krumbein, 1966b, p. 25).

First, several arbitrary terms are selected and a regression is performed. Next, based on the results from the regression, new arbitrary terms are added and some of the original terms are removed. An equation is eventually derived that fits the observations nearly or exactly at the data points, which is the chief advantage of this method. However, outside the data points, this equation may behave poorly. It may yield absurdly high or low values in regions of real data and act even worse for predictions beyond the net of data points. Other disadvantages of this method are that it makes no attempt to relate the final equation to a model of the real world, the developed function has no physical interpretation, and it requires a large amount of art and computer time. A better way to obtain a detailed representation of data—contouring by linear interpolation within grid points—is explained in the next subsection.

Contouring by Linear Interpolation within Grid Points

The final type of map analysis to be introduced is surface fitting by linear interpolation within grid points. The computations are essentially those familiar to the geologist who has made topographic contour maps by plane-table surveying (for instance, see Compton, 1962, p. 149). Although several ready-made computer programs are available, written by either users or manufacturers, only the one used by the U.S. Geological Survey in 1966 is described here.

In the U.S. Geological Survey program, Miesch (1966, personal communication) states that "for any type of data, the actual plotting process always employs linear interpolation in two directions within grid points of a rectangular array of points. These points are either given or computed over a grid." The plotting is done on a roll of paper by a printer controlled by the computer. The map is of unlimited length in one direction (y in the U.S.G.S. program) and is either 12 or 13.2 inches wide in the other direction (x in the U.S.G.S. program). Thus, "given a fixed y-value, a complete line in the x-direction is scanned for all possible contour values between grid points. Symbols are printed according to the particular contour value and corresponding x-value derived from the interpolation process." The paper is then advanced one line in the y-direction and the process repeated. If the required distance in the x-direction is more than the width of the print line, the area to be contoured has to be subdivided and the resulting partial maps pasted together. If the data are not on a grid to start with, a grid is constructed by a weighted average method.

When the U.S. Geological Survey computer program is applied to the 90 points on James' (1966) grid for the McClure, Pennsylvania, data, the map in figure 9.27 is obtained, closely resembling figure 9.25 obtained by the Fourier method.

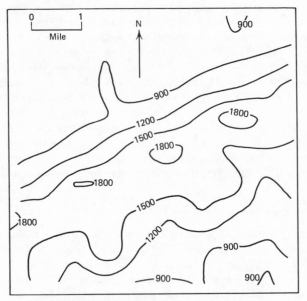

Fig. 9.27. Contour map made by linear interpolation within grid points for the McClure quadrangle, Pennsylvania, topographic data.

9.10 FURTHER USES OF TREND ANALYSIS

The uses of trend analysis previously explained in this chapter are for summarizing data, exemplified by observations from the Rangely oil field in Colorado, from veins in the Fresnillo and Frisco mines, and from a simulated petrological problem. In this section are examined uses of trend analysis for prediction, improved estimates of means, location of extrema, and calculation of volumes.

Regression Analysis for Prediction

Prediction by regression analysis has become popular in geology, presumably because everyone would like to be able to prophesy from the known to the unknown. When a geologist predicts with the aid of a hand-drawn contour map, he closely interacts with the geological situation and has a good grasp of what is happening. But when he predicts by a computer-implemented method, unless he is careful he loses touch with the data and imputes to the computer an anthropomorphic power that it lacks.

When predicting by regression analysis, one should carefully think out a suitable model for the observations at hand or to be obtained and should then keep the model in mind. To return once more to the Rangely data, the model for the elevation of the key bed is a dome. If this model is correct, the elevation of the key bed can be predicted within the net of data points and outward beyond them. The prediction will succeed to the extent that the geological and statistical models represent the real world. In some geological situations, no geologist expects a prediction to succeed, although to a statistician or other layman it may appear to be excellent. For instance, at Rangely, the key bed may not even be present, perhaps because it was never deposited, was eroded away, or was displaced by a fault or an igneous body. Or the key bed may be present but may have been faulted to an entirely unexpected elevation. Such geological phenomena may be easy to recognize in a simple structural or stratigraphic situation, but are otherwise often difficult to discern. At the least, one should be suspicious of predicted values of chemical elements of less than zero or more than the value corresponding to 100 percent of the pure mineral or minerals, and of predicted subsurface elevations above the surface elevation.

It is helpful to distinguish between trend analysis for interpolation and for extrapolation. Interpolation, seeking predictions among observed data points, is relatively easy to evaluate. First, the behavior of the residuals at the observed data points can be inspected. Then a few data points can be omitted, the regression analysis repeated, and the new prediction compared with the original one. Usually little change occurs unless too many points are omitted. So, if the data points are more or less evenly spaced, the original pattern of residuals shows how reliable interpolation will be. Because interpolation is straightforward, we do not examine it further.

Extrapolation, in contrast to interpolation, is predicting outward from areas with observations to those without observations. At symposia on electronic computing in applied geology, the enthusiasts of trend-surface analysis exchange anecdotes about whether or not certain extrapolations have worked. Successful extrapolations are hard to prove, just as it is difficult to fix responsibility for a producing oil well or mine, on the one hand, or for a dry well or noncommercial mineral deposit, on the other. Does the fact that a drill hole finds oil or ore indicate that the man who spotted the drill hole was perceptive or only lucky?

For an example of prediction by extrapolation, we return to the analysis by two-dimensional quadratic regression of data from the 2137 vein of the Fresnillo mine (sec. 9.5). An extrapolation eastward from the surface fitted to silver in the west part of the vein (fig. 9.11) predicts an increase in silver content (fig. 9.12). When this predicted increase is compared with the silver data in the east part of the vein (fig. 9.28), fair agreement is found. And a predicted increase in silver content downward below the 920-meter level to

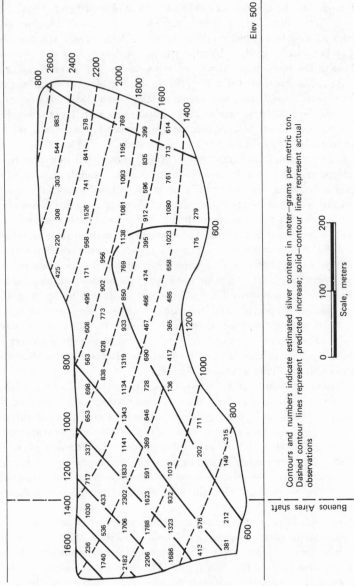

Fig. 9.28. Comparison of predicted increase in silver content on upper levels of the 2137 vein, Fresnillo mine, with actual data.

the northwest (lower left in fig. 9.11) has been confirmed by mine workings driven after the study was made (G. K. Lowther, personal communication).

Thus quadratic regression surfaces may be useful for prediction. In the past eight years we have made several hundred quadratic regression analyses of mine assay and other types of geological data. Predictions, when checked by known data, or data obtained subsequent to our investigation, have been sometimes good, other times not. Often predictions made using only the data closest to the place to be predicted are better than those done with regression analysis.

Higher-order polynomial equations are often dangerous for use in prediction, as discussed in an excellent article by Krumbein (1966b) and illustrated in a revealing map, reproduced in this book as figure 9.29. The Fourier model is seldom of value for extrapolation, because it is a periodic function that can only repeat itself, as figure 9.30 illustrates.

Other extrapolative methods might be used to advantage for certain types of situations. J. W. Tukey (personal communication) suggests that an equation could include higher-order polynomial terms close to the known data and lower-order terms at a distance from them, while J. Hartigan (personal communication) suggests that a polynomial equation could be fitted to data with the points on the margin weighted to increase their influence on the fitted

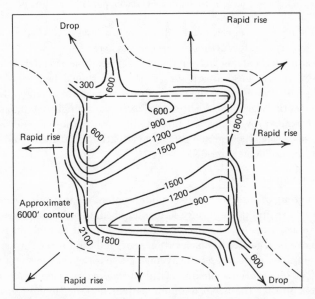

Fig. 9.29. Sixth-order polynomial surface fitted to the McClure quadrangle, Pennsylvania, topographic data (after Krumbein, 1966b, p. 37).

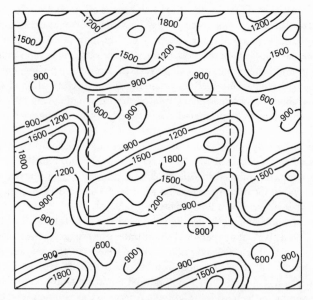

Fig. 9.30. Fourier trend surface based on 49 coefficients fitted to the McClure quadrangle, Pennsylvania, topographic data (after Krumbein, 1966b, p. 37).

equation. A similar concept, a weighted moving average, has been advanced by Krige (1966). A field for further research exists here.

Regardless of what extrapolative method is devised, the crucial difficulty is deciding to what geological area a statistical model should be applied for a meaningful prediction, because the range of prediction is always limited. This problem is essentially geological. Often the area is vanishingly small, because the adjacent points provide the best prediction, but sometimes the areas may be fairly large (Krige, 1966). Each geological situation is different, and more study of specific problems is needed.

Improved Estimates of Means

Consider a group of observations of uranium content made in mine workings driven to develop a uranium-bearing sedimentary bed. In order to estimate the grade and tonnage of ore, the average uranium content is required. Each observation can be regarded as an elevation above a datum plane corresponding to zero uranium content, and the average grade is then the mean of these elevations. The observations must constitute a random sample in some sense of the phrase, so that the mean uranium content may be thought of as an unbiased estimate of the average uranium content of the bed. If there is no trend in the observations, a confidence interval for the mean may be obtained by using the statistic $t_{10\%}(n-1)\sqrt{s^2/n}$, where s^2 is the sample variance and n is the sample size. The model is that the average grade corresponds to a horizontal plane above the datum plane, with the residual values from the horizontal plane, $w - \overline{w}$, used to calculate s^2, the estimate of σ^2.

Through regression analysis, other models can be formulated in which the mean grade is represented by an inclined surface obtained through linear regression, or a curved surface obtained through quadratic or higher-order regression. The trend-surface residuals, $w - \hat{w}$, are used to calculate s_e^2, the residual mean square, which is the appropriate estimate of σ^2. On the average, these residuals are smaller if the trend is reliable, because they are measured from a surface that fits the data points more closely than does the horizontal plane; accordingly the confidence interval is narrower. [Confidence limits are obtained by using the relation $t_{10\%}(n-k)\sqrt{s_e^2/n}$, where s_e^2 is the residual mean square and k is the number of fitted parameters.]

These methods are applied to a set of data from the Mi Vida mine, Utah, which worked an ore deposit of uranium and vanadium minerals disseminated in certain sedimentary beds of the Triassic Chinle Formation. The basic data (Koch, Link, and Hazen, 1964) are assay results for uranium (U_3O_8), vanadium (V_2O_5), and lime from 225 channel samples cut at 79 sample points; only the uranium data are discussed in this book. According to the pattern of uranium observations, the mine was divided into two areas: the northern area, with strong quadratic trend, and the southern area, without quadratic trend. Table 9.19 lists confidence limits for the mean grade, computed by taking trend into account and computed without regard for trend. In the northern area, confidence limits are narrowed if trend is recognized, while in the southern area there is no appreciable change. The confidence limits for the entire sampled area are also calculated—the pooled mean squares are used to estimate σ^2, and the degrees of freedom for residual sums of squares (number of observations minus 6) are used as weighting factors.

TABLE 9.19. CONFIDENCE INTERVALS, TAKING INTO ACCOUNT TRENDS, FOR URANIUM DATA FROM THE MI VIDA MINE

Area	Number of observations	Mean	Confidence limits taking trend into account		Confidence limits without regard for trend	
			Lower	Upper	Lower	Upper
Entire mine	79	0.510	0.452	0.568	0.441	0.579
Northern area	47	0.468	0.387	0.549	0.367	0.569
Southern area	32	0.570	0.487	0.653	0.485	0.655

Interval estimates taking trend into account can be calculated directly as described in the previous paragraphs only when the mean of the weighted assays is an appropriate estimate of the desired mean. In particular, this requirement implies a sampling scheme whose "center" is at the right location, and whose sample points are spaced so that all areas of interest have an equal chance of being sampled. If the Mi Vida sample points were concentrated in one working place rather than spread out more or less uniformly, the mean might be inappropriate, and a more complicated formula (Link, Koch, and Gladfelter, 1964, p. 24) would be needed in order to calculate confidence limits.

Location of Extrema

Regression analysis can readily locate extrema (secs. 9.4 and 9.6), which may be maxima, such as the highest point of a key sedimentary bed on an oil dome, or may be minima, such as the lowest point in a basin-shaped aquifer. Another method of locating extrema, by evaluation of sequentially obtained observations, is explained in section 12.5.

Maxima for the Rangely and Mi Vida example data were located by quadratic regression. For the Rangely data, the estimated location of the highest point on the dome is plotted in figure 9.2 by the letter M; the estimated elevation is -246 feet. This should be a good place to drill for oil, and indeed several wells are nearby. At the Mi Vida mine, the highest point of the dome (which corresponds to a quadratic equation fitted to uranium for the northern area) is within a few feet of the place where the discovery diamond-drill hole intersected 14 feet of high-grade ore. The location, selected by Charles A. Steen in July 1952, made his fortune.

Calculation of Volumes

Regression equations can be integrated to obtain the area under a curve or the volume between two surfaces. Full details of several petrologic applications have been published by Whitten (1962). For his example data of the modal

TABLE 9.20. COMPARISON OF ESTIMATED MEANS CALCULATED BY ARITHMETIC AVERAGING AND DOUBLE INTEGRATION FOR PETROLOGIC DATA FROM THE "OLDER GRANITE" OF DONEGAL, EIRE*

Variable	Arithmetic mean of 118 observations	Double integration mean
Quartz percentage	21.77	23.38
Color index	13.07	11.37
Potash feldspar/total feldspar ratio	0.50	0.60
Total feldspar percentage	65.16	65.14

* After Whitten, 1962, p. 557.

composition of the "older granite" of Donegal, Eire, the results, shown in table 9.20, do not differ substantially from the arithmetic means. (The figures have been corrected by us and checked by Whitten from the values published on page 557 of the original paper.)

Integration is a method of obtaining a weighted average, but it is seldom useful. Arithmetic averaging of the observations will yield nearly the same mean if the points are fairly evenly spaced. If they are not, the best procedure is usually to weight points by their area of influence or by the number of points per unit area, as is commonly done in mining geology (Chap. 13) and in contouring points in stereographic projection (sec. 10.7).

9.11 STRATEGY IN TREND ANALYSIS

Three topics are reviewed in this section: evaluation of the reliability of fitted trend surfaces, choice of geological and statistical models, and pertinence of trend-surface analysis of geological data.

Evaluating the Reliability of Fitted Trend Surfaces

This subsection explains three methods of evaluating the reliability of trend surfaces that have been fitted by linear and quadratic regression; the methods also apply in some degree to evaluating trend surfaces fitted by other statistical models.

One way to determine the reliability of a trend surface is by an analysis of variance. The total variability of a trend surface is divided into parts that are associated with the linear and quadratic terms and into an unexplained residual part. Two quantities in particular should be inspected: (1) the proportion of the total sum of squares accounted for by the linear and quadratic terms, and (2) the F-statistics. If the "explained" proportion of the total sum of squares is large, one can be reasonably well satisfied. How large is large depends on the purpose of the investigation. In the determination of a simple over-all trend and the pattern of residuals, the proportion need not be very large. Interpretation of the F-statistic in an analysis of variance for a trend surface is no different than for any other analysis of variance. However, it is quite common to have so many observations that the slightest trend yields a "significant" result, even though no purpose can be achieved with the trend. The next step is to compare the residual variability, which is unexplained by the linear or quadratic model, with the local variability, as was done for the Fresnillo silver data in table 9.8. A well-designed experiment for obtaining observations for a trend-surface analysis must include a provision for estimating local variability.

A second method for evaluating a trend surface is to see how well it predicts, for both interpolation and extrapolation. To do this, the trend analysis is repeated several times, leaving out first a few points in the middle of the area or volume and then a few on the margins. If the new equations predict the omitted points fairly well, the investigator can have some confidence in the analysis.

A third way to test the reliability of a surface is to divide the points in half, randomly or systematically, so that each set of points more or less covers the original area while the density of points is cut in two. Then a new surface is fitted to each set, and the new surfaces are overlaid on the original one and compared with it. This technique provides a severe test (more rigorous if the points are divided randomly) that is particularly instructive because the investigator realizes how sensitive an analysis may be to the number and pattern of points.

Table 9.21. Analysis of variance for two randomly selected sets of one-half the original zinc data from the 2137 vein of the Fresnillo mine

Source of variation	Sum of squares	Degrees of freedom	Mean square	F	$F_{10\%}$
Linear terms	1008.71	2	504.36	33.50	2.37
Quadratic terms	14.92	3	4.97	0.33	2.15
Residual variation	1445.32	96	15.06		
Total	2468.95				

Source of variation	Sum of squares	Degrees of freedom	Mean square	F	$F_{10\%}$
Linear terms	553.75	2	276.88	23.87	2.37
Quadratic terms	93.28	3	31.09	2.68	2.15
Residual variation	1113.46	96	11.60		
Total	1760.49				

Table 9.21 and figure 9.31 present the results of such an analysis for the zinc data at Fresnillo, previously reviewed in table 9.10 and figure 9.13. Analyses of variance (tables 9.21 and 9.10) show that, while the residual mean square is about the same in both, the linear mean square is smaller in table 9.21, so that the corresponding F-value is also smaller. In both tables the explained proportion of the total sum of squares is about the same. Figure 9.31, which plots the three fitted surfaces, shows that they correspond well in some areas

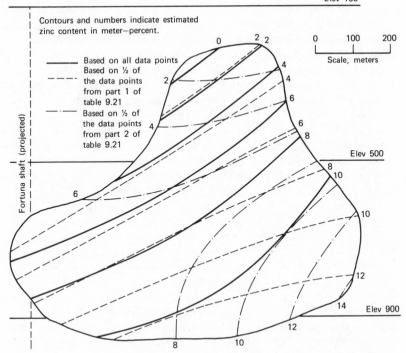

Fig. 9.31. Comparison of three quadratic regression surfaces based on one-half the data points randomly selected, with the surface based on all the data points. Zinc data are from the 2137 vein of the Fresnillo mine.

and poorly in others. (This test gave much better results when applied to the silver data in figure 9.12.) We have performed this test on many sets of data and have found the results always instructive, if often dismaying.

Choice of Models

When choosing a model, the question to consider is: What geological model is appropriate, and how is it related to the real world as well as to a statistical model that can be implemented at the present state of statistical knowledge and of the computer art? The choice of a geological model should be based on theoretical knowledge, on experience with similar geological situations, and on internal evidence in the data if they have already been collected.

For some problems, theoretical knowledge is a primary determinant. For instance, one of the simple geological structures, such as an anticline, dome, syncline, basin, or monocline, may be expected from theory; sand sizes and shapes may follow well-defined physical laws in a delta, alluvial fan, or sand dune; and volcanic ashes are expected to be dispersed radially outward from a vent. Experience with similar geological situations may be the basis for choosing a model, as when repetition of certain evidence on the Gulf Coast suggests a salt dome, or a set of subsurface data from the Los Angeles region indicates an oil pool even though the theory for these phenomena may be controversial. In such cases geology provides a positive guide to the selection of a model.

For existing data, the choice of a model is aided by the internal evidence obtained after grouping the observations into cells on a two- or three-dimensional grid. The local variability, which is independent of whatever model is chosen, should be calculated. A rough plot of the raw or grouped data may also suggest trends, or clusters of high or low values.

A statistical model should be chosen in relation to a geological model. A simple model, such as a linear or quadratic polynomial, is to be preferred if the geological model is poorly known or unknown as well as when its form is linear or quadratic. A more complicated model, such as higher-order polynomial or Fourier series, may be useful for a well-defined geological model that corresponds to that mathematical form, or if local variability is low and well controlled.

Here, as elsewhere in statistical analysis of geological data, choice of a model depends on taste, judgement, and luck. The choice must be made subjectively. A single "correct" trend surface cannot be found by fitting more and more complicated equations. Although a sufficiently complicated function will fit any set of data, the function is not unique. Therefore there is no reason to try more and more complicated functions with the hope that a "correct" surface will eventually be found.

In the straight-line example (fig. 9.3), 22 data points were fitted. One way to get an exact fit is to use an equation with 22 terms. Another way is to start with a less complicated equation of 11 terms by cutting the number of data points in half and using every other point. If the number of points is progressively reduced, eventually, when only three data points remain, a parabola will exactly fit them, and when only two remain, a straight line will provide an exact fit. Yet a perfect fit would not represent reality at any of these stages, because the silver is known to be at least as variable in detail as is represented by the plot of the 22 original points. This discussion illuminates the situation that arises when data are sparse to begin with, as in exploration drilling, and one is tempted to fit all the data exactly, through the choice of a complicated function. Yet even such a function will not truly portray the data, any more than do functions only slightly more complicated than the straight line in the example of figure 9.3.

The fitting of a surface in either two or three dimensions is analogous to the fitting of a line. With observations exhibiting as much local fluctuation as those from the Frisco and Fresnillo mines, a detailed fit would not be pertinent. Whether or not most geological data do exhibit this much local fluctuation is a real question, the answer to which is not yet available.

If low-order polynomial surfaces are selected, the question remains of what to do with areas or volumes that are so grossly complicated that a quadratic surface will not afford a satisfactory fit. The best procedure usually is to break up the complicated areas into smaller pieces (overlapping them if desired) that can be fitted individually with a quadratic polynomial. Initial identification of the complete surface can be made through ordinary methods of hand contouring or through fitting a preliminary quadratic surface and examining it in relation to its residuals. It is easier to use subjective judgment in choosing an approximate area to be fitted than in choosing an appropriate statistical model. The areas to be fitted can be selected by hand contouring, in which subjective knowledge can be rapidly applied by an experienced geologist. The reliability of the final analysis is undisturbed since, for one thing, the whole area to be contoured had to be selected subjectively in the first place.

Areas should also be broken up if the data are discontinuous, because fitting a smooth function to discontinuous data makes little if any sense. For instance, if a major fault cuts through an area to be contoured, it is better to fit the data on either side of it than to lump all the data together. Faults and other discontinuities can often be located by preliminary hand contouring or machine contouring by linear interpolation within grid points (sec. 9.9).

Pertinence of Trend-Surface Analysis of Geological Data

Trend-surface analysis requires particularly thoughtful use in order to be pertinent to the analysis of geological data, because ready-made computer programs make the method easy to use and do not require the understanding of the investigator and because the resulting maps superficially resemble contour maps already familiar to all geologists.

Chayes and Suzuki (1963) and Whitten (1963) have written interesting and thoughtful critiques of trend-surface analysis. Two of Chayes and Suzuki's comments are especially pertinent. First, they state that "without suitable explanation of basic concepts, most of us simply will not take seriously a contour map which leaves more than a small fraction of the observed variation in the 'unexplained' or 'residual' category. . . . There is a vast difference between detecting significant sources of variation in a body of data and reducing the 'residual' or 'unexplained' sum of squares to a negligible size. It is the latter condition which most geologists take for granted when they attempt to read contour maps." All of this is true, and these are the basic concepts that are stressed in this chapter. However, geologists have many different things in mind when they read contour maps, and this is why we stress local variability, without knowledge of which there is little point in drawing a detailed contour map (of nontopographic data) by subjective hand methods, by statistical methods, or at all. The large conceptual jump from a topographic contour map to a contour map of other data is one that is often glossed over.

Chayes and Suzuki's second criticism, about integrating volumes under a trend surface in order to calculate means, points out the importance of thorough understanding of the model. Most aspects of trend-surface analysis can be better understood if one thinks first about a simple case. Suppose that a geologist has traversed four miles across a rock body and has taken 40 observations at milepost zero, and four each at mileposts 1, 3, and 4, with the results plotted in figure 9.32. Now he wants to estimate the mean at milepost 2.

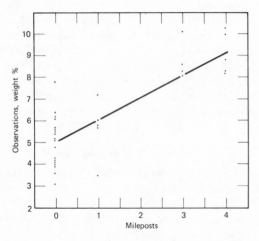

Fig. 9.32. Linear regression line fitted to fictitious data to illustrate model formation.

Should he use the arithmetic average of 6.23 percent? Clearly not. The best estimate is either 7.12 percent from the linear regression line (if the appropriate model is linear) or 7.04, the mean of the observations at mileposts 1 and 3 (if the appropriate model is a smooth but not necessarily linear curve). The analysis hinges on evaluating local variability, choosing a model, devising a sampling plan so that the observations are spread out evenly, or making adjustments, as with the diamond-drill-hole deflection data (sec. 7.4), if the observations cannot be evenly spread out.

By extrapolating from this trivial one-dimensional case to data in two or three dimensions, one can see why it is wise to plot and examine the actual data points before fitting a surface in order to help reveal mistakes in the data and to learn whether a trend-surface analysis is likely to be appropriate. A trend-surface analysis for the fictitious data in section 5.12, for instance, yields misleading results. In that section it was supposed that a geologist wished to investigate potash content in a gneiss exposed in a linear outcrop belt and that he made ten traverses at right angles to the foliation with five geological samples from each traverse. The 50 fictitious potash analyses, listed in table 5.40, may be assumed to come from traverses spaced 2000 feet apart with the samples taken at 100-foot intervals within each traverse. If a quadratic trend-surface analysis is now performed on these data, the analysis of variance in table 9.22 shows that only 10 percent of the variability is "explained," and that the F is not significant. The plotted surface, in which little

TABLE 9.22. ANALYSIS OF VARIANCE OF FICTITIOUS POTASH DATA FROM TABLE 5.40 TO ILLUSTRATE MISUSE OF QUADRATIC REGRESSION ANALYSIS

Source of variation	Sum of squares	Degrees of freedom	Mean square	F	$F_{10\%}$
Linear terms	1.756	2	0.878	0.25	2.43
Quadratic terms	15.057	3	5.019	1.42	2.22
Residual variation	154.985	44	3.522		
Total	171.798				

or no confidence can be placed, is an elongated saddle trending north-south and rising to the north and south. Through an inappropriate analysis, only misleading results have been obtained. The pattern shown by the traverse means in table 5.40 is hidden. By contrast, the multiple-comparison analysis in section 5.12 illuminates, instead of obscuring, these same data.

In summary, the usual advantage of statistically fitting a surface rather than drawing contours by hand or by machine is that it affords an objective determination of the reliability of the resulting map. Thus problems of interpretation arising in the analyses of hand-drawn maps, which can be excellent or incompetent, depending on the skill of the mapmaker, can be avoided. If a surface is fitted statistically, the amount of variability explained by the fitted surface may be compared with the amount unexplained and with the local variability. If the amount of explained variability is large, one concludes that the fitted surface is a good approximation that, other things being equal, may be accepted with confidence. If the amount of explained variability is small, one concludes that, although there may be an underlying trend, it has not been found with the chosen statistical model.

REFERENCES

Agterberg, F. P., 1969, Interpolation of areally distributed data, *in* Spear, S. W., ed., Operations research and computer applications in the mineral industries, 7th Internat. Symposium, Golden, Colo., Quart. Colo. Sch. Mines, v. 64, no. 3, p. 217–237.

Albert, Adrian, 1949, Solid analytic geometry: New York, McGraw-Hill, 162 p.

Berkson, J., 1950, Are there two regressions?: Jour. Am. Statistical Assn., v. 45, p. 169–180.

Chayes, F., and Suzuki, Y., 1963, Geological contours and trend surfaces: Jour. Petrology, v. 4, p. 307–312.

Compton, R. R., 1962, Manual of field geology: New York, John Wiley & Sons, 362 p.

Conolly, H. J. C., 1936, A contour method of revealing some ore structures: Econ. Geology, v. 31, p. 259–271.

Davis, J. C., 1969, Distribution of hydrocarbons in three dimensions, *in* Merriam, D. F., ed., Symposium on computer applications in petroleum exploration: Kansas Geol. Survey, Computer Contr. 40, p. 34–40.

Dixon, W. J., and Massey, F. J., Jr., 1969, Introduction to statistical analysis: New York, McGraw-Hill, 638 p.

Fraser, D. A. S., 1958, Statistics—an introduction: New York, John Wiley & Sons, 398 p.

Grant, Fraser, 1957, A problem in the analysis of geophysical data: Geophysics, v. 22, p. 309–344.

Grenander, U., and Rosenblatt, M., 1957, Statistical analysis of stationary time series: New York, John Wiley & Sons, 300 p.

Griffiths, J. C., 1967, Scientific method in analysis of sediments: New York, McGraw-Hill, 508 p.

Harbaugh, J. W., 1964, A computer method for four-variable trend analysis illustrated by a study of oil–gas gravity variations in southeastern Kansas: Kansas State Geol. Survey, Bull. 171, 37 p.

Harbaugh, J. W., and Merriam, D. F., 1968, Computer applications in stratigraphic analysis: New York, John Wiley & Sons, 259 p.

Hewlett, R. F., 1964, Polynomial surface fitting using sample data from an underground copper deposit: U.S. Bur. Mines Rept. Inv. 6522, 27 p.

James, W. R., 1966, The Fourier series model in map analysis: Evanston, Ill., Northwestern Univ. Dept. of Geology, Tech. Rept. 1, ONR Task No. 388-078, 37 p.

Koch, G. S., Jr., and Link, R. F., 1967, Geometry of metal distribution in five veins of the Fresnillo mine, Zacatecas, Mexico: U.S. Bur. Mines Rept. Inv. 6919, 64 p.

Koch, G. S., Jr., Link, R. F., and Hazen, S. W., Jr., 1964, Statistical interpretation of sample assay data from the Mi Vida uranium mine, San Juan County, Utah: U.S. Bur. Mines Rept. Inv. 6550, 40 p.

Krige, D. G., 1966, Two-dimensional weighted moving average trend surfaces for ore valuation, *in* Symposium on mathematical statistics and computer applications in ore valuation: South African Inst. Mining Metall., p. 13–79.

Krumbein, W. C., 1963, Confidence intervals on low-order polynomial trend surfaces: Jour. Geophys. Research, v. 68, p. 5869–5878.

————, 1966a, unpublished lectures.

————, 1966b, A comparison of polynomial and Fourier models in map analysis: Evanston, Ill., Northwestern Univ. Dept. Geology Tech. Rept. 2, ONR Task 388-078, 45 p.

Krumbein, W. C., and Graybill, F. A., 1965, An introduction to statistical models in geology: New York, McGraw-Hill, 475 p.

Li, J. C. R., 1964, Statistical inference, v. 1: Ann Arbor, Michigan, Edwards Bros., 658 p.

Lindley, D. V., 1953, Estimation of a functional relationship: Biometrika, v. 40, p. 47–49.

Link, R. F., Koch, G. S., Jr., and Gladfelter, G. W., 1964, Computer methods of fitting surfaces to assay and other data by regression analysis: U.S. Bur. Mines Rept. Inv. 6508, 69 p.

Link, R. F., Yabe, N. N., and Koch, G. S., Jr., 1966, A computer method of fitting surfaces to assay and other data in three dimensions by quadratic regression analysis: U.S. Bur. Mines Rept. Inv. 6876, 42 p.

McKinstry, H. E., 1948, Mining geology: Englewood Cliffs, N.J., Prentice-Hall, 680 p.

Miesch, A. T., and Connor, J. J., 1968, Stepwise regression and nonpolynomial models in trend analysis: Lawrence, Kans., State Geol. Survey, Computer Contr. 27, 40 p.

Peikert, E. W., 1962, Three-dimensional specific-gravity variation in the Glen Alpine stock, Sierra Nevada, California: Geol. Soc. Am. Bull., v. 73, p. 1437–1448.

Preston, F. W., and Harbaugh. J. W., 1965, BALGOL programs and geologic application for single and double Fourier series using IBM 7090/7094 computers: Lawrence, Kans., State Geol. Survey, Spec. Distrib. Pub. 24, 72 p.

Stone, J. B., and McCarthy, J. C., 1948, Mineral and metal variations in the veins of Fresnillo, Zacatecas, Mexico: Am. Inst. Mining Metall. Engineers Trans., v. 178, p. 91–106.

Thomas, C. R., et al., 1945, Structure contour map of the Rangely anticline: U.S. Geol. Survey Oil and Gas Inv. (Prelim.), Map 41.

Tukey, J. W., 1951, Components in regression: Biometrics, v. 7, p. 33–69.

Whitten, E. H. T., 1959, Compositional trends in a granite: modal variation and ghost-stratigraphy in part of the Donegal granite, Eire: Jour. Geophys. Research, v. 64, p. 835–848.

————, 1961a, Systematic quantitative areal variation in five granitic massifs from India, Canada, and Great Britain: Jour. Geology, v. 69, p. 619–646.

————, 1961b, Quantitative areal modal analysis of granitic complexes: Geol. Soc. Am. Bull., v. 72, p. 1331–1360.

————, 1962, A new method for determination of the average composition of a granitic massif: Geochimica et Cosmochimica Acta, v. 26, p. 545–560.

————, 1963, A reply to Chayes and Suzuki: Jour. Petrology, v. 4, p. 313–316.

Wilks, S. S., 1962, Mathematical statistics: New York, John Wiley & Sons, 644 p.

Wisser, Edward, 1941, Discussion of a paper by H. E. McKinstry: Trans. Am. Inst. Mining Engineers, v. 144, p. 87–93.

Wylie, C. R., Jr., 1960, Advanced engineering mathematics: New York, McGraw-Hill, 640 p.

Chapter 10

Analysis of Multivariate Data

10.1 INTRODUCTION

All of the statistical methods introduced in the previous chapters were devised to analyze *univariate* observations, each consisting of a single number. For instance, observations of quartz percentage from a batholith are univariate if the location of each sample point is recorded as three independent variables in three-dimensional space and its quartz percentage is recorded as the univariate dependent variable. Even when there was more than one dependent variable at each data point, as in the analysis of the 2200 vein of the Fresnillo mine in section 9.5, the different metals were treated one at a time as univariate variables. Now, in this chapter, analysis of *multivariate* data is introduced. An example is petrographic modal analysis data; each thin section yields a single observation consisting of several numbers, specifying percentages of the minerals counted. An observation written as $(10, 22, 5)$ might stand for 10 percent quartz, 22 percent feldspar, and 5 percent opaque minerals—the other minerals having been disregarded.

Because the concept of a single observation consisting of more than one number may be new, a few more examples of multivariate observations are mentioned; others will doubtless occur to the reader. Familiar to all is a strike-and-dip observation, composed of two numbers, such as 310, 85-N, designating a strike of 310 degrees (N 50° W) and a dip of 85 degrees northeastward. Or the two numbers $(170, 20)$ may be the multivariate observation for a line whose bearing is 170 degrees (S 10° E) and whose plunge is 20 degrees southward.

Most mineralogical and petrological observations are multivariate. Composition data, whether obtained by chemical, modal, or other analysis, are multivariate. So are physical property measurements, including specific gravity, index of refraction, color, and electrical conductivity. Multivariate data also occur because the science of geology often involves the dimensions of space and time.

All of the methods explained previously for univariate data can be applied to multivariate data if the variables are taken one at a time. However, the interrelationships of the variables can also be investigated. In Chapter 9 the relationship of a single dependent variable, such as elevation of a key bed or silver content, to independent variables that designated data-point locations was studied. In that chapter only the single dependent variable was of interest, and the independent variables were ancillary to determining its behavior. Here, in Chapter 10, ways to study interrelationships among variables, several or all of which may be of interest, are developed.

For instance, if several sandstones are studied, it is natural to inquire if they are alike or dissimilar, whether the data are univariate or multivariate. But if multivariate observations are obtained—say porosity, permeability, specific gravity, organic content, fossil assemblages, and oil content—many additional questions can be posed. One might investigate the possibility that oil content is related to porosity but not to permeability, a possibility that might suggest that oil was formed in the rock rather than introduced. Or sandstones could be classified according to their fossil assemblages.

Some Geometry of Multivariate Data

Univariate observations can be plotted on a straight line (fig. 10.1). If the straight line is divided into class intervals, the number of observations in each interval can be counted, and a histogram can be formed (sec. 2.3). Plotting the histogram requires two dimensions.

$$w_1$$

Fig. 10.1. Plot of univariate observations on a straight line.

Bivariate observations can be plotted on a scatter diagram in two dimensions (fig. 10.2). If the diagram is divided into cells, the number of observations in each cell can be counted and plotted on a histogram. This histogram takes three dimensions to plot and may be represented by a contour map, like those commonly made by geologists to represent points plotted in stereographic projection (sec. 10.7).

Trivariate observations can be plotted on a scatter diagram in three dimensions (fig. 10.3). The most usual configuration of points defines an

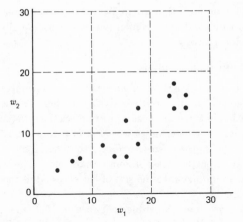

Fig. 10.2. Plot of bivariate observations on a scatter diagram in two dimensions.

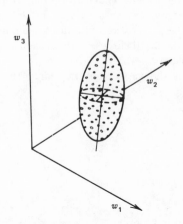

Fig. 10.3. Plot of trivariate observations on a scatter diagram in three dimensions.

ellipsoid. [The geometrical properties of ellipsoids, familiar to geologists from the study of structural geology and optical crystallography, are not reviewed in this book; Albert (1949) provides details.] If the three-dimensional space is divided into cubes or other solid figures, the number of observations in each cube can be counted and a frequency distribution made. A histogram would require four dimensions to plot. Trivariate observations of constant sum can also be plotted on triangular coordinates (Chap. 11).

Observations with four or more variables cannot be plotted with ordinary methods, although Mertie (1949) has devised complicated hypertetrahedra plots. The algebraic methods that are explained in this chapter for geometric interpretations of three and fewer dimensions will work equally well for four and more dimensions, however.

Procedures for Multivariate Analysis

The following is an explanation of how the organization of this chapter relates to discussions of multivariate analysis elsewhere in the book. A geometric interpretation in three dimensions introduces the different procedures. For each procedure, the cluster or swarm of data points defines an ellipsoid.

Section 10.2 is about *generalized variance*, the analog of the variance of univariate statistics for the case of more than one variable. Generalized variance involves an inquiry into the size of the ellipsoid defined by a cluster of points (fig. 10.3).

Multiple linear regression, the subject of section 10.3, was introduced in Chapter 9 as the special method for trend-surface analysis. This procedure considers several variables, in contrast to simple linear regression (sec. 9.2), in which only one variable is independent. In geometrical terms (fig. 10.4), a plane (for a linear regression) or a curved surface (for a higher-order regression) is fitted to the observations in order to minimize the sum of the squared distances from the fitted surface to the reference plane, as measured perpendicular to that plane.

Generalized variance is the measure of variability that is used in three

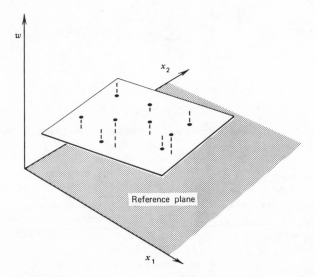

Fig. 10.4. Plane fitted to observations by multiple linear regression.

different methods of multivariate analysis: generalized analysis of variance (sec. 10.4), discriminant analysis (sec. 10.5), and factor analysis (sec. 10.6). A *generalized analysis of variance* investigates whether two or more groups of multivariate observations come from populations with the same mean, or, in geometrical terms, whether two or more ellipsoids have the same center. Of the four ellipsoids in figure 10.5, two have the same center within the limits of expected statistical fluctuation.

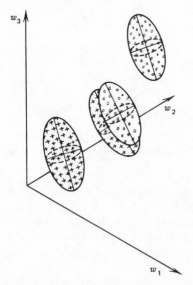

Fig. 10.5. Four ellipsoids representing clouds of observations whose four means are compared by the generalized analysis of variance.

Discriminant analysis (sec. 10.5) yields a function for classifying multivariate observations into groups. Through the version presented in this book, ellipsoid centers are located and the distances between paired ellipsoids are calculated. Figure 10.6 shows three ellipsoids and three dashed lines denoting paired distances.

Factor analysis (sec. 10.6) ascertains whether multivariate observations occupy a number of dimensions equal to the number of measured variables or instead may be contained in a smaller number of dimensions, implying fewer variables present than those measured. For example, the two variables plotted in figure 10.7-a may be essentially contained in the one dimension represented by the dashed line and can therefore be expressed as one variable. Similarly, figure 10.7-b implies that data in three dimensions lie in one plane and can be expressed as two rather than three variables. Figure 10.7-c implies that data in three dimensions lie in one dimension. Ellipsoid axes may be rotated and

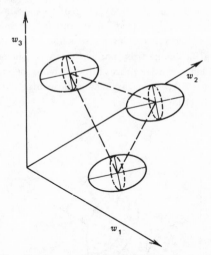

Fig. 10.6. Three ellipsoids representing clouds of observations that are classified by discriminant analysis.

translated to place the ellipsoid center at the center of coordinates, so that the ellipsoid axes correspond to the Cartesian coordinate axes. Factor analysis specifies the rotation and translation. It also indicates whether there are fewer dimensions than directions, as, for example, in figure 10.7-b, in which the shortest major axis of an ellipsoid was collapsed to make the ellipsoid an ellipse.

Several other multivariate methods, important in some disciplines, are omitted from this book because they require data more nearly normal than those usually found in geology. These omitted methods include generalizations of the chi-square and F-tests to inquire whether two or more groups of observations come from populations with the same variability (Wilks, 1962, p. 595).

Another method omitted because its application is limited is canonical correlation (Wilks, 1962, p. 587), which generalizes the correlation coefficient (sec. 9.3) to the correlation among groups of variables. For problems with several dependent as well as several independent variables, questions such as which linear combinations of the independent variables best depict the dependent variables, and vice versa, are investigated.

Besides these principal methods, many procedures exist for special problems. The analysis of directional data in three dimensions, usually plotted in stereographic projection, is the only one of these discussed (sec. 10.7), because of its geological importance.

Especially interesting multivariate data are those that add to a constant

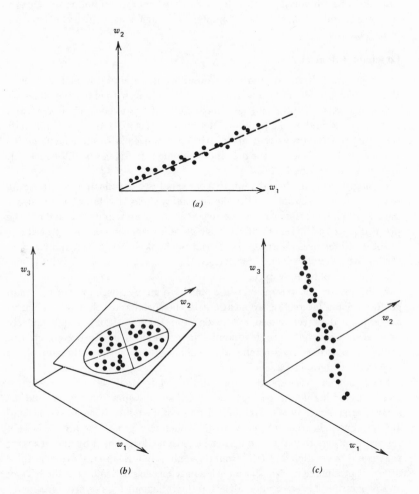

Fig. 10.7. Examples of dimensionality of observations investigated by factor analysis. In figure 10.7-a, although two variables are measured, they may be essentially contained in the one dimension represented by the dashed line. In figure 10.7-b, although three variables are measured, they may be essentially contained in two dimensions represented by the plane. In figure 10.7-c, the three measured variables may be essentially contained in the one dimension represented by the line.

sum and those involving ratios. Because these data are common in geology, they are examined separately in Chapter 11, which is logically a continuation of this chapter.

Concluding Remarks

To explain multivariate analysis in detail would expand this book too much. Wilks (1962, pp. 540–601) gives a complete treatment with extensive literature references. Also helpful and less mathematical are a book by Cooley and Lohnes (1962) and a long paper by McCammon (1969), both of which include the computer programs that are nearly always needed to implement multivariate methods. Because the field is developing rapidly, many of the methods are found only in periodicals.

Geologists have made many multivariate analyses. A collection of interesting papers in a special supplement to the *Journal of Geology*, volume 74, number 5, part 2, 1966, has references to the recent literature. Also instructive are the proceedings of a University of Kansas colloquium on classification procedures in the earth sciences (Merriam, 1966) and the book by Krumbein and Graybill (1965), who devote several chapters to the subject. Miller and Kahn (1962) reference the older literature.

Multivariate analysis has been applied in many disciplines other than geology. Some interesting applications in the social and behaviorial sciences are explained or referenced by Cooley and Lohnes (1962). A particularly interesting book on a problem similar in principle to many encountered in geology was written by Mosteller and Wallace (1964) on the authorship of the disputed "Federalist Papers."

Some current uses suggest that multivariate analysis will become increasingly valuable in geology as geologists develop experience and as statisticians devise new techniques. Studies of coal data (Gomez and Hazen, 1970) show that some of the measured chemical properties are so closely related to others that they would not have to be determined by actual assay; they could be predicted instead from those already measured. Petroleum from different producing horizons or oil fields may be classified and identified. Time and geographical area of mineralization (metallogenic eras or provinces) may be distinguished.

As with all statistical analyses of geological data, one should think carefully about the purposes before embarking on a study of multivariate data. Suppose that an investigator wishes to distinguish sandstones derived from two source areas. If one variable can be recognized that will do the job, it is better to measure it carefully than to do a multivariate analysis on many variables. More often than not, however, the problem cannot so easily be solved, and a multivariate analysis may be necessary to solve it or at least identify the pertinent variables that demand concentrated attention.

For real geological data, it is difficult enough to analyze univariate observations. The problems are compounded with multivariate observations, because of uncertainties about the number of variables actually present and the difficulties of specifying exact objectives. Clearly formulating the problems becomes even more important than with univariate data. If that is not possible, it may be better to treat the data, even if multivariate, as univariate, taking one variable at a time, at least until some preliminary understanding has been achieved. We cannot overemphasize that thoughtless trying of the techniques explained in this chapter will do nothing but run up a large computing bill and lead to frustration, confusion, and unhappiness.

10.2   GENERALIZED VARIANCE

In the analysis of variance of univariate observations, the means of two or more populations are often compared. For this comparison a measure of the underlying variability is needed; the usual measure is the population variance σ^2 as estimated by the sample variance s^2, the measure used in the F-test (sec. 5.3).

To compare means of multivariate observations, an analog to the variance σ^2 of univariate observations is needed. This analog is the *variance-covariance* matrix. If $w_1, w_2, w_3 \ldots, w_k$ is a multivariate observation with k variables, each variable w_i has a population variance $\sigma_{w_i}^2$ and each pair of variables has a population covariance definied as $\rho_{w_i w_j} \sigma_{w_i} \sigma_{w_j}$, where $\rho_{w_i w_j}$ is the population correlation coefficient of the paired variables. The variance-covariance matrix is the symmetric matrix

$$\mathbf{V} = \begin{bmatrix} \sigma_{w_1}^2 & \rho_{w_1 w_2} \sigma_{w_1} \sigma_{w_2} & \rho_{w_1 w_3} \sigma_{w_1} \sigma_{w_3} & \cdots & \rho_{w_1 w_k} \sigma_{w_1} \sigma_{w_k} \\ & \sigma_{w_2}^2 & \rho_{w_2 w_3} \sigma_{w_2} \sigma_{w_3} & & \\ & & \sigma_{w_3}^2 & & \\ & & & & \sigma_{w_k}^2 \end{bmatrix}$$

The diagonal elements of this matrix are the variances, and the off-diagonal elements are the covariances. Because the matrix is symmetric, only the upper half is displayed. The element in the ith row and the jth column is

$$\rho_{w_i w_j} \sigma_{w_i} \sigma_{w_j}.$$

If i and j are equal,

$$\rho_{w_i w_j} \sigma_{w_i} \sigma_{w_j} = \sigma_{w_i}^2$$

because $\rho_{w_i w_i}$ is equal to 1. In the corresponding matrix for sample observations, the parameters $\rho_{w_i w_j}$ and σ_{w_i} are replaced by the corresponding statistics $r_{w_i w_j}$ and s_{w_i}.

The variance-covariance matrix is illustrated through forming matrices for two fictitious populations, each consisting of five bivariate observations (table 10.1). The first population of correlated observations yields the variance-covariance matrix

$$\begin{bmatrix} 2.00 & 2.40 \\ 2.40 & 3.20 \end{bmatrix}$$

because $\sigma^2_{w_1} = 2.00$, $\sigma^2_{w_2} = 3.20$, and $\rho_{w_1 w_2} = 0.95$. However, the second population of uncorrelated observations yields the variance-covariance matrix

$$\begin{bmatrix} 2.00 & 0.00 \\ 0.00 & 0.56 \end{bmatrix}$$

because $\sigma^2_{w_1} = 2.00$, $\sigma^2_{w_2} = 0.56$, and $\rho_{w_1 w_2} = 0.00$.

TABLE 10.1. TWO FICTITIOUS POPULATIONS TO ILLUSTRATE THE VARIANCE-COVARIANCE MATRIX

Observation number	Correlated population		Uncorrelated population	
	w_1	w_2	w_1	w_2
1	1	1	1	3
2	2	1	2	2
3	3	3	3	1
4	4	5	4	2
5	5	5	5	3

The variance-covariance matrix has many uses, of which several are developed later in this chapter. Sometimes a modified matrix is used, such as the matrix of sums of squares and cross-products. Similarly, in univariate statistics, the sum of squares often replaces the sample variance.

In univariate statistics, equal variance (homoscedasticity) is often assumed (sec. 6.3); testing the correspondence of observations to this assumption requires the further assumption that the observations are normally distributed. Similarly, in multivariate statistics, equality of variance-covariance matrices is often assumed; testing the correspondence of observations to this assumption also requires the same further assumption. Fortunately, variance-covariance matrices need be only approximately equal in order to make valid comparisons of means, just as the variances in univariate statistics need be only approximately equal for this comparison.

10.3 MULTIPLE REGRESSION

Given multivariate data, one may wish to learn how the mean of a particular variable changes as the other variables change. For example, multivariate data from the Robena coal mine in Pennsylvania consisted of measurements in some 500 drill holes of the following variables: ash, sulfur (total), organic sulfur, and sulfur in pyrite. These variables were measured on samples taken at different strata within the seam, on different size fractions of coal, and on float and sink fractions. The principal use made of the data was to investigate the behavior of the variable sulfur to learn whether it could be predicted by these other variables, some of which may be easier to recover in drill core or to measure in the chemical laboratory than the sulfur itself. Sulfur is important because it must be cleaned from the coal to make a marketable product. Similarly, the concentration of gold in ore may be related to many variables (Pretorius, 1966).

In multiple regression one variable, designated w, is selected as the dependent variable, while the remaining variables, designated $x_1, x_2, ..., x_k$, are the independent variables. The designation of w is made arbitrarily to suit the investigator's purpose. The linear statistical model is

$$w = \alpha + \beta_1 x_1 + \beta_2 x_2 + \cdots + \beta_k x_k + e,$$

where α is the general coefficient and β_1 to β_k are coefficients for the independent variables. The coefficients must then be estimated.

The problem is similar to those already discussed in Chapter 9 for trend-surface analysis, in which a variable such as silver content was measured at various places and the change in location of the mean of this variable was estimated. Indeed, trend-surface analysis is nothing more than a special case of multiple regression, which was introduced separately in Chapter 9 for convenient exposition, although in statistics the two topics are one. Therefore the independent variables x, y, x^2, y^2, xy, etc. could have been designated instead as x_1, x_2, x_3, x_4, x_5, etc.

Two basically different conceptual frames are used in multiple regression. In the first, the variables are interrelated by a reasonably well-known theory so that those to be included in the model are recognized, and the possible need to transform any of them is understood. In this frame, *multiple linear regression*, as explained in the next subsection, is used to estimate the coefficients and the residual variance. In the second frame, the variables are known to be interrelated somehow, but it is not clear which ones should be included to express the relationship, nor is it clear which variables should be transformed and by what transformations. Therefore the exploratory method of *stepwise multiple linear regression*, explained in a subsequent subsection, is appropriate, for not only must the coefficients be estimated but the terms to include must be found. In this kind of regression, variables are introduced into the linear model one at a time in order of their evident importance.

As the name implies, multiple linear regression, whether stepwise or not, is only for linear models. For most geological problems, in the present state of knowledge, this is certainly good enough. However, as geologists become more adept in using models, they may increasingly require nonlinear ones (James, 1967). Because nonlinear models require special techniques that must be carefully applied, they are not discussed in this book; the reader is referred to Draper and Smith (1966).

Multiple Linear Regression

Multiple linear regression is a method for estimating how the mean of the dependent variable w changes as the values of the independent variables α and $x_1, x_2, \cdots, x_k$ change. In the equation

$$w = \alpha + \beta_1 x_1 + \beta_2 x_2 + \cdots + \beta_k x_k + e,$$

relating these variables, the random fluctuations e are assumed to have a mean μ_e of zero, to have a variance σ_e^2 equal to an unknown constant, and to be statistically independent. The parameters are estimated by the least-squares technique (sec. 9.2) to find values of $a, b_1, b_2, \cdots, b_k$ that minimize the expression

$$\sum (w - [a + b_1 x_1 + b_2 x_2 + \cdots + b_k x_k])^2.$$

The arithmetic details are similar to those explained in section 9.4 for two independent variables (fitting a plane) and for five independent variables (fitting a quadratic surface).

When the previous expression is differentiated with respect to a and $b_1, b_2, \ldots, b_k$, and each derivative is set equal to zero, the following linear equations, $(k+1)$ in number, are obtained:

$$na + b_1 \Sigma x_1 + b_2 \Sigma x_2 + \cdots + b_k \Sigma x_k = \Sigma w,$$
$$a \Sigma x_1 + b_1 \Sigma x_1^2 + b_2 \Sigma x_1 x_2 + \cdots + b_k \Sigma x_1 x_k = \Sigma w x_1,$$
$$\vdots$$
$$a \Sigma x_k + b_1 \Sigma x_k x_1 + b_2 \Sigma x_k x_2 + \cdots + b_k \Sigma x_k^2 = \Sigma w x_k.$$

The constant a is next eliminated. As in linear regression (sec. 9.2), a can be expressed by the equation

$$a = \overline{w} - (b_1 \overline{x}_1 + b_2 \overline{x}_2 + \cdots + b_k \overline{x}_k).$$

When a is eliminated from the above equations by algebraic manipulations, the $(k+1)$ linear equations are reduced to the following linear equations, k in number:

$$b_1 \, \mathrm{SS}_{x_1 x_1} + b_2 \, \mathrm{SS}_{x_1 x_2} + \cdots + b_k \, \mathrm{SS}_{x_1 x_k} = \mathrm{SS}_{w x_1},$$
$$b_1 \, \mathrm{SS}_{x_2 x_1} + b_2 \, \mathrm{SS}_{x_2 x_2} + \cdots + b_k \, \mathrm{SS}_{x_2 x_k} = \mathrm{SS}_{w x_2},$$
$$\vdots$$
$$b_1 \, \mathrm{SS}_{x_k x_1} + b_2 \, \mathrm{SS}_{x_k x_2} + \cdots + b_k \, \mathrm{SS}_{x_k x_k} = \mathrm{SS}_{w x_k}.$$

The number of equations is reduced because it is easier to solve k rather than $(k+1)$ equations and also because writing the equations in terms of the sums of squares and cross-products about the means provides fewer arithmetic difficulties for an electronic computer. These k equations are solved for the values of the b coefficients, which are substituted into the first equation of the set of $(k+1)$ equations to obtain a.

In matrix notation, these equations for the values of the b coefficients may be represented as

$$\mathbf{S} \cdot \mathbf{b} = \mathbf{P},$$

and the solution as

$$\mathbf{b} = \mathbf{S}^{-1} \cdot \mathbf{P},$$

where $\mathbf{P}$ is a $k \times 1$ matrix of the sums of products of the w- and x-values, $\mathbf{S}$ is a $k \times k$ matrix of the sums and products of the x-values, and $\mathbf{b}$ is a $k \times 1$ matrix of the coefficients $b_1, b_2, \ldots, b_k$. The regression sum of squares is obtained by evaluating the product

$$\mathbf{b}' \cdot \mathbf{P},$$

where $\mathbf{b}'$ is the transpose of $\mathbf{b}$. The residual sum of squares is obtained by subtracting the regression sum of squares from SS_{ww}.

For the n data points, the variation in the dependent variable w has $(n-1)$ degrees of freedom. The total sum of squares can be partitioned into two parts: a regression sum of squares with k degrees of freedom, explained by the variables $x_1, x_2, \ldots, x_k$, and a residual sum of squares with $(n-k-1)$ degrees of freedom. The analysis of variance has the following format, in which the residual mean square estimates the variance of the random fluctuation e:

Source of variation	Sum of squares	Degrees of freedom	Mean square	F	$F_{10\%}$
Regression		k			
Residual variation		$n-k-1$			

Stepwise Multiple Linear Regression

Being an exploratory method, stepwise multiple linear regression usually furnishes *indications* (sec. 4.9) that suggest further investigations. One dependent variable w and a set of independent variables $x_1, x_2, \ldots, x_k$ are selected. The independent variables, unlimited in number, may be observed or may be derived by transformations.

The regression is then carried out in an iterative fashion. Each of the independent variables is examined in turn and the linear regression of the variable w with it is calculated. Then the variable with the largest regression sum of squares is chosen as the *first entering variable*. Next a residual from $\hat{w}$ is calculated for each w observation, and each residual becomes an observation of the new dependent variable, still designated as w. For each x observation, a residual from $\hat{x}$ is calculated and becomes the new x-value. Thus the linear regression of the first entering variable is removed from each of the other independent variables as well as from the dependent variable. This calculation is made by treating each x variable as a dependent variable and by calculating a residual by using the regression equation. The residuals define $(k-1)$ new independent variables, the first entering variable being omitted.

The iterative process is repeated until stopped by applying one of the following rules:

1. The reduction in the sum of squares effected by the entering variable is so small that the ratio of the regression sum of squares of the entering variable to the residual mean square is less than the 10 percent point of the F-statistic with one and infinity degrees of freedom.

2. The regression sum of squares for the entering variable becomes equal to or less than the residual mean square; that is, the F-statistic computed by dividing the regression mean square by the residual mean square becomes equal to or less than 1.

3. The degrees of freedom of the residual sum of squares becomes equal to zero because $(n-1)$ regression coefficients have been removed, where n is the number of observations.

For most problems the first rule is the most reasonable one to apply, because all variables are removed that appear to be meaningful in the context of several hypothesis tests. The second rule removes each variable whose contribution to the current sum of squares is at least as large as the residual mean square, which is a kind of error variance. The last rule allows all variables to enter the model.

More than in most statistical methods, the numerical results of stepwise linear regression depend on details of computing. The computer programs (sec. 17.2) can introduce one or more variables at a time and can remove variables previously introduced. Moreover, the first variable or variables introduced, and the variables introduced in intermediate stages, may be different with different programs. Only if all variables are introduced are the final results identical (aside from round-off and similar errors) regardless of the program.

Observations may be transformed or combined to derive new variables before a regression analysis is made. The dependent variable is often transformed. For instance, a lognormal transformation of gold observations often causes the residuals to have a more nearly constant variance. Independent variables may be transformed to see if some function of them is a more sensitive predictor than the original variable. It is always desirable to scrutinize the residuals. If one or a few of them are surprisingly large, they may indicate mistakes in the data, interesting anomalies, or unexpected clusters that point to nonrandom observations or a poor fit of the model.

As in trend-surface regression (sec. 9.4), if replicate observations of the dependent variable are available, comparison of the replication mean square with the residual mean square will indicate how good the fit is.

Example of Multiple Linear Regression

For an example of multiple linear regression in geology, we use petrographic data from the Lacorne, La Motte, and Preissac granitic complex, Canada, published by Dawson and Whitten (1962) for a study in trend-surface analysis. The purpose is not to comment on Dawson and Whitten's excellent paper, but rather, for a convenient set of data, to demonstrate how specific gravity can be predicted from the other variables by multiple linear regression.

Dawson and Whitten (1962, p. 26) write that "the Lacorne, La Motte, and Preissac granitic complex constitutes one lithologic unit within the structurally complex greenstone belt at the south side of the Superior Province of the Canadian shield . . . in north-western Quebec." From their analysis, they conclude (p. 1) that "marked patterns of variation emerge; the more clear-cut patterns relate to massifs with mafic- and potash feldspar-rich centers, and quartzose peripheries."

The original data were measured on specimens taken by Dawson. "The collecting procedure was ideally to collect one 'representative' specimen at mile-grid intersections where outcrops are abundant, and from all outcrop areas where drift cover is extensive. The samples were hammered from available edges free of xenoliths, pegmatites, and veins, and they were trimmed to remove weathered and obviously inhomogeneous material" (Dawson and Whitten, 1962, p. 13). The original multivariate observations (Dawson and Whitten's table 4, p. 14) consist of the seven variables: quartz percentage, color index (percentage of dark-colored silicate minerals), ratio of potash feldspar to plagioclase, total feldspar percentage, specific gravity, north-south geographic coordinate, and east-west geographic coordinate. All seven variables were measured for 89 of the data points.

To learn if specific gravity could be predicted from the other variables by multiple regression, we randomly selected one-half of the observations. Each of the 44 observations listed in table 10.2 contains six of Dawson and Whitten's

TABLE 10.2. RANDOM SELECTION OF 44 OBSERVATIONS FOR SIX VARIABLES FROM
THE LACORNE, LA MOTTE, AND PREISSAC GRANITIC COMPLEX, QUEBEC, CANADA*

Variable	Quartz (%)	Color index (%)	Total feldspar (%)	Specific gravity	North–south coordinate	East–west coordinate
Our notation	x_1	x_2	x_3	w	U	V
Dawson and Whitten's notation	x_1	x_2	x_4	x_5	U	V
	21.3	5.5	73.0	2.63	0.920	6.090
	38.9	2.7	57.4	2.64	1.150	3.625
	26.1	11.1	62.6	2.64	1.160	6.750
	29.3	6.0	63.6	2.63	1.300	3.010
	24.5	6.6	69.1	2.64	1.400	7.405
	30.9	3.3	65.1	2.61	1.590	8.630
	27.9	1.9	69.1	2.63	1.750	4.220
	22.8	1.2	76.0	2.63	1.820	2.420
	20.1	5.6	74.1	2.65	1.830	8.840
	16.4	21.3	61.7	2.69	1.855	10.920
	15.0	18.9	65.6	2.67	2.010	14.225
	0.6	35.9	62.5	2.83	2.040	10.605
	18.4	16.6	64.9	2.70	2.050	8.320
	19.5	14.2	65.4	2.68	2.210	8.060
	34.4	4.6	60.7	2.62	2.270	2.730
	26.9	8.6	63.6	2.63	2.530	3.500
	28.7	5.5	65.8	2.61	2.620	7.445
	28.5	3.9	67.8	2.62	3.025	5.060
	38.4	3.0	57.6	2.61	3.060	5.420
	28.1	12.9	59.0	2.63	3.070	12.550
	37.4	3.5	57.6	2.63	3.120	12.130
	0.9	22.9	74.4	2.78	3.400	15.400
	8.8	34.9	55.4	2.76	3.520	9.910
	16.2	5.5	77.6	2.63	3.610	11.520
	2.2	28.4	69.3	2.74	4.220	16.400
	29.1	5.1	65.7	2.64	4.250	11.430
	24.9	6.9	67.8	2.70	4.940	5.910
	39.6	3.6	56.6	2.63	5.040	1.840
	17.1	11.3	70.9	2.71	5.060	11.760
	0.0	47.8	52.2	2.84	5.090	16.430
	19.9	11.6	67.2	2.68	5.240	11.330
	1.2	34.8	64.0	2.84	5.320	8.780
	13.2	18.8	67.4	2.74	5.320	13.730
	13.7	21.2	64.0	2.74	5.330	12.450
	26.1	2.3	71.2	2.61	5.350	1.430
	19.9	4.1	76.0	2.63	5.610	4.150

Table 10.2. Continued

Variable	Quartz (%)	Color index (%)	Total feldspar (%)	Specific gravity	North–south coordinate	East–west coordinate
	4.9	18.8	74.3	2.77	5.850	13.840
	15.5	12.2	69.7	2.72	6.460	11.660
	0.0	39.7	60.2	2.83	6.590	14.640
	4.5	30.5	63.9	2.77	7.260	12.810
	0.0	63.8	35.2	2.92	7.420	16.610
	4.0	24.1	71.8	2.77	7.910	14.650
	23.4	12.4	63.1	2.79	8.470	13.330
	29.5	9.8	60.4	2.69	8.740	15.770

* Data of Dawson and Whitten, 1962.

Table 10.3. Numerical results of multiple linear regression to estimate specific gravities of rocks from a Quebec granitic complex

Part A: Four equations in four unknowns from differentiation:
$$118.380 = 44.0a + 925.6b_1 + 624.2b_2 + 2821.8b_3$$
$$2252.148 = 925.6a + 22410.73b_1 + 6830.43b_2 + 55132.93b_3$$
$$1830.762 = 624.2a + 6830.43b_1 + 18351.35b_2 + 40710.79b_3$$
$$7697.565 = 2821.8a + 55132.93b_1 + 40710.79b_2 + 188423.27b_3$$

Part B: Three equations in three unknowns after elimination of a:
$$-34.5198 = 6040.4644b_1 - 5963.7225b_2 - 42.2144b_3$$
$$43.6207 = -5963.7225b_1 + 8352.1025b_2 - 2411.2476b_3$$
$$-9.5322 = -42.2144b_1 - 2411.2476b_2 + 2458.2640b_3$$

Part C: Inverse matrix of b-coefficients:
$$\begin{bmatrix} 0.0720 & & \\ 0.0689 & 0.0669 & \\ 0.0711 & 0.0684 & 0.0708 \end{bmatrix}$$

Part D: Three b-values:
$$-0.01882$$
$$-0.01316$$
$$-0.01711$$

Regression sum of squares: 0.2389476.

variables (feldspar ratio omitted) in a notation different from theirs. If the sum of the three mineralogical variables—quartz, color index, and feldspar—were exactly 100 percent for all 44 observations, two variables at most would be independent (sec. 11.4) and only two variables could be used in the regression. However, all three variables could be used because they were calculated independently and their sum for most of the observations was far enough from a constant 100 percent to indicate that all of them contained distinct information.

The prediction is tested by applying the resulting regression equation to the other one-half of the observations. However, for reasons that are discussed later, a perfect prediction is not expected.

Through multiple regression, we obtained the four equations in four unknowns listed in part A of table 10.3 and, by eliminating the constant term a, the three equations in part B. In part C of the table is given the inverse matrix of the coefficients of the b-values, whose solution, in part D, concludes the necessary algebra. To evaluate the success of the regression, we made the analysis of variance in table 10.4, as performed in Chapter 9, and found that the fit was excellent, the residual mean square of 0.0008 corresponding to a standard deviation of only about 0.03.

TABLE 10.4. ANALYSES OF VARIANCE FOR MULTIPLE LINEAR REGRESSION TO ESTIMATE SPECIFIC GRAVITIES OF ROCKS FROM A QUEBEC GRANITIC COMPLEX

Source of variation	Sum of squares	Degrees of freedom	Mean square	F	$F_{10\%}$
Regression	0.2389	3	0.0796	99	2.23
Residual variation	0.0321	40	0.0008		

TABLE 10.5. COMPARISON OF PREDICTED AND OBSERVED SPECIFIC GRAVITIES OF ROCKS FROM A QUEBEC GRANITIC COMPLEX

Specific gravity				
Observed	Predicted	Residual	U	V
2.63	2.63214	−0.00214	1.010	6.510
2.62	2.60632	0.01368	1.165	5.400
2.62	2.62941	−0.00941	1.230	7.550
2.63	2.63890	−0.00890	1.350	6.700
2.62	2.64705	−0.02705	1.805	2.700
2.65	2.65919	−0.00919	1.810	8.500

TABLE 10.5. CONTINUED

| | Specific gravity | | | |
Observed	Predicted	Residual	U	V
2.63	2.62985	0.00015	1.960	5.360
2.61	2.62391	− 0.01391	2.030	7.550
2.73	2.67471	0.05529	2.130	13.650
2.63	2.62056	0.00944	2.130	11.600
2.75	2.73194	0.01806	2.250	14.470
2.63	2.63439	− 0.00439	2.300	4.405
2.78	2.73408	0.04592	2.400	15.295
2.89	2.91418	− 0.02418	2.550	10.860
2.63	2.63328	− 0.00328	2.890	11.730
2.65	2.64788	0.00212	2.900	10.800
2.73	2.75971	− 0.02971	3.020	16.260
2.76	2.75971	0.00029	3.020	16.260
2.63	2.61313	0.01687	3.190	12.840
2.74	2.65704	0.08296	3.600	7.950
2.87	2.89606	− 0.02606	3.820	8.640
2.78	2.76403	0.01597	3.840	15.890
2.69	2.65511	0.03489	3.930	19.160
2.67	2.71193	− 0.04193	4.040	8.800
2.75	2.75577	− 0.00577	4.300	13.240
2.62	2.63741	− 0.01741	4.540	1.350
2.63	2.63737	− 0.00737	4.560	0.600
2.75	2.76521	− 0.01521	4.600	16.920
2.62	2.61959	0.00041	4.790	1.060
2.64	2.64196	− 0.00196	4.890	6.910
2.63	2.62681	0.00319	4.900	2.270
2.62	2.65423	− 0.03423	4.900	5.545
2.73	2.77794	− 0.04794	4.910	8.710
2.62	2.63489	− 0.01489	4.950	3.890
2.75	2.89733	− 0.14733	5.000	8.560
2.62	2.65588	− 0.03588	5.190	1.090
2.62	2.64774	− 0.02774	5.760	2.400
2.73	2.67386	0.05614	5.880	7.450
2.64	2.63309	0.00691	5.900	5.670
2.71	2.73615	− 0.02615	6.040	6.390
2.66	2.67931	− 0.01931	6.330	3.070
2.75	2.81388	− 0.06388	6.700	13.710
2.81	2.80375	0.00625	7.250	14.600
2.77	2.78383	− 0.01383	8.120	13.680

To test how well the regression equation predicted specific gravity, we applied the coefficients to the other one-half of the observations and obtained the results in table 10.5. For each observation the measured values of quartz, color index, and total feldspar were multiplied by the regression coefficients to yield predicted values for specific gravity, which were compared with the observed values by calculating residuals. One way to appraise the prediction was to test the hypothesis H_0, that the population mean of the differences between observations and residuals is equal to zero, against the alternative hypothesis H_1, that it is greater or less than zero. When this test was performed, with t as the test statistic, the computed calue of t was -1.32. Because this value is outside the critical region defined by the tabled t-value of 1.68 (10 percent significance level, two-tailed t), we accepted hypothesis H_0.

Although the over-all prediction was excellent, table 10.5 lists large residuals for observations with U-values of 3.600 and 5.000 and rather large residuals for U-values of 2.130, 5.880, and 6.700. The sizes of these five residuals suggest a faulty measurement of one or more variables or some peculiarity in mineralogy. If one had the specimens, they should be examined again.

Although the prediction is excellent, it is worth considering some of the causes that make it imperfect. Most obviously, because the rocks presumably contained accessory minerals, the sum of quartz, color index, and total feldspar should be less than 100 percent, and surely could not exceed 100 percent; actually, however, because of measurement errors, the sum for some specimens is greater than 100 percent. The other causes are mineralogical. The color index includes various minerals with different specific gravities; the feldspar varies in composition and therefore in specific gravity; and the quantities and specific gravities of accessory minerals are not accounted for. In addition, voids are doubtless present in some specimens.

It is interesting to perform a stepwise regression on the same set of data. While we expected the geographic variables U and V to have little or no

TABLE 10.6. COEFFICIENTS OBTAINED IN STEPWISE MULTIPLE REGRESSION TO ESTIMATE SPECIFIC GRAVITIES OF ROCKS FROM A QUEBEC GRANITIC COMPLEX

			Coefficients			
Step	a	b_1	b_2	b_3	b_4	b_5
1	2.6156		0.00522			
2	2.5894		0.00467		0.00879	
3	2.6401	-0.00164	0.00354		0.00825	
4	4.0065	-0.01528	-0.01014	-0.01376	0.00767	
5	4.0605	-0.01584	-0.01058	-0.01426	0.00802	-0.00060

relation to specific gravity, we included these variables with the mineralogical ones in a stepwise regression. Table 10.6 shows how the five independent variables entered stepwise into the regression: first color index, followed in turn by U, quartz, total feldspar, and finally V. The coefficients changed unpredictably as the variables entered.

TABLE 10.7. ANALYSES OF VARIANCE FOR FIVE STEPS OF MULTIPLE REGRESSION TO ESTIMATE SPECIFIC GRAVITY OF ROCKS FROM A QUEBEC GRANITIC COMPLEX

Step	Source of variation	Sum of squares	Degrees of freedom	Mean square	F	$F_{10\%}$
1	Regression	0.22782	1	0.22782	221	2.84
	Residual variation	0.04326	42	0.00103		
2	Regression	0.24091	2	0.12046	163	2.44
	Regression added at step 2	0.01237	1	0.01237	17	2.84
	Residual variation	0.03017	41	0.00074		
3	Regression	0.24562	3	0.08187	128	2.23
	Regression added at step 3	0.00471	1	0.00471	7	2.84
	Residual variation	0.02546	40	0.00064		
4	Regression	0.24864	4	0.06216	107	2.09
	Regression added at step 4	0.00302	1	0.00302	5	2.84
	Residual variation	0.02244	39	0.00058		
5	Regression	0.24878	5	0.04976	84	2.00
	Regression added at step 5	0.00014	1	0.00014	0.2	2.84
	Residual variation	0.02230	38	0.00058		

Table 10.7 gives the five analyses of variance for the five steps of the regression. After the first step, the size of the residual mean square is reduced very slowly. The criteria of the previous subsection may be used to decide where the regression should have been stopped. Under the 10 percent significance criteria, it would have been stopped at step 4 (the F-value of 0.2 at step 5 is smaller than 2.84). With the criterion that reduction in regression mean square becomes less than reduction in residual mean square, it would also have been stopped at step 4 (0.00014 < 0.00058). The residual mean square increases between steps 4 and 5, because inclusion of the last entering variable V causes only a small reduction in the regression sum of squares.

10.4 GENERALIZED ANALYSIS OF VARIANCE

In univariate analysis of variance, the basic principle is to partition the total sum of squares into a part that measures variation among groups and a part that measures variation within groups. Thus (sec. 5.3)

total SS = among-group SS + within-group SS.

In multivariate analysis of variance, three multiples of the variance-covariance matrix play the same roles as these three sums of squares. The corresponding relationship is

$$\mathbf{T} = \mathbf{A} + \mathbf{W},$$

where $\mathbf{T}$ is the matrix of the sums of squares and cross-products of the observations grouped into a single sample. $\mathbf{A}$ is the matrix of the among-group sums of squares and cross-products, and $\mathbf{W}$ is the matrix of the within-group sums of squares and cross-products.

In univariate analysis of variance, the among-sample sum of squares is compared with the within-sample sum of squares by the F-test. Alternatively, the within-sample sum of squares could be compared with the total sum of

TABLE 10.8. NOTATION FOR GROUPS OF MULTIVARIATE OBSERVATIONS

		Variable number	
Group	Observation number	1	2
	1	w_{111}	w_{121}
	2	w_{211}	w_{221}
1	3	w_{311}	w_{321}
	4	w_{411}	w_{421}
	5	w_{511}	w_{521}
	1	w_{112}	w_{122}
	2	w_{212}	w_{222}
2	3	w_{312}	w_{322}
	4	w_{412}	w_{422}
	5	w_{512}	w_{522}
	1	w_{113}	w_{123}
	2	w_{213}	w_{223}
3	3	w_{313}	w_{323}
	4	w_{413}	w_{423}
	5	w_{513}	w_{523}

squares through a statistic algebraically related to the F-statistic. This alternative formulation is paralleled in multivariate analysis of variance by comparing the within-group sum-of-squares matrix with the total sum-of-squares matrix. The analysis requires that, for each group, the number of observations exceed the number of variables. The rather complicated algebra is given by Cooley and Lohnes (1962, p. 62).

Digital computer programs are needed to calculate most real problems. Cooley and Lohnes devised the program explained in this book. In order to illustrate the calculations, a fictitious numerical example is analyzed, and the results of an analysis of petrographic data are given. First, table 10.8 indicates a system of notation for groups of multivariate observations. Each variable w_{ijm} has three subscripts in order to denote the number of observations within a group, the number of variables within an observation, and the number of groups. The number of observations within a group ranges from $i = 1, n$; the number of variables within an observation ranges from $j = 1, k$; and the number of groups ranges from $m = 1, g$.

TABLE 10.9. BIVARIATE OBSERVATIONS FOR
FICTITIOUS ILLUSTRATION OF GENERALIZED
ANALYSIS OF VARIANCE

Group	Observation	
	w_1	w_2
	0	1
	2	1
1	4	2
	4	3
	5	3
	2	2
	4	2
2	6	3
	6	4
	7	4
	1	3
	3	3
3	5	4
	5	5
	6	5

The fictitious data of table 10.9 consists of three groups with five observations per group and two variables per observation; the means and variances are given in table 10.10. The numerical values for the matrices,

$$\mathbf{T} = \mathbf{A} + \mathbf{W},$$

are

$$\begin{bmatrix} 58 & 26 \\ 26 & 22 \end{bmatrix} = \begin{bmatrix} 10 & 5 \\ 5 & 10 \end{bmatrix} + \begin{bmatrix} 48 & 21 \\ 21 & 12 \end{bmatrix}.$$

TABLE 10.10. MEANS AND VARIANCES OF THE THREE GROUPS OF BIVARIATE OBSERVATIONS FROM TABLE 10.9

Group	Mean, $\overline{w}_i$		Variance, $s^2_{w_i}$	
	w_1	w_2	$s^2_{w_1}$	$s^2_{w_2}$
1	3	2	4	1
2	5	3	4	1
3	4	4	4	1

TABLE 10.11. VALUES OF SOME VARIABLES FOR FICTITIOUS GENERALIZED ANALYSIS OF VARIANCE

Item	Value
$\mid\mathbf{W}\mid$	135
$\mid\mathbf{T}\mid$	600
Δ	0.225
g	3
N_g	5,5,5
N	15
p	2
s	2
q	2
m	11.5
n	14
λ	-0.5
r	2
y	0.52
F	6.10
d.f. for numerator	4
d.f. for denominator	22
$F_{10\%}$	2.22

The pooled within-group variance-covariance matrix is calculated from W by dividing the elements of W by the number of degrees of freedom. For these data, the number of degrees of freedom is 12 (3×4), so that the variance-covariance matrix is

$$\begin{bmatrix} 4.00 & 1.75 \\ 1.75 & 1.00 \end{bmatrix}.$$

An F-statistic to test the hypothesis H_0 that the three samples come from populations with the same mean is obtained from the values in table 10.11, which are presented in order to provide the reader with intermediate results produced by Cooley and Lohnes' computer program. When the calculated F-value of 6.10 with 4 and 22 degrees of freedom is compared with the tabled $F_{10\%}$-value of 2.22, the hypothesis H_0 is rejected, and the conclusion is reached that the population means are different. If the hypothesis H_0 is true, this F-statistic approximately follows the F-distribution with the indicated numbers of degrees of freedom.

To illustrate the generalized analysis of variance for real data, we use rock analyses published by Washington (1917). The rocks are granites, nepheline syenites, and gabbros. There are three groups, ten observations in each group, and six variables per observation. Table 10.12 gives the means, and table 10.13 gives the pooled within-group variance-covariance matrix. As before, the

TABLE 10.12. MEANS OF THREE GROUPS OF GRANITES, NEPHELINE SYENITES, AND GABBROS*

| | \multicolumn{6}{c}{Oxide means} | | | | | |
Rock	SiO_2	Al_2O_3	MgO	CaO	Na_2O	K_2O
Granite	72.34	13.40	0.53	1.55	3.12	5.21
Nepheline syenite	56.06	22.22	0.61	1.10	9.20	5.11
Gabbro	49.34	15.26	7.96	10.79	2.28	0.83

* From Washington (1917), U.S. Geol. Surv. Prof. Paper 99.

TABLE 10.13. POOLED WITHIN-GROUP VARIANCE–COVARIANCE MATRIX FOR ROCK DATA OF TABLE 10.12

	SiO_2	Al_2O_3	MgO	CaO	Na_2O	K_2O
SiO_2	6.1540					
Al_2O_3	0.4772	2.3020				
MgO	-0.1724	-0.3668	1.3271			
CaO	-0.7224	0.6864	-0.2165	1.2660		
Na_2O	0.3366	0.1806	-0.0715	-0.3233	0.3201	
K_2O	-0.1195	-0.6907	-0.0290	-0.1768	-0.1620	0.9123

hypothesis H_0 is that the means of the three groups are the same. The calculated F-value of 194 with 12 and 44 degrees of freedom is compared with a tabulated $F_{10\%}$-value of 1.70. The hypothesis H_0 is therefore rejected, and we conclude that the means are different. Inspection of the means, table 10.12, suggests that this conclusion could have been reached from an analysis of fewer than the six variables, as is indeed true.

10.5 DISCRIMINANT ANALYSIS

Discriminant analysis is a set of methods for classifying multivariate observations into two or more groups through developing one or more linear combinations (sec. 5.8) of the variables that comprise the observations. In principle, the directions and lengths of the distances connecting ellipsoid centers (fig. 10.6) are examined.

Two methods of discriminant analysis are explained in this section. The first is a method of classifying groups of observations; the second is a method of classifying a single observation into one of two predefined groups.

Classification of Groups of Multivariate Observations

In this subsection a procedure of linear discriminant analysis to distinguish groups of multivariate observations by comparing paired groups is explained. The *generalized t-statistic*, also named *Mahalanobis' d^2-statistic*, is calculated. The approach is analogous to proceeding directly to single-degree-of-freedom tests in the analysis of variance of univariate data (sec. 5.9), skipping the general F-test. The d^2-statistic is closely related to the discriminant analysis first introduced by Fisher (1938) and is equivalent to Hotelling's T^2 (Wilks, 1962, p. 557). The statistic is introduced through an explanation of the distance concept, and examples are given of its use.

THE DISTANCE CONCEPT. The distance concept may be introduced by considering two populations, each with the same standard deviation. The simple difference of population means $(\mu_{w_1} - \mu_{w_2})$ is insufficient to describe the distance between the two populations, because the inherent variability within each population, as measured by the standard deviation, must also be considered. An example is given in figure 10.8; the numerical difference of 3 between the two means μ_{w_1} and μ_{w_2} is the same in cases A and B; however, the data of case A are distinctly different, whereas those of case B are close together. Cases A and B may be contrasted by dividing the simple difference of the means by the standard deviation to give the distance measure

$$\frac{|\mu_{w_1} - \mu_{w_2}|}{\sigma}.$$

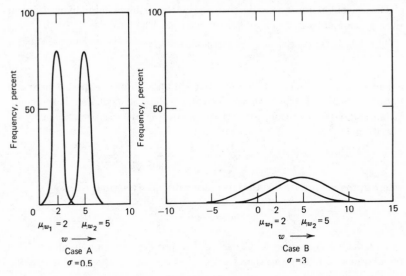

Fig. 10.8. Frequency distributions for paired samples from two hypothetical cases to illustrate the distance concept in discriminant analysis.

With this measure, the distances for cases A and B are no longer both 3, but become 6 and 1, respectively, corresponding to an intuitive assessment of the situation.

Next this concept of distance is developed from univariate distance through bivariate distance to multivariate distance. In univariate analysis the problem of whether different statistical samples come from populations with the same mean is attacked by calculating a statistic such as Student's t. For two samples, the basic measure of the difference between sample means is a distance

$$d = \frac{\overline{w}_1 - \overline{w}_2}{s},$$

where $\overline{w}_1$ and $\overline{w}_2$ are the sample means and s is the estimated standard deviation. The sample sizes, n_1 and n_2, in the formula for Student's t (sec. 4.3),

$$t = \frac{\overline{w}_1 - \overline{w}_2}{s\sqrt{(1/n_1) + (1/n_2)}},$$

are used only to measure the reliability of d. This value d, the ratio of the difference of the two means to the estimated standard deviation of the underlying population, is the dimensionless distance between two means drawn from populations with unit variance.

The concept of univariate distance is presented in more detail by writing out the algebra and by giving a numerical example. The formula

$$u = \frac{\overline{w}_1 - \overline{w}_2}{\sigma \sqrt{(1/n_1)+(1/n_2)}},$$

where u is a statistic that follows a distribution with mean equal to zero and variance equal to 1, where $\overline{w}_1$ and $\overline{w}_2$ are sample means, where σ is the population standard deviation, and where n_1 and n_2 are sample sizes, may be rewritten as

$$|u| = d \frac{1}{\sqrt{(1/n_1)+(1/n_2)}},$$

where d, the univariate distance, is obtained by the relation

$$d = \frac{|\overline{w}_1 - \overline{w}_2|}{\sigma}.$$

The statistic d is simply the Cartesian distance between sample means from populations with unit variance.

Fig. 10.9. One-dimensional distance.

Figure 10.9 presents a numerical example illustrating univariate distance. If w is the unit in which the observations are made and σ is the population standard deviation, a dimensionless quantity, w', may be defined as

$$w' = \frac{w}{\sigma}.$$

In figure 10.9 the points $\overline{w}_1'$ and $\overline{w}_2'$ represent two samples with means of 15 and 25 from a population with σ equal to 10. The dimensionless distance between these two points is 1, exactly the number calculated by the formula

$$d = \frac{|\overline{w}_1 - \overline{w}_2|}{\sigma}.$$

In practice, σ, usually unknown, is replaced by an estimate of the standard deviation, s. Then the distance, d, is estimated by the ratio $|\overline{w}_1 - \overline{w}_2|/s$; and, with introduction of the sample sizes, n_1 and n_2, the formula for Student's t is obtained:

$$|t| = \frac{|\overline{w}_1 - \overline{w}_2|}{s \sqrt{(1/n_1)+(1/n_2)}}.$$

TABLE 10.14. FICTITIOUS UNCORRELATED BIVARIATE DATA

Item	Sample 1	Sample 2
$\overline{w}_1$	10	50
$\overline{w}_2$	20	80
$\overline{w}_1'$	1	5
$\overline{w}_2'$	1	4

Assumptions for the data: $\sigma_{w_1} = 10$, $\sigma_{w_2} = 20$, $\rho_{w_1 w_2} = 0$.

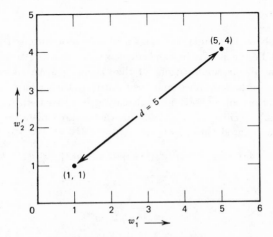

Fig. 10.10. Two-dimensional distance.

The concept of univariate distance between two sample means may be generalized for multivariate data. For the special case of bivariate data presented in this subsection, the subcase of two independent variables is given first, followed by the subcase of two correlated variables.

For the bivariate uncorrelated subcase, two variables, w_1 and w_2, are measured for each sample. Then $w_1' = w_1/\sigma_{w_1}$; $w_2' = w_2/\sigma_{w_2}$; and, because w_1 and w_2 are uncorrelated, the correlation coefficient $\rho_{w_1 w_2} = 0$, and their covariance is zero. Table 10.14 and figure 10.10 present a numerical example. The Cartesian distance, d, between the two sample points $(1, 1)$ and $(5, 4)$ is 5, as calculated by the familiar distance formula

$$d = \sqrt{(\overline{w}_{11}' - \overline{w}_{12}')^2 + (\overline{w}_{21}' - \overline{w}_{22}')^2},$$

$$= \sqrt{(1-5)^2 + (1-4)^2},$$

$$= 5,$$

where the second subscripts refer to samples 1 and 2. Alternatively, d may be calculated by the formula

$$d = \sqrt{\frac{(\overline{w}_{11} - \overline{w}_{12})^2}{\sigma_{w_1}^2} + \frac{(\overline{w}_{21} - \overline{w}_{22})^2}{\sigma_{w_2}^2}}.$$

If the numerical values from table 10.14 are substituted in this formula, the result is again

$$d = \sqrt{\frac{(10 - 50)^2}{100} + \frac{(20 - 80)^2}{400}},$$

$$= 5.$$

For the bivariate correlated case, the simple procedure outlined above is insufficient. Before it can be applied, new uncorrelated variables must be calculated as linear combinations of the original correlated variables. The calculations are first outlined and then illustrated by a numerical example. Taking a square root is avoided by calculating d^2 rather than d.

The formula (Wilks, 1962, p. 556) for the squared distance between two samples of correlated variables is

$$d^2 = a(\overline{w}_{11} - \overline{w}_{12})^2 + 2b(\overline{w}_{11} - \overline{w}_{12})(\overline{w}_{21} - \overline{w}_{22}) + c(\overline{w}_{21} - \overline{w}_{22})^2,$$

where

$$a = \frac{1}{\sigma_{w_1}^2(1 - \rho_{w_1 w_2}^2)},$$

$$b = \frac{-\rho_{w_1 w_2}}{\sigma_{w_1}\sigma_{w_2}(1 - \rho_{w_1 w_2}^2)},$$

and

$$c = \frac{1}{\sigma_{w_2}^2(1 - \rho_{w_1 w_2}^2)}.$$

Significantly, a, b, and c are, in fact, elements of the inverse matrix

$$\mathbf{A}^{-1} = \begin{bmatrix} a & b \\ b & c \end{bmatrix}$$

of the covariance matrix

$$\mathbf{A} = \begin{bmatrix} \sigma_{w_1}^2 & \rho_{w_1 w_2}\sigma_{w_1}\sigma_{w_2} \\ \rho_{w_2 w_1}\sigma_{w_2}\sigma_{w_1} & \sigma_{w_2}^2 \end{bmatrix}.$$

The concept of squared distance between two samples may be generalized for more than two variables with the expression of d^2 in matrix notation.

A 1×2 matrix may be formed as

$$\mathbf{V}' = [\overline{w}_{11} - \overline{w}_{12}, \; \overline{w}_{21} - \overline{w}_{22}],$$

whose transpose

$$\mathbf{V} = \left[\begin{array}{c} \overline{w}_{11} - \overline{w}_{12} \\ \overline{w}_{21} - \overline{w}_{22} \end{array} \right]$$

is a 2×1 matrix. Thus it is easily verified that

$$d^2 = \mathbf{V}' \cdot \mathbf{A}^{-1} \cdot \mathbf{V}.$$

These results, generalized to the case of k variables, require only that matrix $\mathbf{A}$ become a $k \times k$ matrix, matrix $\mathbf{V}'$ become a $1 \times k$ matrix, and matrix $\mathbf{V}$ become a $k \times 1$ matrix.

TABLE 10.15. FICTITIOUS CORRELATED BIVARIATE DATA

Item	Sample 1	Sample 2
$\overline{w}_1$	1	2
$\overline{w}_2$	1	3

Assumptions for the data: $\sigma_{w_1}{}^2 = 16$, $\sigma_{w_2}{}^2 = 9$, $\rho_{w_1 w_2} = 0.5528$.

The procedure is illustrated through the numerical data of table 10.15. For these data, because $\rho_{w_1 w_2} \sigma_{w_1} \sigma_{w_2} = (0.5528)\sqrt{16}\sqrt{9} = 6.633$, the result obtained is that

$$\mathbf{A} = \left[\begin{array}{cc} 16 & 6.633 \\ 6.633 & 9 \end{array} \right].$$

Then

$$\mathbf{A}^{-1} = \left[\begin{array}{cc} 0.09 & -0.06633 \\ -0.06633 & 0.16 \end{array} \right].$$

Hence

$$d^2 = (0.09)(1-2)^2 - 2(0.06633)(1-2)(1-3) + (0.16)(1-3)^2,$$

or

$$d^2 = 0.465.$$

Any set of correlated variables can be transformed into a set of uncorrelated variables by a linear transformation that can be interpreted geometrically as a rotation of the coordinate system. That the d^2 calculated in the last paragraph is indeed the same as the d^2 calculated by the simple formula for uncorrelated variables is shown by rotating the correlated variables of table 10.15 into uncorrelated variables. This rotation is defined by an orthonormal matrix. The device used to find the rotation is also used in one method of factor analysis (sec. 10.6) to investigate the basic dimensionality of a set of multivariate data.

The new variables are defined as

$$u_1 = \alpha_1 w_1 + \beta_1 w_2$$

and

$$u_2 = \alpha_2 w_1 + \beta_2 w_2.$$

Therefore two pairs of numbers, (α_1, β_1) and (α_2, β_2), must be found so that

$$\alpha_1^2 + \beta_1^2 = 1,$$

$$\alpha_2^2 + \beta_2^2 = 1,$$

$$\alpha_1 \alpha_2 + \beta_1 \beta_2 = 0,$$

and

$$\alpha_1 \alpha_2 \sigma_{w_1}^2 + \beta_1 \beta_2 \sigma_{w_2}^2 + \alpha_1 \beta_2 \rho_{w_1 w_2} \sigma_{w_1} \sigma_{w_2} + \alpha_2 \beta_1 \rho_{w_1 w_2} \sigma_{w_1} \sigma_{w_2} = 0.$$

These two pairs of numbers are obtained by finding the roots of the determinantal equation

$$|\mathbf{A} - \lambda \mathbf{I}| = 0,$$

where $\mathbf{A}$ is the covariance matrix and $\mathbf{I}$ is the identity matrix. In general the equation has two roots, λ_1 and λ_2. Once the roots are found, the pairs are the solutions of the equations

$$[\alpha_1, \beta_1] \cdot \mathbf{A} = [\lambda_1 \alpha_1, \lambda_1 \beta_1]$$

or

$$[\alpha_1, \beta_1] \cdot [\mathbf{A} - \lambda_1 \mathbf{I}] = [0, 0]$$

and

$$[\alpha_2, \beta_2] \cdot \mathbf{A} = [\lambda_2 \alpha_2, \lambda_2 \beta_2],$$

where λ_1 and λ_2 are the two roots of the equation

$$|\mathbf{A} - \lambda \mathbf{I}| = 0.$$

The process is illustrated for the data of table 10.15. For these data

$$\left| \begin{bmatrix} 16 & 6.633 \\ 6.633 & 9 \end{bmatrix} - \lambda \begin{bmatrix} 1 & 0 \\ 0 & 1 \end{bmatrix} \right| = 0,$$

which may be rewritten as

$$\begin{bmatrix} 16 - \lambda & 6.633 \\ 6.633 & 9 - \lambda \end{bmatrix} = 0$$

and reduces to

$$(16 - \lambda)(9 - \lambda) - 44 = 0.$$

The above equation becomes

$$\lambda^2 - 25\lambda + 100 = 0,$$

whose roots are

$$\lambda_1 = 20, \qquad \lambda_2 = 5.$$

The numerical values for the quantities (α_1, β_1) and (α_2, β_2) may now be obtained. As was shown, (α_1, β_1) must satisfy the expression

$$[\alpha_1, \beta_1] \cdot [\mathbf{A} - \lambda_1 \mathbf{I}] = [0, 0],$$

which, upon insertion of numerical values, becomes

$$[\alpha_1, \beta_1] \cdot \left[\begin{bmatrix} 16 & 6.633 \\ 6.633 & 9 \end{bmatrix} - 20 \begin{bmatrix} 1 & 0 \\ 0 & 1 \end{bmatrix} \right] = [0, 0]$$

and reduces to

$$[\alpha_1, \beta_1] \cdot \begin{bmatrix} -4 & 6.633 \\ 6.633 & -11 \end{bmatrix} = [0, 0].$$

The last equation may be written as the two linear equations

$$(-4)\alpha_1 + (6.633)\beta_1 = 0$$

and

$$(6.633)\alpha_1 + (-11)\beta_1 = 0,$$

from which it is evident that

$$\alpha_1 = 1.6583\beta_1.$$

However, $\alpha_1{}^2 + \beta_1{}^2 = 1$. Therefore $\beta_1 = 0.5164$, and $\alpha_1 = 0.8563$.

If the second root $\lambda_2 = 5$ is inserted into the previous expression, the result is that $\alpha_2 = -0.5164$ and $\beta_2 = 0.8563$. Thus the transformation is

$$u_1 = 0.8563w_1 + 0.5164w_2,$$

$$u_2 = -0.5164w_1 + 0.8563w_2.$$

The variance of the variable u_1 is

$$(0.8563)^2 \sigma_{w_1}^2 + (0.5164)^2 \sigma_{w_2}^2 + 2(0.8563)(0.5164)\rho_{w_1 w_2}\sigma_{w_1}\sigma_{w_2},$$

which is equal to 20. A similar calculation shows that the variance of u_2 is 5, and the covariance of u_1 and u_2 is 0.

Now the simple distance formula may be applied to the u_1 and u_2 variables by substituting the new means as well as the new variances. The new mean of u is calculated by applying the transforming coefficients to the corresponding means. For example, for the first sample,

$$\bar{u}_1 = 0.8563\bar{w}_1 + 0.5164\bar{w}_2,$$

and from table 10.15 it is seen that

$$\overline{w}_1 = 1, \qquad \overline{w}_2 = 1;$$

hence

$$\overline{u}_1 = 0.8563 + 0.5164,$$

$$= 1.3727.$$

The complete set of means and variances is given in table 10.16.

TABLE 10.16. TRANSFORMED DATA FROM TABLE 10.15

Item	Sample 1	Sample 2
$\overline{u}_1$	1.3727	3.2618
$\overline{u}_2$	0.3399	1.5361

Assumptions for the data: $\sigma_{u_1}^2 = 20$, $\sigma_{u_2}^2 = 5$, $\rho_{u_1 u_2} = 0$.

The simple squared-distance formula based on the values from table 10.16 gives

$$d^2 = \frac{(3.2618 - 1.3727)^2}{20} + \frac{(1.5361 - 0.3399)^2}{5},$$

from which

$$d^2 = 0.465.$$

The numerical value for d^2 is the same value given by the squared-distance formula for correlated variables.

Through a direct extension of the method used for bivariate distance, the distance may be calculated between two samples, each with k correlated variables. If $\mathbf{A}$ denotes the k-variate covariance matrix, $\mathbf{A}^{-1}$ is its inverse. If the k variables are designated $w_1, w_2, \ldots, w_k$, the means of the first sample may be designated $\overline{w}_{11}, \overline{w}_{12}, \ldots, \overline{w}_{1k}$, those of the second sample $\overline{w}_{21}, \overline{w}_{22}, \ldots, \overline{w}_{2k}$, etc. Then the $1 \times k$ matrix,

$$\mathbf{V}' = [\overline{w}_{11} - \overline{w}_{21}, \overline{w}_{12} - \overline{w}_{22}, \ldots, \overline{w}_{1k} - \overline{w}_{2k}],$$

may be formed. The squared distance, d^2, is calculated by forming the product of the matrix $\mathbf{V}$ with $\mathbf{A}^{-1}$ and with the transpose matrix $\mathbf{V}'$; that is,

$$d^2 = \mathbf{V}' \cdot \mathbf{A}^{-1} \cdot \mathbf{V}.$$

This distance is equal to that which would be calculated by generalizing the simple distance formula applied to the uncorrelated variables that were previously obtained by a rotation of the original variables.

An F-value may be calculated for each d^2-value through the formula

$$F = \frac{n_1 n_2 (\text{d.f.} - k + 1) d^2}{k (n_1 + n_2) (\text{d.f.})},$$

where n_1 is the number of observations in the first group, n_2 is the number of observations in the second group, and d.f. is the number of degrees of freedom in the covariance matrix **A**. This F-value has k and $(\text{d.f.} - k + 1)$ degrees of freedom.

EXAMPLE DISCRIMINATION AMONG VEINS. For an example of linear discriminant analysis, we turn again to the Frisco mine (sec. 1.4). The data are assay results from 19,050 mine samples cut at 2-meter intervals in all drifts of mine levels 7, 10, 13, 14, and 15. The assays were divided into 102 groups,

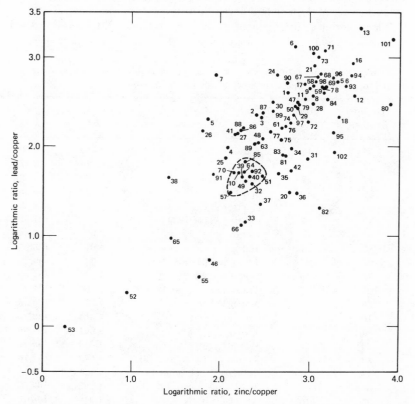

Fig. 10.11. Means of logarithms of Pb/Cu and Zn/Cu ratios for 91 vein groups of the Frisco mine.

each corresponding to one drift on a vein, except for drifts more than 1 kilometer long, which generally were subdivided into more than one group. Examples of discriminating bivariate and trivariate observations are given.

The linear discriminant analysis of bivariate data may be introduced by a plot of group means in a scatter diagram. Figure 10.11 is such a diagram for 91 mean logarithmic ratios of Pb/Cu versus mean logarithmic ratios of Zn/Cu. (The groups are numbered from 1 to 102 rather than from 1 to 91, because 11 groups from short drifts on structurally indistinct veins were omitted from this particular analysis.) The figure shows that several of the points are outliers and also that there are several clusters of points. A quantitative measure of cluster size is obtained by calculating d^2 for each of the group pairs, in order to indicate which groups are similar. In this calculation the intrinsic variability and covariability of these data are considered, as well as the mean differences; therefore an advance is made over plotting a simple scatter diagram; just as in figure 10.8 use of the standard deviation distinguishes cases A and B.

When the linear discriminant analysis is performed, 91 lists of ordered distances are obtained, one for each vein. Table 10.17 is a partial list of

TABLE 10.17. Some distances from level 10 Auras vein, Frisco mine, based on logarithms of Pb/Cu and Zn/Cu ratios

Group number	Mine level	Vein name	d^2	F
53	10	Auras	0.000	0.00
52	7	Auras	0.578	23.92
65	7	North Frisco	1.746	46.82
38	10	San Luis	2.632	90.55
55	7	Guerrero	2.950	78.22
46	7	Cobriza (S. of 19,500)	3.251	121.52
91	13	Rosario	3.627	80.84
57	7	Marte	4.014	126.65
25	7	Transvaal (S. of 18,500)	4.321	159.22
70	15	Don Tomás	4.356	189.62
26	7	Transvaal (18,500–19,000)	4.496	121.87
66	10	North Frisco	4.513	120.35
10	14	East Footwall	4.515	129.79
4	14	Frisco	4.601	162.05
39	10	Boreál	4.619	195.31
33	13	Transvaal (N. of 19,500)	4.672	172.76
49	7	Cobriza (N. of 20,500)	4.721	168.11
64	7	T	4.793	167.27
85	10	Atenas	4.949	106.43
5	7	Brown	5.011	191.07

distances for group 53, which is made up of 117 observations from the Auras vein on level 10 (fig. 1.2, at N-18,200) and which is plotted for the two variables considered in the lower left-hand corner of the scatter diagram, figure 10.11, as an outlier. Group 52 (283 observations from the Auras vein, level 7) and group 65 (99 observations from the North Frisco vein, level 7) plot closest to group 53 on the scatter diagram. Conversely, group 38 (167 observations from the San Luis vein, level 10) is closer in a d^2 sense than group 55 and group 46, which plot closer, because the d^2 calculation includes the interrelationship between the variable pairs as expressed by the covariance matrix. Group 101 (201 observations from the Chispa vein, level 13) in the upper right-hand corner of the scatter diagram is also farthest from group 53 in the complete list (Link and Koch, 1967) of d^2-values.

Table 10.17 also lists the F-value calculated for each d^2-value. If appropriate assumptions could be justified, these F-values could be used in statistical significance tests. Even though these assumptions are not rigorously justified for the example data, the values provide indications (sec. 4.9). For these paired

TABLE 10.18. SOME DISTANCES FROM LEVEL 7 COBRIZA VEIN (NORTH OF N-20,500) BASED ON LOGARITHMS OF Pb/Cu AND Zn/Cu RATIOS

Group number	Mine level	Vein name	d^2	F
49	7	Cobriza (N. of 20,500)	0.000	0.00
40	10	Footwall Two (West Footwall)	0.006	0.19
32	13	Transvaal (18,500–19,500)	0.006	0.35
39	10	Boreál	0.009	0.54
92	14	Rosario	0.012	0.39
64	7	T	0.017	0.79
57	7	Marte	0.028	1.12
10	14	East Footwall	0.036	1.27
51	7	East Madroños	0.042	2.11
70	15	Don Tomás	0.059	3.51
85	10	Atenas	0.065	1.61
35	14	Transvaal (18,500–19,500)	0.175	9.27
63	13	East Bronces	0.182	5.29
83	14	Footwall (N. of 20,000)	0.183	6.75
89	15	Santa Cruz	0.185	8.23
48	7	Cobriza (20,000–20,500)	0.201	10.86
81	13	Footwall (N. of 20,000)	0.208	11.87
37	15	Transvaal (19,309–19,561)	0.210	8.12
75	13	West Brown	0.225	5.51
77	7	Footwall (N. of 20,000)	0.270	5.05

comparisons with only two variables, each F-ratio would have two degrees of freedom in the numerator and an effectively infinite number of degrees of freedom in the denominator, because the number of observations in each group is large. Hence an F larger than 2.5 would indicate a risk level of about 10 percent. [An investigator who believed that all the assumptions were met would use an F-value of 2.30 for the exact 10 percent level with two and infinity degrees of freedom. He would have an additional problem of multiple comparisons (sec. 5.12).] The list of F-values makes clear that group 53 is indeed an outlier, because the closest other group has an associated F-value of 24.

In contrast with the outlying group 53, group 49 (level 7 Cobriza vein, north of N-20,500) is now considered. Table 10.18, a partial list of distances from group 49, indicates that the first groups in the list have effectively the same relation between the two variables Pb/Cu and Zn/Cu. Under a criterion of an F less than 2.5, the first 11 groups, excluding group 70 with an F of 3.51, comprise a cluster encircled by a dashed line in figure 10.11. All other groups have d^2-values with corresponding F-values larger than 2.5.

Table 10.19. Some distances from level 7 Cobriza vein (north of N-20,500) based on logarithms of Ag/Pb, Ag/Zn, and Ag/Cu ratios

Group number	Mine level	Vein name	d^2	F
49	7	Cobriza (N. of 20,500)	0.000	0.00
51	7	East Madroños	0.053	1.77
64	7	T	0.170	5.05
48	7	Cobriza (20,000–20,500)	0.201	7.24
63	13	East Bronces	0.214	4.15
75	13	West Brown	0.229	3.75
25	7	Transvaal (S. of 18,500)	0.274	8.70
61	7	El Mante	0.322	10.23
77	7	Footwall (N. of 20,000)	0.347	4.33
40	10	Footwall Two (West Footwall)	0.501	9.99
39	10	Boreál	0.518	19.69
97	10	Corona	0.545	21.30
41	7	Aurora	0.578	25.40
76	7	Footwall (S. of 20,000)	0.579	22.26
2	10	Frisco	0.616	25.25
79	10	Footwall (N. of 20,000)	0.658	21.94
27	7	Transvaal (N. of 19,000)	0.707	25.60
65	7	North Frisco	0.725	15.50
31	13	Transvaal (S. of 18,500)	0.727	26.92
66	10	North Frisco	0.762	16.18

TABLE 10.20. DISTANCES BETWEEN PAIRS OF LAMPROPHYRES BASED ON THE LINEAR COMBINATION $(SiO_2 \cdot MgO \cdot K_2O)$*

Number of analyses		Kersantite	Minette	Vogesite	Spessartite	Camptonite	Monchiquite	Ouachitite	Alnöite
103	Kersantite								
66	Minette	2.164							
30	Vogesite	0.064	2.171						
45	Spessartite	0.757	5.423	0.794					
78	Camptonite	3.237	7.449	2.808	2.927				
61	Monchiquite	5.150	7.592	4.301	5.692	0.985			
10	Ouachitite	18.468	16.246	17.237	22.069	10.409	6.315		
19	Alnöite	28.416	26.421	26.418	31.538	16.954	10.548	2.331	

* After Chayes and Métais, 1964, p. 184.

The second example of linear discriminant analysis is a trivariate one, based on the three silver/base metal ratios calculated from these Frisco data. The ratios are expressed as the mean logarithmic ratios of Ag/Pb, Ag/Zn, and Ag/Cu. Although for three variables it is inconvenient to plot a three-dimensional scatter diagram, the d^2-values can readily be calculated. Table 10.19, for group 49 (level 7 Cobriza vein, north of N-20,500), presents a partial list of groups obtained with the variable silver introduced into the analysis. With silver included, table 10.19 shows that, unlike table 10.18, most of the groups close to group 49 are on level 7 and that none is on level 14 and 15. This difference in the two tables occurs because silver mineralization in the mine is very sensitive to elevation.

Lamprophyre Example. A third example of linear discriminant analysis is a classification of lamprophyre dikes by Chayes and Métais (1964). Table 10.20 lists distances calculated by these investigators using the method explained in this section. The variables were SiO_2, MgO, and K_2O. Chayes and Métais state (1964, p. 183) that "d^2 values are invariably small if based on variables other than SiO_2, MgO, and K_2O, whereas little further enlargement of d^2 over the values yielded by this set results if other variables are added to it." Through averaging these distances, Chayes and Métais obtained the four clusters defined by the average distances in table 10.21. (For example, the distance 3.252 between cluster I and cluster II is the mean of the three distances 2.164, 2.171, and 5.423 from table 10.20, corresponding respectively to the distances from minette of kersantite, vogesite, and spessartite.) The values on the diagonal are cluster diameters.

Table 10.21. Average within- and among-cluster distances calculated from the distances between pairs in table 10.20*

Lamprophyre	Cluster	I	II	III	IV
Minette	I	0.000	3.252	7.520	21.334
Kersantite Vogesite Spessartite	II		0.538	4.019	24.024
Camptonite Monchiquite	III			0.985	11.056
Ouachitite Alnöite	IV				2.331

* After Chayes and Métais, 1964, p. 184.

Chayes and Métais conclude from this analysis that lamprophyres cannot satisfactorily be classified from chemical analyses alone. They write,

There is thus a strong suggestion that for distinguishing between closely related varieties mineralogy is more useful than chemical composition. Kersantites and vogesites almost certainly cannot be efficiently distinguished by means of discrinant functions based on chemical composition, for instance, but if one is told the names of the minerals in a specimen one can decide at once, without benefit of discriminant functions, whether it is a kersantite or vogesite. In view of the fact that the basic definitions are mineralogical this is hardly a surprising result.

CONCLUDING REMARKS. The discriminant analysis just described may be generalized in order to compare several groups of observations together rather than in pairs. An explanation of this method, too lengthy for this book, is offered by Cooley and Lohnes (1962). Recall that the procedure explained in this section finds for each pair of groups the line that best separates them. If more than two groups are treated simultaneously, more than one function may be developed for linear discrimination. Additional functions are often needed, because three or more groups may not lie on or near one line.

Classification of a Single Multivariate Observation into One of Two Groups

A familiar geological problem is to classify a single multivariate observation into one of two groups. The observation may consist of the variables in a rock analysis, the measurements of fossil dimensions, the sizes and shapes of sand grains, etc. A discriminant function is a linear combination (sec. 5.8) of the variables in existing observations formed to maximize the chance of classifying a new observation correctly. The problem arises only when the groups are similar. To distinguish a centipede from a cow, one has only to count the number of legs, but to assign a sandstone to one of two formations with similar lithology may be difficult.

To illustrate the method, a discriminant function is formed for the fictitious data of table 10.9, which consists of three groups of bivariate observations. The discriminant function to compare the second and third groups is calculated in table 10.22. Line 1 gives the pooled within-group variance-covariance matrix $\mathbf{A}$ for the three groups. In line 2 the coefficients of the discriminant function

$$L = \sum c_i w_i$$

are calculated. $\mathbf{C}$ is the 2×1 coefficients' matrix, $\mathbf{A}^{-1}$ is the inverse of $\mathbf{A}$, and $\mathbf{V}$ is the 2×1 matrix for the differences of the means of the variables in groups 2 and 3. Thus the discriminant function is

$$L = 2.93w_1 - 6.13w_2,$$

where the subscripts refer to variables 1 and 2. In lines 2 and 3 the discriminant

TABLE 10.22. CALCULATION OF THE FUNCTION TO DISCRIMINATE BETWEEN GROUPS 2 AND 3 OF THE FICTITIOUS DATA IN TABLE 10.9

Line	Item	Values
1	$\mathbf{A}$	$\begin{bmatrix} 4.00 & 1.75 \\ 1.75 & 1.00 \end{bmatrix}$
2	$\mathbf{C} = \mathbf{A}^{-1} \cdot \mathbf{V}$	$\begin{bmatrix} 2.93 \\ -6.13 \end{bmatrix} = \begin{bmatrix} 1.0667 & -1.8667 \\ -1.8667 & 4.2667 \end{bmatrix} \cdot \begin{bmatrix} 1 \\ -1 \end{bmatrix}$
3	$L = c_1 \overline{w}_{12} - c_2 \overline{w}_{22}$	$2.93(5) - 6.13(3) = -3.73$
4	$L = c_1 \overline{w}_{13} - c_2 \overline{w}_{23}$	$2.93(4) - 6.13(4) = -12.80$
5	Mean of lines 3 and 4	-8.27
6	$d^2 = \mathbf{V}' \cdot \mathbf{C} = \mathbf{V}' \cdot \mathbf{A}^{-1} \cdot \mathbf{V}$	$\begin{bmatrix} 1 & -1 \end{bmatrix} \cdot \begin{bmatrix} 2.93 \\ -6.13 \end{bmatrix} = 9.06$

functions are calculated for values of the variables equal to the means of groups 2 and 3, respectively. The subscripts of the coefficients c are the variable numbers; the first subscript of the means $\overline{w}$ is the variable number, and the second is the group number. The mean of lines 3 and 4 is a *cutting value* (line 5). If the value of L calculated from an observation is larger than the cutting value, the observation is assigned to group 2; if the calculated value is smaller than the cutting value, the observation is assigned to group 3. For comparison with the previous subsection, d^2 is calculated in line 6.

To apply the discriminant function, assume an observation ($w_1 = 3$, $w_2 = 5$). By substitution in the function,

$$L = 2.93(3) - 6.13(5) = -21.8,$$

and the observation is assigned to group 3. For an observation ($w_1 = 4$, $w_2 = 3$) $L = -6.67$, and the observation is assigned to group 2.

The discriminant function treats the groups or populations by pairs, except that the variance-covariance matrix used is the pooled within-groups matrix for all groups with similar variance-covariance matrices (three in the fictitious illustration). Alternatively, all groups can be treated simultaneously in order to form more than one linear combination for discrimination. Thus, if two linear combinations are developed, the resulting points can be plotted on a plane, which can be divided into regions. This alternative method, beyond the scope of this book, is explained by McCammon (1969, p. RME-10 ff.)

In section 10.4 three groups of rocks—granites, nepheline syenites, and gabbros—are compared by the generalized analysis of variance, and the conclusion is drawn that their compositions, as defined by six variables, are different. Discriminant functions for these three groups yield the cutting values in table 10.23. Several individual rocks are classified in table 10.24 by

Table 10.23. Cutting values for discriminant functions comparing the granites, nepheline syenites, and gabbros of table 10.12

| | | Classification | |
Rocks compared	Cutting value	Greater than cutting value	Less than cutting value
Granite versus nepheline syenite			
(1 versus 2)	− 146.4	1	2
Granite versus gabbro (1 versus 3)	75.8	1	3
Nepheline syenite versus gabbro			
(2 versus 3)	222.2	2	3

comparing calculated L-values with the cutting values in table 10.23. Three rocks—the quartz monzonite, trachyte, and basalt—are classified with the granites, nepheline syenites, and gabbros, respectively, whenever one of these groups appears in the classification. These results are expected from a comparison of the oxides with the mean oxide values for the granites, nepheline syenites, and gabbros (table 10.12). Some ambiguity arises in the classification of the other rocks. Diabase, for instance, is classified as granite, although one might expect it to be more nearly like a gabbro. The attempt to force it into one of these two pigeonholes is foolish, since diabase closely resembles neither granite nor gabbro.

10.6   FACTOR ANALYSIS

Regardless of the number of measured (or derived) variables, the number of statistically independent variables is often smaller than the number measured. Thus, in data of constant sum, at least one variable, chosen at will, is not statistically independent because it can be obtained by subtracting the sum of the others from 100 percent (sec. 11.4). Or one variable may be a linear combination of all or some of the others, again reducing by one the number of statistically independent variables.

TABLE 10.24. CLASSIFICATION OF SIX ROCKS BASED ON THE CUTTING VALUES OF TABLE 10.23

Name	Observations, w_i							Classification		
	SiO$_2$	Al$_2$O$_3$	MgO	CaO	Na$_2$O	K$_2$O		1 vs. 2	1 vs. 3	2 vs. 3
Quartz monzonite (246, 4)*	64.31	15.44	2.21	4.22	2.71	4.09		1	1	3
Syenite (272, 23)	62.87	17.45	0.76	2.81	4.23	8.87		1†	1	2
Trachyte (274, 45)	64.58	17.52	0.22	0.39	6.41	6.23		2†	1	2
Anorthosite (303, 1)	53.43	28.01	0.63	11.24	4.85	0.96		2	3	2†
Diabase (378, 60)	57.21	12.99	1.59	5.97	3.07	1.61		1	1†	3
Basalt (596, 9)	49.05	12.88	8.20	6.96	3.42	3.81		1	3	3

* Page and analysis numbers for Washington's (1917) tables.
† Calculated value of L close to the cutting value of table 10.23.

From this specific situation, in which the number of statistically independent variables is distinctly smaller than the number of measured ones, we pass to the more general case, in which some or all of the variables are partly rather than completely dependent on one another.

The inclusive name for a set of techniques to investigate the number of independent variables and the dimensionality (sec. 10.1) of observations is *factor analysis*. Factor analysis was first applied in psychology by various investigators who had different aims and who often had to use certain methods, not because they were the best, but because they were relatively easy to calculate before the days of digital computers. For a clearly written, elementary account of factor analysis as developed in psychology, the reader is referred to Cooley and Lohnes (1962, pp. 151–185), and for mathematical details to Wilks (1962, p. 564). Geological applications have been made by many investigators, notably Imbrie (Imbrie and Purdy, 1962; Imbrie, 1963; Imbrie and Newell, 1964; Imbrie and van Andel, 1964), Klovan (1968), and McCammon (1969). These authors also reference the voluminous literature.

A basic concept held by early workers in psychology was that humans possess general underlying "factors," for instance "intelligence." Various test scores, such as the IQ test, provide multivariate numerical data that can be analyzed to yield one or a few "factors" that measure intelligence. This measurement is done by obtaining values named *components*; then each student's test scores are multipled by coefficients of these components to rank him on an "intelligence" scale.

In other disciplines, "factors" are also recognized, expressed in terms including socioeconomic status, teachability, receptivity to advertising, susceptibility to disease, accident proneness, etc. And geologists speak of grade of metamorphism, complexity of structure, favorability of rocks for petroleum or ore, rock entropy, etc. All these terms express generalizations perhaps so broad as to be of little or no value. But that depends on one's point of view. Commenting on one of these terms, Joralemon (1967) suggested in a perceptive paper that the most important factor in ore search is intensity of mineralization and that even though geologists do not know exactly what that is, if they keep searching some day they may find out.

Elements of Eigenvalue Analysis

Eigenvalue analysis, one of the techniques of factor analysis, is explained through the study in this section of two sets of simple fictitious data. Subsequently, in sections 10.7, 11.3, and 11.5, eigenvalue analyses of real data are made.

A set of 16 fictitious observations, each with two variables, is listed in the first two columns of table 10.25 and plotted in figure 10.12. Because the observations lie nearly on a straight line, their variables are highly correlated

TABLE 10.25. SET OF 16 FICTITIOUS OBSERVATIONS, EACH WITH TWO VARIABLES, TO ILLUSTRATE EIGENVALUE ANALYSIS

Original observations		Transformed observations	
w_1	w_2	u_1	u_2
1.0	1.5	1.49	1.02
1.5	2.5	2.33	1.76
2.5	2.0	3.07	0.92
2.5	2.8	3.36	1.66
3.0	3.5	4.09	2.12
3.5	4.0	4.74	2.40
3.5	3.0	4.37	1.47
4.0	2.5	4.64	0.82
4.5	3.7	5.56	1.75
5.0	4.5	6.32	2.30
5.5	3.0	6.22	0.72
6.0	4.0	7.06	1.46
6.0	4.5	7.24	1.93
7.0	3.5	7.80	0.63
7.5	5.0	8.82	1.83
8.0	4.5	9.10	1.18

(sec. 9.3), a fact which suggests that the two variables can be replaced by a single variable.

Figure 10.12 graphs the 16 points plotted with respect to the original w_1- and w_2-axes. Also plotted are the axes u_1 and u_2 obtained by rotating the original axes to a position that concentrates as much as possible of the variability into a single variable. When these rotated axes are translated so that their origin coincides with the mean of the observations, their positions are those shown by the dashed lines. Then the sum of the squared distances from the observations to the u_1-axis, measured perpendicular to this axis, is a minimum, and the sum of the squared distances to the u_2-axis is a maximum. Notably, the translated axis u_1 does not coincide with the regression line, plotted as a short-dashed line. Although the regression line also passes through the mean of the observations, it is defined in order to minimize the sum of the squared distances from the observations to the w_1-axis, measured perpendicular to this axis, rather than perpendicular to the u_1-axis.

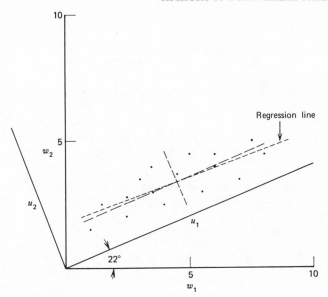

Fig. 10.12. Original axes w_1 and w_2 and rotated axes u_1 and u_2 obtained by eigenvalue analysis of 16 bivariate observations. The rotated axes as translated to the mean are shown by long-dashed lines, and the linear regression line is shown by a short-dashed line.

TABLE 10.26. RESULTS OF EIGENVALUE ANALYSIS OF 16 FICTITIOUS OBSERVATIONS FROM TABLE 10.25

Eigenvalue		Eigenvector	
Rank	Size	w_1	w_2
1	5.136	(0.927,	0.374)
2	0.319	(−0.374,	0.927)

The method of rotation is that described in section 10.5. The rotation provides the eigenvalues and eigenvectors listed in table 10.26. The values of the transformed variables u_1 and u_2 can be calculated from the eigenvectors by the formulas

$$u_1 = 0.927w_1 + 0.374w_2$$

and

$$u_2 = -0.374w_1 + 0.927w_2.$$

The transformed variables are listed in table 10.25. The variance of u_1 is the larger eigenvalue, 5.1359, and the variance of u_2 is the smaller eigenvalue,

0.3192. The correlation of u_1 and u_2 is 0. The angle of rotation, θ, is given by the formula

$$\theta = \tan^{-1}\left(\frac{0.374}{0.927}\right) = \tan^{-1}(0.403) = 22.0°.$$

The slope of the dashed line is 0.403. On the other hand, the same calculation for the regression line yields

$$\theta = \tan^{-1}\left(\frac{\text{SP}}{\text{SS}_{w_1}}\right) = \tan^{-1}\left(\frac{1.67}{4.46}\right) = \tan^{-1}(0.374) = 20.5°,$$

where SP is the sum of products and SS_{w_1} the sum of squares for w_1.

Eigenvalue analysis is a stepwise procedure. In the first step, the individual variables are placed in a linear combination u, which has the following properties:

$$u = \Sigma c_i w_i,$$

$$\Sigma c_i^2 = 1,$$

$$s_u^2 \text{ is a maximum.}$$

These equations define the terms *eigenvalue* and *eigenvector*; s_u^2 is an eigenvalue, and the coefficients c_i are the components of the corresponding eigenvector. The term s_u^2 accounts for a percentage of the total variance $V = \Sigma s_{w_i}^2$, expressed by the relation

$$100 \frac{s_u^2}{V}.$$

In the second step, another linear combination, u', is found, which has the following properties, similar to those in step 1:

$$u' = \Sigma c'_i w_i,$$

$$\Sigma c'^2_i = 1,$$

$$s_{u'}^2 \text{ is a maximum,}$$

and the restriction

$$r_{uu'} = 0,$$

where $r_{uu'}$ denotes the correlation between u and u'. Now the percentage of the total variance accounted for is larger, being expressed by the relation

$$100 \frac{(s_u^2 + s_{u'}^2)}{V}.$$

The stepwise process is continued until all of the total variance V is accounted for, which occurs when the number of linear combinations becomes equal to d, where d is the number of *dimensions*.

For the original 16 fictitious observations (table 10.25), the variance-covariance matrix is the 2×2 matrix

$$\begin{bmatrix} s_{w_1}^2 & r_{w_1 w_2} s_{w_1} s_{w_2} \\ r_{w_2 w_1} s_{w_2} s_{w_1} & s_{w_2}^2 \end{bmatrix}$$

with the numerical values

$$\begin{bmatrix} 4.4625 & 1.6704 \\ 1.6704 & 0.9926 \end{bmatrix}.$$

When this matrix is operated on, the two eigenvalues and associated eigenvectors, listed in table 10.26, are obtained for the original variables. The eigenvalues are the variances of the new variables into which the original variables were transformed by multiplying the original variables by the coefficients that form the components of the eigenvectors. The sum 5.455 of the eigenvalues is the sum of the variances.

The variance-covariance matrix is operated on as follows. The *dimensionality* of a set of k variables is the number of algebraically independent variables necessary to express the k variables. For example, if a set of (x, y) points falls on a straight line, the set is one-dimensional because the points can be represented by $(x, a + bx)$, where $y = a + bx$, the equation of a line. Bivariate data are nearly one-dimensional if they are highly correlated, as in the fictitious illustration. This correlation explains why, instead of measuring these two variables, one of them alone, or another single variable, could in principle be measured.

If two variables are not perfectly correlated, they can be transformed into two algebraically independent variables whose correlation is zero. If there are k original variables, but the dimensionality d of the data is less than k, because the variables are linearly related, only d variables will be obtained from the k variables for which there are positive eigenvalues; the remaining $(k - d)$ eigenvalues equal zero. Because in real observations correlations among the variables are imperfect, several small eigenvalues are usually obtained, and an investigator must use judgment to decide how many dimensions to consider.

The matrix manipulations are as follows. If $\mathbf{A}$ is a $k \times k$ matrix used to measure the variability of multivariate data, the eigenvalues (characteristic roots) of $\mathbf{A}$ may be found by solving the determinantal equation

$$|\mathbf{A} - \lambda \mathbf{I}| = 0,$$

where $\mathbf{I}$ is the identity matrix and λ is an eigenvalue. This procedure leads to a polynomial in λ of the kth degree, and the roots of such polynomials can in general be found only by iterative methods.

A $1 \times k$ matrix, the eigenvector, is associated with each eigenvalue. These matrices, k in number, are found by solving equations that may be written as

$$\mathbf{X} \cdot |\mathbf{A} - \lambda_i \mathbf{I}| = \mathbf{0},$$

where $\mathbf{X}$ is a $1 \times k$ matrix, λ_i is one of the roots, and $\mathbf{0}$ is a $1 \times k$ matrix of zeros.

Once the eigenvalues and eigenvectors are obtained, a mathematical manipulation may be performed to make some of the components of each eigenvector as large as possible and the others as small as possible. The purpose is to aid interpretation by reducing the number of original variables associated with each eigenvalue. The manipulation is a rotation, in principle like that made in structural geology to rotate plunging fold axes to a horizontal plane so that their directions can be designated by a single number for azimuth rather than by one number for azimuth and a second number for plunge. Two of these procedures, named quartimax and varimax, are explained by Cooley and Lohnes (1962, p. 161). If one wishes only to identify new variables, this rotation is unnecessary.

Other Modes of Eigenvalue Analysis

The previous illustration is for the *variance mode* of eigenvalue analysis. However, other modes can be used. The following are two other modes of analysis commonly used by geologists, the R- and Q-modes.

In the *R-mode* of analysis, the matrix formed to express the variability is a correlation matrix. For the fictitious observations (table 10.25), the 2×2 correlation matrix is

$$\begin{bmatrix} 1 & r_{w_1 w_2} \\ r_{w_2 w_1} & 1 \end{bmatrix}.$$

TABLE 10.27. RESULTS OF THREE MODES OF EIGENVALUE ANALYSIS OF FICTITIOUS OBSERVATIONS FROM TABLE 10.25

	Eigenvalue		Eigenvector	
Mode	Rank	Size	w_1	w_2
Variance	1	5.136	(0.927,	0.374)
	2	0.319	(−0.374,	0.927)
R	1	1.794	(0.707,	0.707)
	2	0.206	(−0.707,	0.707)
Q	1	15.569	Not displayed	
	2*	0.431	,,	

* Other 14 eigenvalues are zero except for round-off errors.

Table 10.27 lists the one large and one small eigenvalue obtained in the R-mode of analysis, compared with the results for the variance mode copied

from table 10.26. The components are interpreted as before. Because in the
R-mode the original variables are standardized to variables with means of 0
and variances of 1, the components for the two variables are numerically
identical.

In the *Q-mode* of eigenvalue analysis, the observations and variables are
interchanged, so that, for the fictitious illustration, the original 2×16 matrix
becomes a 16×2 matrix of 16 variables and two observations. A 16×16
matrix to measure the variability is then formed, the diagonal elements of
which are 1 and the off-diagonal elements of which are essentially uncorrected
correlation coefficients for the 16 variables. Two positive eigenvalues, one
large and one small, are listed in table 10.27; the other 14 eigenvalues, not
listed, are zero, except for computer round-off errors. The 16 components for
each of the two eigenvectors associated with the two eigenvalues are not
displayed. The Q-mode of analysis obtains only two eigenvalues from the
16×16 matrix, because there can never be more nonzero eigenvalues than the
smaller number of either rows or columns in the original data matrix. In this
fictitious illustration, there were only two original variables.

Discussion of the Variance, R-, and Q-Modes

At present, eigenvalue analysis is pursued by geologists for two main
purposes. For the first, identifying the number of independent variables in a
set of observations, either the variance or the R-mode is appropriate. For the
second, classifying multivariate observations, the Q-mode is suitable.

In the variance mode, because a variance-covariance matrix is used, each
variable is weighted according to its variance, and too much weight may be
given to one or a few of the variables if their variances are much larger than the
others'. In the R-mode, because a correlation matrix is formed, each variable
receives equal weight. Although the variance mode is advocated by some
statisticians because the scale of the original measurements is preserved, the
R-mode is more commonly used. In practice, the two modes of analysis often
give similar results.

In the Q-mode of eigenvalue analysis, each off-diagonal element is obtained
from the cosine of the angle between two vectors corresponding to two
observations, which is the quantity

$$\frac{\Sigma w_1 w_2}{\sqrt{\Sigma w_1^2 \; \Sigma w_2^2}}.$$

This quantity forms a correlation-like measure of association especially
appropriate for proportional data. (Proportional data are those for which
$w_1 = aw_2$ in contrast to nonproportional data, for which $w_1 = a + bw_2$, a and b
being nonzero constants.) Each eigenvalue has an associated eigenvector,
whose component for each observation is a coefficient, which may be refined

by rotation if desired. The total number of eigenvalues is chosen arbitrarily; usually selected are those making up 80 to 90 percent of the total sum of eigenvalues. For the chosen eigenvalues, observations with similar coefficients are grouped by inspection into a cluster that defines the classification. Imbrie and Purdy (1962) apply the technique to a problem of classifying recent sediments.

Eigenvalue Analysis of a Fictitious Illustration with 20 Observations and 10 Variables

We continue this explanation of eigenvalue analysis with a fictitious illustration of 20 observations and ten variables of constant sum, of which the previous illustration with 16 observations and two variables is a subset. This additional illustration further explains the eigenvalue procedure and provides the reader with a convenient data set with known properties for practice.

TABLE 10.28. FORMULAS TO CALCULATE
FICTITIOUS OBSERVATIONS TO DEMONSTRATE
EIGENVALUE ANALYSIS

Variable		Formula
w_1	=	$0.3x_1 - 5$
w_2	=	$w_1 + 0.25w_4 + 0.05x_4 - 2.5$
w_3	=	$0.1x_2$
w_4	=	$0.2x_3 - 2$
w_5	=	$w_1 - w_3 + 0.05x_5 - 2.5$
w_6	=	$0.5x_1 + w_3 + 0.05x_6 - 2.5$
w_7	=	$w_1 + w_3 - 0.625w_4 + 0.05x_7 - 2.5$
w_8	=	$0.2w_1 + w_3 + w_4$
w_9	=	$1.67w_4 - 1.67w_3$
w_{10}	=	$100 - (w_1 + w_2 + \cdots + w_9)$

Table 10.28 explains how the fictitious data were calculated. Each observation consists of 10 variables w_1 to w_{10}; the variables are calculated by the formulas in the table. The x_1 to x_7 values are obtained from a set of random normal numbers with a mean of 50 and a standard deviation of 7. The formulas are devised so that three variables account for most of the variability; four others are linear combinations of these three, with some additional random error; and the other three are perfect linear combinations of the others. Specifically, variables w_1, w_3, and w_4 comprise practically all of the inherent variability; variable w_2 and variables w_5 to w_7 are linear combinations of variables w_1, w_3, and w_4, with a small amount of random error introduced by

Table 10.29. Fictitious data to demonstrate eigenvalue analysis

Observa-tions	Variables									
	1	2	3	4	5	6	7	8	9	10
1	9.700	11.150	4.100	6.200	5.200	8.700	10.125	12.240	3.500	29.085
2	8.800	11.300	6.000	10.600	2.100	10.150	8.275	18.360	7.667	16.748
3	11.800	15.400	4.000	11.000	7.700	10.600	9.725	17.360	11.667	0.748
4	11.500	14.400	5.300	8.000	5.600	10.550	11.600	15.600	4.500	12.950
5	8.500	9.950	4.000	5.400	4.350	7.900	9.125	11.100	2.333	37.342
6	11.200	12.800	3.700	7.200	8.400	9.300	11.000	13.140	5.833	17.427
7	10.600	12.450	5.200	7.600	5.800	10.150	11.200	14.920	4.000	18.080
8	12.400	14.550	5.400	9.000	7.150	11.400	11.775	16.880	6.000	5.445
9	11.200	13.050	6.300	7.800	5.750	11.200	12.675	16.340	2.500	13.185
10	10.900	12.700	6.200	8.200	5.400	11.300	11.225	16.580	3.333	14.162
11	11.800	13.650	4.500	6.600	7.600	10.250	11.975	13.460	3.500	16.665
12	8.800	11.700	5.300	9.600	4.000	9.650	8.550	16.660	7.167	18.573
13	6.100	7.700	5.800	7.200	0.150	8.750	7.000	14.220	2.333	40.747
14	6.100	9.150	3.700	8.800	2.450	6.350	4.150	13.720	8.500	37.080
15	13.600	14.750	4.900	7.400	8.200	11.150	14.425	15.020	4.167	6.388
16	9.400	11.150	3.900	8.000	6.150	8.250	8.800	13.780	6.833	23.737
17	12.700	14.000	6.000	8.400	6.200	11.650	14.300	16.940	4.000	5.810
18	10.600	12.250	5.100	10.400	5.850	9.900	8.150	17.620	8.833	11.297
19	7.000	9.100	5.600	8.200	0.650	8.950	7.625	15.200	4.333	33.342
20	9.700	12.050	5.700	7.200	4.150	10.350	10.600	14.840	2.500	22.910

the contributions of the x_4 to x_7 values; and variables w_8 to w_{10} are linear combinations of the other variables.

The resulting fictitious data are listed in table 10.29. This 20×10 table may be considered as a list of rock analyses summing to 100 percent by rows, with some constituents correlated more or less with others.

The three modes of eigenvalue analysis produce the eigenvalues ranked in table 10.30. As is to be expected from the way the observations were constructed, all three modes of analysis yield three large eigenvalues corresponding

TABLE 10.30. EIGENVALUES FOR THREE MODES OF
ANALYSIS OF FICTITIOUS DATA

Rank	Variance mode	R-mode	Q-mode
1	149.10	5.38	18.012
2	12.21	2.60	1.731
3	5.13	1.91	0.178
4	0.24	0.05	0.071
5	0.19	0.04	0.004
6	0.05	0.02	0.003
7	0.04	0.01	0.001
8	0.00	0.00	0.001
9	0.00	0.00	0.000
10	0.00	0.00	0.000

Note: Q-mode eigenvalues 11 to 20 are all zero.

to the three variables that account for most of the variability, four small eigenvalues corresponding to the four variables that are linear combinations of those three plus random errors, and three zero eigenvalues (except for 0.001 appearing in the Q-mode analysis through round-off error). Moreover, the Q-mode eigenvectors, rank numbers 11 to 20, are also all zero.

Because the eigenvalues must sum to the original total variance, their relative sizes in the three modes of analysis differ from one another. Those in the variance mode are the largest; those in the R-mode are the smallest because all variables were standardized to give them equal variances.

For these data, the Q-mode analysis is inappropriate, because the observations were derived from a uniform set of random numbers, and no clusters should exist. The analysis is made only to demonstrate the method. However, the same distribution of eigenvalues is found as in the previous analyses in the other modes, because the data have the same inherent variability, whether the matrix is taken by rows or by columns. The extra eigenvalue, arising from round-off error, appears because the matrix is larger, and problems develop in manipulating large matrices with electronic computers.

Classification

We first comment on classification itself and then explain and compare two methods of classification: discriminant analysis and factor analysis.

Classification of *univariate* or *bivariate observations* is straightforward because boundaries can readily be set. For instance, students' examination grades can be plotted on a straight line (fig. 10.1). Anyone can group these univariate observations into categories of A, B, C, etc., although defining the cutting points between categories may be difficult, and the examination papers may have been graded wrongly in the first place. Bivariate observations can be plotted in two dimensions (fig. 10.2) on rectangular coordinates, and trivariate observations of constant sum can be plotted in two dimensions on triangular coordinates (fig. 11.1); the observations can then be classified by forming groups of clusters.

Multivariate observations with more than two or three variables are harder to classify because reasonable conceptual schemes become more and more difficult to devise as the number of variables increases. For instance, there are many schemes for computing grade-point averages, which are used to classify students. Some schools do not count grades in physical education; others exclude grades in one course a year; still others allow a course to be repeated to raise a grade, etc. In geology, rocks may be classified according to age, mineralogy, oil content, texture, and many other properties, and one variable can be weighted more than another; a petrologist is likely to weight feldspar composition and content high, while a gold miner is likely to emphasize gold.

The methods of discriminant analysis presented in this book superimpose univariate classifications on multivariate observations. Mahalanobis' d^2-statistic classifies groups of multivariate observations by pairs, so that d^2-values, $m(m-1)/2$ in number, can be calculated for m groups of observations. From the lists of d^2-values, the investigator can choose the list or lists that he prefers. The discriminant function classifies single multivariate observations into one of two established groups. Although a particular observation may always be classified, the assignment may not be meaningful if the particular observation has little or nothing in common with the observations in either group.

The R- and Q-modes of factor analysis may also be used to classify multivariate observations (McCammon, 1969). In R-mode analysis, the eigenvectors associated with the two largest eigenvalues are used to define two new variables. These variables are plotted in two dimensions, and a classification is defined from the clusters evident in the plot. In Q-mode analysis, there is an eigenvector for each observation. The number of components of each eigenvector is equal to the number of variables. Clusters are formed by grouping together observations with similar components.

The choice of a classification method is highly subjective and must contain many arbitrary elements. Geologists concerned with classification have investigated criteria for choice; McCammon (1969) explains some criteria and gives a good bibliography of other works.

10.7 ANALYSIS OF DIRECTIONAL DATA

Among the most common data in geology are directional observations. They are essential in structural geology to represent attitudes of planes and lines, and are often made in crystallography, paleomagnetism, structural petrology, sedimentology, and many other fields.

Geologists have devised several conventions for measuring directional data. We adopt one of the systems of angular measurements and relate it to direction cosines, required for statistical analysis. Horizontal angles are measured by an azimuth rotated *clockwise* from north at zero degrees, so that east is 90 degrees, south is 180 degrees, and west is 270 degrees. Thus the azimuth 280 degrees designates the strike of a plane or the direction of a line northwestward on a bearing of north 80 degrees west. Vertical angles are measured by the inclination from the horizontal, with a plus or minus sign if necessary for clarity. Thus the vertical angle 20 degrees designates the dip of a plane or the plunge of a line inclined 20 degrees *below* the horizon; the vertical angle -20 degrees denotes an inclination *above* the horizon. (The sign of the vertical angle follows the convention usual in paleomagnetism and provides a right-handed coordinate system without which the direction cosines would not follow the usual mathematical convention.) Alternatively, directions may be expressed as direction cosines defined as

$$L = \cos\phi\cos\theta,$$

$$M = \cos\phi\sin\theta,$$

$$N = \sin\phi,$$

where θ is the azimuth and ϕ is the inclination. Table 10.31 summarizes these definitions by giving the values for six directions on a sphere defined both ways.

As with any other data, it is advisable to plot a set of directional observations

TABLE 10.31. TRANSFORMATION OF ANGULAR MEASUREMENTS TO DIRECTION COSINES

| Direction | Azimuth | Inclination | Direction cosines | | |
			L	M	N
General	θ	ϕ	$\cos\phi\sin\theta$	$\cos\phi\cos\theta$	$\sin\phi$
Down	0	90	0	0	1
Up	0	-90	0	0	-1
North	0	0	1	0	0
East	90	0	0	1	0
South	180	0	-1	0	0
West	270	0	0	-1	0

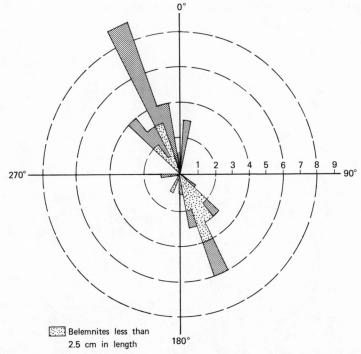

Fig. 10.13. Orientation of belemnites on a single bedding plane in Upper Jurassic rock in eastern Scotland (after Ager, 1963, p. 78). The direction of the apex of the rostrum is plotted in each case.

so that their pattern may be observed. Univariate directional data in a plane can be plotted in two dimensions. Thus belemnite orientations in a sedimentary bed can be graphed in plan (fig. 10.13), or slickensides can be plotted in a fault plane. Multivariate data can also be plotted on a map or section by symbols, with the inclinations (dips or plunges) written out on the drawing. Because it is difficult if not impossible to integrate these observations by scanning the strike and dip, or other symbols, the observations are commonly plotted in stereographic projection. Familiarity with how to read and plot the different varieties of stereographic projections (Schmidt and Wulff nets, etc.) is assumed; full details are furnished by Phillips (1960), Billings (1954), and others. For many observations, computer plotting, explained by Louden (1964), is convenient.

Procedures for averaging data to be contoured appear in the cited works and in the periodical literature that they list. These methods are needed to summarize many points for the display of preferred orientations but can be replaced by the statistical methods explained in this section. Original data are usually better for statistical analysis than averaged data.

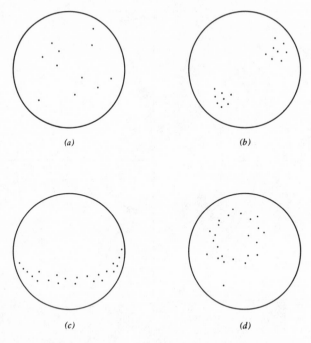

(a)

(b)

(c)

(d)

Fig. 10.14. Typical configurations of multivariate directional observations plotted in stereographic projection. (a) Randomly distributed points. (b) Two clusters of points. (c) A girdle of points on a great circle. (d) A girdle of points on a small circle.

Figure 10.14 shows typical configurations of multivariate directional observations plotted in stereographic projection. Many different classifications of the pattern of points have been devised for different reasons (for instance, Billings, 1954; Hills, 1963); we adopt a geometrical classification suitable for explaining the statistical analysis. Figure 10.14-a plots randomly distributed points (to be distinguished from haphazardly distributed points, sec. 3.4). Figure 10.14-b shows two clusters of points, figure 10.14-c a girdle of points on a great circle, and figure 10.14-d a girdle of points on a small circle.

Once a set of observations is plotted in stereographic projection, outliers should be identified and inspected because they may indicate blunders in field measuring, data transcription, and plotting, or, on the other hand, they may suggest something significant. Then the general configuration of points should be inspected and such questions should be asked as: Do the points define a pattern or are they scattered ? If there is a pattern, what kind is it ?

Problems in Statistical Analysis

Satisfactory statistical methods for directional data have been devised for points distributed on a sphere. For geological observations that are directed lines (vectors), principal among which are paleomagnetic measurements, the sphere is the proper representation and statistical analysis is straightforward. Procedures are well developed for points in one cluster (point maximum) on a sphere and for points in a girdle. At present the field is one of active study in statistics. Some important works are those by Fisher (1953), Watson (1966), and Bingham (1964). Irving (1964) applied the work of the first two of these authors in an excellent study of paleomagnetic data.

Unfortunately, most directional observations in geology are of undirected lines, such as strike and dip measurements, plunges and pitches of fold and other axes, and compass directions. These data are conventionally plotted on a hemisphere; no better way to plot them for statistical analysis has evidently been devised. However, statistical analyses for data on a hemisphere are in their infancy because the equatorial boundary poses a difficult statistical problem. In two dimensions, the problem was recognized by Krumbein (1939) for data that should be plotted on a semicircle rather than on a circle. His solution was to double the angles and study the doubled distribution with statistical methods that are available for the circle (fig. 10.15). While this procedure indeed produces a full-circular distribution, it also doubles the dispersion by spreading out the points, as the reader can verify by plotting a few. Batschelet (1965) has published an excellent review of the two-dimensional problem, which we recommend because the problems are easier to understand in two dimensions than in three. Similarly, it is not satisfactory to plot randomly one half of a set of observations on the lower hemisphere and to ignore the other one half. Nor is it satisfactory to plot them all, because, if for instance they define a cluster, the points will often plot as two antipodal clusters (depending on the conventions used for recording the angles).

Thus, although statistical methods for observations on a hemisphere are needed, they are undeveloped. The difficulty is that statistics that have been suggested are not isotropic with respect to a hemisphere. In other words, the details of the orientations of the points with respect to the hemisphere must be taken into account. This fact is illustrated in stereographic projections when one recognizes that points plotted in two clusters 180 degrees apart on the equator (perhaps one at the north and the other at the south pole) really define only one cluster.

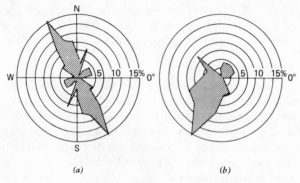

(a)   *(b)*

Fig. 10.15. Distribution of pebble directions in glacial drift in Sweden. In figure 10.15-a the original centrally symmetric distribution is shown. To get figure 10.15-b from figure 10.15-a, the angles must be doubled (after Batschelet, 1965, who references the original paper by Köster, 1964).

In summary, points on a sphere that are distributed in clusters or girdles can be analyzed by methods applied to paleomagnetic data by Irving (1964) and Watson (1966). These methods are explained in the next two subsections. They differ in the extent to which they apply to analysis of data plotted on a hemisphere, as is discussed in a subsequent subsection through the analysis of some representative lineation data from structural geology.

Statistical Analysis of Clustered Points on the Sphere

If points plotted in stereographic projection appear to define one or more clusters (point maxima) on the sphere, one or more of the following statistical procedures, which are described in turn in this subsection, may be applied:

1. Estimate the azimuth and inclination of the center (mean direction) of a cluster, and estimate the precision (closeness) with which points are clustered.

2. Determine whether two or more clusters have the same mean direction.

3. Determine whether two clusters have the same precision of clustering.

4. Determine whether points are clustered or are randomly distributed on the sphere.

Statistical analysis of clustered points depends on Fisher's (1953) frequency distribution of points on a sphere. Properties of Fisher's frequency distribution have been further explored by Watson (1966, with references to his earlier works) and in part applied to geological data by Irving (1964). This frequency distribution is symmetrically distributed about the mean direction of a cluster, just as the normal distribution is symmetrically distributed about the mean. In the normal distribution, the measure of tightness of clustering of observations about the mean is the standard deviation (fig. 10.8). Similarly, in Fisher's distribution, the measure of tightness of clustering is a statistic named the "estimate of precision" (k in Irving's notation, 1964, p. 58). The estimate of precision is defined in the reciprocal sense to the standard devi-

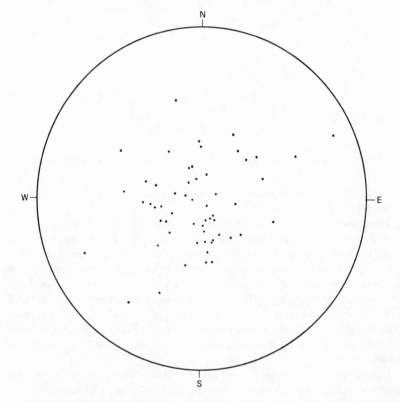

Fig. 10.16. Stereographic projection of 50 random points with a precision of 10.

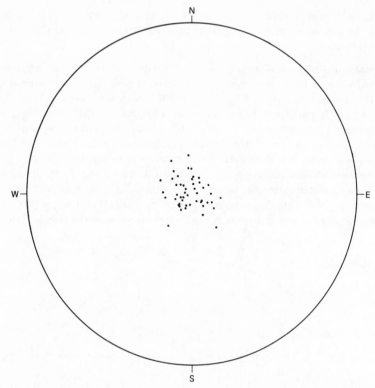

Fig. 10.17. Stereographic projection of 50 random points with a precision of 50.

ation; a large value indicates close clustering of points, and a small value indicates little or no clustering.

The concept of precision is illustrated by plotting two groups of points, one with a precision of 10 and the other with a precision of 50. Figures 10.16 and 10.17 are each plots of 50 randomly drawn points, whose mean direction is vertically downward, but whose precision is different. Notice that the points are much less clustered in figure 10.16, with a precision of 10, than in figure 10.17, with a precision of 50. If the precision were further reduced to 2, there would be about a 10 percent chance that some points would be so far dispersed from the mean that they would plot on the upper rather than on the lower hemisphere. The reader wishing a mathematical explanation of precision is referred to Irving.

In order to explain the calculations for the various statistical procedures for clustered points, the calculations are applied to directions of magnetization observed at four sites in the Tasmanian dolerites of lower to middle Jurassic

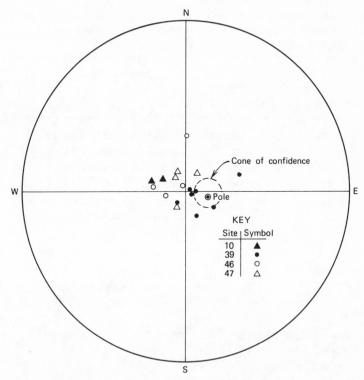

Fig. 10.18. Directions of magnetization observed at four sites in the Tasmanian dolerites (after Irving, 1964).

age (Irving, 1964, p. 65). The observations are plotted in stereographic projection in figure 10.18.

First, the mean azimuth, mean inclination, and precision are estimated. The seven observations from site 39, listed in table 10.32 with the corresponding direction cosines, are chosen in order to demonstrate the calculations. In lines 8 and 9 the sums, together with the sums of squares of the direction cosines, are accumulated, and in lines 10 and 11 the vector resultant R is calculated. In line 12 the mean direction cosines are given, and in lines 13 to 15 the desired statistics are listed. Details of the algebra are given in any standard text on advanced algebra [for example, MacLane and Birkhoff (1965) and Fisher (1953)].

A confidence interval for the mean direction of a cluster of points, named a *cone of confidence*, may be calculated as outlined in table 10.33. Fisher (1953) showed that, for a 90 percent confidence level, the true mean direction lies within a circular cone whose axis is the estimated mean direction and whose

TABLE 10.32. CALCULATION OF MEAN AZIMUTH, MEAN INCLINATION, AND PRE-
CISION FOR SEVEN DIRECTIONS OF MAGNETIZATION OBSERVED AT SITE 39 IN THE
TASMANIAN DOLERITES*

	Angles		Direction cosines			
Line	Azimuth, θ	Inclination, ϕ	L	M	N	
1	75	-62	0.1215	0.4535	-0.8829	
2	99	-75	-0.0405	0.2556	-0.9659	
3	120	-74	-0.1378	0.2387	-0.9613	
4	213	-83	-0.1022	-0.0664	-0.9925	
5	157	-77	-0.2071	0.0879	-0.9744	
6	127	-87	-0.0315	0.0418	-0.9986	
7	86	-78	0.0145	0.2074	-0.9781	
8	$\Sigma L, \Sigma M, \Sigma N$		-0.3831	1.2185	-6.7538	
9	$(\Sigma L)^2, (\Sigma M)^2, (\Sigma N)^2$		0.1468	1.4847	45.6138	
10	$R^2 = (\Sigma L)^2 + (\Sigma M)^2 + (\Sigma N)^2$					47.2453
11	R					6.8736
12	$\bar{L} = \Sigma L/R, \bar{M} = \Sigma M/R,$ $\bar{N} = \Sigma N/R$		-0.0557	0.1773	-0.9826	
13	Estimate of azimuth $\theta = \tan^{-1}(\bar{M}/\bar{L})$				107°	
14	Estimation of inclination $\phi = \sin^{-1}(\bar{N})$				$-79°$	
15	Estimate of precision				47	

* After Irving, 1964.

TABLE 10.33. CALCULATION OF A CONE OF CONFIDENCE FOR
SEVEN DIRECTIONS OF MAGNETIZATION FROM TABLE 10.32

Line	Value	Symbol	Explanation
1	7	n	Number of observations
2	0.468		Value from table A-2
3	6.8736	R	From table 10.32
4	0.99139	$\cos A$	$1.0 - 0.468(n - R)/R$
5	7.52°	A	

semivertical angle A is given by the formula

$$\cos A = 1 - \frac{n-R}{R} \left\{ \left(\frac{1}{0.9}\right)^{1/(n-1)} - 1 \right\},$$

where R has the same meaning as before, and n is the number of observations. Values of the bracketed quantity,

$$\left\{ \left(\frac{1}{0.9}\right)^{1/(n-1)} - 1 \right\},$$

are tabulated in table A-2.

The second statistical procedure is the calculation to determine whether two or more clusters have the same mean direction within limits of statistical fluctuation, shown with the example data. The 17 observations are divided into four groups and the groups are compared to determine whether they have the same mean direction. The groups are compared by an F-test, whose principle, similar to that of the F-test in the one-way analysis of variance (sec. 5.3), and given by Irving (1964), is not reviewed here. In table 10.34, lines 1 to 4 list for the four groups the resultant R-values and the mean

TABLE 10.34. CALCULATIONS TO DETERMINE WHETHER CLUSTERS OF POINTS ON A SPHERE HAVE THE SAME MEAN DIRECTION, APPLIED TO DIRECTIONS OF MAGNETIZATION OBSERVED AT FOUR SITES IN THE TASMANIAN DOLERITES*

Line	Site number	Sample size, n	Resultant, R	Mean direction Azimuth, θ	Mean direction Inclination, ϕ
1	10	2	1.998	290	-76
2	39	7	6.874	107	-79
3	46	4	3.899	318	-80
4	47	4	3.961	340	-86
5	Σn_i				17
6	ΣR_i				16.732
7	R_T				16.492
8	k (number of groups)				4
9	$F = \dfrac{(\Sigma n_i - k)(\Sigma R_i - R_T)}{(k-1)(\Sigma n_i - \Sigma R_i)}$				3.85
10	d.f. in numerator $= 2(k-1)$				6
11	d.f. in denominator $= 2(\Sigma n_i - k)$				26
12	$F_{10\%}$ with 6 and 26 d.f.				2.01

* After Irving, 1964

directions; the latter are not needed for the calculations but are supplied for a direct comparison of the different directions. Lines 5 to 8 record the sum of observations, the sum of R-values, the R-value for all 17 points taken as a single group, and the number of groups. The value of F is calculated in line 9. In lines 10 and 11 are calculated the degrees of freedom required to look up the tabled F-value of line 12. Because the calculated F-value of 3.85 is larger than the tabled F-value of 2.01, the conclusion is that the mean directions of all clusters are different.

Procedure 3 is for determining whether two clusters have the same precision of clustering within limits of statistical fluctuation. For sites 46 and 47, table 10.35 gives the sample size and the numbers of degrees of freedom

TABLE 10.35. CALCULATIONS TO DETERMINE WHETHER TWO CLUSTERS OF POINTS ON A SPHERE HAVE THE SAME PRECISION OF CLUSTERING, APPLIED TO DIRECTIONS OF MAGNETIZATION OBSERVED AT TWO SITES IN THE TASMANIAN DOLERITES*

Line	Site number	Estimate of precision, k	Sample size, n	Degrees of freedom
1	46	30	4	6
2	47	78	4	6
3	$F = \dfrac{\text{larger estimate of precision}}{\text{smaller estimate of precision}}$			2.6
4	$F_{5\%}$			4.28

* After Irving, 1964.

[equal to $2(n-1)$]. An F-statistic is equal to the larger value of k divided by the smaller value. Because the calculated value of 2.6 is smaller than the tabled $F_{5\%}$-value of 4.28 for a two-tailed test at the 10 percent significance level, the conclusion is that the two clusters have the same precision of clustering. More than two groups can be compared by an extension of this method explained by Irving (1964).

Finally, one can determine whether the points are clustered or randomly distributed by comparing the calculated value of R with a value for the same sample size tabulated in table A-3. For the seven points at site 39 (table 10.32), the calculated R-value of 6.78 is compared with the tabulated value of 3.78, and the conclusion is drawn that the points are not randomly distributed.

Statistical Analysis of Points Distributed in a Girdle on a Sphere

The statistical model thus far described was devised by Fisher and is useful for points that form a single cluster. However, if the points form a girdle, the

Table 10.36. Azimuths, inclinations, and direction cosines for 35 directions of natural remanent magnetization observed in the Keuper marls, England*

Number	Angles		Direction cosines		
	Azimuth, θ	Inclination, ϕ	L	M	N
1	28	41	0.6664	0.3543	0.6561
2	19	70	0.3234	0.1114	0.9397
3	23	52	0.5667	0.2406	0.7880
4	35	40	0.6275	0.4394	0.6428
5	28	47	0.6022	0.3202	0.7314
6	24	60	0.4568	0.2034	0.8660
7	28	49	0.5793	0.3080	0.7547
8	29	67	0.3417	0.1894	0.9205
9	12	76	0.2366	0.0503	0.9703
10	17	66	0.3890	0.1189	0.9135
11	22	56	0.5185	0.2095	0.8290
12	28	60	0.4415	0.2347	0.8660
13	24	44	0.6571	0.2926	0.6947
14	27	60	0.4455	0.2270	0.8660
15	341	81	0.1479	-0.0509	0.9877
16	213	85	-0.0731	-0.0475	0.9962
17	235	64	-0.2514	-0.3591	0.8988
18	240	67	-0.1954	-0.3384	0.9205
19	192	77	-0.2200	-0.0468	0.9744
20	193	49	-0.6392	-0.1476	0.7547
21	215	31	-0.7021	-0.4917	0.5150
22	216	55	-0.4640	-0.3371	0.8192
23	219	45	-0.5495	-0.4450	0.7071
24	234	51	-0.3699	-0.5091	0.7771
25	198	67	-0.3716	-0.1207	0.9205
26	229	45	-0.4639	-0.5337	0.7071
27	230	70	-0.2198	-0.2620	0.9397
28	231	37	-0.5026	-0.6207	0.6018
29	224	75	-0.1862	-0.1798	0.9659
30	217	19	-0.7551	-0.5690	0.3256
31	237	84	-0.0569	-0.0877	0.9945
32	276	58	0.0554	-0.5270	0.8480
33	30	73	0.2532	0.1462	0.9563
34	78	86	0.0145	0.0682	0.9976
35	13	76	0.2357	0.0544	0.9703

* Data from Creer, 1957, p. 131.

Fisher model does not reveal it unless a transformation is used, such as one to obtain the poles of the original observations. For girdle data, Watson (1966) devised the method of analysis explained in this subsection. Watson's method is also useful for clustered points.

Watson's method is explained through the analysis of a set of paleomagnetic data from the Keuper marls in England (table 10.36). These points define the girdle plotted in figure 10.19. First, the direction cosines L, M, and N are

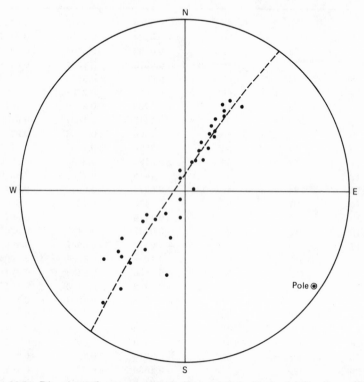

Fig. 10.19. Directions of magnetization in Keuper marls, England (after Creer, 1957).

calculated for each observation (table 10.36). Then the cross-products of the direction cosines are summed to form the matrix

$$\left[\begin{array}{ccc} \Sigma L^2 & & \\ \Sigma LM & \Sigma M^2 & \\ \Sigma LN & \Sigma MN & \Sigma N^2 \end{array} \right],$$

whose numerical values are given in table 10.37. Next the eigenvalues and eigenvectors are calculated for this matrix.

Table 10.37. Example of analysis of points distributed on a sphere; Watson's model applied to 35 directions of magnetization observed in the Keuper marls, England*

Sum of cross-products matrix of direction cosines:

	L	M	N
L	6.70657		
M	4.12005	3.42493	
N	1.84181	-1.30868	24.86850

Eigenvalues and eigenvectors:

Eigenvalues		Eigenvectors		
Rank	Size	L	M	N
1	25.09			
2	9.46			
3	0.45	-0.56538	0.82027	0.08660

Azimuth and inclination:

Estimate of azimuth $\theta = \tan^{-1}(0.82027/-0.56538) = 125°$

Estimate of inclination $\phi = \sin^{-1}(0.08660) = \qquad 5°$

* Data from table 10.36.

The relative sizes of the eigenvalues indicate whether the points form a girdle or a cluster. If two of the eigenvalues are large and roughly equal in size, the observations lie on a girdle or great circle. The direction of the line perpendicular to the girdle is obtained by substituting the eigenvector components for the smallest eigenvalue in the formulas

$$\text{estimate of azimuth} = \theta = \tan^{-1}\left(\frac{M}{L}\right)$$

and

$$\text{estimate of inclination} = \phi = \sin^{-1}(N).$$

The numerical results for the Keuper marls example are listed in table 10.37, and the pole of the direction perpendicular to the girdle is plotted in figure 10.19.

If one eigenvalue is large and the other two are small, the observations lie either in one cluster or in two antipodal clusters. The direction of the cluster or clusters is given by substituting the eigenvector components for the largest eigenvalue in the previous formulas. The numerical values obtained by applying Watson's method to the Tasmanian dolerite data of table 10.32 are given in table 10.38. The azimuth and inclination are essentially the same as those calculated in table 10.32 by Fisher's method.

TABLE 10.38. EXAMPLE OF ANALYSIS OF POINTS DISTRIBUTED ON A SPHERE; WATSON'S MODEL APPLIED TO SEVEN DIRECTIONS OF MAGNETIZATION OBSERVED AT SITE 39 IN THE TASMANIAN DOLERITES*

Sum of cross-products matrix of direction cosines:

	L	M	N
L	0.08992		
M	0.00213	0.38486	
N	0.38478	−1.14115	6.52521

Eigenvalues and eigenvectors:

Eigenvalues		Eigenvectors		
Rank	*Size*	*L*	*M*	*N*
1	6.752	−0.05670	0.17611	−0.98274
2	0.214			
3	0.034			

Azimuth and inclination:

Estimate of azimuth $\theta = \tan^{-1}(0.17611/ - 0.05670) = 108°$

Estimate of inclination $\phi = \sin^{-1}(- 0.98274) = \quad -79°$

* Data from table 10.32.

If all three eigenvalues are of about the same size, the observations are neither particularly clustered nor arranged in a girdle. However, this is not evidence that the points are distributed randomly.

Thus far no statistical tests have been devised for Watson's method and it is therefore essentially descriptive, but extremely useful nonetheless, especially for points in a girdle. The advantage of Fisher's method for clustered points is that tests can be performed.

Statistical Analysis of Clustered Points on the Hemisphere

Some of the methods developed for observations on the sphere may be used more or less successfully for points on the hemisphere. In this subsection these methods are illustrated by analysis of a set of 49 *b*-axis lineations measured in Precambrian gneisses and other crystalline rocks in the Gray Rock–Livermore Mountain area, Larimer County, Colorado, by Connor (1962). The original data are plotted in stereographic projection in figure 10.20.

When Fisher's model is applied to Connor's data, the results in table 10.39 are obtained. The estimated mean direction is plotted as a cross in figure 10.20. Connor's mean direction, estimated by contouring points averaged in 1 percent

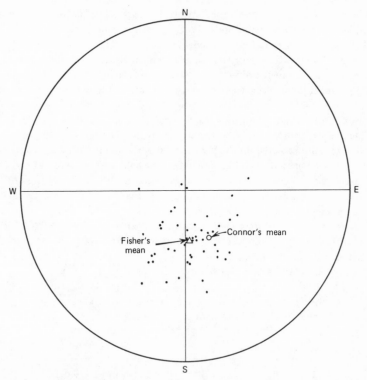

Fig. 10.20. Directions of 49 b-axis lineations in Precambrian gneisses and other crystalline rocks in Colorado (data from Connor, 1962).

TABLE 10.39. CALCULATION OF MEAN AZIMUTH, MEAN INCLINATION, AND PRECISION FOR 49 b-AXIS LINEATIONS

Item	Value
Number of points	49
Mean direction cosines:	
L	-0.421
M	0.036
N	0.907
R	46.42
Estimate of azimuth	$175°$
Estimate of inclination	$65°$
Estimate of precision	19

circles (azimuth 158 degrees, inclination 64 degrees), is plotted as a circle. Considering that the two methods of estimation are conceptually different, we find the agreement remarkably good.

The preceding estimate of the azimuth and inclination of the center of a cluster is sensitive to the direction of the mean relative to the equator of the hemisphere. If points are clearly clustered on or near the equator, they should be rotated so that the approximate center of the cluster is far removed from the equator. Under these conditions the preceding estimates of the azimuth and inclination are valid, but the estimate of precision is too large because the number of points plotted on the hemisphere to represent a certain direction is twice the number that would be plotted on the sphere. The relative precisions of clusters can be compared, however, provided both that the centers of the clusters are well removed from the equator (either originally or by rotation) and that the precision of clustering is fairly high.

A cone of confidence for the 49 lineations is calculated in table 10.40. This calculation and the calculation to determine whether two or more clusters have the same mean direction should work fairly well for data on the hemisphere, with the qualification stated in the preceding paragraph.

Table 10.40. Calculation of a cone of confidence, applied to 49 b-axis lineations

Line	Value	Symbol	Explanation
1	49	n	Number of observations
2	0.049		Value from table A-2
3	46.42	R	From table 10.39
4	0.99728	$\cos A$	$1 - 0.049(n - R)/R$
5	4.23°	A	

When Watson's model is applied to the lineation data, the results in table 10.41 are obtained. They are in good agreement with those obtained by Fisher's model.

The test of randomness described previously for points on the sphere is clearly invalid for observations on the hemisphere, since the points, just because they are restricted to one hemisphere, cannot be randomly distributed with respect to the sphere. A valid way to determine whether points are randomly distributed on a hemisphere is to test whether the number of points per unit area follows the Poisson distribution (sec. 6.1). This test can readily be performed with the aid of an equal-area Schmidt net or a computer program. In table 10.42 this test is applied to the fictitious observations of figure 10.16, which are 50 points weakly clustered (with a precision of 10) about a line

TABLE 10.41. EXAMPLE OF ANALYSIS OF POINTS DISTRIBUTED ON A
HEMISPHERE; WATSON'S MODEL APPLIED TO 49 b-AXIS LINEATIONS*

Sum of cross-products matrix of direction cosines:

	L	M	N
	L	*M*	*N*
L	9.538		
M	− 0.410	2.934	
N	− 16.165	1.547	36.528

Eigenvalues and eigenvectors:

Eigenvalues		Eigenvectors		
Rank	Size	L	M	N
1	44.15	− 0.42321	0.03819	0.90522
2	2.96			
3	1.89			

Azimuth and inclination:

Estimate of azimuth $\theta = \tan^{-1}(0.03819/-0.42321) = 175°$

Estimate of inclination $\phi = \sin^{-1}(0.90522) = \qquad 65°$

* Data from Connor, 1962.

TABLE 10.42. CALCULATION TO DETERMINE
WHETHER POINTS ARE RANDOMLY DISTRIBUTED
ON THE HEMISPHERE, APPLIED TO 50 CLUS-
TERED POINTS FROM FIGURE 10.16

Number of points per cell	Number of cells	
	Observed	Expected
0	51	43.7
1	21	27.0
2	4	8.3
3	1	1.7
4	2 ⎫	0.3
5	2 ⎭	

directed vertically downward. In order to obtain a mean number of about
one point per cell, the hemisphere was divided into 20-degree squares, 81 in
number. The observed and expected numbers of cells with a specified number
of points are listed in table 10.42. When the chi-square goodness-of-fit test

(sec. 6.1) is applied, the observed value of 50.7 is much larger than the tabled chi-square value of 6.25 with 3 degrees of freedom at the 10 percent significance level; therefore these points are not randomly distributed. (Of course, this fact was already known for these data.)

Concluding Remarks

The reader is cautioned once again that the methods in this section for the analysis of directional data are valid only for directed lines, as plotted on the sphere. For undirected lines, as plotted on the hemisphere, the methods can yield only approximations, pending the development of more rigorous statistical procedures.

REFERENCES

Ager, D. V., 1963, Paleoecology: New York, McGraw-Hill, 371 p.

Albert, Adrian, 1949, Solid analytic geometry: New York, McGraw-Hill, 162 p.

Batschelet, Edward, 1965, Statistical methods for the analysis of problems in animal orientation and certain biological rhythms: Am. Inst. Biol. Sci., Monograph, 57 p.

Billings, M. P., 1954, Structural geology: Englewood Cliffs, N.J., Prentice-Hall, 514 p.

Bingham, C., 1964, Distributions on a sphere and on the projective plane: Yale Univ., Ph.D. thesis.

Chayes, Felix, and Métais, D., 1964, Discriminant functions and petrographic classification, *in* Carnegie Institution, Annual report of the director of the geophysical laboratory, 1963–1964, p. 179–186.

Connor, J. J., 1962, Precambrian petrology and structural geology of the Gray-Rock— Livermore Mountain area, Larimer Co., Colo.: Colo. Univ., Ph.D. thesis.

Cooley, W. W., and Lohnes, P. R., 1962, Multivariate procedures for the behaviorial sciences: New York, John Wiley & Sons, 211 p.

Creer, K. M., 1957, The remnant magnetism of unstable Keuper marls: Philosophical Trans. Royal Soc. London, ser. A., no. 974, v. 250, p. 130–143.

Dawson, K. R., and Whitten, E. H. T., 1962, The quantitative mineralogical composition and variation of the Lacorne, La Motto, and Preissac granitic complex, Quebec, Canada: Jour. Petrology, v. 3, no. 1, p. 1–37.

Draper, N. R., and Smith, Harry, 1966, Applied regression analysis: New York, John Wiley & Sons, 407 p.

Fisher, R. A., 1938, The statistical utilization of multiple measurements: Ann. Eugenics, v. 8, p. 376–386.

Fisher, R. A., 1953, Dispersion on a sphere: Proc. Royal Soc. London, ser. A., v. 217, p. 295–305.

Gomez, Manuel, and Hazen, Kathleen, 1970, Evaluating sulfur and ash distribution in coal seams by statistical response surface regression analysis: U.S. Bur. Mines, Rept. Inv. 7377, 120 p.

Hills, E. S., 1963, Elements of structural geology: New York, John Wiley & Sons, 483 p.

Imbrie, John, 1963, Factor and vector analysis programs for analyzing geologic data: U.S. Office of Naval Research, Tech. Rept. 6, 83 p.

Imbrie, John, and Newell, Norman, eds., 1964, Approaches to paleoecology: New York, John Wiley & Sons, 432 p.

Imbrie, John, and Purdy, E. G., 1962, Classification of modern Bahamian carbonate sediments, *in* American Association of Petroleum Geologists, Classification of carbonate rocks—A symposium, Mem. 1, p. 253–272.

Imbrie, John, and van Andel, T. H., 1964, Vector analysis of heavy mineral data: Geol. Soc. Am. Bull., v. 75, p. 1131–1156.

Irving, E., 1964, Paleomagnetism: New York, John Wiley & Sons, 399 p.

James, W. R., 1967, Nonlinear models for trend analysis in geology, *in* Merriam, D. F., and Cocke, N. C., eds., Computer applications in the earth sciences—Colloquium on trend analysis, Kansas State Geol. Surv., Lawrence, Computer Contr. 12, p. 26–30.

Joralemon, Peter, 1967, The fifth dimension in ore search: Soc. Mining Engineers, Am. Inst. Mining Engineers, fall mtg., preprint, 18 p.

Journal of Geology, 1966, v. 74, no. 5, pt. 2, p. 653–830 (special supplement devoted to papers on applications of multivariate statistics in geology).

Klovan, J. E., 1968, Selection of target areas by factor analysis, *in* Proceedings of a symposium on decision-making in mineral exploration, Vancouver, B.C., 1968, p. 19–27.

Krumbein, W. C., 1939, Preferred orientation of pebbles in sedimentary deposits: Jour. Geology, v. 47, p. 673–706.

Krumbein, W. C., and Graybill, F. A., 1965, An introduction to statistical models in geology: New York, McGraw-Hill, 475 p.

Link, R. F., and Koch, G. S., Jr., 1967, Linear discriminant analysis of multivariate assay and other mineral data: U.S. Bur. Mines, Rept. Inv. 6898, 25 p.

Louden, T. V., 1964, Computer analysis of orientation data in structural geology: Northwestern Univ., Tech. Rept. 13, ONR Task No. 389-135, 129 p.

MacLane, Saunders, and Birkhoff, Garrett, 1967, Algebra: New York, Macmillan, 598 p.

McCammon, R. B., 1969, Multivariate methods in geology, *in* Fenner, Peter, ed., Models of geologic processes, Washington, D.C., Am. Geol. Inst. (pages not sequentially numbered, McCammon's article is about 141 p. long).

Merriam, D. F., ed., 1966, Computer applications in the earth sciences—Colloquium on classification procedures: Kansas Geol. Surv., Lawrence, Computer Contr. 7, 79 p.

Mertie, J. B., Jr., 1949, Charting five and six variables on the bounding tetrahedra of hypertetrahedra: Am. Mineralogist, v. 34, p. 706–716.

Miller, R. L., and Kahn, J. S., 1962, Statistical analysis in the geological sciences: New York, John Wiley & Sons, 483 p.

Mosteller, Frederick, and Wallace, D. L., 1964, Inference and disputed authorship: The Federalist: Reading, Mass., Addison-Wesley, 285 p.

Phillips, F. C., 1960, The use of stereographic projection in structural geology, 2nd ed.: London, Edward Arnold, 86 p.

Pretorius, D. A., 1966, Conceptual geological models in the exploration for gold mineralization in the Witwatersrand basin, *in* Symposium on mathematical statistics and computer applications in ore valuation, South African Inst. Mining Metall., p. 225–275.

Washington, H. S., 1917, Chemical analyses of igneous rocks: U.S. Geol. Surv. Prof. Paper 99, 1201 p.

Watson, G. S., 1956, A test for randomness of directions: Monthly Notices, Royal Astron. Soc., Geophysical Suppl., v. 7, p. 160–161.

————, 1966, The statistics of orientation data: Jour. Geology, v. 74, no. 5, pt. 5, p. 786–797.

Wilks, S. S., 1962, Mathematical statistics: New York, John Wiley & Sons, 644 p.

Chapter 11

Ratios and Variables of Constant Sum

11.1 INTRODUCTION

Ratios and variables of constant sum are of more interest in theoretical than in applied geology, particularly in petrology, where the ratio of one oxide, chemical element, or mineral to another is often studied, and where rock-analysis data sum to a constant 100 percent. Because the subject of ratios and the subject of variables of constant sum have historically been intertwined in geology, they are presented together in this chapter. Sections 11.2 and 11.3 are about ratios, and sections 11.4 and 11.5 are about variables of constant sum.

Ratios are a natural way to express many geological data. For example, the quartz/feldspar ratio may be used to characterize a granite, and the C^{14}/C^{12} ratio may be used to date a specimen containing organic material. A comparison of ratios may also be used to illuminate a geological process, such as ore formation. A silver/lead ratio, for instance, would vary from place to place if all or part of these metals were introduced into an ore deposit at different times. But if the metals were introduced simultaneously and the galena were saturated with silver, the ratio would remain constant.

A geologist often has a well-defined purpose in mind when calculating means and standard deviations. He may want to estimate the porosity of an oil sand or the specific gravity of a rock. It is not clear that all the geologists who calculate ratios have such well-defined purposes, however. Problems are difficult to formulate sharply; much of the formulation depends on tradition and on groping, as is revealed by the diverse ways to plot and interpret ratio data that abound in the literature of petrology. We therefore avoid discussing any particular theoretical basis in this chapter and explain instead some empirical methods that are useful in formulating and solving problems.

Variables of constant sum have the property of adding up to a certain total. Because many geological data, especially rock analyses, are expressed in percentages, the constant sum is often 100 percent. Whenever the sum of variables is constant, their statistical properties are peculiar and deserve attention. Standard statistical tests contain little or no information on constant-sum data.

11.2   RATIOS

Frequency distributions of ratios have already been introduced in the study of the t- and F-distributions (secs. 4.3 and 5.2). These ratios were formed to compare differences between or among means, with the variance of the difference as a yardstick. The ratios calculated from the samples are dimensionless quantities. The ratios studied in this chapter, however, are themselves the subject of interest, rather than being quantities derived secondarily for the purpose of making comparisons.

A ratio is formed from two numbers, a numerator w and a denominator x. Depending on the investigator's purpose, w and x may be defined as observations, or the ratio w/x may be the observation.

Two formulations of a ratio problem may be made, either of which may be more appropriate to a particular situation. In the first, the *ratio formulation*, the new variable R, equal to w/x, is defined and the relation to be studied is

$$R = \mu_R + e,$$

where the mean of the fluctuation e is assumed to be zero.

The second, the *regression formulation*, assumes a closer relationship between the two variables; in statistical terms, more "structure" is introduced, and the appropriate relation to be studied is

$$w = \beta x + e.$$

With this equation, if the mean of the fluctuation e is assumed to be zero, it can be shown that

$$\beta = \frac{\mu_w}{\mu_x}.$$

The two formulations can be contrasted through an example. Consider that silver and lead assays have been determined for a set of geological samples. If the silver assay is designated w and the lead assay x, an estimate of the average silver/lead ratio may be interesting. Under the first formulation, both

w and x may be assumed to be distributed lognormally, and the problem becomes how to measure the ratio for the means of these two distributions. Alternatively, if one supposes that galena can contain a constant proportion of silver, the silver content of galena should be constant within the limits of statistical fluctuation, independent of the lead assay. In this case the second formulation, a special situation for a regression equation, is more appropriate. Analysis of observations under both formulations is discussed in section 11.3.

In practice, it is often difficult to choose between the two formulations, although fortunately both lead to estimating nearly the same quantity. The value μ_w/μ_x in the ratio formulation equals the value of β in the regression formulation. One would intuitively suppose that the value of μ_w/μ_x in the first formulation should be nearly or exactly equal to the mean of R, so that R would also be equal to β. In fact, the mean of R is close to but not exactly the value of μ_w/μ_x. The regression formulation can always be used if enough assumptions can be satisfied about the frequency distributions and the relation between the variables w and x. If fewer assumptions can be satisfied, the approximate methods of calculating a ratio formulation are appropriate.

Behavior of Statistics Used to Estimate Ratios

Little is known about how statistics used to estimate ratios behave. The exact mathematical theory for calculating the distribution of R in the ratio formulation has not been developed, except for a few cases. An approximate theory, explained in section 11.3, is available. For the regression formulation, ordinary regression analysis (sec. 9.2) applies.

Most statistics used to estimate μ_R are biased, but a moderate amount of bias is not troublesome because ratios are usually estimated for descriptive purposes. The bias is even less troublesome when comparing the means of R for two or more groups of ratios, because it tends to be a slowly varying function of the values of the means and to disappear when the comparisons are made. If sufficient data are available so that the coefficients of variation are all small, the bias must always be small.

Properties of Ratios

Certain mathematical properties of ratios are now briefly reviewed. In a ratio w/x, small changes in w and x have very different effects on the ratio, depending on the sizes of w and x. In table 11.1, four cases are illustrated in a 2×2 table, with each cell partitioned into nine subcells. In the upper left-hand cell, the initial value of w is 2.00 and that of x is 1.00, yielding a ratio of 2.00 in the central subcell. The effect of changing numerator and denominator by 0.25 is shown by the entries in the other eight subcells. In this case, with small values of both w and x, a small change in the denominator has a greater effect than a small change in the numerator.

Entries in the other three cells show the effects of making the same change of 0.25. In the upper right-hand cell, with a large value of w and a small value of x, a small change in the numerator has almost no effect, but a small change in the denominator has a substantial effect. In the lower left-hand cell, with a small value of w and a large value of x, a small change has little effect if made in the denominator but a substantial one if made in the numerator. Finally, in the lower right-hand cell, with large values of both w and x, small changes in either numerator or denominator have virtually no effect on the ratio.

TABLE 11.1. EFFECTS ON RATIO w/x, OF SMALL CHANGES IN w AND x, FOR DIFFERENT INITIAL VALUES OF w AND x

			w			
	1.75	2.00	2.25	59.75	60.00	60.25
0.75	2.33	2.67	2.99	79.67	80.00	80.33
1.00	1.75	2.00	2.25	59.75	60.00	60.25
1.25	1.40	1.60	1.80	47.80	48.00	48.20
29.75	0.0588	0.0672	0.0756	2.01	2.02	2.03
30.00	0.0583	0.0667	0.0750	1.99	2.00	2.01
30.25	0.0579	0.0661	0.0744	1.98	1.98	1.99

x labels the rows.

Because division by zero is not allowed by mathematical rule, the ratio w/x cannot be computed if the denominator is zero. However, data often contain zero x-values, usually because the amount of a constituent is below the detection limit of the analytical method used (sec. 8.5). If only a few x-values are zero, they can be reset to a small arbitrary value. If many x-values are zero, they should be sorted out and examined separately. The regression formulation may then be used or a modified ratio calculated. Calculation of a modified ratio is developed in the next subsection.

An Example to Demonstrate Some Properties of Ratios

Some of the properties of ratios are demonstrated with data taken from 1000 typical mine samples of the Transvaal vein of level 10 of the Frisco mine. Each of the 1000 observations consists of three numbers—the assays for silver, copper, and lead. Because the veins also contain about 10 percent zinc, gold in trace amounts, and about 85 percent gangue minerals, the silver, copper, and lead assays are not constant-sum data. In table 11.2 the observations are represented on a 2×2 table with values from the Cu/Pb ratio that are obtained after sorting out zero assays in the numerator or denominator. The means of copper and zinc are not sensitive to this classification, and only two observations require special handling because zero appears in the denominator.

Table 11.2. Classification of 1000 observations according to zero and nonzero values in the ratio Cu/Pb, for Frisco mine data

		Lead	
		Zero values	Nonzero values
Copper	Zero values	0 observations Cu = 0 Pb = 0	19 observations Cu = 0 Pb = 4.37
	Nonzero values	2 observations Cu = 0.82 Pb = 0	979 observations Cu = 0.63 Pb = 7.00

Three different ways to treat the two observations with zero lead assays are to disregard them, to set the values of the ratios Cu/Pb and Ag/Pb equal to arbitrarily high numbers, or to define the new ratios Cu/(Cu + Pb) and Ag/(Ag + Pb). The advantage of using the form

$$\frac{w}{w+x}$$

is that its largest possible value is 1, obtained when x is equal to zero but w is not equal to zero, and its smallest possible value is zero, obtained when w is equal to zero. Table 11.3 illustrates the three methods of calculating. If the two observations are omitted, the results in column 1 are obtained. If the two observations are included, with Cu/Pb set equal to 9999 and Ag/Pb set equal to 9,999,999, the means and standard deviations of these ratios are enormously inflated even though the means and standard deviations of the assays of copper and silver are not appreciably changed. This result would be obtained even if lead were equal to a very small value instead of zero. Finally, if the form $w/(w+x)$ is used, the means and standard deviations of both ratios are scarcely changed.

Table 11.3. Means and variances of 1000 and 998 observations of Frisco mine data, to illustrate three methods of calculating

Item	(1) 998 observations		(2) 1000 observations	
	$\overline{w}$	s	$\overline{w}$	s
Cu	0.61	0.48	0.63	0.54
Pb	6.95	5.73	6.95	5.73
Ag	153.15	135.70	154.65	139.62
Cu/Pb	0.23	0.56	20.23	446.94
Ag/Pb	41.67	199.49	20,041.58	446,987.85
Cu/(Cu + Pb)	0.13	0.14	0.13	0.15
Ag/(Ag + Pb)	0.94	0.09	0.94	0.09

Table 11.4. Frequency distribution of Cu/Pb ratios from 1000 observations of Frisco mine data

Class interval	Frequency
0.0–0.2	759
0.2–0.4	135
0.4–0.6	38
0.6–0.8	16
0.8–1.0	10
1.0–1.2	8
1.2–1.4	4
1.4–1.6	5
1.6–1.8	4
1.8–2.0	1
- - - - - - - - - - - - -	
2.0–10.0	18
- - - - - - - - - - - - -	
9999	2

Table 11.3 illustrates that including as few as two values with zero in the denominator can make a big difference in calculations involving ratios, and also shows a way out of this difficulty in the form of the transformation $w/(w+x)$. However, this method will not work if too many numbers in the denominator are nearly or exactly zero. How many is too many cannot be specified exactly, because the number depends on the frequency distribution

of all values, but more than 10 percent would certainly be too many. How near to zero the numbers can be without creating a problem cannot be specified either, but trouble results if a small fraction of the ratios is several orders of magnitude larger than the rest. If the transformation is not successful, the regression formulation may be used.

Three frequency distributions show the effect of the transformation on the ratios. The distribution of Cu/Pb ratios (table 11.4) is highly skewed to the right. Most of the values are less than 1, two arbitrary values are 9999, and 18 large values between 2 and 10 warrant attention because copper is exceptionally large or lead is exceptionally small. Inspection of all data for the 18 corresponding mine samples might suggest mistakes in assaying, copying, or data processing; a geographic concentration of metals; or something else of interest. Table 11.5, for Ag/Pb ratios, is also a distribution highly skewed to

TABLE 11.5. FREQUENCY DIS-
TRIBUTION OF Ag/Pb RATIOS
FROM 1000 OBSERVATIONS OF
FRISCO MINE DATA

Class interval	Frequency
0–20	467
20–40	359
40–60	70
60–80	27
80–100	19
100–120	12
120–140	10
140–160	5
160–180	9
180–200	5
200–1000	15
9,999,999	2

the right; most of the values are larger than 1, two arbitrary values are 9,999,999, and 15 large values between 200 and 1000 require inspection.

The frequency distributions of the two transformed observations are compared in table 11.6. The distribution for Cu/(Cu + Pb) is skewed to the right like that for Cu/Pb and is superficially similar, but that for Ag/(Ag + Pb) is skewed to the left, with the eight entries in class interval 0.00 to 0.04 representing zero silver values. The extreme skewness is not uncommon and suggests that a logarithmic transformation might be appropriate. Of course, if the original variables—copper, lead, and silver—were distributed exactly lognormally, the distributions of the ratios would also be lognormal. Then it would be necessary to change the 29 observations, for which at least one assay was zero, to small arbitrary values or to discard them from the analysis.

TABLE 11.6. FREQUENCY DISTRIBUTIONS OF $Cu/(Cu + Pb)$ AND $Ag/(Ag + Pb)$ RATIOS FROM 1000 OBSERVATIONS OF FRISCO MINE DATA

Class interval	$\dfrac{Cu}{Cu + Pb}$	$\dfrac{Ag}{Ag + Pb}$
0.00–0.04	209	8
0.04–0.08	264	0
0.08–0.12	170	0
0.12–0.16	105	0
0.16–0.20	58	0
0.20–0.24	53	0
0.24–0.28	33	0
0.28–0.32	22	0
0.32–0.36	12	0
0.36–0.40	11	0
0.40–0.44	11	0
0.44–0.48	10	0
0.48–0.52	4	0
0.52–0.56	5	0
0.56–0.60	4	0
0.60–0.64	8	0
0.64–0.68	1	0
0.68–0.72	5	0
0.72–0.76	4	0
0.76–0.80	5	1
0.80–0.84	2	2
0.84–0.88	1	7
0.88–0.92	1	85
0.92–0.96	0	495
0.96–1.00	2	402

11.3 STATISTICAL ANALYSIS OF RATIOS

In this section a statistical analysis of ratios is presented for the two formulations previously explained.

Ratio Estimation

In the ratio formulation, w and x are two random variables. A statistic must be found to estimate μ_R, the mean of the ratio w/x. If the distributions of both w and x are lognormal, the distribution of R is also lognormal, and methods for the lognormal distribution may be applied to the n sample ratios R (sec. 6.2).

If the exact distributions of w and x are unknown, an approximate method must be used. The parameter μ_R is estimated by the statistic $\overline{w}/\overline{x}$, whose mean is approximately correct for moderately large samples (the bias would be about $1/n$). The variance of $\overline{w}/\overline{x}$ is approximately

$$\frac{\overline{w}^2}{n\overline{x}^2}\left(\frac{s_w^2}{\overline{w}^2} + \frac{s_x^2}{\overline{x}^2} - \frac{2\,r_{wx}s_x s_w}{\overline{x}\,\overline{w}}\right),$$

where r is the correlation coefficient between w and x. If $\overline{w}/\overline{x}$ is assumed to be normally distributed, confidence intervals may be calculated in the usual manner (sec. 4.4).

In the regression formulation,

$$w = \beta x + e,$$

the mean value of e is supposed to be zero, and the assumption made about the variance of the e-values will determine the statistic used to estimate β. Many assumptions may be made; we discuss the consequences of only three.

First, if the variance of the e-values is assumed to be a constant, β is estimated by the expression

$$\frac{\sum wx}{\sum x^2}.$$

Second, if the variance of the e-values is assumed to be proportional to x, β is estimated by the expression

$$\frac{\overline{w}}{\overline{x}}.$$

And third, if the variance of the e-values is assumed to be proportional to x^2, β is estimated by the expression

$$\frac{\sum w/x}{n}.$$

An example demonstrates the differences among statistics calculated under these three assumptions about the variance of e-values. In the study of data from the City Deep mine, South Africa (Koch and Link, 1966), we considered three different estimates of variance for width-measurement errors. When gold-bearing conglomerate beds in South Africa are sampled, the bed width is measured, a sample is taken across this width, and an assay is made. The gold value is expressed in inch-pennyweights (in.-dwt), a unit obtained by multiplying width in inches by assay in pennyweights (dwt). Because the measured width is usually wider than the true width of the bed, the estimate of average pennyweight depends on the behavior of the errors of width measurements and is not the same as the assay result.

Table 11.7. Comparison of estimates of average pennyweight values for development and stope data, City Deep mine, South Africa

Assumption about variance of width measurement errors	Mean of pennyweight values	
	Development data	Stope data
$\sigma^2 = a$	11.69	8.21
$\sigma^2 = bx$	15.02	9.23
$\sigma^2 = cx^2$	22.58	12.00

If the bias in width measurements is ignored, the estimates in table 11.7 are appropriate under the previous assumptions about the variance of width-measurement errors. For the City Deep data, σ^2 is the variance of width-measurement errors; a, b, and c are constants; x is the width measurement for a mine sample; w is the inch-pennyweight value for a mine sample; and n is the number of mine samples. Unless the nature of the measurement errors of width is known, the best of the three estimates cannot be identified, and, in general, none is unbiased. However, table 11.7 makes clear that the estimate is sensitive to the assumption made about the measurement errors.

Ratio Comparison

The method used to compare groups of ratios depends on the formulation adopted. For the ratio formulation, assuming that the variance of the e-values is constant, the groups of ratios may be compared in pairs (two at a time). This comparison can be made because the approximate mean and standard deviation of the statistic estimating the ratio are known, and, if the groups are independent, the variances of the difference of the ratios equal the sum of the variances of the individual ratios. The assumption must be made that the difference of the ratios is normally distributed.

For the regression formulation, an analysis of variance can be performed. This procedure is indicated when the variance of the e-values is assumed to be proportional to x. There are k groups of data, consisting of similar w- and x-variables. A ratio statistic $\overline{w}/\overline{x}$ is calculated for each group. A within-ratio sum of squares can then be calculated by the formula

$$\sum \left\{ w - \left(\frac{\overline{w}}{\overline{x}} \right) x \right\}^2 = \sum w^2 + \left(\frac{\overline{w}}{\overline{x}} \right)^2 \sum x^2 - 2 \left(\frac{\overline{w}}{\overline{x}} \right) \sum wx.$$

The variance of $\overline{w}/\overline{x}$ is the within-ratio sum of squares multiplied by the quantity $1/(n\overline{x}^2)$. An among-ratios sum of squares may be calculated directly by finding the variance of the k ratios. This variance has $(k-1)$ degrees of of freedom. The within-ratio sum of squares is the pooled statistic estimating s^2. The usual F-statistic, with $(k-1)$ and $k(n-1)$ degrees of freedom, may then be calculated.

Results of ratio comparisons are approximate for both the first and second formulations because the within-ratio variances do not have the same mean-square values unless the x-values themselves have the unlikely property of being equal in number, mean, and variance. As soon as more than a simple F-test is required, multiple comparisons are needed, and methods similar to those explained in section 5.12 are used, with modifications to account for the unequal variances. A detailed explanation, beyond the scope of this book, is given in the works cited in that section.

Eigenvalue Analysis of Ratio Data

With more than two variables, as in the previous example involving silver, copper, and lead, several ratios may be formed, more ratios sometimes than there are dimensions and independent variables. A set of ratios may be examined by eigenvalue analysis (sec. 10.6) to determine whether the number of dimensions present is equal to the number of ratios. If the number of dimensions is smaller, a new set of independent variables may be defined, equal to the number of dimensions.

TABLE 11.8. EIGENVALUES OF LOGARITHMS OF SIX RATIOS, FORMED FROM FOUR METALS IN VEINS OF THE FRISCO MINE

Dimension	Eigenvalue
1	2.31123
2	2.00198
3	0.96667
4	0.00000
5	0.00000
6	0.00000

As an example, we cite again the study of grouped veins of the Frisco mine introduced in section 10.5. From logarithms of the four metals (silver, lead, copper, and zinc), the six logarithmic ratios $\ln(Ag/Pb)$, $\ln(Ag/Zn)$, $\ln(Ag/Cu)$, $\ln(Pb/Zn)$, $\ln(Pb/Cu)$, and $\ln(Zn/Cu)$ may be formed. However, these six ratios are not independent variables because some of them are linear combinations of the others. Table 11.8 summarizes the results of an eigenvalue analysis applied to the within-group dispersion matrix for six ratios for the data of the 91 groups. Because three of the eigenvalues are zero, the data have only three dimensions, which may be expressed alternatively as any one of the sets of ratios in table 11.9.

TABLE 11.9. BASES FOR LOGARITHMS OF METAL RATIOS, CALCULATED FOR VEINS OF THE FRISCO MINE

Sets of ratios		
Ag/Pb,	Ag/Zn,	Ag/Cu
Pb/Zn,	Pb/Cu,	Ag/Cu
Pb/Zn,	Pb/Cu,	Ag/Pb
Pb/Zn,	Pb/Cu,	Ag/Zn
Pb/Zn,	Ag/Zn,	Ag/Cu
Pb/Zn,	Ag/Pb,	Ag/Cu
Pb/Cu,	Ag/Zn,	Ag/Cu
Pb/Cu,	Ag/Pb,	Ag/Zn
Zn/Cu,	Ag/Pb,	Ag/Cu
Zn/Cu,	Ag/Pb,	Ag/Zn

Only three dimensions are present because any one of the six ratios can be expressed as a linear combination of any of the sets of ratios that appear in table 11.9. For example, if the first set is used,

$$\ln\frac{Pb}{Zn} = \ln\frac{Ag}{Zn} - \ln\frac{Ag}{Pb}.$$

If the second set is used,

$$\ln\frac{Ag}{Zn} = \ln\frac{Ag}{Cu} - \ln\frac{Pb}{Cu} + \ln\frac{Pb}{Zn}.$$

Confidence Intervals and Hypothesis Tests for the Coefficient of Variation

The coefficient of variation, introduced in section 3.3, is developed as a measure of relative geological variability in section 7.8. Because the coefficient of variation is a ratio, that of the standard deviation to the mean, the confidence intervals and hypothesis tests for it are discussed in this chapter. Hald (1952) explains the methods. They apply to the special case of uncorrelated, independent, normally distributed variables, of which the coefficient of variation is one example. The material in this subsection is relevant only to the coefficient of variation and to other ratios with the same statistical properties.

The coefficient of variation C is approximately normally distributed, provided that the observations w are approximately normally distributed and their number is moderately large (greater than $\gamma^2/0.01$). The mean of the distribution of C is approximately

$$\gamma = \frac{\sigma}{\mu},$$

and the variance is approximately

$$\gamma^2 \frac{1 + 2\gamma^2 (\text{d.f.}/n)}{2\,\text{d.f.}}.$$

Approximate confidence limits may be obtained for γ by replacing γ by its statistic C in the previous formula for the variance of γ. The values obtained are listed in tables A-4 and A-5. (An alternative approximation that may be more accurate is also discussed by Hald.)

Two forms of hypothesis tests also depend on the previous formula for the variance of γ. The first hypothesis is that a sample with a coefficient of variation C was drawn from a population with a specified population coefficient of variation γ. This hypothesis might be tested to investigate whether a new set of observations w came from a population whose relative variability was like that of a known population. For example: Do the gold assays from a particular deposit correspond to those from a known mineralization type ? Or has a new sampling method yielded observations whose relative variability γ is consistent with the observations that were set as a standard ? In order to test this hypothesis, the standardized value of z of the difference between C and γ is obtained by the formula

$$z = \frac{C - \gamma}{\sqrt{\gamma^2 (1 + 2\gamma^2 [\text{d.f.}/n])/2\,\text{d.f.}}},$$

where the denominator is the square root of the variance of γ. The calculated value of z is then compared with the critical value of z, obtained from a table for the standardized normal deviate. For a 10 percent significance level, the critical value of z is ± 1.645; for a 5 percent significance level, the critical value is ± 1.960.

The second hypothesis is that two samples are drawn from populations with the same population coefficient of variation, γ. For instance, one might want to investigate whether a particular method of sampling two batholiths yielded results with the same relative variability. The statistic z is computed by the formula

$$z = \frac{C_1 - C_2}{\sqrt{C_1^2 (1 + 2C_1^2[\text{d.f.}_1/n_1])/2\text{d.f.}_1 + C_2^2(1 + 2C_2^2[\text{d.f.}_2/n_2])/2\text{d.f.}_2}},$$

where C_1 is the coefficient of variation calculated for the first sample, C_2 is the coefficient of variation calculated for the second sample, and the denominator is the square root of the sum of the variances. Then, as with the previous test, the calculated value of z is compared with its critical value.

Three example problems are worked for data from the Getchell mine, Humboldt County, Nevada (Koch and Link, 1970). The data are from mine samples taken by two types of drills, wagon and diamond.

The first problem is to calculate confidence intervals for the coefficients of variation for the two types of data. For the wagon-drill-hole cuttings (table 11.10), C is 1.026 and n is 956. The corresponding value from table A-4 is 0.068

Table 11.10. Data for example problems about the coefficient of variation*

Item	Sample 1, wagon-drill-hole cuttings	Sample 2, diamond-drill-hole cores
Mean, $\bar{w}$, ounces/ton	0.152	0.111
Standard deviation, s	0.156	0.139
Coefficient of variation, C	1.026	1.252
Number of observations, n	956	87
Degrees of freedom, d.f.	955	86

* Data from Koch and Link, 1970.

(by interpolation). Thus a 90 percent two-sided confidence interval for γ extends from 0.958 to 1.094 (1.026 ± 0.068). For the diamond-drill-hole cores, C is 1.252 and n is 87. The corresponding value from table A-4 is 0.328 (by interpolation). Thus the confidence interval extends from 0.934 to 1.570.

The second problem is to test the hypothesis that the gold assays of the wagon-drill cuttings come from a population with a specified coefficient of variation of 1.5. This hypothesis would be of interest in order to test whether the Getchell mineralization was consistent with a coefficient of variation that is often typical of fine-grained gold mineralization. Would it be consistent with the 1.58 coefficient of variation that has been calculated from the geologically similar Carlin, Nevada, gold deposits, for instance? The conclusion reached (table 11.11) is that the Getchell mineralization, as sampled

TABLE 11.11. EXAMPLE TEST OF HYPOTHESIS THAT A SET OF OBSERVATIONS COMES FROM A POPULATION WITH A SPECIFIED RELATIVE VARIABILITY

Hypothesis: $H_0 =$ The gold assays of the 956 Getchell wagon-drill cuttings are from a population with a coefficient of variation of 1.5, that is, $\gamma = 1.5$

Alternative hypothesis: $H_1 = \gamma \neq 1.5$

Assumptions: The observations were drawn at random from a normal population

Level of significance: The 10 percent level of significance is chosen

Critical region: The critical region is where z is < -1.645 or > 1.645

Conclusion: Because the calculated value, -5.9, of z is inside the critical region, the hypothesis H_0 is rejected, and the alternative hypothesis H_1 is accepted

by wagon drill, is inconsistent. The relative variability is less than expected, perhaps because the natural variability is small, or perhaps because the variability was reduced by mixing in the process of sample preparation, or perhaps for yet another reason.

The third problem is to test the hypothesis that the gold assays of mine samples from the two types of drilling are from populations with the same relative variability. This hypothesis may be used to determine whether one sampling process introduced more variability than the other. The conclusion reached (table 11.12) is that the two kinds of sampling yield consistent results.

TABLE 11.12. EXAMPLE TEST OF HYPOTHESIS THAT TWO SETS OF OBSERVATIONS COME FROM POPULATIONS WITH THE SAME RELATIVE VARIABILITY

Hypothesis: H_0: The gold assays of the 956 Getchell wagon-drill cuttings are from a population with the same relative variability as that of the gold assays of the 87 Getchell diamond-drill-hole cores, that is, $\gamma_1 = \gamma_2$

Alternative hypothesis: H_1: $\gamma_1 \neq \gamma_2$

Assumptions: The observations were drawn at random from a normal population

Level of significance: The 10 percent level of significance is chosen

Critical region: The critical region is where z is < -1.645 or > 1.645

Conclusion: Because the calculated value, -1.14, of z is outside the critical region, the hypothesis H_0 is accepted

For neither set of data is the assumption satisfied that the samples were drawn at random from a normal population. However, the conclusions and the confidence intervals should be essentially valid.

Finally, it should be noted that, for these data, confidence intervals and hypothesis tests about relative variability are of more interest than corresponding statistical techniques to investigate the variance, because the sampled populations have quite different means and relative variability is of interest rather than variance itself.

11.4   VARIABLES OF CONSTANT SUM

With variables of constant sum, the number of independent variables is always one or more less than the number of measured variables. In the first subsection, the problem for two or three variables of constant sum is outlined and an extension indicated for k variables. Although the presentation is in terms of percentages, the common mode of expression for constant-sum data, it applies to constant-sum data expressed in any form.

In the second subsection, problems arising from correlations induced among variables are considered. These problems stem from the fact that variables in closed arrays must be more or less correlated with one another.

Number of Independent Variables

With two-variable data of *constant* sum, the sum may be exactly or only nearly constant. If *exactly* constant, the percentage of the first constituent plus the percentage of the second must add to 100. One of the two variables is chosen arbitrarily for study; the other need not be considered, because it is equal to 100 percent minus the variable selected. Any method for studying univariate data is applicable.

Two-variable data of *nearly* constant sum require slightly more consideration. If, for example, the percentages of quartz and calcite in sandy limestone are determined separately by independent chemical analyses, the sum of the two analyses will be nearly but not exactly 100, because of random error. Therefore, in a statistical model, with random error included, any univariate method can again be used (because of the random error). If only one constituent is chemically analyzed, the other is determined by difference, and the random error is disregarded.

A further complication is introduced in dealing with trace-element impurities. One's attitude toward the impurities depends on his purpose and on the nature of the trace elements. If the trace were 4 or 5 parts per million of gold, it might well be the most important constituent. From the several ways to handle the situation, an arbitrary choice must be made. The impurities can simply be lumped with the analytical error in the random error. Or they can be added to the list of variables, and the system may then be considered to be a system of three or more components.

With three variables of constant sum, two at most need to be considered. In a sandy limestone, with gold as an impurity, any two constituents may be independent. It is important to recognize, however, that the three-component system is completely defined by one independent variable if the ratio of gold to quartz is constant.

With more than three variables of constant sum, that is, with k variables summing to 100 percent, $(k-1)$ variables at most can be independent because one variable can always be calculated by subtracting the sum of the others from 100. This restriction on the number of variables is closely related to the fact that the kth observation can be calculated, provided that the sample mean and $(k-1)$ observations are known. The variance thus depends on $(k-1)$ observations and has $(k-1)$ degrees of freedom.

Because the number of independent variables is usually less than $(k-1)$, it is highly desirable to attempt to find the number of independent variables by eigenvalue analysis before proceeding to statistical analysis of a multi-component system. The analysis will provide eigenvalues to describe the number of independent variables and will yield eigenvectors whose coefficients can be used to weight the original variables and transform them into independent variables. Sometimes the weighting can be interpreted geologically; at other times no clear-cut interpretation is evident (sec. 10.6).

Induced Correlations

Chayes, who pioneered the study of geological variables of constant sum, was perhaps the first investigator to recognize that variables in closed arrays must be more or less correlated with one another. This subsection summarizes Chayes' work on these induced correlations. Most of the results are given in a paper by Chayes and Kruskal (1966); other results and full references to the literature appear in the annual reports of the Geophysical Laboratory (Carnegie Institution of Washington) from 1959 onward.

Chayes and Kruskal postulate the following model to study induced correlations. Consider a closed array of variables y_i. (An example of a closed array is constant-sum data, such as a rock analysis for several chemical elements, that add to 100 percent.) Define an open array of variables x_i equal in number to the y_i variables, but with a nonconstant sum. The variables x_i have means μ_i and variances σ_i^2 and are uncorrelated. The y_i variables are related to the x_i variables by the formula

$$y_i = \frac{x_i}{\sum x_i}.$$

The algebraic properties of the model allow the variances of the x_i variables to be inferred from the means and variances of the y_i variables and allow the co-variances of the y_i variables to be calculated in terms of the x_i variables.

Thus the expected values of the induced correlations among the y_i variables can be calculated from the covariances of the y_i variables (sec. 10.2). If the observed correlations are different from these expected values, one concludes that there are relationships among the y_i variables that are not due merely to the induced correlations. The complicated arithmetic, given by Chayes and Kruskal (1966), can be performed easily with an electronic computer.

Table 11.13. Data from modal analyses of 15 thin sections of Bellingham, Minnesota, granite[*]

Mineral		Observed value		Variance, $s^2_{x_i}$
No.	Name	y_i	$s^2_{y_i}$	
1	Quartz	29.37	19.5021	21.8262
2	Microcline	34.19	35.7878	68.1950
3	Plagioclase	29.85	20.5712	24.1408
4	Biotite	4.50	6.1757	5.5157
5	Muscovite	2.08	1.0189	1.0082

[*] After Chayes and Kruskal, 1966, p. 698.

Table 11.14. Observed and expected correlations for modal data of table 11.13[*]

Mineral pair	Correlations	
	Observed	Expected
1, 2	− 0.6466	− 0.5781
1, 3	0.0086	− 0.1466
1, 4	− 0.4492	− 0.1173
1, 5	0.5243	− 0.0015
2, 3	− 0.6543	− 0.5967

[*] After Chayes and Kruskal, 1966, p. 699.

Chayes and Kruskal apply their calculations to data from modal analyses of 15 thin sections of the Bellingham, Minnesota, granite (table 11.13). The observed and expected values of the induced correlations are given in table 11.14. They conclude that correlations among the major minerals—quartz, microcline, and plagioclase—are those expected from the correlation induced by the closed system. However, because the correlation between quartz and muscovite is significantly larger than expected, there may be a real relationship between these two minerals in the Bellingham granite.

Certain limitations in the model exist, as pointed out by Chayes and Kruskal. For certain means and variances of y_i, the model fails because negative x_i variances result. The model may also fail if there are fewer underlying variables than the number of y_i variables minus 1 (sec. 11.5).

One important application of Chayes' work is in the plotting of variation diagrams (Harker diagrams, etc.) in petrography. A variation diagram for quartz and muscovite would be the only meaningful representation of data for the Bellingham granite. Variation diagrams for other pairs of minerals, such as microcline and plagioclase against quartz, would be rendered meaningless by induced correlations.

11.5 INTERPRETATION OF THREE-VARIABLE DATA OF CONSTANT SUM

In geology, three-variable data of constant sum are commonly plotted on triangular coordinate paper. Although an ordinary plot of three-variable data on Cartesian coordinates requires three dimensions, the plotted points define a plane. The triangular coordinate paper is one way of projecting onto a plane to begin with. A two-dimensional plot is possible because, with three-variable data of constant sum, two variables at most are independent.

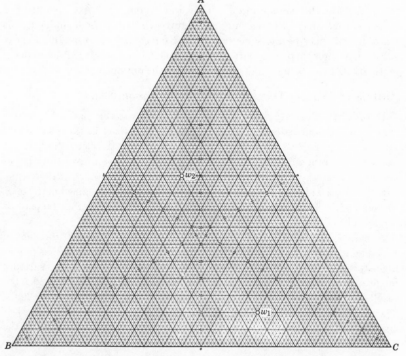

Fig. 11.1. Triangular coordinate paper for plotting three-variable data of constant sum.

Figure 11.1 illustrates how data are plotted on triangular coordinates. The three axes A, B, and C are scaled from zero to 100; thus the constant sum of $A + B + C$ is also scaled to 100, appropriate for percentage data. Any observation can be represented by a point on the graph. For instance, the point w_1 represents an observation with $A = 10$, $B = 30$, and $C = 60$; and the point w_2 represents an observation with $A = 50$, $B = 30$, and $C = 20$. The same paper used to plot triangular coordinates is also used to plot four-dimensional data in a tetrahedron; properties of this plot, not discussed in this book, are explained by Levin (1956–1959).

In an analysis of a set of three-variable data of constant sum, the first step is to plot the observations on triangular coordinates so that the configuration of points can be inspected. For numerous observations, a computer plot is convenient (sec. 17.6). The points may fall on or near a straight line, may form one or more clusters, or may be scattered helter-skelter.

Once the data are plotted, one's interest turns, as always, to effective means to summarize them. Various questions arise. Are there one or two independent variables? Do the observations fall into distinct clusters? Do the observations define a straight or curved line? One way to answer these questions is to investigate the configuration within and among groups of points. For instance, there may be two groups of points, each with a single, independent variable that may be the same or different for the two groups. In the next subsection some fictitious data are constructed and analyzed to indicate methods that can be used and results that can be obtained.

Analysis of Fictitious Three-Variable Data of Constant Sum

In order to illustrate some properties of three-variable data of constant sum, five fictitious cases are presented, as characterized in table 11.15 and illustrated in figures 11.2 to 11.8. In order to obtain the fictitious data for cases 1 to 4, random normal independent numbers were drawn from populations and listed, with their means and variances, in the table. If there is no entry for a variable, the variable was calculated in a way explained in the text. For case 5, the 10 observations are evenly spaced between 10 and 100 percent of variable B.

In case 1 variable C was calculated by subtracting from 100 the sum of variables A and B. In figure 11.2 the 50 randomly chosen observations plot as a cluster near the C-apex. Two eigenvalue analyses, one with the variance mode and the other with the R-mode, were made for these data and yielded the results in table 11.16. For the variance mode, the three eigenvalues are 206.3, 29.9, and 4×10^{-7}. After rounding, all eigenvector components for the last eigenvalue are 0.58; therefore, because the sum of $A + B + C$ is a constant, the value obtained by forming the linear combination whose coefficients are the components of the eigenvector is also a constant.

TABLE 11.15. CHARACTERISTICS OF FIVE FICTITIOUS CASES OF THREE-VARIABLE DATA OF CONSTANT SUM

| | | | Variable | | | | | |
| | | | A | | B | | C | |
Case	Properties	Number of points	μ	σ	μ	σ	μ	σ
1	Ellipsoidal cluster of points	50	25	10	15	5		
2	Two circular clusters of points	25	45	2	45	2		
			10	2	10	2		
3-a	Two linear groups of points. $A/C = 2$, no random variability in the ratio	25			20	5		
		25			80	5		
3-b	Two nearly linear groups of points. $A/C \approx 2$, random variability in the ratio	25			20	5		
		25			80	5		
4-a	One linear group of points. $A/C = 2$, no random variability in the ratio	50			50	15		
4-b	One nearly linear group of points. $A/C \approx 2$, random variability in the ratio	50			50	15		
5	Curved line of points, no random fluctuation	10						

TABLE 11.16. EIGENVALUE ANALYSIS OF DATA FROM FICTITIOUS CASE 1

| | | Eigenvector | | |
Rank	Eigenvalue	A	B	C
1	206.3	$(-0.7027145,$	$-0.00872983,$	$-0.71143127)$
2	29.9	$(-0.41578521,$	$0.81644990,$	$-0.40066471)$
3	4×10^{-7}	$(0.57735026,$	$0.57735028,$	$0.57735027)$

Because of round-off errors, the last eigenvalue is not exactly zero, and the coefficients are not precisely constants. However, the sensible interpretation clearly is that only two independent variables exist. Of the two nonzero eigenvalues, the larger corresponds approximately to the variable $0.7C - 0.7A - 0.009B$ (sec. 10.6). Because $C = 100 - (A + B)$, the larger variable is equal to

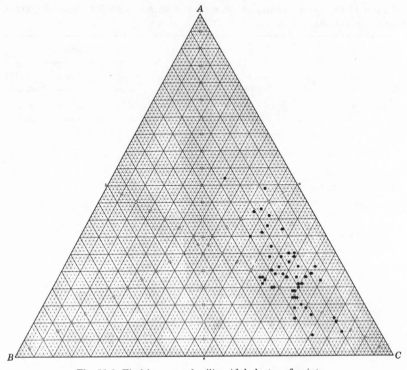

Fig. 11.2. Fictitious case 1, ellipsoidal cluster of points.

$70 - 1.4A - 0.7B$. The smaller eigenvalue of about 30 corresponds approximately to $0.8B - 0.4A - 0.4C$; in terms of the original variables, the smaller variable is equal to $1.2B - 40$.

When logarithms of these data are taken, the logarithms no longer form a constant sum, and an eigenvalue analysis shows three variables, because none of the eigenvalues is equal to zero. This analysis, whose numerical results are not presented, illustrates one reason why caution must be exercised in transforming constant-sum data.

In case 2, variable C was also calculated by subtracting from 100 the sum of variables A and B. The observations lie in two clusters (fig. 11.3). An eigenvalue analysis of all 50 observations yields two eigenvalues. One eigenvalue, of size 4, corresponds to the residual variability measured perpendicular to a line joining the two clusters. The other, of size 2000, corresponds to the between-cluster variability. If the two clusters are analyzed separately, both eigenvalues are about equal in size because in the within-cluster data there are two dimensions with about the same variances of both the variables.

For both cases 3-a and 3-b, the ratio A/C was set equal to 2, and two groups of observations were defined by making the mean of B equal to 20 for half the points and equal to 80 for the other half. After a value B was drawn at random from a normal distribution with a variance of 5 and a mean of 80 or 20, A and C were calculated by the formulas $C = (100 - B)/3$ and $A = 2C$. The resulting observations, too close together to plot individually, lie along two segments of the straight line corresponding to the ratio $A/C = 2$ (fig. 11.4). An eigenvalue analysis of these data indicates only one dimension, stated approximately as $4B - 2A - C$.

For case 3-b, the observations were obtained in the same way as for case 3-a, except that a random error e, with mean equal to zero and variance equal to 1, was introduced. The variable A was obtained by the formula $A = 2[(100 - B)/3] + e$, and the variable C by the formula $C = 100 - (A + B)$. Because of the random error, the observations do not plot exactly on straight-line segments, but in elongated clusters (fig. 11.5). An eigenvalue analysis reveals a second dimension, with variance equal to 2, introduced by the random error. The weights for the primary dimension are the same as in case 3-a.

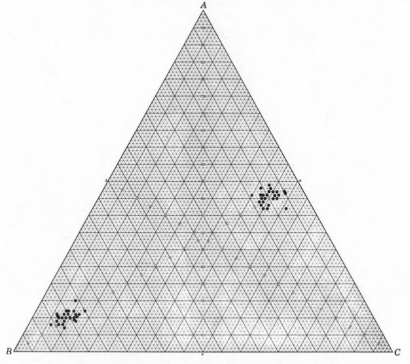

Fig. 11.3. Fictitious case 2, two circular clusters of points.

TABLE 11.17. EIGENVALUE ANALYSIS OF DATA FROM FICTITIOUS CASE 4-b

Rank	Eigenvalue	Eigenvector	
	Ratio set 1	A/B	C/B
1	0.169	(0.897,	0.442)
2	0.609×10^{-3}	(−0.442,	0.897)
	Ratio set 2	A/C	B/C
1	6.267	(0.002,	1.000)
2	0.037	(1.000,	−0.002)
	Ratio set 3	B/A	C/A
1	1.873	(1.000,	0.009)
2	0.002	(−0.009,	1.000)

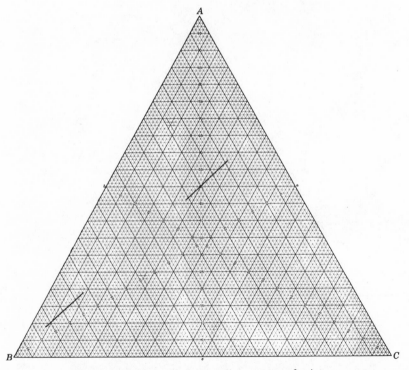

Fig. 11.4. Fictitious case 3-a, two linear groups of points.

Cases 4-a and 4-b are similar to cases 3-a and 3-b, except that the B-variable has a mean of 50 and a variance of 15, instead of having a mean of 20 or 80 and a variance of 5. Therefore figure 11.6, for case 4-a, and figure 11.7, for case 4-b, are similar to figures 11.4 and 11.5. An eigenvalue analysis reveals only one dimension for case 4-a but two dimensions for case 4-b, because the random error e is introduced. The coefficients of the dimension with the larger eigenvalue are roughly the same as for the previous cases.

Table 11.17 presents the results of an eigenvalue analysis for case 4-b. The 50 observations may be described by any one of the three sets of ratios

$$\frac{A}{B}, \quad \frac{C}{B},$$

$$\frac{A}{C}, \quad \frac{B}{C},$$

$$\frac{B}{A}, \quad \frac{C}{A}.$$

Each of these three sets of ratios yields one large and one small eigenvalue.

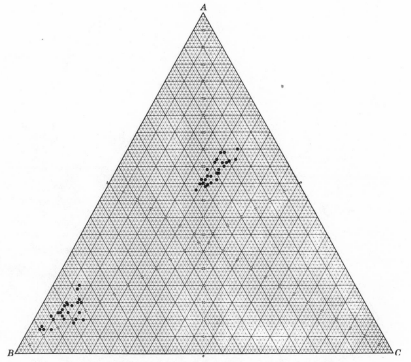

Fig. 11.5. Fictitious case 3-b, two nearly linear groups of points.

For each set, the variable calculated from the eigenvectors that correspond to the small eigenvalue is nearly a constant. For the first set, the variable $-0.442\,(A/B)+0.897\,(C/B)$ is nearly a constant, and the variable $0.817\,(A/B)$ $+0.442\,(C/B)$ is approximately proportional to $2\,(A/B)+(C/B)$. For the second set, the ratio (A/C) is nearly a constant, and the variable corresponding to the large eigenvalue is (B/C). For the third set, the ratio (C/A) is nearly a constant, and the variable corresponding to the large eigenvalue is (B/A).

Case 5 is introduced to demonstrate that eigenvalue analysis treats data linearly. The 10 values for B are equally spaced from 10 to 100. Variable A was calculated by the formula $A=(100-B)^2/100$, and C was calculated by subtracting A and B from 100. Thus, although only a single variable is independent, it is not related linearly to the other variables. Figure 11.8 shows that the observations define a smooth curve. Because it is linear, an eigenvalue analysis shows two dimensions. It is important to note that both the plot of data and the results of statistical analysis must be examined because the plot of a smooth curve reveals the quadratic relation that can readily be made linear by transforming the 10 values for B.

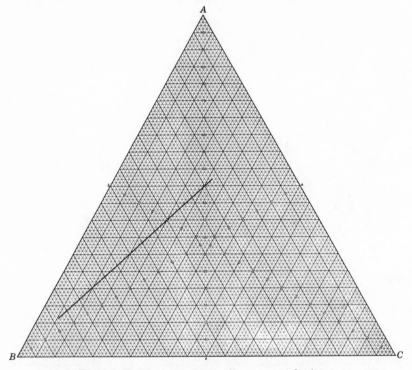

Fig. 11.6. Fictitious case 4-a, one linear group of points.

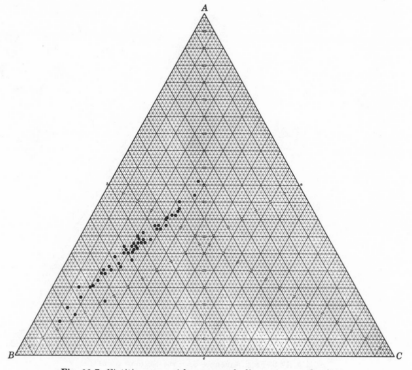

Fig. 11.7. Fictitious case 4-b, one nearly linear group of points.

These five cases illustrate the following suggestion. Plot three-variable data of constant sum on triangular coordinates and examine the pattern. If all the points fall on a single straight or curved line, there is only one independent variable. If the line is straight and passes through the vertex of the triangle corresponding to one variable, the ratio between the other two variables is constant. If the points form more than one cluster, more than one group must be present, and there may be an independent variable defining the groups. An example of such an independent variable is a series of percentage measurements, taken at different times of day, forming clusters, where time is the independent variable defining the groups.

If the clusters are indistinct, eigenvalue and discriminant analyses may be helpful. Group the observations as well as possible by eye. Then perform an eigenvalue analysis on each cluster to extract one or two principal eigenvectors. Apply the corresponding eigenvectors to all the groups to obtain a new set of one or two variables for each of the groups. Use these new variables in a discriminant analysis to support the clustering, if it does in fact exist.

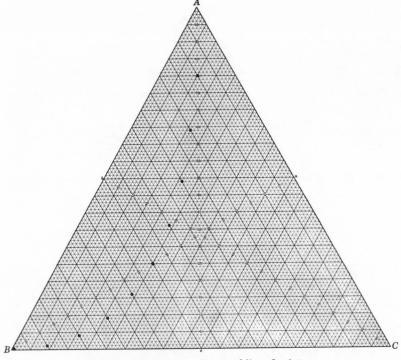

Fig. 11.8. Fictitious case 5, curved line of points.

Conditions for Independence of Two Variables

As already emphasized, with three variables of constant sum, although at most two can be independent, only one need be independent. In this subsection the conditions that permit but do not require two variables to be independent are examined, first for normally distributed variables and then for variables with skewed frequency distributions.

For independence, the sum of the means of the variables must be bounded away from the extremes of 0 and 100 percent, with the standard deviation of the sum as the yardstick. For three normally distributed variables of constant sum, two variables can, but need not, be independent only if

$$\mu_A + \mu_B + 3\sqrt{\sigma_A^2 + \sigma_B^2} < 100$$

and

$$\mu_A + \mu_B - 3\sqrt{\sigma_A^2 + \sigma_B^2} > 0.$$

These inequalities correspond to the requirement that little of the area under the frequency-distribution curve of $(A + B)$ can be nearly equal to either 0 or

100 percent. The constant 3 in the inequalities corresponds to three standard deviations from the mean, which, for normally distributed variables, defines the area under the frequency curve that contains nearly all the observations.

Figures 11.9 and 11.10 plot the maximum values of σ that permit but do not require A and B to be independent, against the given values of μ. Figure 11.9 is for the case of the mean and variance of A equal to those of B. Figure 11.10

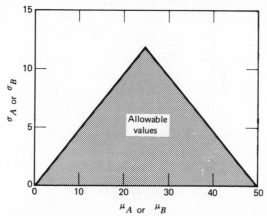

Fig. 11.9. Maximum values of α that permit, but do not require, A and B to be independent, if the means and variances of A are equal to those of B.

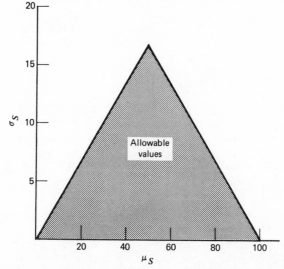

Fig. 11.10. Maximum values of α_S that permit, but do not require, A and B to be independent, if the means and variances of A are not equal to those of B.

is for the case of the mean and variance of A not equal to those of B; in this case the maximum allowable values of σ_S, the standard deviation of $(A + B)$, are plotted for given values of μ_S, the mean of $(A + B)$.

For variables whose distributions are skewed, the inequalities that must be satisfied for independence become

$$\mu_A + \mu_B + K_1\sqrt{\sigma_A^2 + \sigma_B^2} < 100$$

and

$$\mu_A + \mu_B - K_2\sqrt{\sigma_A^2 + \sigma_B^2} > 0,$$

where the constant 3 is replaced by the constants K_1 and K_2. The behavior for distributions skewed to the right and left may be illustrated by the cases for sum of $(A + B)$ that follow the lognormal and negatively skewed lognormal distributions. Table 11.18 gives values of K_1 and K_2 for variables whose sum

TABLE 11.18. TYPICAL VALUES OF K_1 AND K_2 FOR VARIABLES WHOSE SUM FOLLOWS THE LOG-NORMAL DISTRIBUTION

β^2	K_1	K_2
0.01	2.62	3.42
0.05	2.21	4.01
0.10	1.95	4.49
0.25	1.51	5.54
0.50	1.13	6.82
1.00	0.74	8.53
1.50	0.53	9.44
2.00	0.39	9.73

follows the lognormal distribution. Because the skewness of the distribution is determined by the variance of logarithms β^2, both K_1 and K_2 are nearly 3 for small values of β^2, while for larger values of β^2 K_1 and K_2 are very different from 3. If the distribution is skewed to the left, for instance in the negatively skewed lognormal distribution, the roles of K_1 and K_2 are reversed.

REFERENCES

Chayes, Felix, and Kruskal, William, 1966, An approximate statistical test for correlations between proportions: Jour. Geol., v. 74, p. 692–702.

Hald, A., 1952, Statistical theory with engineering applications: New York, John Wiley & Sons, 783 p.

Koch, G. S., Jr., and Link, R. F., 1966, Some comments on the distribution of gold in a part of the City Deep mine, Central Witwatersrand, South Africa, *in* Symposium on mathematical statistics and computer applications in ore valuation: South African Inst. Mining Metall., p. 173–189.

Koch, G. S., Jr., and Link, R. F., 1970, A statistical interpretation of sample assay data from the Getchell mine, Humboldt county, Nevada: U.S. Bur. Mines Rpt. Inv. 7383, 23 p.

Levin, E. M., 1956–1959, Phase diagrams for ceramists: Columbus, Ohio, Am. Ceramic Soc.

SOME PROBLEMS IN APPLIED GEOLOGY

Part V, composed of chapters 12 to 16, is concerned with statistical analysis of selected problems in the discovery, appraisal, and exploitation of natural resources. In applying geology to these problems, the focus is on making money for a company or increasing the national wealth of a country; and economics plays a part. In this book only some topics are emphasized for which statistical analysis can offer insight. Although most examples are drawn from mining, many of the same principles apply to investigation of other natural resources, such as petroleum and water, as well as to problems in academic geology.

In Chapter 12, exploration for natural resources is discussed. In Chapter 13, delineation of ore deposits, once they are found, is reviewed, particularly statistical methods that can markedly improve estimates of grade and amount of ore. In Chapter 14, some methods of operations research for making decisions in the natural resources field are explained to show how to select from the methods in Chapters 12 and 13 those suitable for particular problems, and also to show how to carry through to the exploitation and manufacturing stage some problems in mineral resources. In the study of these large-scale systems, as contrasted to the simple ones considered previously, many arbitrary decisions must be made. However, with statistics, scientists and engineers can obtain additional insight into problems and can provide management with results that, when mingled with experience, provide guides for action.

Chapters 15 and 16 apply techniques developed in Volume 1 to several problems in applied geology. Chapter 15 takes up three kinds of geological sampling—sampling of broken rock, placer sampling, and sampling in exploration geochemistry—and concludes with an account of an investigation to compare several different techniques of geological sampling in a Uganda copper mine. Chapter 16 appraises the relative merits of the normal and log-normal distributions for the analysis of gold assay and other trace-element data. Assays from South African and American gold mines are discussed.

Much of Part V results from articles by and discussions with men in various mining companies—particularly the Kennecott Copper Corporation and its exploration subsidiary the Bear Creek Mining Company and American Metal Climax, Inc. But of course these men, whose work is acknowledged in the text, share no responsibility for misinterpretations that we may have made.

Chapter 12

Exploration

The exploitation of natural resources is a *primary industry* because the material removed from the ground is usually further processed, as are petroleum or metalliferous ores, although for some materials (some coal and industrial mineral products such as concrete aggregate and building stone) the processing may be minimal or unnecessary. In common with other primary industries, including fishing, farming, and logging, mineral industries must be carried on wherever nature has placed the substance to be exploited. And they have the further property of being extractive industries, in that, once removed, the material is gone forever and, once a mineral deposit is worked, it is rarely possible to return to recover more material. Therefore it is desirable to decide at first what is to be done with the resource. Ventura (1964) offers a good further explanation of these concepts.

What is exploration? Bailly (1968, p. 19) writes that "with regard to a new mine, exploration can be technologically defined as all the activities and evaluations necessary before an intelligent decision can be made establishing size, initial flowsheet and annual output of the new extractive operation based on a certain mineral deposit." He recognizes that this engineering definition may not correspond to one made for tax purposes. In the past ore in outcrops and petroleum in oil seeps were discovered by prospectors; today most discoveries are made by technically trained geologists and engineers.

In this chapter we first discuss some risks and rewards inherent in exploration and suggest that they are a function not only of nature but also of the policies of the explorers. We review, in sections 12.2 and 12.3, ways to classify areas of the earth in the search for natural resources, and we introduce some theoretical models of how the prizes sought may be distributed. Exploration

187

on a grid is reviewed in section 12.4; and one other way to explore, the response-surface method, is developed in section 12.5. Section 12.6, on theory of search, adapts methods of military search to exploration for natural resources.

12.1 RISKS AND REWARDS IN EXPLORATION

Before discussing risks and rewards in exploration, we mention that in exploration it is necessary to have a policy or a purpose for searching; otherwise one is not sure what he is risking and whether, if anything is found, he will be rewarded. Purposes or policies may be those of private companies or governments.

By and large, private-company policies are simpler than those of governments, because the companies' principal if not sole business is to make money. Different policies are found, depending on the purposes and resources of the companies. A few large companies explore for anything valuable, but most limit their scope—some to metals, some to petroleum, and others to base metals, for example. Vertically integrated companies, particularly in the steel industry, may be interested only in the raw materials that go into their final product.

Some companies explore all over the world, while others limit their activities to one or a few countries. The extent and location of a company's foreign involvement may be controlled in part by governmental policies, such as those regarding the disposal of nonconvertible currency or special taxes on which the profit or loss of a future development could depend. These policies do not require the ignoring of a deposit not sought for—one that is found by accident—but they do rule that some targets will not be the specific focus of exploration and, if found, may be turned over to another company better fitted to exploit them.

The policies of a government are diverse and often diffuse, especially if responsibility is shared by several agencies. One of the simplest policies is resource appraisal, which is of interest in estimating how long resources will last and, especially in an underdeveloped country, for determining what new industries can best be supported by the country's natural resources. Of great concern are a government's conservation policies, perhaps to determine whether only the best natural resources should be mined or whether any should be mined at all. For example, one of the questions that arises in the planning for oil-shale development in western Colorado is to what cutoff grade the shale should be mined.

Concern with the balance of payments may play a part, as in some mines owned by the Finnish government that work subcommercial ores to produce metals that are sold in order to gain foreign exchange. Or a nation may wish to produce a commodity even at a loss to use as a base line from which to regulate the cost of imports, as in the case of the Riddle, Oregon, nickel deposits, which were brought into production through substantial government financing. These are only a few of the ways in which governmental policies are diverse and diffuse.

Risks in Exploration

Exploration is a gambling game in which the odds in favor of winning are small and not well known, although the prizes may be very valuable; thus the risks in exploration are great, and little is known about them. The game is like playing a slot machine that pays only jackpots or like a roulette game on the single numbers that pay off at 35 to 1, with the difference that, whereas in these gambling games the odds favor and are known to the house (although not necessarily to the player), the odds in exploration are unknown.

In exploration the odds for failure are great, as illustrated by table 12.1, an interesting compilation by Bailly (1967). For some governmental and some private exploration ventures carried on in recent years, mostly in the United States, Bailly shows the progression of exploration from the detailed reconnaissance stage through later steps to the establishment of new mineral deposits (which he defines as occurrences that can be mined either at present or in the future with normally expected technological improvements)—finally to the stage of new ore deposits, which are mineral deposits that can be mined with present technology. In the table, the question marks denote lack of information, not an uncertain classification. For instance, the Bear Creek Mining Company (last line in the table) performed detailed reconnaissance on 1649 prospects in the four-year period under discussion but had carried only one of them through to the stage of probably being an ore deposit at the time the chart was compiled.

Elementary probability calculations, discussed by statisticians (e.g., Feller, 1968, p. 313) under the name "gambler's ruin," can be made to determine how long a run of bad luck a gambler can stand before going broke. The subject, evidently first introduced into the exploration literature by Slichter (1960), is fully explained by Feller. One simple case is that of a gambler with a certain capital that he bets one unit at a time on a game with given odds; the question is: How many times can he play for a given chance of being ruined ? This case is illustrated by table 12.2, which is in a form published by Arps (1961) but recalculated for probabilities that seem to us more appropriate to exploration. Explanation of one entry shows how to read the table. For a 1 percent probability of successfully developing a prospect into a mine, at least 229 prospects must be examined to reduce the risk of complete failure to 10 percent, 298 to reduce this risk to 5 percent, and 458 to reduce this risk to 1 percent.

Rewards in Exploration

The risks in exploration are large, and entering this game usually costs a lot of money. Investigating a target thoroughly enough to determine that it is a mine costs from a few hundred thousand to a few million dollars, aside from costs for land acquisition or purchase, although not all this money need be spent on a target if it is turned down during the investigation. In this subsection a few examples of rewards are introduced.

Table 12.1. Numbers of properties surviving successive stages in exploration*

Project name and description	Detailed reconnaissance	Detailed surface investigations	Physical sampling	New mineralization discovered	New mineralization with some tonnage	New mineral deposits	New ore deposits
U.S. Strategic Minerals Development Program, 1939–1949		10,071		1,342	1,053	?	?
U.S. Atomic Energy Commission, 1948–1965	15,000			4,317	643	?	?
U.S. Defense Minerals Exploration Administration, 1951–1958		3,888		?	374	45	?
Canada, estimated total prospecting activity/year in recent years		6,000		?	?	5	?
Phelps-Dodge Corp., 1962	73	?	few	?	?	0	0
Phelps-Dodge Corp., 1966	313	107	16	?	?	?	?
International Nickel Co., 1958		>100	>100	?	?	1	?
Texas Gulf Sulfur Co., 1959–1961	Several 1,000	Several 1,000	>66	?	?	>1	1
Five exploration companies in southwestern U.S., dates not given		352	23	?	?	2	?
Bear Creek Mining Co., 1963–1966	1,649		60	15	8	5	1

* After Bailly, 1967, p. 10.

TABLE 12.2. PROBABILITY OF GAMBLER'S RUIN

Probability of success	Probability of ruin through a run of bad luck		
	10%	5%	1 %
0.1%	2301	2994	4603
0.5%	459	598	919
1%	229	298	458
2%	114	148	228
5%	45	58	90
10%	22	28	44
20%	10	13	21

Mines are still occasionally found by individuals. Charles A. Steen, a geologist-prospector, found the famous Mi Vida mine in Utah in 1952 by drilling exactly one hole in an exploration effort whose costs were a few thousand dollars. Before the original operation closed in 1967, it produced about $40 million of high-grade ore.

More commonly, a natural resource target is found by a group of geologists and engineers financed by a substantial company. An example is the Carlin gold deposit in Nevada (*Mining Engineering*, 1965), which, at the start of mining, was announced to contain reserves of 11 million tons of ore with a grade of 0.32 ounce of gold per ton, worth $123 million, with the $10 million investment to be returned within five years of the 15-year life.

Another recent success story is the Timmins ore deposit in northern Ontario, Canada, reported at its discovery (Sheehan, 1964) to contain 25 million tons of ore averaging 1.18 percent copper, 8.1 percent zinc, and 3.8 ounces of silver per ton, based on information from seven drill holes in ore, which delineated an ore body 800 feet long, 400 feet wide, and more than 800 feet deep below 20 feet of overburden. This great deposit was found by drilling 66 holes on geophysical anomalies at a total exploration cost of about $3 million; the value of the ore is about $850 million, and the profit, calculated at $10 per ton, is about $250 million.

But perhaps more common than these three ore deposits, which, once recognized, were clearly valuable natural resources, is the natural resource that falls in the category of a mineral deposit in Bailly's sense and stands on the edge of being an exploitable ore deposit. One example is the Tyrone, New Mexico, copper deposit, which was first unsuccessfully developed and mined at great expense only to close in the 1920s and which had to wait until 1967 for a resumption of active mining. Another example is the White Pine, Michigan, copper deposit, for which Boyd (1967) has prepared an interesting tabulation (table 12.3) comparing cash investment, cash return from operations, and net annual return over the 36 years from 1929 to 1965. By 1965 cash return from operations was still $8.7 million less than the $93.2 million investment, without regard for interest. And the present value in 1965 of the cash investment at

TABLE 12.3. CASH FLOW FROM 1929 TO 1965 FOR THE WHITE PINE, MICHIGAN, COPPER MINE*

Activity	Year	Cash flow (millions of dollars)		
		Cash investment	Cash return from operations	Net annual return
Property purchased	1929	0.1		− 0.1
Exploration and development	1930–1943	0.1		− 0.1
	1944–1945	0.1		− 0.1
	1946	0.4		− 0.4
	1947	0.4		− 0.4
	1948	0.1		− 0.1
	1949	0.1		− 0.1
	1950	0.2		− 0.2
	1951	0.1		− 0.1
Construction period	1952	9.6		− 9.6
	1953	31.6		− 31.6
	1954	18.9		− 18.9
Production started	1955	12.5	9.7	− 2.8
	1956	1.8	11.9	10.1
	1957	1.7	4.2	2.5
	1958	0.7	5.2	4.5
	1959	1.6	5.2	3.6
	1960	2.2	3.6	1.4
Operation reached maturity	1961	1.5	6.0	4.5
	1962	2.8	7.5	4.7
	1963	2.1	8.3	6.2
	1964	2.0	9.4	7.4
	1965	2.6	13.5	10.9
Totals		93.2	84.5	− 8.7

* After Boyd, 1967.

4 percent interest was $139.6 million, suggesting that the money might have better been put in tax-free municipal bonds.

In summary, risks in exploration are large, and so also may be the rewards. It is not at all clear that the game is worth the candle (Groundwater, 1968), although presumably someone is making money. In the rest of this chapter we explain some ways to improve the odds by statistical analysis of the geological data obtained in an exploration campaign.

12.2   GEOLOGICAL CLASSIFICATION OF AREAS

If the policy of an exploration organization is systematic exploration of large areas rather than examination of properties submitted by prospectors or investigation near the home mine, the first step is to classify areas as favorable or unfavorable for exploration according to their geology. The necessary geological knowledge is usually obtained through library research, map compilation, and study of aerial photographs and existing exploration reports —all of which is perhaps supplemented by reconnaissance such as airborne geophysics or airborne visual examination.

Classification of areas may be easy for substances like coal, whose general distribution is well known, may be difficult for substances whose geology is poorly known (as for uranium in the United States when intensive exploration started near the end of World War II), and may be unnecessary for sand, gravel, and crushed rock, which are valuable only if found close to a market because of the high shipping costs in proportion to their value. Thus for coal one can readily find a map (fig. 12.1), but for a porphyry copper one might have to do intensive research after eliminating such areas as those where sedimentary rocks or basalt flows are exceptionally thick.

Early attempts by geologists to classify areas according to favorability for exploration were based on generalized hypotheses about metalliferous (Lindgren, 1933, pp. 878–894) and petroliferous (Levorsen, 1967, pp. 14–46, 627–657) provinces and times in geologic history. Today most explorationists believe that these concepts are suspect because they are too closely tied to the highly uncertain genetic theories of origin of the resources. Thus close adherence to these concepts makes one liable wrongly to exclude areas from exploration because of geology that is not favorable according to the genetic theories currently fashionable.

However, some recent studies are bringing forth ideas that may eventually make possible areal classification based on quantitative evidence evaluated statistically. An excellent example is the studies of Cannon and Pierce (1967), who, from isotopic analyses of lead, postulate three major metallogenic eras:

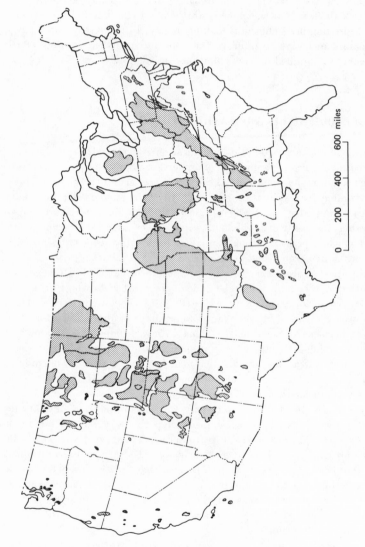

0   200   400   600 miles

Fig. 12.1. Coal fields of the conterminous United States (after Averitt, 1961, p. 4).

Phanerozoic (Paleozoic, Mesozoic, and Cenozoic) time, mid-Precambrian time, and early Precambrian time. Gold and lead appear to be unevenly distributed among these times. Scarcely any lead occurs in early Precambrian deposits, about 30 percent occurs in mid-Precambrian deposits, and about 70 percent occurs in Phanerozoic deposits. On the other hand, more than 50 percent of gold deposits evidently were formed in early Precambrian time. Clearly, as studies like these are extended and further interpreted, strong bases of classification may be developed.

Present-day thinking is well expressed by Jerome (1959), who proposes for the first stage of exploration that "all the facts are gathered on at least two map scales and preferably three, assimilated, imaginatively interpreted and projected." Jerome also points out that the available factual information possesses different reliability in various places. Thus, for North America, the geological map of the continent (Goddard, 1965) might be taken and overlays prepared on transparent sheets to show tectonic features, intrusives of various types, commodities, etc. If information is detailed, these maps may be re-compiled on larger scales. The maps will pinpoint certain regions which can be selected for additional map work on larger scales and for geological examination and physical exploration in turn if they are indicated.

In the remainder of this section some statistically guided methods for geological classification of areas in exploration for metals, petroleum, and industrial minerals are first discussed, and then a generalized statistical approach is reviewed.

Classification of Areas in Exploration for Metals

Not all areas have the same potential value of metalliferous ores; this well-known fact is demonstrated by figure 12.2, which shows the variability in value of total mineral and nonferrous metal production for the western United States in 1964. On the assumption that these states, except Alaska, have been explored with about equal attention, the variability must show a relation to the classification of these areas in exploration for metals. The differences in the illustrations demonstrate that metals are only a small part of the total mineral natural resources in many states. Maps like these can be used in several ways. One can say that Oregon must be a prime exploration target because the non-ferrous metal production is distinctly lower than that in any other state (this is the argument we advanced while employees of that state), or one can say that it must be a poor target because almost nothing is mined there today. Geo-logical arguments have been marshaled by the dozen to support both points of view.

Among the western states, Nevada has received great attention in explor-ation recently, and Joralemon (1967) states that, in 1967, 85 percent of exploration was concentrated in the Walker Lane and Carlin Belts (fig. 12.3).

He plots the metal-mining districts of the state without regard to size because "great ore bodies in the future may be developed in once unimportant districts, as at Carlin and Yerrington," and he "seriously question[s] whether we know enough about regional ore control to say that the concentration of nearly six out of seven Nevada exploration programs within these belts is justified."

Joralemon's conclusion may be investigated by a statistical analysis. In table 12.4 are listed the mean numbers of mines per thousand square miles in the three areas of Nevada. Under the assumption that the entire state has been

TABLE 12.4. MEANS AND CONFIDENCE INTERVALS FOR THE NUMBER OF MINES PER THOUSAND SQUARE MILES IN THREE AREAS OF NEVADA*

		90% Confidence interval	
Area	Mean	Lower	Upper
Carlin Belt	3.3	2.4	4.2
Walker Lane	5.2	4.0	6.4
Rest of Nevada	2.6	2.4	2.9

* Data compiled from a map by Joralemon, 1967.

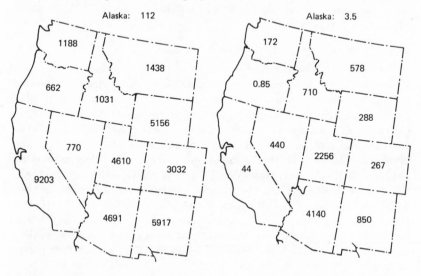

(a) Total mineral production   (b) Nonferrous metal production

Fig. 12.2. Total mineral production and nonferrous metal production, in dollars per square mile, for the western United States in 1964 (after Bailly, 1967).

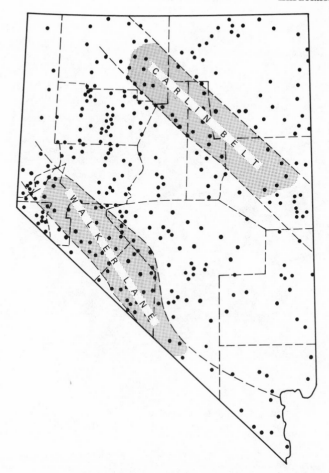

Fig. 12.3. Mining districts of Nevada (after Joralemon, 1967).

explored with the same thoroughness, these means are estimates of the amount of mineralization in the three areas, and the number of mines observed represents a Poisson variable with some underlying characteristic mean that is a function of the intensity of exploration. Confidence limits for this mean can be calculated: for the Carlin Belt and the rest of Nevada, the overlapping of the 90 percent confidence limits implies that the amount of mineralization is not necessarily different for these two areas. However, the limits for the Walker Belt do not overlap those for the rest of Nevada and only slightly overlap those for the Carlin Belt, a fact suggesting either that the Walker Belt actually contains more mineralization or that it has been more thoroughly explored.

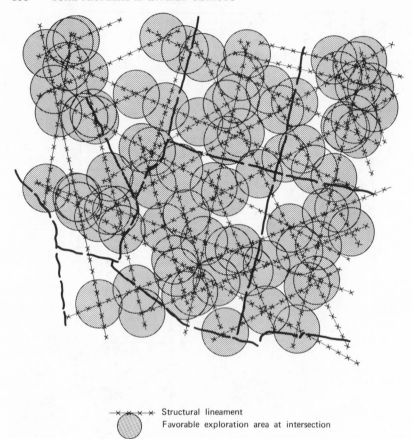

Structural lineament
Favorable exploration area at intersection

Fig. 12.4. Suggested major structural trends of the southwestern United States (after Joralemon, 1967).

Joralemon (1967) also presents an interesting map (fig. 12.4) locating major structural trends that have been mapped in the southwestern United States, with circles of 50 miles' radius drawn to delimit ground that spatial relations between known ore bodies and these structural trends show may be favorable. He states "that over three-quarters of the area studied falls within the favorable circles. With this bewildering array of favorable areas to choose from, the geologist is no better off than he would be without this study." Having rejected genetic, lithlogic, structural, and mineralogic ore guides, Joralemon concludes that "intensity of mineralization" is a "fifth dimension" to guide ore search.

These few illustrations are intended to leave the reader with an uneasy feeling, only some of which is dispelled later in this section. Some areas can be ruled out on clear-cut factual grounds if the exploration targets are sufficiently restricted. For instance, if a company seeks only porphyry copper deposits near the surface, states such as Kansas (Bailly, 1964, p. 8), Louisiana, and eastern Oregon, where Columbia River basalt is thick, can be eliminated. On the other hand, to rule out parts of Nevada because they are outside the Walker Lane and Carlin Belts is in the category of geological controversy— that is, in the range of opinion.

Classification of Areas in Exploration for Petroleum

It is well known that most petroleum is found in rocks formed at certain times in the geologic past. Moreover, these rocks have special characteristics that make them *reservoir rocks*, and they are contained in specific stratigraphic or structural environments that constitute *reservoir traps*.

TABLE 12.5. WORLD PRODUCTION AND RESERVES OF PETROLEUM, 1947*

Geologic age	Total cumulative production to date		Total reserves		Total ultimate production	
	Billion barrels	Per-cent	Billion barrels	Per-cent	Billion barrels	Per-cent
Tertiary	32.0	58.1	40.2	58.0	72.2	58.0
Cretaceous	10.8	19.6	11.7	17.0	22.5	18.3
Jurassic-Triassic	2.4	4.3	9.2	13.0	11.6	8.6
Paleozoic	9.9	18.0	8.3	12.0	18.2	15.0
Totals	55.1	100.0	69.4	100.0	124.5	100.0

* After Levorsen, 1967, who adopted the table from Gester, *World Oil*, November, 1948.

Table 12.5 shows that all petroleum, except for insignificant freak occurrences, has been found in rocks of Paleozoic age or younger and that the great bulk has been found in Tertiary rocks. Commenting on this table, Levorsen (1967) pointed out that "lack of drilling and testing is probably the chief reason why geologists have a low opinion of rocks of some geologic ages, notably the Cambrian and the Triassic." If this statement is correct, reserves in rocks of these ages should tend to increase faster than those in better explored rocks.

Not only may areas be classified as favorable or not for petroleum according to the age of the rocks, but, if Levorsen's belief that only Precambrian igneous rocks should be ruled out is accepted, a classification of areas according to rock lithology is even more interesting. The concept, too broad to summarize in this book, is well explained by Levorsen (1967, Chap. 14), who writes,

If we compare the geologic conditions that are found in the known petroleum provinces, we see that they are extremely diverse. Each province has its own geologic history, its own characteristic deformation, its own stratigraphy, and its own peculiar types of petroleum accumulation. . . . Yet there are certain empirical characteristics that seem to be present in most productive areas, and that seem to carry more weight than others in a pre-discovery evaluation of the prospects of a region. These characteristics might be classified as (1) sediments, (2) evidences of oil and gas, (3) unconformities, (4) wedge belts of permeability, (5) regional arching, (6) nature of local traps.

A statistically based study to learn which of these characteristics are most significant for classifying areas similar to the classification model of Harris reviewed later in this section should be fruitful but does not seem to have been made.

Classification of Areas in Exploration for Industrial Minerals and Rocks

In exploration for industrial minerals and rocks, classification of areas depends foremost on the economic factor of whether the commodity sought has a high or low *place value* ["If a large part of the value of a rock or mineral results from its geographic position, it is said to have a high place value" (Bates, 1960, p. 7).] Thus sand and gravel have high place values, and diamonds and fluorspar have low place values. Bates classifies industrial minerals into two categories, with place value an important distinguishing criterion. For commodities with high place values, most areas can simply be eliminated as being too far from markets.

A search for industrial minerals or rocks with low place values is not restricted geographically. The search is focused by the fact that, for a particular commodity, the geologic environment is often extremely restricted, particularly if physical properties are important, in contrast to metalliferous ores and petroleum, which may occur in many different environments. For instance, asbestos suitable for the manufacture of such products as roofing is rather narrowly restricted to serpentine bodies in ultramafic rocks, and if the additional requirement is made that the asbestos have a low electrical conductivity, one's search is further restricted to the few serpentine bodies associated with limestone.

In the search for industrial minerals, while classification schemes for areas are interesting to develop for particular commodities, we can discern no general principles for all industrial minerals. For further reading on this subject, the reader is referred to Bates (1960) or to Gillson, ed. (1960).

Harris' Geologic Occurrence Model

For geological classification of areas, the models previously described in this section are highly subjective and intuitive. In the past few years, a mathematical or statistical model to analyze data derived from geologic maps, which may be named a *geologic occurrence* model, has been advocated by various investigators and first implemented by Harris (1966). Because its purpose is the screening of large geologic areas to select those most favorable for exploration, this model is considered in this subsection; in a sense it is also one of the models of prize distribution, which are taken up in section 12.3.

Taking the point of view that natural-resource prizes are a result of the geologic processes reflected in the areal geology as well as in the prize itself, Harris relates, by a statistical model, the probability of occurrence of the prize to the areal geology.

$$V = f_1(L, S, F, A);$$

hence

$$P(V) = f_2(L, S, F, A),$$

where V is a measure of mineral wealth, L is the lithology, comprising the age and type of rock, S comprises structural forms, F is rock fracturing, A comprises the age of igneous activity and the contact relationships, and $P(V)$ is the probability of a prize of a certain value.

Harris translates the generalized equation into operational form by measuring from geologic maps the 26 variables listed in table 12.6. To test the model, Harris made these measurements in 243 cells, each 20 miles square, located in parts of New Mexico and Arizona. He also calculated the value of mineral production for each cell by assigning each to one of six value groups. To relate the measured variables to the known values in the various cells, Harris used a Bayesian type of multivariate statistics, in the form of a multiple discriminant analysis (a generalization of the discriminant analysis of section 10.5). In this analysis, rather than summarizing the variables by a single linear combination, a vector combination is formed; the basic concept is that the space over which a variable is defined is carved up into regions corresponding to the classifications developed. Then, using the relative frequency of classes as weights in a Bayesian context (sec. 4.8), and presuming that the variables follow a multivariate normal distribution, one can calculate the probability that a new area will belong to one of the six classes by calculating the value of the discriminant vector for the region. Harris (1966) gives mathematical details, and the general procedure is outlined by Wilks (1962).

After obtaining statistics for the area in New Mexico and Arizona, Harris tested the model by repeating the measurements for another area consisting of most of Utah. He arbitrarily set up the decision rule to retain for exploration any cell with a probability equal to or greater than 20 percent of having a

TABLE 12.6. HARRIS' 26 GEOLOGIC VARIABLES*

Age and type of rock
1. Percent of cell consisting of pre-Tertiary sedimentary rocks
2. Percent of cell consisting of Precambrian igneous intrusive rocks (includes diabase)
3. Percent of cell consisting of Laramide and Nevadan igneous intrusive rocks (includes injection gneiss and contact schists)
4. Percent of cell consisting of igneous extrusive rocks
5. Percent of cell consisting of Precambrian metamorphic rocks, undifferentiated

Rock fracturing
6. Number of high angle faults 0 to 8 miles in length
7. Number of high angle faults greater than 8 miles in length
8. Number of thrust faults 0 to 8 miles in length
9. Number of thrust faults greater than 8 miles in length
12. Number of high angle fault intersections
13. Number of low-high angle fault intersections
24. Number of igneous dikes

Structural forms
10. Number of anticlines 0 to 8 miles in length
11. Number of anticlines greater than 8 miles in length
25. Distance from northeast-trending structural lineament as measured normal to the trace
26. Distance from northwest-trending structural lineament as measured normal to the trace

Age of igneous activity and contact relationships
14. Length of the contact of Laramide and Nevadan intrusive rocks with sedimentary rocks
15. Number of exposures of the contact in variable 14
16. Length of the contact of Laramide and Nevadan igneous intrusive rocks with Precambrian igneous intrusive rocks
17. Number of exposures of the contact in variable 16
18. Length of the contact of Laramide and Nevadan igneous intrusive rocks with Precambrian metamorphic rocks
19. Number of exposures of the contact in variable 18
20. Length of the contact of Laramide igneous intrusive rocks with Cretaceous igneous intrusive rocks
21. Number of exposures of the contact in variable 20
22. Length of the contact of Precambrian igneous intrusive rocks with Precambrian metamorphic rocks
23. Number of exposures of the contact in variable 22

---

* After Harris, 1966.

value of more than \$1 million. Under this rule, 19 of the original 144 cells in Utah were sélected as ones for which exploration would have been done were Utah an unexplored state. These 19 cells comprise nine that were failures, having recorded values less than \$1 million, and ten that were successes. These ten comprise all five cells in Utah with values of more than \$100 million, two of the four cells with values of between \$10 and 100 million, and three of the eight cells with values between \$1 and \$10 million. Thus ten of the most valuable 17 cells were selected.

An explorationist may take one of two points of view toward using Harris' model as a guide to action. The first, which is attuned to a fixed exploration budget, is that enough money is available to explore a certain percentage of cells in the search area, regardless of their merit. The second, corresponding to a flexible exploration budget with money allocated as needed, is that all cells of merit should be explored. Regardless of which viewpoint is adopted, decision rules can be developed through statistical theory, although Harris did not investigate this problem analytically. With the first viewpoint, an exploration-ist would need a rule to allocate exploration effort among the cells in the different value classes. With the second viewpoint, an explorationist would need a rule on how high the probability must be of a cell being correctly assigned to a value class to make spending money for exploration worthwhile. With the second viewpoint, an ordered list of cells according to their expected values is also required, values that are calculated by multiplying the value assigned to each cell by the probability that the cell will contain that value or a greater one. This probability is the sum of the probabilities given by Harris.

Harris' approach to geological classification of areas is both interesting and provocative. Admittedly, the 26 geological variables that he picks are arbi-trary, but, clearly, at least some of them contain information that is evaluated by his model. As usual, when one measures something—even if he does not have the best set of variables and then makes an analysis—he makes progress beyond purely intuitive classical geology.

12.3 MODELS OF PRIZE DISTRIBUTION

In this section, first some empirical models of distribution of various geological prizes are discussed, and then some theoretical models. The dis-cussion is incomplete and unsatisfactory because the necessary data are not available for the empirical models, and, without a good understanding of them, construction of theoretical models is not likely to be profitable. Therefore purposes for models of prize distribution and some ways to devise them are stressed, rather than the models themselves.

In the preceding section geological classification of areas was reviewed as a guide to some "favorable" places for the explorationist to seek prizes. Consider now what the target may be like in the selected areas and how it may be found. One obvious comparison is with military or animal targets, about which Griffiths (1966, p. 191) has written extensively in work drawn on here, pointing out three differences in the two types of targets. First, unlike the animal or submarine, which takes evasive action, and therefore generally renders itself more rather than less vulnerable to detection, the geological target ignores the searcher. Our readers will agree that rattlesnakes are easier to find than fossils, mainly by virtue of their moving. Second, Griffiths points out that "a natural resource is not bounded by such marked discontinuities as a military target; . . . the natural resource target in its environment is not like a needle in a haystack but more like a needle surrounded by iron filings." Although this description holds true for the petroleum targets Griffiths had most in mind, it may not be true for the massive sulfide or magnetite ore body, the coal bed, or the salt dome. Third, Griffiths states that "the change from environment to military target is quite abrupt (except perhaps where camouflage is adopted), whereas the change from barren to natural-resource target is usually continuous, and the boundaries of the resource are quite often subject to change when—at a later date—improved technology permits exploitation of a 'lower grade' deposit."

Probably the most important peculiarity of natural-resource targets is the problem of recognition. To recognize the snake, deer, or submarine is trivial and requires no special training, although admittedly the second phase of evaluation—whether the snake is poisonous, the deer is a buck or a doe, or the submarine is a friend or an enemy—may require some investigation. (Griffiths recognizes but does not emphasize this point.) Of natural-resource targets, some may be easily recognized, such as when Sutter found gold in California in 1849; but others, for example the gold of submicroscopic size at the Getchell and Carlin mines in Nevada, and the copper ore body within shouting distance of the San Manuel ore body in Arizona (Lowell, 1968), may defy detection for years. From the bewildering variety of responses to the search mechanisms—whether human eye, microscope, drill, or other device—the geologist must seek out those that tell him he has found or is closing in on the target..

When World War II began, the size of the German navy and its complement of vessels was well known to Great Britain; moreover, the positions of these ships were also known, together with auxiliary information including cruising ranges, locations of refueling and rewatering stations, which vessels were submarines that could submerge, etc. With this information in hand, the British navy took action against these ships and some were promptly hunted down, for instance the *Graf Spee* (Churchill, 1948, p. 511 f.). In seeking geological targets, the corresponding sort of information, that is, empirical distributions of number or prizes arranged in order of size and value, ideally should first be obtained. These would then be stratified according to geologic nature of the target, geologic time, depth below the surface, "favorable" geological areas, rock formations, commodity, and anything else that might be pertinent.

Unfortunately, however, this strategy cannot be followed because sufficiently detailed information is lacking even for the targets thus far discovered, much less for those of most interest, which are those that have not yet been found. In the next two subsections the problem is discussed, and some of the few data that are available for petroleum and ore deposits are reviewed.

As expected from the discussion in section 6.1, the distributions are highly skewed. This fact is emphasized in table 12.7, which lists the values of the 23

TABLE 12.7. VALUES OF THE 23 MOST VALUABLE NATURAL-RESOURCE TARGETS IN THE UNITED STATES*

Target number	Target name	Value (billions of dollars)	Resource
1	Mesabi Range, Minnesota	25	Iron
2	East Texas	10	Oil
3	Butte, Montana	5	Copper
4	Bingham Canyon, Utah	3	Copper
5	Midway, California	3	Oil
6	Wilmington, California	3	Oil
7	Elk Hills, California	3	Oil
8	Climax, Colorado	2.5	Molybdenum
9	Bradford, Pennsylvania	2.5	Oil
10	West Texas	2.0	Oil
11	East Bay, Louisiana	2.0	Oil
12	Coeur d'Alene, Idaho	2.0	Lead, silver
13	Oklahoma City, Oklahoma	1.5	Oil
14	Bisbee, Arizona	1.5	Copper
15	Globe-Miami, Arizona	1.5	Copper
16	Buena Vista, California	1.5	Oil
17	Michigan Copper	1.5	Copper
18	Southeast Missouri	1.5	Lead, copper
19	Homestake, South Dakota	1.0	Gold
20	Tri-State, Missouri-Kansas-Oklahoma	1.0	Lead, Zinc
21	Santa Rita, New Mexico	1.0	Copper
22	Leadville, Colorado	0.8	Zinc, lead, silver
23	Hugoton gas field, Kansas	> 2.0	Gas, oil

* After Griffiths, 1967.

most valuable natural resource targets found thus far in the United States. The values range widely, as do the sizes (which are not given); for instance, target 4, Bingham Canyon, worth $3 billion, is only $1\frac{2}{3}$ mile long, $1\frac{1}{3}$ mile wide, and $\frac{1}{2}$ mile deep, whereas at least one major oil field is 28 miles across (Griffiths, 1967).

We pause here to say that the distribution of natural-resource targets is a fruitful field for research and that an obvious starting point would be to make size, shape, and value distributions for coal fields of the world. As pointed out previously (sec. 12.2), the locations of all coal fields are essentially known, and the data would not be difficult to obtain and record, although a reliability index would have to be included. The resulting distributions should be highly instructive, particularly with reference to petroleum, which has similarities in origin but differences in that, being a fluid, it can move. Perhaps this work has been done, but reports that we have seen are too confounded with political boundaries, too unevaluated in reliability, and too incomplete to be conclusive.

Geological Models for Petroleum Distribution

Figure 12.5 gives the results of a study by Arps (1961) of the size distribution of petroleum fields in the Denver-Julesburg basin of Colorado and Nebraska and of some other petroleum distributions in the United States. All of the distributions are highly skewed, and Arps suggests that at least some of them may be lognormal (sec. 6.2). Hendricks (1965), in a discussion of petroleum resources of the world, reaches much the same conclusion and also gives a good bibliography of supporting works. Thus most petroleum distributions are evidently highly skewed, but details of these distributions are not well known.

Geological Models for Distributions of Ore Deposits

As for the other distributions discussed in this section, nearly all size distributions of ore deposits are highly skewed. Figure 12.6, a histogram of the amount of gold produced from districts of the United States, drawn from data published by Koschmann and Bergendahl (1968), is a good example. More than half the districts fall in the lowest class interval, with less than 100,000 ounces production, but the two largest ones, Cripple Creek, Colorado, and Lead, South Dakota, are too large to plot on the linear scale of the drawing.

The same sort of relation is almost always found—a few large deposits and many small ones—no matter whether size is reckoned in tonnage, value, total reserves, total production, or some other measure. This relation is seen for grade of gold mines in South Africa (sec. 16.4) and for tonnage and grade of lead mines of the world (sec. 13.5). Within one district the relation usually holds; for example, figure 12.7 is a histogram of the tonnages of 44 iron-ore bodies in the Knob Lake area of the Labrador-Quebec peninsula in Canada. The relation also occurs on the scale of a single mine, and often within one ore body.

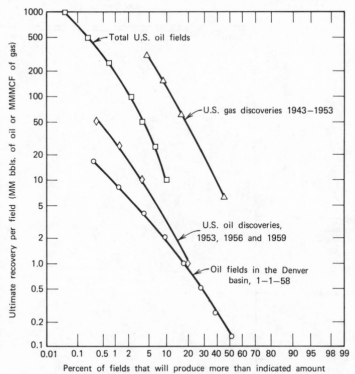

Fig. 12.5. Distribution of some petroleum fields of the United States (after Arps, 1961, p. 157).

Theoretical Distributions of Prizes

Many statistical models for theoretical distributions of prizes have been devised. One kind of empirical distribution is a count of the number of prizes per unit area. The prizes may be defined in any purposeful way, such as mines, prospects, oil fields, or all natural resources. An empirical distribution may be approximated by a theoretical frequency distribution. Thus Slichter (1960) postulated that mines in Ontario and the Basin and Range province in the United States are Poisson-distributed and therefore were formed by random processes, although these distributions were later shown to be nearly approximated by the negative binomial distribution (sec. 6.1). It has been argued by Allais (1957) and by others that, if one makes a frequency distribution for prizes in an explored area, the information might be put to use in exploring an unexplored area in order to estimate such items as the number of prizes per unit area, the total number of prizes in a total area such as a country or a continent, and the average value per unit area.

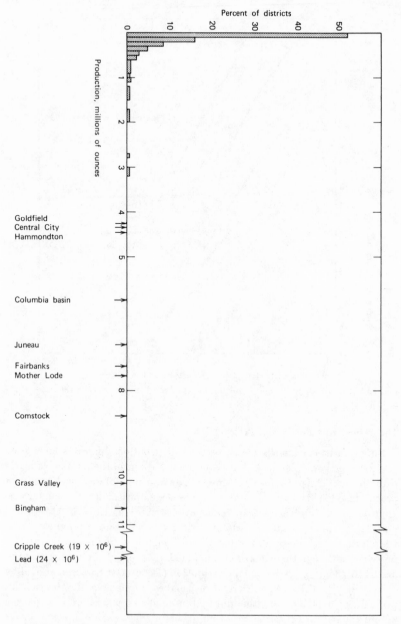

Fig. 12.6. Amount of gold produced from districts of the United States (data from Koschmann and Bergendahl, 1968).

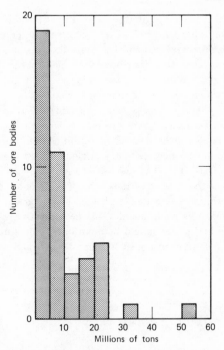

Fig. 12.7. Tonnages of 44 iron-ore bodies in the Knob Lake area of the Labrador, Quebec, Canada, peninsula (after Choubersky, 1957).

Other distributions besides Slichter's have been presented, some of which appear previously in this section. If enough distributions were made for many areas of the world, it might be possible to organize this information for predicting that, for an unexplored area of sufficient size, say of continental or subcontinental proportions, one might expect to find a certain number and value of prizes. But thus far nowhere near enough information has been collected, nor have good enough definitions been given of what constitutes an explored area and what constitutes an unexplored area to afford a useful guide to action.

We can, however, suggest a rather far-fetched illustration of how theoretical distributions of prizes might be used in exploration, noting that no convincing monetary interpretation can be given, although a generalized cost benefit (sec. 14.1) might be calculated. Suppose that an unexplored country, say in South America, wants to make a systematic resource appraisal. As advisor to the minister of the interior, a geologist might be able to discern that the geology seems similar to that of Ontario, Canada. [This geologist would be not only statistically informed but also remarkably lucky, because Ontario is one of the few areas for which a prior distribution (sec. 4.8) is available.] The geologist notes that the mean number of mines per 1000 square kilometers in Ontario is about 0.867 (table 6.4) and that the distribution is approximately a negative binomial. Assuming that this distribution applies to the South American country, the geologist explores some part of it, say one-third, and then calculates a new mean, which can be used to predict what will be found in the rest of the country. The geologist can also assume that if the distribution observed in the South American country is very different from the negative binomial, for instance, if two out of 100 cells contain four mines and the other 98 contain none, the exploration was incompetently carried out. Our enthusiasm for this scheme is so little that, if we had the job of this hypothetical geologist and expected this advice to be taken seriously, we would get another job lined up and our return ticket in hand before the field exploration started.

Finally, theoretical distributions of value per unit area without regard to value of the individual prizes should be mentioned. Some unpublished work (Wilmot, personal communication, 1968) suggests that these distributions tend to be lognormal.

12.4 EXPLORATION ON A GRID

In section 8.6, several experimental designs suitable for exploration were introduced, some based on gridded and others on nongridded sample points. Some of the gridded patterns are examined further in the present section. For simplicity, the method of data collection is referred to throughout as drilling, although many other types of sampling, including geochemical, trenching, etc., are also done on a grid.

Consider a specific problem for which exploration on a grid would be suitable. Assume that the object of search is Spanish galleons sunk at sea in the sixteenth century. Historical records would define the size, shape, and physical properties of the galleons; would indicate the general locations of where they sank; and might tell whether they were surrounded by other clues to their location, perhaps cannonballs. These data would suggest detection devices useful in the search, such as metal detectors, and a general area for search, and would specify those variables, like the cannonballs, that are valueless in themselves but that would indicate when the search was nearing the target. The available resources for the search, comprising detection gear, divers, ships, etc., would be inventoried, as well as the cost of moving these resources from place to place. With this much known, a highly specific and therefore meaningful statistical model could be devised to guide the search.

In geological exploration, information comparable to that used in a search for a Spanish galleon is not known, nor is it likely to be known in the future. Targets do exist, but their presence, size, and shape in any area of search are all uncertain, as is their likelihood of being surrounded by other response variables, for instance, a zone of rock alteration.

Rather than consider specific problems in this section, we shall develop some relationships among targets, area of search, and grid that may be investigated, and we shall explain the sort of assumptions that must be made. While developing this section, we performed many computations, of which only a small fraction appear here. The reader with a specific exploration problem is advised that incisive statistical guidance can be supplied for some real problems but that much thought must be applied in both statistics and geology if more than a perfunctory analysis is to be obtained, because the necessary parameters involved are difficult to evaluate.

The necessary assumptions relate to the number of targets; their shapes, geometric relationships, and overlap of one with another; the grid spacing; and the randomness of the relationship between the target or targets and the grid. For simplicity, the discussion that follows is restricted to a square grid of equally spaced points; a target is defined as a discrete unit; and the statement that a certain percentage of the total area of search is target means that this percentage is the total area in surface projection of the target or targets in question.

Target, Area of Search, and Grid Interrelationships

Cases of only one target in the area of search and cases of more than one target are considered in turn.

Regardless of the size of the area of search, if only one target is present, the number of grid intersections is inversely proportional, disregarding boundary effects, to GS^2, where GS is the grid or mesh spacing, in the same sense as used in sieve and screen sizing. If the grid spacing is cut in half, 4 times as many bore holes are required to search an area.

Choice of grid spacing is related to the size of the expected target. If the target is circular, it will always be found, provided that the grid spacing is smaller than $r\sqrt{2}$, where r is the target radius; for larger grid spacings, the probability of detection is graphed in figure 12.8. If the target is elliptical, the probability of detection may be obtained from a formula by Drew (1967, p. 701) or from tables by Savinskii (1965) or Singer and Wickman (1969). In figure 12.9 are given probabilities of detection for representative ellipses with major axes either 2 or 4 times the length of the minor axis. Similar graphs can be more or less readily drawn for targets bounded by curves susceptible to mathematical analysis and can be developed by simulation for targets whose boundaries do not readily yield to analysis.

If more than one target is present in the area of search, the number of

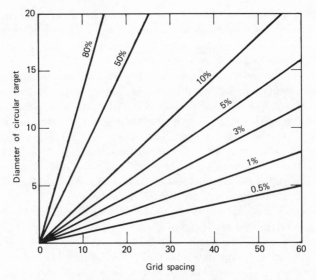

Fig. 12.8. Probabilities for detection of circular targets with grid spacings of various sizes.

targets, the percentage of the total area in target, and whether targets may overlap must all be considered. For simplicity, only a few simple concepts are developed for square targets. In order to suggest the complications that arise, it suffices to mention that, if the entire area to be searched is in target, one borehole will find it; and if there are ten targets that completely overlap one another, because they are stacked like a pile of dinner plates, the borehole that finds one target finds them all. Therefore, for multiple targets, details must be carefully specified.

For multiple targets, two subcases are considered. The first depends on the results (hits or misses) of successive boreholes being independent in a probability sense. Independence is satisfied if the targets in an area are randomly distributed throughout. Figure 12.10 graphs the probability of *at least* one hit for the percentage of area in target plotted against the number of holes drilled. For instance, if 45 holes are drilled on an equidimensional grid, and if targets comprise 5 percent of the total area, there is a 90 percent chance that at least one of the holes will hit a target. One practical application of this graph is to show a geologist who feels that the assumptions are satisfied how many holes he must drill in order to have a specified probability of locating at least one prize to present the board of directors as a productive result of the drilling campaign. (Figure 12.10 was plotted by applying the binomial distribution, where p is the fraction of the area in target and n is the number of holes drilled.)

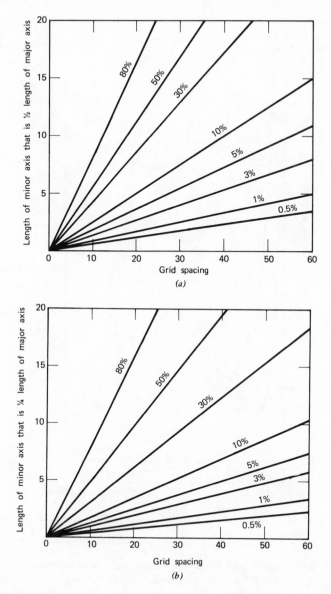

Fig. 12.9. Probabilities for detection of ellipses with grid spacings of various sizes.

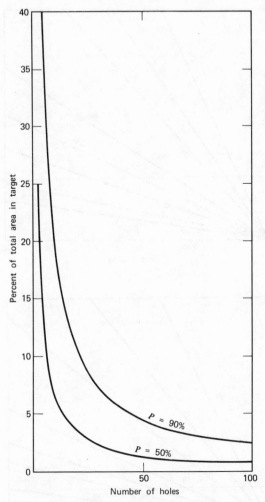

Fig. 12.10. Probability of at least one hit for the percentage of area in target, plotted against the number of holes drilled.

For only a few targets and a few holes drilled on a grid, the results depend intimately on the details of the assumptions made. For instance, the targets may be square ones that do not overlap and are located in a rectangular array like a checkerboard, but at random rather than in a regular alternating pattern, with a given square being either target or not. If five squares are investigated, one of which is target, and if all five are drilled, the probability is 100 percent that the target will be found. If two of ten squares are target and if only five randomly chosen ones are drilled, the probability of finding one target is only about 90 percent. If five of 25 squares are target and if five holes are drilled, the probability of missing all of the targets is 29 percent; in contrast, if the same 20 percent of the area is divided among many targets rather than among five out of 25 squares, and if the same number of five holes is drilled, the probability of all of them missing the target is 33 percent.

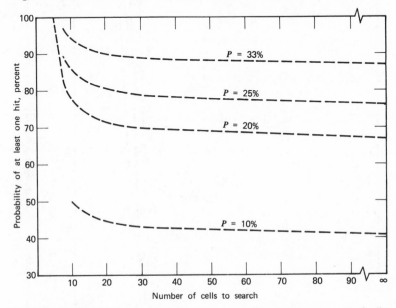

Fig. 12.11. Probability of hitting at least one target graphed against number of cells to search for several ratios of targets to cells.

Figure 12.11 graphs these relationships for the case of drilling five holes. For several ratios of targets to cells from 10 percent to 33 percent, the probability of hitting at least one target is graphed versus the number of cells. For instance, the graph shows that, for the probability of at least one hit to be 72 percent in 25 cells searched, 20 percent of the area must be in target. For a small number of either cells or targets, the probability of at least one hit is sensitive to the number of cells and targets, not merely to their ratio. This sensitivity diappears for a large number of cells, for which the limiting probabilities are plotted on the right-hand edge of the graph. (Figure 12.11 was constructed for a finite number of cells with the use of the hypergeometric distribution and for an infinite number of cells with the use of the binomial distribution.)

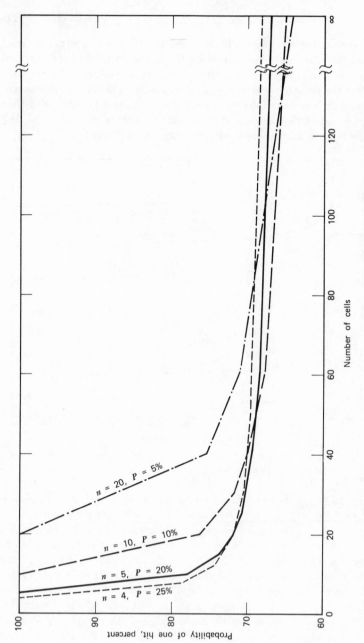

Fig. 12.12. Probability of hitting at least one target graphed against number of cells for a variable number of holes.

Figure 12.11 illustrates the general point that if there are only a few targets, the details of their geometry, relative configuration, and other fine points must be taken into account for exact probability calculations. It is unlikely that these details could ever be known well enough in a practical problem to make an exact analysis. However, if the assumption of independence in hitting the target from borehole to borehole is reasonable, an examination, using the binomial model, of the effects of the proportion of target in the area versus the number of holes drilled will always give the minimum probability of a hit. Thus the assessment is conservative, because, if all details are included, the probability of a hit can be calculated to be somewhat higher than this minimum, but is not overly conservative, because of the uncertainty involved in making the key assumption about the proportion of total cells that are target.

Figure 12.12 is similar to figure 12.11, except that the number of holes is not kept constant at five but is allowed to vary, and the probability of at least one hit is plotted against the number of cells to be searched. The curves relate the number of independent holes drilled and the probabilities of each hole being a hit; for instance, the curve labeled 10 holes and 10 percent probability means that 10 independent holes are drilled, each with a 10 percent probability of success. As before, all the curves flatten once the number of cells becomes large.

Finally, the question of drilling pattern arises. Equidimensional holes may be spaced either on a square or on a hexagonal grid (also named a rhombic or diamond-shaped grid), as illustrated in figure 12.13. If all targets are circular or nearly so, a hexagonal drilling pattern is more efficient than a square one. For instance, to be assured of hitting a circular target 1 mile in radius (fig. 12.13), a square grid would have to be 1.414 miles on a side, whereas a hexagonal grid could be 1.75 miles on a side. Thus, for a large area, hexagonal drilling requires only 80 percent (1.414/1.750) as many holes as square drilling. If targets are not exactly circular, the relative efficiency of the hexagonal pattern declines to approach or equal that of the square one. For real targets, it is doubtful that the hexagonal pattern has enough advantage to overcome the drawbacks of superimposing it on the square grid formed in many parts of the world by roads and property lines.

12.5 RESPONSE SURFACES

Consider a subsurface area of rectangular shape known to contain a maximum point, perhaps a center of mineral content, a volcanic vent from which pyroclastic material was dispersed, or an oil pool. In order to explore the area, holes can be drilled on a rectangular grid across it, and appropriate variables can be analyzed statistically by two-dimensional quadratic regression (sec.

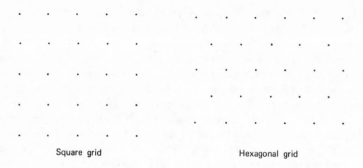

Square grid   Hexagonal grid

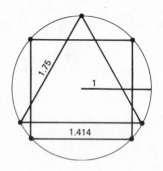

Comparison of square and hexagonal grids

Fig. 12.13. Comparison of square and hexagonal grids.

9.1) or by a factorial design (Cochran and Cox, 1957; Scheffé, 1959). However, either of these analyses may be inefficient if only the center of the phenomenon of interest is to be located and if information is not required across the entire area.

The response-surface method was developed by statisticians in the early 1950s to seek such a maximum point. The area can be defined by any variables of interest; in fact, the first problems were in chemistry, with the area defined by temperature and pressure axes, and the purpose was to find the optimum temperature and pressure at which to manufacture a product. Such applications may also be of interest in experimental and other fields of geology, but in this section the axes will be assumed to be spatial ones. An important distinction between chemistry and exploration geology is that in chemical problems at least one maximum point almost certainly is present whereas in exploration none need exist. In section 14.4 the two methods of exploration, response-surface drilling and grid drilling the entire area, are compared.

The problem resolves itself into a determination of the arrangement of sample points. The response-surface method requires the assumption that the surface sought is more or less quadratic—a dome, a basin, or a saddle. When a few observations are made, a "response" is obtained at each sample point; these responses are a guide to the discovery of the maximum or minimum point (if one exists) without the need to explore the entire area systematically.

Many procedures have been devised for response-surface analysis. Cochran and Cox (1957, Chap. 8-a) provide an excellent summary, Wilde and Beightler (1967) have written an entire book, and Hill and Hunter (1966) review methodology. We explain only a single method that should be useful in geology.

A Design Based on the Work of Box and Hunter

In this subsection a modified central-composite-rotatable design based on the work of Box and Hunter (Cochran and Cox, 1957, p. 345) is explained through a hypothetical problem to find the center of the Rangely oil-field dome. Although in practice the dome would be delineated through geophysics rather than through drilling, pretend that key-bed elevations are the basis for locating the center. In Chapter 9 the data from the 26 wells drilled in a haphazard pattern provide a good quadratic fit (table 9.12). With a response-surface design, the center can be located with fewer wells.

Because a quadratic equation in two dimensions has six coefficients (sec. 9.4), six data points are necessary for estimating them through a quadratic response model, without allowing any points for estimating residual variability. Box and Hunter suggest locating a center point at the maximum or minimum point of the quadratic surface, predicted from any prior information, and then taking the other points on the circumference of a circle centered on that point. If the minimum number of six points is taken, one is at the center and five are at the corners of a pentagon inscribed in the circle; if seven points are taken, one is at the center and six are at the corners of an inscribed hexagon; if nine points are taken, one is at the center and eight are at the corners of an inscribed octagon; etc. We discuss only the design for nine points, although we repeated the analysis for some other numbers of points with similar results.

First, an appropriate center-point location and circle radius are chosen by geological judgment. For the Rangely data, figure 12.14 plots the chosen radius of 2 miles and the center point located at coordinates East 7 miles, North 3 miles. (These choices are arbitrary for the purpose of exposition rather than realistic, because prior information such as the outcrop pattern would be used.)

Next nine wells would be drilled at the location plotted in figure 12.14. From each hole a response, which is the elevation of the key bed, would be obtained, and a quadratic trend equation would be fitted to the nine responses to predict the location and the size of the maximum or minimum point of the

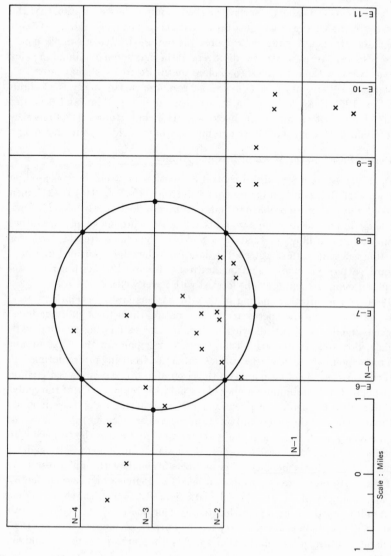

Fig. 12.14. Drilling pattern, corresponding to a quadratic-response-model design, for the Rangely, Colorado, oil field.

fitted surface. To simulate the responses, we estimated the variability in key-bed elevation from the observations in the 26 wells (sec. 9.5) and randomly added it to the observations corresponding to the original equation fitted to the 26 wells.

When the outlined calculations were performed, the estimated maximum was found to be at East 6.5 miles, North 2.1 miles, at an elevation of −241 feet. These numbers agree closely with the corresponding values of East 7.1 miles, North 2.0 miles, and an elevation of −246 feet, which were derived from the quadratic equation fitted to the 26 points. An analysis of variance, table 12.8, shows a highly reliable fit. Table 12.9 compares the simulated key-bed elevations with those predicted by the quadratic equation fitted by the response-surface design.

If a maximum or minimum point is not located closely enough from the responses at the original nine points, a second experiment can be performed

TABLE 12.8. ANALYSIS OF VARIANCE FOR QUADRATIC EQUATION FITTED BY RESPONSE SURFACE DESIGN TO KEY-BED ELEVATIONS OF THE RANGELY, COLORADO, OIL-FIELD DOME

Source of variation	Sum of squares	Degrees of freedom	Mean square	F	$F_{10\%}$
Linear terms	728,714	2	364,357	728	5.46
Quadratic terms	114,339	3	38,113	76	5.39
Residual variation	1,503	3	501		

TABLE 12.9. COMPARISON OF SIMULATED AND PREDICTED KEY-BED ELEVATIONS OF THE RANGELY, COLORADO, OIL-FIELD DOME

Hole number	Response elevation (ft)		Difference in response, observed minus predicted	Coordinates, miles	
	Observed	Predicted		East	North
1	− 961.3	− 951.9	− 9.4	8.00	4.00
2	− 1128.6	− 1128.3	− 0.3	7.00	4.41
3	− 774.3	− 778.5	4.2	6.00	4.00
4	− 354.7	− 354.8	0.1	5.59	3.00
5	− 263.7	− 253.8	− 9.9	6.00	2.00
6	− 267.7	− 287.3	19.6	7.00	1.59
7	− 310.6	− 287.2	− 23.4	8.00	2.00
8	− 481.9	− 501.1	19.2	8.41	3.00
9	− 380.7	− 380.7	0.0	7.00	3.00

with a new circle of points centered on the maximum point that is first found. The number of points would not necessarily have to be the same as in the first experiment, and presumably the circle radius could be decreased, conventionally to one-half the previous radius. The new data could be analyzed separately or together with the old data; usually one would make both analyses. The disadvantage of combining all the data is that, if the quadratic model is inadequate, some distortion may be introduced, whereas the new data, being from points closer together, will suffer less from inadequacy of the model. If the results from separate analyses of new data agree with the results from combined old and new data, evidently the quadratic model is suitable, and the answers are reliable.

Concluding Remarks

Although for the Rangely data the variable of interest, the key-bed elevation, was the response variable, it need not be. Another response variable may be chosen if the variable of interest is impossible or difficult to measure. For instance, a search may be made for a porphyry copper deposit below the depth of surface sampling. If copper had been leached from the surface rocks, a correlated response variable, such as lead that had not been leached, might be measured. Or it might be better to measure a response variable that is not correlated directly with the copper but that signals its approach, such as sericite in an envelope of alteration surrounding the ore deposit. Or several kinds of alteration might be measured and combined in a multivariate model (Chap. 10) to yield a single response variable.

12.6 THEORY OF SEARCH

Whether the hunt is for a needle in a haystack, for an honest man, or for evidence bearing on continental drift, a theory of search can illuminate the path. Much of the recent development-of-search theory was done for the U.S. Navy during World War II in order to locate life rafts, disabled planes, enemy submarines, etc. In particular, B. O. Koopman, working with the Operations Evaluation Group of the U.S. Navy, pioneered methods which he reviewed in three papers (1956–1957). Koopman considers both stationary and movable targets; in this section only parts of the second and third papers that are particularly relevant to geology are reviewed. The reader requiring more information is referred to Koopman's papers and to a review article by Brown (1960).

Previously in this chapter it was assumed that if a target is encountered, say by a drill hole, it also is detected with certainty. But this assumption is not necessarily fulfilled; for instance, the probability that petroleum, even if present, will actually be recognized in a particular drill hole may be only 25 percent. If the probability of recognition p is the same for each drill hole and if the target is intersected by a number of drill holes n, each with an independent chance of recognition, the probability P of detecting the target is

$$P = 1 - (1-p)^n.$$

Thus, for the probability p of 25 percent associated with each hole, the probabilities of two, three, or four holes detecting the target are 44, 58, and 68 percent, respectively. The assumption that the different holes are independent is essential. Otherwise, if for instance one hole is deflected, some dependence may be introduced (sec. 7.4); and if n is the number of deflections, the formula will overestimate the probability of finding the target.

The previous formula shows that the average number of drill holes necessary to detect a given target is proportional to $1/p$, and therefore that the best of several different methods of search may be found by comparing ratios of the cost of a sampling method with the probability of finding a target by using that method. In practice these costs and especially the probabilities may be difficult to establish, but if they can be estimated, a guide for action is afforded.

Consider again the case in which, if a target is encountered, it is detected. If an area of size A is searched over a fraction of the area, A_f, the probability of finding a target is proportional to the area searched divided by the entire area, that is, to A_f/A. If the area searched contains k targets, more or less uniformly and independently distributed across the area and small enough so that any overlaps of targets can be ignored, the probability P of finding the number s of the total number of k targets is

$$P = \binom{k}{s} p^s (1-p)^{k-s},$$

where p is equal to A_f/A. The assumption must be made that there are k independent targets, each with the same probability p of being found in the area searched.

If the size of the target is large with respect to the size of the area searched, all these relationships break down because they are exact only for point targets being searched for by a device that covers a small area. The primary consideration in using this method is that the individual target is presumed to be very small relative to the size of the area searched, so that these targets may be considered to be small areas being searched for by pinpricks, that is, by drill holes. In preceding sections the calculations were for particular simple cases. However, for exact solutions to real problems, the details of target shape are important and the results presented in this book serve only as illustrations to give the reader an idea of what to expect.

Up to this point the assumption has been made that the target sought is as likely to be in one part of the search area as in another. But often there is reason to believe that the target is more likely to be in one sector, and Koopman again shows how to proceed. Suppose that Wyoming and Nebraska are to be searched for uranium, and a geologist believes that the probability of finding a deposit in Wyoming is 75 percent and that of finding one in Nebraska is 25 percent. If n holes are drilled, they may be allocated so that the number x are drilled in Wyoming and the number $n - x$ are drilled in Nebraska. Now each hole has a probability p of detecting the target, and if the holes are independent, the probability P_W of finding the target in Wyoming, given that a target exists in Wyoming, is

$$P_W = [1 - (1-p)^x],$$

while the probability P_N of finding the target in Nebraska, given that a target exists in Nebraska, is

$$P_N = [1 - (1-p)^{n-x}],$$

both these probabilities being conditional (sec. 2.5). The total probability of finding the target is

$$P = 0.75P_W + 0.25P_N,$$

and x must be chosen to maximize this expression.

We may illustrate the application of this formula for a few cases. If the probability of the target being in Wyoming is 75 percent and that of its being in Nebraska is 25 percent, if these is a 10 percent probability of detecting the target through one borehole, and if ten holes in all are to be drilled, all ten holes should be drilled in Wyoming to maximize the probability of finding the target, which is 49 percent. On the other hand, if 20 holes in all are to be drilled, 15 should be drilled in Wyoming and the remaining five in Nebraska, the probability of finding the target now being 70 percent.

With the probabilities for Wyoming and Nebraska specified, the best plan is first to drill all holes in Wyoming and then, after a certain number of holes have been drilled there, to alternate drilling between the two states. To find the certain number of holes to drill first in Wyoming, one must assume the probability of discovery of the target by a single hole, the probability of which is plotted in figure 12.15 against the certain number of holes to drill in Wyoming. For instance, the figure shows that, for a probability of discovery by a single hole of 10 percent, 11 holes should first be drilled in Wyoming, then the 12th hole in Nebraska, the 13th in Wyoming, etc.

How best to distribute the total number of holes to be drilled between the two states depends, of course, on the assumed prior probabilities of discovery of the target in them. Figure 12.16 plots the proportion of holes that should be

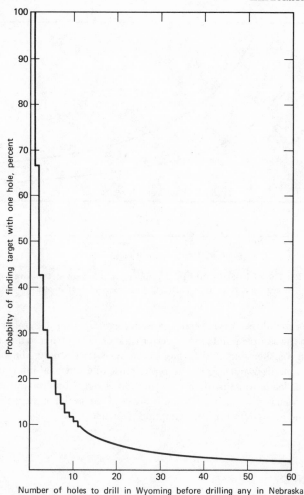

Fig. 12.15. Probability of finding a target with one hole, graphed against the number of holes to drill in Wyoming before drilling any in Nebraska.

drilled in Wyoming against the total number of holes to be drilled; the probability of discovery of the target by each hole is assumed to be 10 percent. Curves are given for assumed probabilities that the target is in Wyoming of 90 percent, 75 percent, and 50 percent. For instance, if the assumed probability of the target being in Wyoming is 75 percent and if 50 holes are to be drilled, 60 percent of them should be drilled in Wyoming. The plotted curves have been smoothed; the exact curves have some vertical segments in them.

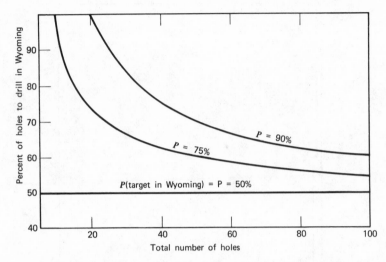

Fig. 12.16. Percent of holes to drill in Wyoming plotted against the total number of holes to drill, assuming a 10 percent probability for discovery of the target by each hole.

If the probabilities have been assessed correctly, the suggested scheme distributes the sample points much better than one of putting half in one area and half in another; but if they have been assessed wrongly, the suggested scheme can be disastrous. As usual, making use of additional information on probabilities leads to gains if the information is right, but to trouble if it is wrong. Koopman also discusses the scheme as revised for more than two subareas and gives details of the sampling sequence.

REFERENCES

Allais, M., 1957, Method of appraising economic prospects of mining exploration over large territories: Management Sci., v. 3, p. 285–347.

Arps, J. J., 1961, The profitability of exploratory ventures, *in* Economics of petroleum exploration, development, and property evaluation: Dallas, Tex., Internat. Oil and Gas Educ. Center, p. 153–173.

Averitt, Paul, 1961, Coal reserves of the United States—a progress report: U.S. Geol. Survey, Bull. 1136, 116 p.

Bailly, P. A., 1964, Methods, costs, land requirements, and organization in regional exploration for base metals: Preprint of paper presented at mtg. of Am. Inst. Mining Engineers, Alaska sec., Fairbanks, Alaska, March 18–21, 1962, 31 p.

————, 1967, Mineral exploration and mine developing problems: Statement presented to the Public Lands Law Conf., Univ. of Idaho, Boise, 44 p.

————, 1968, Exploration methods and requirements, *in* Pfleider, E. P., ed., Surface mining: New York, Am. Inst. Mining Engineers, p. 19–42.

Bates, R. L., 1960, Geology of the industrial minerals and rocks: New York, Harper & Brothers, 441 p.

Boyd, James, 1967, The influence of the minerals industry on general economics: Mining Engineering, v. 19, March, p. 54–59.

Brown, A. A., 1960, Search theory and problems of exploration drilling: Bull. Mineral Industries Expt. Sta., Pennsylvania State Univ., No. 72, p. 33–37.

Cannon, R. S., and Pierce, A. P., 1967, Lead-isotope data and occurrence of ore deposits: U.S. Geol. Survey, Prof. Paper 575-A, p. A-4.

Choubersky, A., 1957, The operation of the Iron Ore Company of Canada: Trans. Inst. Mining Metall., v. 67, p. 34–88.

Churchill, W. S., 1948, The gathering storm: Boston, Houghton Mifflin Co., 763 p.

Cochran, W. G., and Cox, G. M., 1957, Experimental designs: New York, John Wiley & Sons, 611 p.

Drew, L. J., 1967, Grid-drilling exploration and its application to the search for petroleum: Econ. Geology, v. 62, p. 698–710.

Feller, William, 1968, An introduction to probability theory and its applications, v. 1: New York, John Wiley & Sons, 461 p.

Gillson, J. L., ed., 1960, Industrial minerals and rocks: New York, Am. Inst. Mining Engineers, 934 p.

Goddard, E. N., ed., 1965, Geological map of North America: U.S. Geol. Survey.

Griffiths, J. C., 1966, Exploration for natural resources: Jour. Operations Research Soc. Am., v. 14, p. 189–209.

————, 1967, Mathematical exploration strategy and decision-making: Proc. Seventh World Petroleum Cong., p. 599–604.

Groundwater, T. R., 1968, review of a book by S. H. Frankel: Trans. Inst. Mining Metall., sec. A, v. 77, No. 740, p. 39–40.

Harris, D. P., 1966, A probability model of mineral wealth: Trans. Soc. Mining Engineers, Am. Inst. Mining Engineers, v. 235, p. 199–216.

Hendricks, T. A., 1965, Resources of oil, gas, and natural-gas liquids in the United States and the world: U.S. Geol. Survey, Circ. 522, 20 p.

Hill, W. J., and Hunter, W. G., 1966, A review of response surface methodology, a literature survey: Technometrics, v. 8, p. 571–590.

Jerome, S. E., 1959, Exploration of large areas: Mining Cong. Jour., v. 45, p. 37–40.

Joralemon, Peter, 1967, The fifth dimension in ore search: Soc. Mining Engineers, Am. Inst. Mining Engineers, Fall mtg., preprint, 18 p.

Koopman, B. O., 1956–1957, The theory of search: Operations Research Soc. Am., Jour., v. 4, p. 324–346, 503–531; v. 5, p. 613–626.

Koschmann, A. H., and Bergendahl, M. H., 1968, Principal gold producing districts of the United States: U.S. Geol. Survey, Prof. Paper 610, 283 p.

Levorsen, A. I., 1967, Geology of petroleum, 2nd ed.: San Francisco, W. H. Freeman, 673 p.

Lindgren, Waldemar, 1933, Mineral deposits: New York, McGraw-Hill, 894 p.

Lowell, J. D., 1968, Discovery and exploration of the Kalamazoo orebody: New York, Am. Inst. Mining Engineers, Fall mtg., preprint, 11 p.

Mining Engineering, 1965, Newmont brings in Nevada gold—the modern way: Mining Eng., v. 17, p. 138–141.

Savinskii, I. D., 1965, Probability tables for locating elliptical underground masses with a rectangular grid: New York, Consultants Bureau, 110 p.

Scheffé, Henry, 1959, Analysis of variance: New York, John Wiley & Sons, 476 p.

Sheehan, Robert, 1964, Great day in the morning for Texas Gulf Sulfur: Fortune, v. 70, no. 7, p. 137–140, 254–258.

Singer, D. A., and Wickman, F. E., 1969, Probability tables for locating elliptical targets with square, rectangular, and hexagonal point-nets: Mineral Sci. Expt. Sta., Pennsylvania State Univ., Spec. Pub. 1-69, 100 p.

Slichter, L. B., 1960, The need of a new philosophy of prospecting: Mining Eng., v. 12, p. 570–576.

Ventura, E. M., 1964, Operations research in the mining industry, *in* Hertz, D. B., and Eddison, R. T., eds., Progress in operations research: New York, John Wiley & Sons, p. 299–328.

Wilde, D. J., and Beightler, C. S., 1967, Foundations of optimization: Englewood Cliffs, N.J., Prentice-Hall, 480 p.

Wilks, S. S., 1962, Mathematical statistics: New York, John Wiley & Sons, 644 p.

Chapter 13

Delineation of Ore Deposits

This chapter explains some applications of methods developed in previous chapters to the narrow and specific field of delineation of ore deposits. One reason for introducing the topic is that it is important in its own right, and many of the issues raised are pertinent as well to oil and gas delineation. Also it shows how statistical procedures can be put to work on one kind of geological data and, by analogy, how they may be applied to other geological problems. Much of this chapter results from work done at the Mine Systems Engineering Center of the U.S. Bureau of Mines in Denver, Colorado, published in part as Bureau of Mines' Reports of Investigations.

By delineation of an ore deposit, we mean all of the facets comprising the size, shape, location, tonnage, grade, and environment of an ore body, or, if there are several ore shoots or ore bodies close together, the over-all pattern, including their interrelationships. In former times, when mining was done predominantly by hand methods so that typically a narrow vein or other small ore body was mined underground very selectively, delineation of ore deposits was concerned mainly with tonnage and grade. Today the tendency is more and more toward large-scale and highly mechanized mining in open pits or by block caving or like methods underground. Many more aspects of delineation must be considered and tied together now that such problems come to the fore as stripping ratios or choosing methods of mining that can be highly mechanized but may pay a penalty of lower-grade ore or less complete extraction (for instance, auger mining of coal).

That delineation of ore bodies is not easy is demonstrated by the interesting results obtained by Wright (1959), who compared the estimates of uranium ore made by eight geologists with production results. Wright selected five mines whose ore bodies were nearly or completely mined out, located in the Salt Wash member of the Morrison Formation in the Gateway mining district, Mesa County, Colorado, which is a portion of the Uravan Mineral Belt. The mines chosen were those relatively difficult to evaluate rather than average ones. Data given the geologists were maps, published in Wright's article, showing locations of vertical rotary-drill holes and assays, but not the mine names.

Table 13.1, calculated from Wright's table 1, lists the highly variable results obtained. For instance, line 1 shows that for mine 1, geologist 1 estimated 13,920 tons of 0.87 percent ore; estimated results were 154 percent of the produced tonnage, and 335 percent of the produced grade. Not only were there many estimates of tonnage, grade, and pounds of uranium far afield, but the fact that the geologists disagreed among themselves in evaluating each deposit shows that no uniform methods were applied. Wright's comments on the results may be condensed and paraphrased:

(1) This mine was unprofitable. Although tonnage was larger than expected, grade was much lower because high-grade ore pockets intersected by the drill holes were small and discontinuous. Low-grade ore was blended with high grade to maintain tonnage, at the expense of extreme dilution.

(2) This mine was a large and profitable producer, with more tonnage than expected, although grade was lower than estimated because the method of mining chosen led to greater dilution.

(3) The orebody worked was small but larger than expected; mined grade was lower than estimated grade because of increased dilution partly caused by the thinness of the orebody.

(4) Although grade of this deposit was fairly well estimated by all the geologists, tonnage was grossly overestimated. Not enough holes were drilled to demonstrate the lack of continuity of the ore.

(5) For this deposit, grade was slightly overestimated, and tonnage was underestimated, because holes intersecting ore were evidently surrounded by holes in waste.

The plan of this chapter is to sketch briefly the steps in delineation of ore deposits and to expand on those to which statistical analysis offers help. The first section reviews general principles of delineation, and the second discusses evaluation of grade, with special attention to choice of metal variable. Section 13.3 explains how to estimate the amount of ore in dimensions that may be length, area, volume, or weight. If estimated by weight, volume must be converted to weight by one of several methods discussed in this section. In section 13.4, ore estimation at the Giant Yellowknife gold mine, Northwest Territories, Canada, is reviewed as an example. In section 13.5, the previous sections on grade and amount of ore are related to one another in order to investigate the

TABLE 13.1. COMPARISON OF ESTIMATES OF GRADE OF URANIUM ORE WITH PRODUCTION RESULTS FOR FIVE MINES IN THE MORRISON FORMATION, COLORADO*

| Mine number | Geologist number | Production results | | Ratio of geologists' estimates to production results | | |
		Tons of ore	U_3O_8 (%)	Tons of ore	U_3O_8 (%)	U_3O_8 (lb)
1	1	9,063	0.26	154	335	514
	2			66	673	442
	3			58	442	255
	4			88	638	560
	5			133	608	808
	6			27	642	176
	7			106	562	598
	8			38	569	215
2	1	93,000	0.28	42	168	70
	2			19	132	25
	3			27	136	36
	4			33	132	44
	5			36	132	47
	6			30	132	40
	7			44	136	60
	8			31	132	41
3	1	1,111	0.34	90	82	74
	2			45	141	64
	3			50	126	63
	4			46	132	61
	5			61	132	80
	6			23	142	32
	7			69	123	85
	8			48	150	71
4	2	573	0.32	340	91	308
	3			248	94	232
	4			612	81	498
	5			623	81	507
	6			266	94	250
	7			593	81	482
	8			555	94	520
5	1	985	0.21	42	129	54
	2			52	114	60
	3			70	109	77
	4			68	119	81
	5			68	114	78
	7			68	138	93
	8			76	119	90

* After Wright, 1959.

well-known relationship that, as amount of ore increases, its grade usually decreases. The last section of the chapter is about classification of ore.

13.1   SOME REQUIREMENTS FOR DELINEATION OF ORE DEPOSITS

Some requirements for delineation of ore deposits are sketched in figure 13.1, a schematic cross-section distinguishing four components, labeled A to D, of an ore body and overburden in relation to their physical and economic environment.

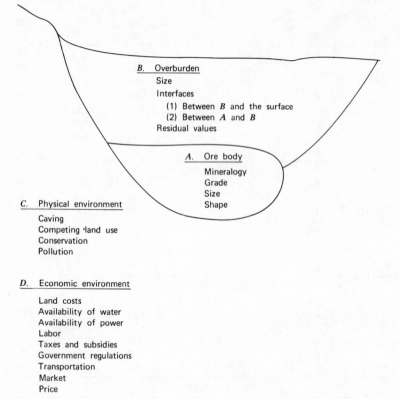

Fig. 13.1. Schematic cross-section of an ore body in relation to its physical and economic environment.

For the ore body itself, grade, size, and shape (component A) are the important requirements for delineation. Size refers to the amount of ore as reckoned in weight, volume, or area, or in a mixture of these units. The variety in units of size evidently mainly reflects tradition in the industry rather than fundamental considerations. In deposits of a commodity sold by volume such as sand and gravel, size is measured by volume, but placers, which yield minerals sold by weight, are also measured by volume. In underground mines, on the other hand, size is measured by weight, probably because in the past weight was the determining factor in hoisting, hand mucking, tramming, etc. In open-pit mines the inconsistent practice is to measure stripping of waste by volume and removal of ore by weight. In mining tabular ore bodies, size may be measured by area, as are reefs in South African gold mines or narrow veins in North American lead and silver mines.

Another important property of an ore body is grade of ore, which may be measured either by volume or by weight, as discussed in the next section. Besides these attributes, the chemical composition of the ore and its amenability to extraction and beneficiation must be considered; decisions that must be made about these subjects are raised in sections 14.6 and 14.7. The shape of the ore body governs sampling and valuation (secs. 13.2 and 13.3) and also influences the mining method. The required decisions are reviewed in section 14.6.

Component B is the overburden, which must be removed to recover the ore if mining is by open pit. This component plays no part in an underground mine unless subsidence takes place, and it is absent from an ore body exposed at the surface, as in a placer deposit, or some ore bodies in glaciated regions. Amounts of overburden and ore are often compared according to stripping ratio (ratio of weight of waste to weight of ore), and the usual statistical measures can be applied (sec. 14.5). Besides the size, there are two interfaces to consider: (1) that between the overburden B and the surface, and (2) that between the overburden B and the ore body. These interfaces govern pit slopes, the steepness of which may lead to slope-stability problems in rock mechanics (sec. 14.6). The interface (2) is also that between ore and waste, separating that rock which goes to the mill from that which goes to leach or the dump; the location of the interface may be set by computer programs for pit design (sec. 14.5). Finally, the overburden may contain some material of sub-ore grade having residual values that may be recovered, as by leaching of copper-bearing minerals.

Component C, the physical environment, may be critically important or of little significance. Caving, either of an open pit or in subsidence above an underground mine, is usually important, even if the mine is in an unpopulated area, because surface structures and lives may be endangered. In cities, land uses often conflict, and many decisions must be made. For instance, in quarrying in cities, decisions are needed about how long to mine before the land becomes more valuable for building lots, or how many claims for blasting damage can be afforded. In a coal deposit, how much to strip mine and how much to underground mine and how to restore the land must all be studied (sec. 14.6). Questions of competing land use arise again in the domain of conservation, with the added problem of complete versus partial extraction of the ore; the interfaces surrounding A shrink or expand. Pollution, especially of water and air, may be a serious problem, particularly if the physical environment is populated.

Finally, component D, the economic environment, comprises a number of economic factors that affect whether a certain mineral deposit is or is not ore. Land cost may be insignificant for a uranium deposit in New Mexico but may be predominant for a sand and gravel pit in New York City. Labor may be readily available for an open-pit mine but may be difficult to find and train for a deep underground mine. Taxes, subsidies, and government regulations vary from country to country and from state to state in the United States. Transportation from mine to market may be a large part of the cost of production for a copper mine in central Africa but may be inconsequential for a diamond mine in the same continent. Markets and prices may be fixed, as for gold in some countries, but may range widely for other commodities.

All of the issues raised in this section concern geology influenced by engineering and economics in relation to the nature of the materials to be moved and the method of moving them. The costs involved can be divided, and this process, first introduced in Chapter 12, is continued through Chapter 13 and into Chapter 14. In the stage of exploration, costs depend on the thickness of the overburden B and on the size, shape, and continuity of the ore body A. In the development stage, the costs depend mainly on the size of the ore body A, but also to some extent on other factors in the exploration stage. In mining, costs depend on the size and shape of both the ore body A and the overburden B, and also on the physical environment C. In beneficiation, costs depend on the size, mineralogy, and grade of the ore body A and on the physical environment C.

13.2   EVALUATION OF GRADE

Grade, a numerical measure of the amount of metal contained in an ore body, is of no interest in itself but is pertinent to the central question, which is how much metal can be produced from a certain ore body. Although grade is usually calculated in weight units, perhaps as percent of copper, it may also be expressed in a volume unit, perhaps as cents of gold per cubic yard, or in a mixed unit, perhaps as meter-percent of lead per square meter. In principle, a mathematical function could be found to relate grade and other variables such as dilution to amount of metal actually produced from the ore body; in practice, grade and the other variables are seldom if ever well enough known to do this.

Evaluation of grade depends closely on geological sampling and variability, as discussed in Chapter 7, particularly on the shape and accessibility of ore bodies for sampling and on the costs of different sampling methods. Several different kinds of grades must be considered (McKinstry, 1948, p. 463):

Sampled grade or *in-place grade*, that of the ore body as sampled in place.

Mined grade, that actually mined, nearly always lower than sampled grade because of dilution by low-grade or barren rock.

Mill-head grade, that received by the mill, which may differ from mined grade because of selective losses in transportation from mine to mill or because of a different sampling scheme.

Recoverable grade, that recovered after losses in milling.

Liquidation grade, that at which the smelter or other purchaser pays after making reductions conventional in the industry.

After taking into account these various kinds of grades, one realizes that the final metal sold is based on a lower grade than that sampled, and appropriate discounts and measures of statistical variation must be applied.

For those ore bodies that have a "normal grade," the main problem may be to identify it. Prescott (1925, p. 670) writes that

Every replacement orebody has a certain "normal grade" of ore which might be defined as the average of all ores above that critical point where the grade commences to drop very suddenly to traces only. The sharp demarcation in the outlines of the orebodies and the absence of grading off of mineralization into the country rock are among the most pronounced characteristics of replacement deposits; and in the same way, there is nearly always a definitely establishable point below which the tonnage for each unit drop in grade is insignificant until the realm of traces is reached, while the tonnage for each unit rise above that point is considerable until the maximum is approached. This critical point and the resulting normal grade are independent of commercial considerations, costs, metal values, prices, etc., and should not be confused with the economic limit of lowest grade which such commercial considerations permit being handled for a suitable profit. In many types of ore deposits, the tonnage increases inversely as the grade; and an improvement in costs, metal market, treatment charges, etc., immediately results in a large increase in tonnage available for production. This is not usually true with replacement deposits providing that the normal grade is equal to or better than the cost of production plus the required profit. This is the first point to be determined by sampling, and it is an essential one. . . .

Similarly, some rocks, such as basalts, have a "normal" composition that varies little from place to place even in a large rock body.

The next subsection takes up the choice of variable (or units) in which to express the metal of interest; this unit depends on the composition of the rock and on the shape of the ore body. Then evaluation of grade of ore bodies of different shapes and appraisal of reliability of evaluation of grade are considered. In the last subsection Matheron's theory of regionalized variables is outlined.

Choice of Metal Variable

No one metal variable is inherently more fundamental than another: what is desired is a measure that is a guide to producing tons of metals contained in concentrates, which is what a mine sells. These remarks also apply to theoretical geology, in which rock composition may be expressed in weight or volume percent of minerals, oxides, or chemical elements; no one of these units is more fundamental than another, although one may be more functional for a particular purpose. What one must do is evaluate critically what the different units really measure and what variances are associated with them and then choose a good one.

One fictitious illustration of choice of metal variable serves to demonstrate the problems. The case presented is one that has caused wide misunderstanding because people have not kept in mind that the goal is a measure of mined grade. The choice is between a weight unit and a mixed unit. The latter is obtained by multiplying assay by ore-body width. The assay width may be expressed in meter-percent, foot-ounces, inch-pennyweights, or other dimensions; for convenience the meter-percent unit is discussed here. Consider a thin, tabular ore body, perhaps a vein or a sedimentary bed. Assume that the minimum width that can be mined (minimum mining width or height) is wider than the ore body and that mineralization is confined to the ore body, with the wall rock being barren. Under these conditions an assay-width unit provides an unbiased estimate of mined grade of ore, rather than an assay unit that yields a biased estimate, unless the width of the ore body can be measured precisely—and it usually cannot be.

This concept is worth examining in more detail. Consider the instructions given samplers. They are directed to sample into the walls, because locations of the walls in many veins or the tops and bottoms of sedimentary beds are difficult to discern. Under these circumstances different samplers are liable to measure different widths at the same place and therefore sample different amounts of wall rock. However, if properly trained, each sampler will sample at least the full width of the ore body. The resulting assay values will, of course, be low if the sampled widths are wide and will be high if the sampled widths are narrow, even though the amount of metal mined per unit area of ore-body surface, as measured in the plane of the ore body, is identical.

Some numerical data may make these points clearer. Consider the following hypothetical situation. An EX diamond-drill hole is bored across a narrow vein in which all the ore mineralization is confined to a width of 0.5 meter. The core is split longitudinally, and the two halves are logged by two samplers. Because of the difficulty in identifying vein boundaries in a small-diameter drill core, sampler A measures the vein width as 0.5 meter, while sampler B measures it as 1 meter. The two halves are assayed, with the results listed in table 13.2,

Sampler	Measured width of orebody (m)	Volume of EX core (cc)	Assay (%)	Metal content (m-%)
A	0.5	479	10	5
B	1.0	958	5	5

and although one measured width is double that of the other, the metal content is the same in both. Now, provided that the minimum stoping width is wider than 0.5 meter, 5 meter-percent is clearly an unbiased estimate of the amount of metal to be mined, in contrast to the unweighted estimate, which fluctuates from 5 to 10 percent, depending on whether sampler A or B does the sampling.

The relationship of sampled grade in meter-percent to mined grade in percent must now be considered, because the mined grade rather than the grade of the ore body itself is the objective of commercial interest. If the entire width of the ore body is stoped,

$$\text{mined grade of ore} = \frac{\text{sampled grade in meter-grams}}{\text{ore body width in meters}}.$$

The one exception to this equation arises when mineralization below profitable grade is disseminated through a section of an ore body that both is wider than the minimum stoping width and also must be stoped to reach a profitable section of the ore body. In this case, a minimum rather than the full width of the ore body is stoped, and sampled grade in meter-percent is a biased estimate of mined grade.

Consider one more fictitious illustration. Assume that a sample is taken that represents a certain volume of ore computed by multiplying the ore body width by unit area. Table 13.3 compares sample data and calculations for five different situations. Samples 1 and 2 differ in both assay and in sampled width, but have the same meter-percent value and mined grade. Samples 2, 3, and 4 yield the same assays but are from samples taken over different widths. As width increases, so does mined grade. Of these first four samples, the sampled percent assay is an unbiased estimate of mined grade only for sample 4; yet

TABLE 13.3. FICTITIOUS SAMPLE DATA FROM FIVE SITUATIONS

Sample number	Sampled grade (%)	Sampled width (m)	Sampled grade (m-%)	Volume per unit area of ore body (m³)	Metric tons of ore per unit area of ore body*	Kilograms of metal per unit area of ore body	Width stoped (m)†	Tons stoped per unit area of ore body	Mined grade (%)
1	5	1.0	5	1.0	3.0	150	1.5	4.5	3.33
2	10	0.5	5	0.5	1.5	150	1.5	4.5	3.33
3	10	1.0	10	1.0	3.0	300	1.5	4.5	6.67
4	10	1.5	15	1.5	4.5	450	1.5	4.5	10.00
5	1	2.0	2	2.0	6.0	60	1.5	4.5	1.00

* Assuming 3 metric tons per cubic meter.

† Assuming a minimum stoping width of 1.5 meters.

sampled grade in meter-percent yields an unbiased estimate of mined grade for all four because 1.5, the width stoped, is substituted in the previous equation to give

$$\text{mined grade of ore} = \frac{\text{sampled grade in meter-grams}}{1.5}.$$

Only for the uncommon case of sample 5 is sampled grade in percent rather than in meter-percent an unbiased estimate of mined grade.

Shape of Ore Bodies

As discussed for narrow tabular ore bodies in the preceding subsection and as introduced in general in Chapter 7, the shape of ore bodies strongly influences the evaluation of their grade. In the next subsections, the effect of shape is reviewed for elongated, tabular, and irregular ore bodies. [By "irregular" ore bodies, we mean those nonuniform in shape, including stockworks, "disseminated deposits" such as the "porphyry" coppers, and replacement deposits nonconformable to rock strata and of no very consistent shape. For the history of use of the term, see Lindgren (1933, p. 204).] It is important to recognize that one ore body may be viewed in different ways; for instance, an irregular ore body may be considered to be a stack of horizontal tabular ore bodies, each of a certain thickness, if it is to be mined by benches in an open pit.

Ore-reserve blocks are usually formed as an aid to ore evaluation. The purpose may be to plan mining, to evalute a mine for a sale price, to determine royalty payments, etc. A block is a part of an ore body of a size suitable for mining, and the grade of ore is usually consistent throughout each block. Delineation of blocks may be routine for ore bodies of well-known shape and consistently minable grade, but difficult for bodies of poorly known shape and with grades that are inconsistent and near the cutoff grade.

Evaluation of Grade for Elongated Ore Bodies

Evaluation of grade is discussed for several representative types of elongated ore bodies. One type, confined to modern or ancient stream channels, includes some placer and some uranium deposits. In order to evaluate grade in these deposits, methods of sampling described in Chapter 15 are followed. Often, most if not all of the ore is concentrated at the extreme bottom of the channel and may be difficult to find and impossible to evaluate satisfactorily. Another type of elongated ore body is a tabular one that is to be mined in only one plane, which is likely to be the surface outcrop. In this case variability across the trace of the tabular ore body is usually much greater than along it, and traverses across the ore body are made with a sampling design like that discussed in section 8.6. For essentially vertical elongated ore bodies, comprising chimneys, breccia pipes, etc., which may be only a few feet across, an economical way to sample ahead of mining can seldom be found, and consequently evaluation before mining is impossible.

Evaluation of Grade for Tabular Ore Bodies

For evaluation of tabular ore bodies, two cases are important: that with sample points distributed more or less uniformly, whether on a grid or in a haphazard or random pattern, and that with sample points clustered on more or less straight lines in the ore body. These are considered in turn.

If the sample points are distributed rather uniformly, grade is best arrived at by summing the assays and dividing by the number of sample points, unless, as discussed in a subsequent subsection, trends exist in the data. Alternative methods are to consider that the sample points lie at the centers of polygons or the vertices of triangles and to weight the assays by the areas of the polygons or triangles. Numerous investigations show that little or no advantage can be demonstrated in the more cumbersome methods of calculating grades, provided that the sample points are distributed more or less uniformly.

In tabular ore bodies, sample points are commonly clustered because they are taken in tunnel-shaped openings in the ore body, comprising drifts, raises, and winzes. In this second case of clustered sample points, illustrated in figures 4.11 and 5.3, the usual procedure is to evaluate the ore in rectangular or triangular blocks for which closely spaced observations are available on one or more sides. For a simple situation when data are at hand for four sides of a rectangular block, say from two drifts and two raises, the best procedure is, as before, to sum the assays and divide by the number of assays, provided that there is no trend in the data and that the reliability of sampling is the same on all four sides of the block. This method is explained in the standard books, for instance by McKinstry (1948, p. 59). If data are available on fewer than three sides, the same procedure is followed but the reliability decreases.

Dilution, although important for ore bodies of any shape, must always be taken into account for tabular ore bodies narrower than the minimum stoping width, a point already introduced in the discussion of the choice of metal variable. Even if the tabular ore body is wider than the minimum stoping width, some dilution is almost inevitable because it is almost impossible to mine cleanly. Dilution usually is expressed as a percentage of the weight of ore mined, and the mined grade is reduced according to the formula

$$\text{percentage grade after dilution} = \frac{100k_1 + k_2 k_3}{100 + k_3},$$

where k_1 is the percentage of metal in the undiluted ore, k_2 is the percentage of metal in the wall rock, and k_3 is the percentage of dilution by weight. Figure 13.2, which graphs the ratio of grade mined after dilution to sampled grade against percentage dilution for various ratios of percentage metal in wall rock

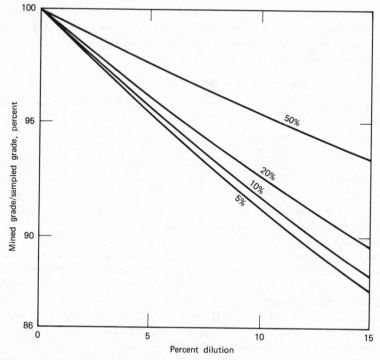

Fig. 13.2. Ratio of grade mined after dilution to sampled grade, for various ratios of percent metal in wall rock to percent metal in ore.

to percentage metal in ore, shows that some small allowance has to be made if the wall rock contains any valuable metal. For example, if the dilution is 10 percent, the metal grade in ore is 10 percent, and the metal grade in wall rock is 2 percent, then the mined grade is 9.3 percent.

The percentage of dilution, based on experience, is usually about 5 to 10 percent. Although it should be a function of ore-body width or thickness for a linear or tabular ore body, this refinement is seldom made. Dilution may be extremely important economically; a good example of how to profit by controlling it is given by Ensign (1964).

Discounting assays is another common practice for all shapes of ore bodies, but especially for the vein type of tabular ore bodies. Discounting is the practice of reducing all original assays by some amount, often 5 to 20 percent. Another common practice, especially for gold assays (sec. 16.2), is to reduce all assays above a certain value to the mean or to some other arbitrary value in order to make the sampled grade correspond more closely to the mined grade. Because both dilution and discounting usually introduce unevaluated "safety factors," these practices are ordinarily undesirable. It is better to evaluate the grade and amount of ore in a straightforward way and then at the end to apply experience factors that may be necessary, rather than to hide the factors here and there in the calculations.

Evaluation of Grade for Irregular Ore Bodies

For irregular ore bodies, more decisions must be made in the evaluation of grade than for linear or tabular ones, as may be illustrated by outlining the operation of a computer program devised by D. G. Mickle (personal communication). In this program, the irregular ore body is divided into cubes of a certain size, say 50 feet on a side, corresponding usually to a bench height. For each cube, grade is estimated by direct observations from drill holes, mine workings, trenches, etc.; or, if there are no observations in a particular cube, grade is derived by one or another type of interpolation from known data. Other information can also be considered, including hardness of rock, tonnage-volume relations, cost of extraction, and need to preserve the cube for some such purpose as a haulage road. Through a statistical model, each cube is examined in turn to determine whether it is minable. Average grades are then computed for the group of selected cubes that define the mine as a whole and for selected subgroups.

Mickle's program is one of several that have been devised for pit design; this field is one of the most successful computer applications in mining because the problem is large in size and tedious to calculate, but it is not conceptually too difficult. The calculations can be repeated with different cube sizes and different statistical models, and the different results can then be evaluated (sec. 14.6).

If a computer method is not used, the best procedure usually is to average the observations of grade at the different sample points, or perhaps at summary sample points (sec. 9.5) if the original points display marked clustering. As for tabular ore bodies, there is little or no purpose in weighting by polygons or triangles unless well-defined trends exist in the data, as discussed in the next subsection.

For irregular ore bodies, dilution may be little or no problem. If selective mining is possible, as in an open pit, percentage dilution for an entire ore body should be low, although it may be locally large on the borders. On the other hand, dilution may be large if the ore is mined by a nonselective method such as block caving and if grade control is not closely maintained at the draw points.

Trend-Surface Analysis and Evaluation of Grade of Ore

In section 9.10, trend-surface analysis of assay data from the Mi Vida uranium mine, Utah, obtains narrower confidence limits for estimates of grade of ore. In this subsection we briefly discuss the general application of trend-

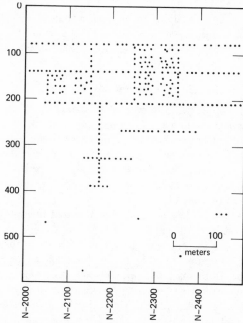

Fig. 13.3. Longitudinal section of a fictitious tabular ore body, with sample points irregularly distributed.

surface analysis to evaluation of grade of ore for a tabular ore body, although the same principles can be applied to any other shape of ore body. Figure 13.3, a longitudinal section of a fictitious tabular ore body, with sample points distributed as shown, is the basis for this discussion.

If a strong linear trend, say an increase in grade vertically downward, is displayed by the assays at the sample points, a trend-surface analysis might be appropriate for the data in figure 13.3. On the other hand, if no trend is found, or if only a relatively high one (higher than, say, quadratic) is found, it would be foolish to use the abundant data above the 210-meter level to estimate grade below, especially below the 330-meter level, where the only information is from the short drift on the 390-meter level and from scattered crosscuts or drill holes below it.

These remarks illustrate the basic principle that, unless the trend is both strong and low-order, trend-surface analysis cannot improve an estimate of the mean of poorly spaced points. In figure 13.3 the trouble is in the uneven distribution of sample points; if they were evenly spaced, either a trend-surface or an arithmetic average would necessarily yield the same mean, although the confidence interval might be narrowed by the trend-surface analysis (sec. 9.10). Of course, this figure portrays where, in the real world of geology, sample points are liable to be, rather than in the places that the statistician would prefer. One must accept the fact that no statistical manipulation can avoid widely variable reliability in the different squares, with observations so unevenly spaced.

Appraisal of Evaluation of Grade of Ore

The best way to appraise the success of an evaluation of grade of ore is to compare the sampled grade with the mined grade, although in practice this may be difficult if not impossible to do objectively because usually the same men in the mine engineer's or mine geologist's office calculate both sets of grades. There is a tendency to adjust one or the other or both to make them agree. Section 13.4 presents an example of an excellent appraisal of evaluation of grade under careful control.

One phenomenon that is almost always observed in mine sampling is a tendency to overestimate the grade of ore in the high-grade parts of a deposit and to underestimate the grade in the low-grade parts of the same deposit. The bias thus introduced is caused by the *regression effect*, well known in other fields (Wallis and Roberts, 1956, pp. 258–263). Some data from a South African gold mine (Storrar, 1966) provide an example of the regression effect. The data consist of 112 observations, which are paired values for stope grades estimated by sampling compared to grades actually mined. The plot of the observations (fig. 13.4) shows that they diverge from the straight dashed line with the equation, $x = y$, that would be obtained if the sampled and mined grades agreed perfectly. Where sampled grade is low, mined grade is always the same or higher; where sampled grade is high, mined grade is usually lower.

Figure 13.4 also shows the linear regression line (sec. 9.2) with the equation

$$y = 4.097 + 0.606x$$

fitted to the data. The equation shows that, with a regression coefficient of 0.606, an increase in sampled grade of almost 2 pennyweights must be attained to predict an increase in mined grade of 1 pennyweight. For all mined grades, the sampled grade must also be raised by a constant 4.097 pennyweights. Thus, for grades below about 10 percent, the sampled grade will tend to be lower than the mined grade, and above this figure the sampled grade will tend to be higher. We fitted the line by assuming homoscedasticity, which is evidently not met, but clearly the fit is about as close as a linear equation will give. Storrar (1966) fitted a curved line to these data, evidently because he wanted to obtain a line that passes through the origin.

Matheron's Theory of Regionalized Variables

Georges Matheron, a French professor of mining engineering, has devised a body of theory for calculating ore reserves and for similar analyses. In principle,

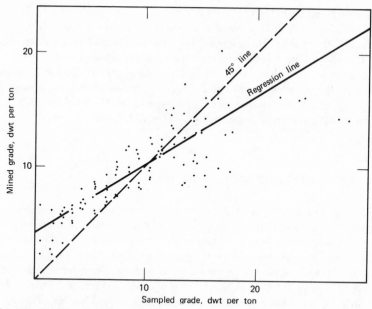

Fig. 13.4. Paired values for stope grades estimated by sampling, compared to grades actually mined (after Storrar, 1966).

Matheron quantifies area-of-influence concepts through *regionalized variables* and the *variogram. Regionalized variables* are variables that assume values dependent on their position in space. The *variogram* is a generalization of mean-square successive differences to the spatial distribution of a metal or of another variable. [A mean-square successive difference is calculated as follows (Hazen, 1967, p. 56). If n number of observations w are ordered, $w_1\, w_2, \dots, w_n$, the mean-square successive difference, is

$$\sum_{i=1}^{n-1} \frac{(w_{i+1} - w_i)^2}{n-1},$$

which can be generalized to an arbitrary lag interval $h = 1, 2, \dots$ to yield

$$\sum_{i=1}^{n-h} \frac{(w_{i+h} - w_i)^2}{n-h}.$$

A plot of this quantity versus h is the estimate of the variogram in one dimension. Clearly, this method can be generalized to more than one dimension.]

For deposits whose metal values exhibit strong area-of-influence trends for substantial distances, Matheron's methods may offer significant advantages. Unfortunately, few if any studies comparing evaluations by Matheron's methods with those by other methods have yet been published. An interesting field for study of ore deposits and other geological phenomena exists here.

Matheron's work has been influential, especially in French-speaking countries, His principal works (1962, 1963a) have not yet been published in English, but he has written a short English summary (1963b). Also, Blais and Carlier (1968) have written an extended review in English.

13.3 EVALUATION OF AMOUNT OF ORE

Grade of ore is always, or nearly always, more important to a mining operation than the amount. Unless the value of the metal represented by the grade per unit weight or unit volume is more than the cost of producing it, a mine is bound to lose money from the day it opens. On the other hand, if the amount of ore is sufficient, say, for only nine years rather than ten, the only consequence is that the mine shuts down one year earlier; and because the present value of the tenth year of production is only a small fraction of the total value (sec. 14.5), this consequence may not be too serious. Yet the approximate amount of ore to be mined must be known if one is to devise the best methods and rate of extraction.

In evaluating the amount of ore, the main guide that statistics can offer is to focus attention on the different degrees of uncertainty that may exist in estimating volume or tonnage and in converting volume to tonnage. It may also offer some help in fixing the natural unit for estimating the amount of ore for a particular deposit. These items are taken up in sequence in the following subsections.

Volume of Ore

To estimate volume, one must first obtain measurements, which may be made directly as drill-hole depths, chip-sample widths, trench lengths, etc., or may be made indirectly on maps and sections by scales or by planimeter. With the measurements in hand, calculations are performed with the aid of everyday trigonometry and geometry, because experience shows that computational refinements are seldom warranted, although occasionally integration of the volumes of geometric figures developed by trend-surface analysis (sec. 9.10) may be useful.

For elongated ore bodies, length is multiplied by cross-sectional area. The best procedure may be to form slices perpendicular to the direction of elongation and evaluate each piece separately.

For irregular ore bodies, volume can usually be estimated rather well, as for a porphyry copper or similar ore body for which information is quite complete down to the depth of drilling. Sometimes volume may be known only poorly, as for an underground ore body to be mined by block caving. There are no particular statistical issues to raise.

However, for tabular ore bodies, such as veins or sedimentary beds, although the area is usually known without significant error, the thickness may not be; and statistical considerations arise. Table 13.4 compares statistics for a coal bed where the thickness is precisely known (with a coefficient of variation of 0.26) with statistics for three veins where the measured thicknesses are less precisely known (with coefficients of variation of 0.43, 0.66, and 0.73). The uniformity of the Pittsburgh coal seam, based on some 500 thickness measurements obtained over an area of some 260 square kilometers, is all the more striking when compared to the veins, whose areas are each less than 1 square kilometer.

For the veins, table 13.4 also compares the means, standard deviations, and coefficients of variation based on the measured thicknesses with those based on a minimum stoping width of 1.2 meters, as calculated by increasing width measurements narrower than 1.2 meters to that value. Necessarily, the coefficients of variation decline markedly, especially for the narrowest vein.

Once calculated, volume often is increased by allowing for dilution either on an unadjusted volume basis, as by adding a band of waste 1 or 2 feet thick to a tabular ore body, or on a percentage basis, as by adding 5 to 10 percent of the total volume of ore. The former method seems more realistic. A dilution factor can hide a multitude of confused calculations and usually is still another unevaluated "safety factor."

Tonnage-Volume Conversion

Volume is converted into tonnage by multiplying or dividing by a constant. In the metric system, the constant is simply the specific gravity; if the specific gravity of an ore is 3.2, a cubic meter of it weighs 3.2 metric tons. In the English system, the volume in cubic feet is divided by the number of cubic feet per short ton, named the *tonnage-volume factor*, to obtain the number of short tons. McKinstry (1948, p. 60) explains four ways to calculate the tonnage-volume constant: by weighing the ore from an excavation of measured size, by weighing cars of broken ore and estimating the volume of ore in place, by measuring specific gravity of ore samples, and by calculating specific gravity from the mineral composition.

Each of these four ways of calculating a tonnage-volume constant is subject to errors. However, sufficient empirical information can be obtained from an ore body to calculate a factor that is correct within narrow limits, unless the

Table 13.4. Comparison of means, standard deviations, and coefficients of variation for a coal seam and several tabular ore bodies

Item	Mean (m)		Standard deviation		Coefficient of variation	
	Measured	MSW = 1.2*	Measured	MSW = 1.2*	Measured	MSW = 1.2*
Pittsburgh coal seam, Robena mine	2.16	2.18	0.38	0.32	0.18	0.15
Level 13 Don Tomás vein, Frisco mine	1.52	1.66	0.66	0.49	0.43	0.30
Level 13 Sirio vein, Frisco mine	1.60	1.85	1.05	0.80	0.66	0.43
2630 vein, Fresnillo mine	0.67	1.27	0.49	0.21	0.73	0.17

* Minimum stoping width (MSW) of 1.2 meters assumed.

Table 13.5. Calculation of the tonnage–volume factor

Line	Item	Calculation	Numerical value	Comments
1	Molecular weight galena	$207.2 + 32.1$	239.3	Handbook value
2	Percent of Pb in galena	$100 \, (207.2/239.3)$	86.586	
3	Grams of galena in 100 grams of ore	$100 \, (5/86.586)$	5.7746	
4	Grams of quartz in 100 grams of ore	$100 - 5.7746$	94.2254	
5	Specific gravity of galena		7.5	Handbook value
6	Specific gravity of quartz		2.7	Handbook value
7	Volume of galena	$5.7746/7.5$	0.76995	Volume = weight/specific gravity
8	Volume of quartz	$94.2254/2.5$	34.89289	”
9	Volume of ore	$0.76995 + 34.89289$	35.66824	
10	Specific gravity of ore		2.8036	Specific gravity = weight/volume
11	Pounds per cubic foot	2.8036×62.5	175.225	Weight of a cubic foot of water = 62.5
12	Cubic feet per short ton	$2000/175.225$	11.4	

variation in specific gravity from place to place is too large. Two main geological phenomena may make the variation too large: the first is a large and variable number of voids; the second is a significant proportion of minerals, whether valuable or not, whose specific gravity is higher than that of the rock and whose amount is variable. A method of calculating the tonnage-volume factor from mineralogical composition is outlined in table 13.5. Being a nonlinear relation, the tonnage-volume factor cannot be calculated as a weighted average of the specific gravities of the various minerals; in table 13.5 calculations are detailed for an ore with 5 percent lead contained in galena and with a quartz gangue. A simple way to work the problem is to consider 100 grams of ore. In lines 1 to 4 are found the number of grams of galena and quartz in 100 grams, and in lines 5 to 9 is found the volume that 100 grams of this ore occupies. In lines 10 to 12 the specific gravity and tonnage-volume factor are determined. The calculations can be easily expanded for ores containing more than two minerals.

If the calculation for converting volume to tonnage is repeated for different grades of ore, a tonnage-volume curve is obtained extending from pure gangue to pure ore-bearing minerals. Figure 13.5, a curve for magnetite and quartz,

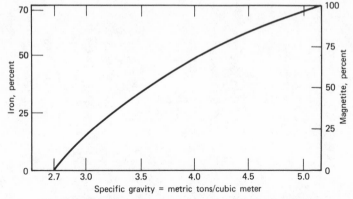

Fig. 13.5. Tonnage–volume curve for magnetite and quartz.

clearly demonstrates that the relation is nonlinear, because a ton of magnetite occupies much less volume than a ton of quartz.

In a magnetite mine, the specific gravity of the ore is obviously important because a given volume of high-grade ore may weigh almost twice as much as the same volume of low-grade ore. But even for a metal whose weight in ore is negligible, the specific gravity of the ore is important; if a gold ore has a grade of 10 grams per metric ton (0.29 ounce per short ton), the yield is 27 grams per cubic meter if the ore has a specific gravity of 2.7; whereas if it has a specific gravity of 4.5, the yield is 45 grams per cubic meter.

Because the tonnage-volume curve is nonlinear, the mean specific gravity of an ore cannot be obtained from its mean metal content unless the frequency distribution of different grades of ore is known. For instance, figure 13.5 shows that, for a uniform magnetite ore of 30 percent iron by weight, the specific gravity is 3.37; but if half the ore is barren and the other half contains 60 percent iron, the average specific gravity is 3.72 [(2.7 + 4.74)/2], a difference of almost 10 percent.

As previously stated, the variation in specific gravity of an ore significantly affects mining recoveries only when the specific gravities of the minerals constituting the ore are quite different from one another. In order to investigate this problem, two assumptions are made: (1) an ore can be considered to be composed of two constituents, one an ore mineral of a certain specific gravity, and the other a gangue mineral or a rock of another specific gravity, and (2) changes in specific gravity of ore of less than 10 percent will not in

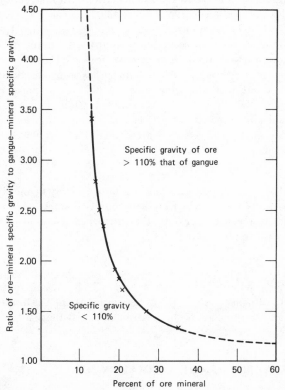

Fig. 13.6. Ratio of ore–mineral specific gravity to gangue specific gravity plotted against the percentage of ore mineral.

practice require the tonnage-volume factor to be adjusted for changes in grade of ore. Under these assumptions, figure 13.6 is a graph of the ratio of ore-mineral specific gravity to gangue specific gravity plotted against the percentage of ore mineral. The curve defines two regions: one above and to the right, where the specific gravity of the ore is more than 110 percent that of the gangue, and the other below and to the left, where the specific gravity of the ore is less than 110 percent that of the gangue. The ratio of specific gravities rather than their individual values defines the curve.

Figure 13.6 was constructed as follows: figure 13.5 shows that a specific gravity of 2.97 (110 percent of the specific gravity of 2.7 for the quartz gangue) corresponds to 19 percent magnetite. In figure 13.6, the ratio of magnetite specific gravity to quartz specific gravity, $5.15/2.7 = 1.91$, is plotted against 19 percent magnetite. Repeating this process for all nine combinations of rocks and ore minerals with the representative specific gravities 2.2 (sandstone), 2.7 (granite), 3.0 (basalt), 4.0 (sphalerite), 5.15 (magnetite), and 7.5 (galena) yields the solid-line curve of figure 13.6. The curve was extended by taking some other combinations of rock specific gravities from 1.6 (loess) to 3.4 (eclogite) and ore-mineral specific gravities up to 10.8 (uraninite); the curve is dashed because actual ores in this region are scarce, if they even exist.

The problem of converting volume to tonnage has been of particular interest to those mining ore with a specific gravity falling in the upper right-hand region of figure 13.6. Bray (1964) gives an interesting account of how the problem is taken into account at the Geco mine, Ontario, Canada, where both massive sulfide and disseminated ores are worked. Dadson (1968) discusses the general problem.

Tonnage of Ore

In the metric system, tonnage of ore is obtained simply by multiplying the volume of ore in cubic meters by its specific gravity. If variances are available for both volume and specific gravity, the variance of the resulting tonnage is calculated by the approximate formula

$$\sigma_T^2 = D^2 \sigma_V^2 + V^2 \sigma_D^2,$$

where σ_T^2 is the variance of the tonnage, D is the specific gravity, σ_V^2 is the variance of the volume, V is the volume, and σ_D^2 is the variance of the specific gravity. In the English system, the same method is applied to the tonnage-volume factor. After tonnage is obtained, an empirical dilution factor is often applied.

13.4   AN EXAMPLE OF ORE ESTIMATION

A clearly presented example of ore estimation based on careful work is given in a paper on the Giant Yellowknife gold mine, Northwest Territories, Canada, by Dadson and Emery (1968), also referred to in section 15.1 on broken-ore sampling. In this section some of the data in this paper are reviewed by statistical analysis. First, something about the mine and about Dadson and Emery's methods of ore estimation and grade control is abstracted from their paper. They write that

> To the end of 1966, the Giant Yellowknife mine had produced approximately 5,400,000 tons of ore with an average grade of 0.75 ounce of gold per ton. The ore bodies occur as irregular masses, lenses, and veins within a complex zone system of chlorite and sericite schist. They are extremely diverse in respect to size, shape, and attitude. Stopes may be from 50 to 400 feet in length, and from under 10 feet to more than 100 feet in width. Ore consists of from 40 to 80 percent quartz-carbonate, schist remnants, and about 10 percent fine-grained arsenopyrite and pyrite. Ore estimates and mine planning have been based, almost entirely, on the interpretation of diamond-drill cores. It must be kept in mind that the basic data are rather sketchy at best. The desire for more information has been outweighed by the extra cost of the work, relative to the benefits to be attained.
>
> Current practice is similar to that used in past years, except that more attention is being given to the dilution problem. The ore shoots, as outlined on 20 scale sections, are adjusted to mineable areas, the minimum width being 7 feet. These areas are measured by planimeter. The average grade for each cross-sectional area is the average of its drill-hole intersections, weighted for core length but not for the area of influence of a hole. The area and average grade, so obtained, are taken as being representative of a block extending half way to each of the adjacent sections. Volume factor is 11.5 cubic feet per ton.

Dadson and Emery's remarks on dilution reflect typical good practice and show how geological reasoning leads to adoption of empirical dilution factors. They write,

> Until 1958 a simple dilution factor of 5 percent (10 percent for certain ore bodies) was applied to the ore as estimated. It was then found that the grade of the calculated ore reserves was not being matched by the trammed grade or by the mill-head grade. In small part this could be attributed to the scheduling of production blocks, with relatively more of the lower grade shoots being mined. However, it was evident that dilution was becoming more serious than experienced hitherto. Undoubtedly the main reason was increased structural complexities in the newer ore bodies, such as abrupt changes in strike, dip, and dimensions which could not be forecast from the diamond-drill hole data. For the current ore-reserve figures the dilution factors are 20 percent for the calculated grades and 10 percent for the calculated tonnage.

Besides applying dilution factors, grade is reduced by cutting high assays for most of the mine history to 5 ounces per ton, but in the past two years to 2 ounces per ton.

In several tables, Dadson and Emery compare various estimates of tonnages and grades stope by stope and year by year. Table 13.6 summarizes grades,

Table 13.6. Ore estimates at the Giant Yellowknife mine, Northwest Territories, Canada*

Line	Stopes	Number of stopes or years	Grade (oz Au/ton)		Tonnage (short tons)		Gold content (oz)	
			Sampled	Mined	Sampled	Mined	Sampled	Mined
1	All completed stopes	50	0.85	0.79	3,160,800	3,046,919	2,686,680	2,407,066
2	A-shaft completed stopes	6	0.66	0.63	340,600	245,158	224,796	154,450
3	B-shaft completed stopes	38	0.87	0.81	2,106,500	2,169,197	1,832,655	1,757,050
4	C-shaft completed stopes	6	0.85	0.77	713,700	632,564	606,645	487,074
5	All uncompleted stopes	44	0.77	0.72				
6	B-shaft uncompleted stopes	30	0.78	0.75				
7	C-shaft uncompleted stopes	14	0.77	0.72				
	Years		Mined	Millhead	Mined	Millhead	Mined	Millhead
8	1949–1956	18	0.751	0.762	5,414,264	5,413,106	4,066,112	4,124,787

* After Dadson and Emery, 1968.

tonnages, and ounces of contained gold in the three categories of completed stopes (lines 1 to 4), uncompleted stopes (lines 5 to 7), and yearly production figures (line 8). The table shows that both the sampled grade and gold content are higher than those actually mined; tonnages based on sampling are also somewhat larger than those mined, except for completed stopes of B-shaft (line 3). (Estimated remaining tonnages, given by Dadson and Emery, are omitted from this book.) In line 8 of the table, mined grades and tonnages are compared with the millhead grades and tonnages, which are calculated from bullion recovery less losses in milling. Surprisingly, the mined grades are lower than those of the mill; the over-all tonnage figures are practically identical.

In table 13.7 the original stope and year data that were summarized to obtain table 13.6 are compared by the paired t-test (sec. 4.7). Each group of years or stopes is compared by considering each individual item as an observation; thus the number of degrees of freedom in each line is equal to the number of stopes or years minus 1. For grade of all completed stopes (line 1), because the t-value of 0.41 is lower than the tabled 10 percent value of 1.678, it is outside the critical region, and on the average the sampled and mined grades agree. Less good agreement in grade was obtained for the A- and C-shaft stopes; in the C-shaft stopes in particular, the average estimated grade is on the edge of being different from the mined grade according to the paired t-test. For the uncompleted stopes, agreement in grade is less satisfactory, especially those from the C-shaft group, which might be expected because estimated grade was based on all data, and evidently the richest parts were mined first.

Tonnage comparisons in table 13.7 show that estimated tonnages are higher on the average than those actually mined for all completed stopes together and for those in the A- and C-shaft areas separately. Tonnage comparisons for uncompleted stopes are omitted because the estimated tonnages should obviously be higher than those mined, as they are indeed for practically all stopes. Comparisons of gold content provide essentially the same results as those of tonnage.

13.5 RELATION OF GRADE TO AMOUNT OF ORE

Grade and amount of ore were considered separately in the three previous sections; the interrelationships of these factors are discussed in the present section. For this discussion, amount of ore could be reckoned in either volume or tonnage; because the conveniently available information is on a tonnage basis, tonnage is considered. The relation may be studied on three scales: that smaller than a single mine (an ore body or a stope), that of a single mine or ore deposit, and that of more than one mine or ore deposit.

TABLE 13.7. COMPARISON OF ORE ESTIMATES AT THE GIANT YELLOWKNIFE MINE BY THE PAIRED t-TEST*

Line	Items compared	Degrees of freedom	Calculated t-values			$t_{10\%}$
			Grade	Tonnage	Gold content	
1	Sampled vs. mined ore, all completed stopes	49	0.41	2.37	2.46	1.678
2	,, ,, A-shaft ,,	5	1.43	2.57	5.70	2.015
3	,, ,, B-shaft ,,	37	−0.40	0.63	1.17	1.688
4	,, ,, C-shaft ,,	5	1.92	2.21	2.30	2.015
5	Sampled vs. mined ore, all uncompleted stopes	43	1.74			1.681
6	,, ,, B-shaft ,,	29	0.83			1.699
7	,, ,, C-shaft ,,	13	2.88			1.771
8	Mined vs. millhead ore, 1949–1956	17	−2.46	−0.01	2.49	1.740

* Data from Dadson and Emery, 1968.

The general relation between tonnage and grade is sketched diagramatically in figure 13.7, in which the "break-even" curve defines two regions, labeled "GO" and "NO GO," where a single ore body or mine is worked or not worked. The generalized curve is asymptotic with respect to minimum grades and tonnages worth mining. The minimum grade is that below which ore is not worth mining no matter how large the tonnage may be; for a particular commodity, the minimum grade presumably could be established from such

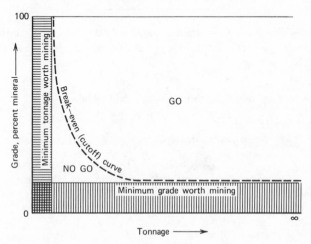

Fig. 13.7. "Break-even" curve relating tonnage to grade.

factors as the costs of the world's lowest-cost producer and the percentage of metal left in tailings by the most efficient recovery process. Similarly, the minimum tonnage is that below which ore is not worth mining no matter how high the grade may be; in general it would be established in the same fashion as for grade, although for a pegmatite mine or a Mexican quicksilver mine this tonnage might be extremely small. Above the minimum tonnage and minimum grade boundaries but below the curve is the region where the combination of tonnage and grade is too low, and here more tonnage must be found or grade must be subdivided to get some ore high enough in grade to rise above the curve. Grades on the "break-even" curve are cutoff grades (sec. 14.5), which are those for which the grade at mining causes a break-even position. Above the curve, in the "GO" region, profitability increases with increasing distance from the curve.

An exponential relationship between the tonnage and grade of metal in a mineral deposit (whether or not ore) has been postulated, linking these two variables in a form expressed in the following table,

Grade	Tonnage with specified or a higher grade
10	1
9	10
8	100
7	1000

which states that one unit of ore has a grade of 10 or higher, 10 units have a grade of 9 or higher, etc. The tabulated relationship can be expressed by the formula

$$T = 10^{(10-G)},$$

where T is tonnage and G is grade. The general formula is

$$T \sim B^{-G},$$

where B is any base, or, equivalently, by

$$T = AB^{-G},$$

where A is a proportionality constant. Conventionally, the base of natural logarithms is used, and the previous expression becomes

$$T = Ae^{-G}.$$

Regardless of the base, A is the amount of contained metal at all grades, because any base raised to the zero power equals 1.

Thus, as either tonnage or grade increases, the other variable decreases. To quantify the relationship, Lasky (1950a, 1950b) studied data from various mines, especially porphyry copper mines of the western United States, with interesting results, although unfortunately his original data are not published in enough detail for statistical appraisal. Figure 13.8 plots the relationship between tonnage and grade that Lasky determined for three ore deposits. For these and other deposits discussed in his report, Lasky believes that tonnage is related to grade by the equation

$$T = e^{(k_1/k_2)}e^{(-G/k_2)},$$

where the constants k_1 and k_2 have different values for different deposits. The previous equation follows the formulation in this book; Lasky writes it as

$$G = k_1 - k_2 \ln T.$$

Lasky (1950a, pp. 83–84) worked out the relation between tonnage and grade in more detail for an unidentified porphyry copper mine. Figure 13.9 plots against grade four variables: tonnage or reserves plus ore mined to date, cutoff grade, ore produced to date, and copper content of the entire ore body.

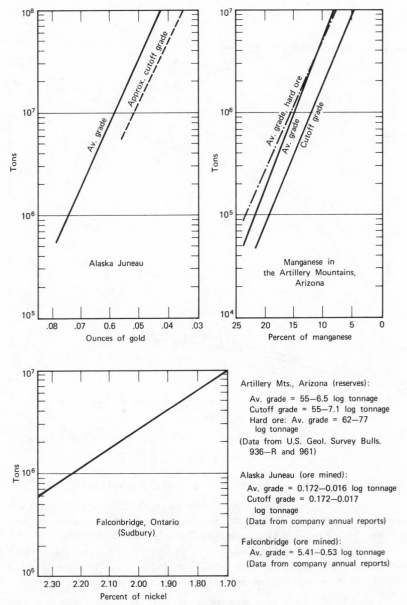

Fig. 13.8. Relationship between tonnage and grade for three ore deposits (after Lasky, 1950b).

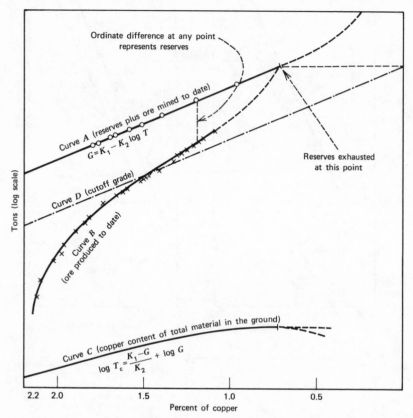

Fig. 13.9. Relation between tonnage and grade for a porphyry copper mine (after Lasky, 1950a).

Relation of Grade to Tonnage for More than One Mineral Deposit

For more than one mineral deposit, the relation of grade to tonnage has also been investigated. Detailed papers reaching rather different conclusions have recently been published by Musgrove (1971) and Lovering (1969).

Earlier, Musgrove (1965) studied the relation of grade to tonnage in lead reserves of the world. His data do not contradict the conclusion that the relationship follows the exponential law, according to the formula

$$T = 26.9 \times 10^6 \exp(-0.1954G)$$

where T is in short tons and G is in percent of lead. The total reserves of lead, 26.9×10^6, compare closely to the sum, 22.6×10^6, of Musgrove's data.

Figure 13.10 plots several data points and the line fitted to Musgrove's data by the least squares procedure for cumulative sum data (Mandel, 1957). We ignored the data below 1 percent lead because these data are perhaps less reliable than the rest.

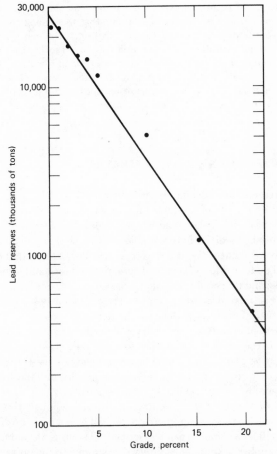

Fig. 13.10. Relation of grade to tonnage in lead reserves of the world (data from Musgrove, 1965).

Thus, these and other tonnage-grade data analyzed by Musgrove (1971) evidently follow the exponential law. Whether or not the postulated relationship holds true for ore deposits as well as for mineral deposits in general is a question that requires further study. Some geologists believe that it does not. Among these is T. S. Lovering (1969, p. 114) who writes:

Although there is no geological reason to conclude that the A/G ratio holds for ore deposits in general, it is true that the closer the minable grade of a metal approaches its clarke the more probable is the existence of an A/G ratio reaching well below the grade of currently commercial ore. For iron and aluminum, it seems probable that A/G relations may hold from ore grade down to (or below) the clarke. Even for them, however, it is unsafe to assume A/G relations for all *individual* deposits. And it is naive to postulate that there must be undiscovered low-grade deposits of astronomical tonnage to bridge the gap between known commercial ore and the millions of cubic miles of crustal rocks that have measurable trace amounts of the various less common metals in them. . . . When cumulative curves of metal content are plotted against volume of rock for elements in the earth's crust, the curves generally show abrupt changes in slope, corresponding to changes from one kind of rock to another.

13.6   CLASSIFICATION OF ORE RESERVES

Once ore reserves are calculated, it is conventional to classify them according to reliability by the use of methods fully explained by McKinstry (1948, pp. 470–473). In the most usual scheme, which was developed for tabular ore bodies but can readily be modified for ore bodies of other shapes, the following three classes of ore are defined: (1) positive ore, or ore blocked out, that ore exposed and sampled on four sides; (2) probable ore, that exposed and sampled on two or three sides; and (3) possible ore, that exposed on only one side, the extent of the ore being estimated by some empirical rule. The purpose of this section is to apply statistical analysis to the classification of ore reserves.

The problem of classifying ore reserves arises because the ore is in place in the ground. If the ore were spread out in a pile on the surface, as is the coal supply of a power plant, there would be no problem because the amount could be determined by ground survey or aerial photography, and the grade could be determined within desired limits by conventional methods of sample-point selection to establish confidence limits as narrow as required. However, when the ore is in place in the ground, there is seldom if ever a measure of its amount to which a confidence interval can be ascribed. Therefore, although it would be tempting to apply the confidence-interval technique to the observations that are available on a particular block of ore and to classify according to the width of the intervals, this procedure would not be valid if divorced from the problem of the extent of the ore.

Measurement of the extent of ore contains many subjective elements and is the main problem. Truscott (1962, p. 5) points out that a measured ore reserve may not even be needed for some mines, for instance some coal mines, although for most mines it is needed for many purposes. We may add that measured ore reserves are certainly needed for some mines for which they are not calculated only because the work is virtually impossible, as for instance chimney deposits in limestones and most mercury deposits.

A thorough coverage of this extent-of-ore problem is given by McLaughlin (1939), who classifies ore by a scheme that is statistically attractive (p. 596):

I. Plenemensurate ore bodies—those capable of being fully measured and sampled at an early stage in the operations;

II. Partimensurate ore bodies—those in which prospects for ore in addition to proved reserves remain a substantial element of value until the later stages of the life of the mines based on them;

III. Extramensurate ore bodies—those difficult to explore and measure much in advance of mining, in which the value of prospects for ore based on geologic evidence exceeds the value based on proved reserves throughout most of the life of mines supported by them. . . . Assignment of particular deposits to one or another of the major groups and to their place in the sequence is determined to a large extent by the following characteristics:

  (1) size, form, and attitude of the deposit,
  (2) accessibility to exploration,
  (3) continuity or regularity in metal content,
  (4) persistence of the mineralization, and particularly of the metal-bearing phase, as revealed by mineral species, textures, and alteration,
  (5) controls determined by structure and environment.

McLaughlin continues with examples of the place, in this classification, of many specific ore deposits and of genetic types of ore deposits. It would be most instructive if he or another experienced geologist familiar with many ore deposits would quantify this treatment with statistical control.

One approach that has been used by various investigators, although not so far as we know in published work, is to compute confidence limits for both the grade and amount of ore and then to construct graphically a joint-confidence region to represent the ore reserve. The region is graphically similar to a joint-confidence region for mean and variance (sec. 6.2), but the interpretation is different because amount and grade of ore are linked by the exponential law or some other relation. Moreover, a joint-confidence region, which should actually be constructed for various cutoff grades, leads to a region in three-dimensional space.

In summary, classification of ore reserves is a subject that needs study. A fruitful approach should be to start with an ore deposit where the total amount of rock to be mined is known accurately, say a porphyry copper deposit with well-defined lateral and downward extent. Here the questions are the division of the total amount of rock between ore and waste and the grade of ore, a division that is calculated at various cutoff grades. These questions are straightforward enough so that it might be possible to construct a joint-confidence region.

REFERENCES

Blais, R. A., and Carlier, P. A., 1968, Applications of geostatistics in ore evaluation, *in* Canadian Institute of Mining and Metallurgy, Ore reserve estimation and grade control, Spec. v. 9, p. 41–68.

Bray, R. C. E., 1964, Techniques used by the geology department, Geco Mines, Ltd.: Canadian Inst. Mining Metall., Trans., v. 67, p. 31–40.

Dadson, A. S., 1968, Ore estimates and specific gravity, *in* Canadian Institute of Mining and Metallurgy, Ore reserve estimation and grade control, Spec. v. 9, p. 3–4.

Dadson, A. S., and Emery, D. J., 1968, Ore estimation and grade control at the Giant Yellowknife mine, *in* Canadian Institute of Mining and Metallurgy, Ore reserve estimation and grade control: Spec. v. 9, p. 215–226.

Ensign, C. O., Jr., 1964, Ore dilution control increases earnings at White Pine: Trans. Am. Inst. Mining Engineers, v. 229, p. 184–191.

Hazen, S. W., Jr., 1967, Some statistical techniques for analyzing mine and mineral-deposit sample and assay data: U.S. Bur. Mines Bull. 621, 223 p.

Lasky, S. G., 1950a, How tonnage and grade relations help predict ore reserves: Eng. Mining Jour., v. 151, no. 4, p. 81–85.

———, 1950b, Mineral-resource appraisal by the U.S. Geological Survey: Colo. School of Mines Quart., v. 45, no. 1-A, p. 1–27.

Lindgren, Waldemar, 1933, Mineral deposits: New York, McGraw-Hill, 894 p.

Lovering, T. S., 1969, Mineral resources from the land, *in* Cloud, P. E., ed., Resources and man: San Francisco, W. H. Freeman, p. 109–134.

Mandel, John, 1957, Fitting a straight line to certain types of cumulative data: Jour. Am. Statistical Assoc., v. 52, p. 552–566.

Matheron, Georges, 1962, Traité de géostatistique appliquée, tome I, Théorie générale: Paris, Editions Technip., 333 p.

———, 1963a, Traité de géostatistique appliquée, tome II, Le krigéage: Paris, Editions Technip., 171 p.

———, 1963b, Principles of geostatistics: Econ. Geology, v. 58, p. 1246–1266.

McKinstry, H. E., 1948, Mining geology: Englewood Cliffs, N.J., Prentice-Hall, 680 p.

McLaughlin, D. H., 1939, Geological factors in the valuation of mines: Econ. Geology, v. 34, p. 589–621.

Musgrove, Phillip, 1965, Lead: grade–tonnage relation: Mining Mag., v. 112, no. 4, p. 249–250.

Musgrove, Philip, 1971, The distribution of metal resources, tests and implications of the exponential grade-size distribution: Paper prepared for presentation at the New York meeting of the Am. Inst. Mining Engineers.

Prescott, Basil, 1925, Sampling and estimating Cordilleran lead–silver limestone replacement deposits: Trans. Am. Inst. Mining Engineers, v. 72, p. 666–676.

Storrar, C. D., 1966, Ore valuation procedures in the Goldfields Group, *in* Symposium on mathematical statistics and computer applications in ore valuation, South African Inst. Mining Metall., p. 276–298.

Truscott, S. J., 1962, Mine economics: London, Mining Publications, 471 p.

Wallis, W. A., and Roberts, H. V., 1956, Statistics: a new approach: Glencoe, Ill., Free Press, 619 p.

Wright, R. J., 1959, An experimental comparison of U_3O_8 ore estimates versus production: Eng. Mining Jour., v. 160, no. 11, pp. 100–102, 188.

Chapter 14

Operations Research

The running of any organization requires continual decision making. Although most of the decision making can be institutionalized by establishing policies and procedures to be followed day by day, new questions arise as soon as answers are found for the old ones, and policies must be made and procedures devised for the new questions. When a new venture is established all the questions are new, and the procedures that worked somewhere else 20 years ago should be carefully examined and probably discarded. In this chapter we explain methods of decision making developed by the new discipline of operations research, as well as by statistics.

14.1 INTRODUCTION

When an anesthetic is about to be administered to a patient, a decision is made about the dosage. An acceptable dosage is one large enough to put the patient under but not so large as to finish him off. In many other situations, procedures have to be developed that on the one hand are effective but on the other hand are not harmful. All practical decisions are judgments based both on available facts and on intuition. The more relevant and usable the facts that are available, the better the decision that can be made by a decision maker with the amount of intuition he possesses on that particular morning. The experienced man has a long sequence of facts from the past available to aid his intuition; it is precisely his experience that makes his intuition valuable. If two people have the same facts, the more experienced one will generally make a better decision than the less experienced one, but an inexperienced person with more facts may make as good a decision as an experienced person.

In order to organize and summarize data so that relevant facts are obtained and clearly displayed for decision making, certain techniques of statistics and operations research have been devised. Crude techniques are often best, offering significant aids to the manager who must exercise judgment, while techniques with mathematical elegance may prove undesirable because they frighten and alienate the decision maker and put him in a frame of mind that freezes his ability to judge.

The use of statistics and operations research to aid formally in decision making is relatively new, although these techniques have been used informally for as long as decisions have been made. Statistics have been used ever since the dawn of written history, as witness the tax records of early Babylon. As a formal activity, operations research got its start during World War II in techniques first consciously used by the British in the Battle of Britain, especially in the deployment of their radar and fighter aircraft. Clark (1962) provides a readable account. These techniques rapidly spread to the United States and were used to plan shipping convoys, bombing techniques, etc. In each application, the main goal was the improved use of existing resources. By and large the available resources were movable, and their distribution could be readily managed.

In the years following World War II, operations research (OR) continued to be influential in military decision making. In the United States Department of Defense, while Robert McNamara was Secretary, new techniques of operations research were developed for organization of military resources as well as for their deployment in the field. These techniques of organization have recently been increasingly applied to other government departments and agencies in the Planning-Programing-Budgeting System (PPBS).

The flexibility in moving resources that was available to the military in World War II is not generally available in established nonmilitary enterprises. For a railroad, an obvious use of operations research is in devising train schedules and in setting fares and freight rates. However, flexibility is lacking because of both intangible and tangible conditions. Intangible conditions include governmental regulations, tax structures, purchase agreements, employee agreements, and leases of land and equipment that may be difficult or impossible to change. And tangible conditions are, for instance, the facts that railroad passengers, unlike servicemen, cannot be summoned to the trains at any time convenient to the railroad; track distances and conditions are more or less set and expensive to move or improve; tunnel clearances are set; and track gauge is set. Even with such a rigid framework as an established railroad, operations research can disclose improvements that may significantly increase profits, and in less regulated industries or those with more movable resources, such as manufacturing industries or airlines, larger improvements can be expected.

A related formalization that came out of World War II was game theory, based on von Neumann and Morgenstern's pioneer investigations made in the 1920s and on work done by A. Wald shortly after World War II. Because these investigations are most pertinent to actual conflict, there are few good applications of game theory to nonmilitary situations. In commerce, application has been attempted, especially in marketing, but the results are not clear-cut because of the complexity of the situations and the limited amount of the necessary information that is available.

Today operations research is an established profession, with its own professional societies (for example, the Operations Research Society of America) and its own journals, of which some of the leading ones in English are *Operations Research*, *International Abstracts in Operations Research*, *Operational Research Quarterly*, and *Management Science*.

The literature in operations research reveals a sharp dichotomy between theory and practice. Most of the published material, particularly that in books, concerns theory, although some journal articles and soft-copy material treat practice. Little has been published on practice because many applications are made with military secrecy, and most successful commercial applications are proprietary, besides being unelegant and uncompleted and therefore unpublishable. In the United States today, the Department of Defense is the largest user of operations research, but commercial use grows year by year, with the largest user in business being the Bell Telephone System.

Of the many books in operations research, several should be particularly interesting to geologists. Two excellent short books written to explain the scope of operations research rather than details of how actually to perform it are by Duckworth (1962) and Ackoff and Rivett (1963). Of the many textbooks, two of the best are by Churchman and others (1957) and Hillier and Lieberman (1967). Case histories of applications to various industries, including mining and petroleum, have been edited by McCloskey and Coppinger (1956) and by Hertz and Eddison (1964). Ackoff (1962) has written a book on the scientific method interpreted through operations-research methods. Coyle (1969) has reviewed the literature on operations research in the mining industry.

In fields closely related to geology, three interesting books on operations research in hydrologic applications by the federal government may be cited; they are by Hufschmidt and Fiering (1966), Maass and others (1962), and Eckstein (1958). Examples of journal articles are one by Crutchfield and others (1967) on benefit-cost analysis and the National Oceanographic Program, and one by Brooks (1966) on economic analysis of benefits and cost of strip-mining reclamation. Because no proprietary or military information is involved in these governmental applications, more details of practice are available than are usually given.

The soft-copy literature is difficult to obtain. This category includes publications of computer manufacturers and computer-user groups, publications that contain computer programs (sec. 17.2) necessary for doing most of the serious work in this field.

The discipline closest to operations research is perhaps industrial engineering. The job of the industrial engineer is efficient utilization of available resources. While he is mainly interested in engineering applications, operations researchers take a more catholic view and are willing to apply their techniques to such things as accounting, manufacturing, and social organization, in addition to engineering.

In the rest of this section we consider a conceptual framework for decision making in the mineral industries and review present and future decision making in theoretical and applied geology. The remainder of the chapter takes up some of these kinds of decision making. Section 14.2 explains three operations-research procedures. Sections 14.3 to 14.8 consider decision making in mining through the stages from pre-search to extraction. Section 14.7 is on decision making in manufacturing; cement production is used as an example. Section 14.8 reviews the decision making involved in geological or natural-resource surveys of large areas.

Although the techniques discussed in this chapter are not complicated, they may assist in the development of new ways of thinking and of using data in theoretical geology as well as in industry and government. If nothing else, formalization of a problem is helpful because one is forced to think about what is relevant and what is not, and about which data are pertinent and which are not. The pause to think about what one is doing is desirable. However, as figure 14.1 suggests, too much thinking and research can be put into a problem, and at some time research should be discontinued before its cost exceeds the savings that it obtains.

A Conceptual Framework for Decision Making in the Mineral Industries

As the remainder of this chapter is about decision making and operations research in developing a mine from start to finish, this is a convenient place to introduce a conceptual framework for the subject. Of first importance is the concept that decision making must ultimately be based on a measure of cost effectiveness. The measure may be one of several. Most simply, the question may be asked: Will a certain decision increase yearly income or profit? Or some other measure may be appropriate, such as *utiles*, a name for units of value based on a scheme peculiar to the corporation or other entity rather than on currency. For instance, if a poor man bets a dollar on the toss of a coin,

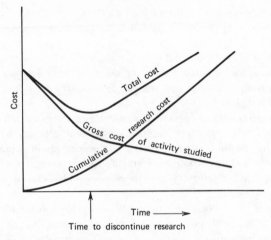

Fig. 14.1. Cost–time graph for research.

the value of a dollar lost may be the same to him as the value of a dollar won. On the other hand, if he bets $10,000 on the same game, the value of $10,000 may be much greater to him lost than won because the loss may force him into a disaster, perhaps bankruptcy, while the win will only allow him a more comfortable life. In terms of utiles, these instances mean that dollars lost, if the sum is substantial, are more valuable to the poor man than dollars won.

Although some decisions are made once and for all, as from the point of view of the condemned murderer when the hangman drops the trap, decision making more commonly involves a sequential framework, as in geological research and in applied geology in the mineral industries, where one makes a decision, sees what happens, and then makes another decision, etc. In fact, the same subject may be repeatedly reviewed with perhaps the same decision reached—for example, a prospect may be turned down year after year. Figure 14.2, illustrating the sequential nature of the decision framework in the mineral industries, also indicates the sections of this chapter that take up the various decisions. Except for crises, the sequence of decision making is typically periodic, whether daily, monthly, yearly, or some other schedule; it is shown on a yearly basis only for convenience.

When profit is the measure of cost effectiveness, the following mathematical model, based on figure 14.2, may be formulated to represent decision making in a sequential program for exploitation of a mineral deposit:

$$\text{profit} = \text{value}\,(P_{\text{dis}}\,P_{\text{eval}}\,P_{\text{devel}}\,P_{\text{extr}}) - (C_{\text{dis}} + C_{\text{eval}} + C_{\text{devel}} + C_{\text{extr}}),$$

where P is the probability of success and C the cost in the sequential stages of "dis" for discovery, "eval" for evaluation, "devel" for development, and

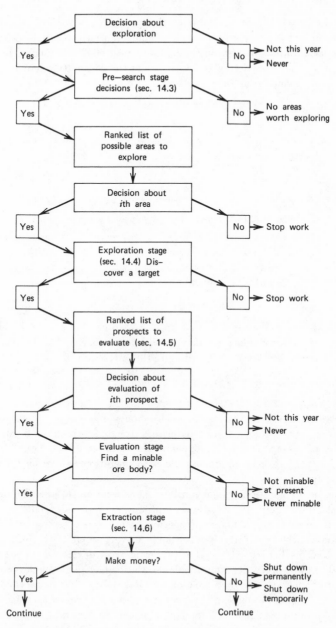

Fig. 14.2. Flow chart for decision making in the mineral industries.

"extr" for extraction. The model states that profit is equal to the inherent value multiplied by the pertinent probabilities minus the corresponding costs.

The several kinds of costs in the mathematical model may be reviewed in the light of the allocation of the resources of the corporation, governmental unit, or other entity; as shown schematically in figure 14.3, these resources

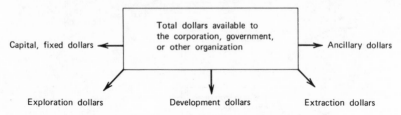

Fig. 14.3. Distribution of the resources of a corporation.

can be spent in different ways. From the resources of the entity, which for simplicity from now on will be named "the corporation," can be extracted liquid dollars, according to the expression

$$\$_{liquid} = \$_{total} - \$_{capital},$$

the $\$_{capital}$ being those dollars tied up in categories that cannot be altered, including fixed assets like mine plants and interest charges. The liquid dollars themselves can be allocated according to the expression

$$\$_{liquid} = \$_{ancillary} + \$_{development} + \$_{extraction} + \$_{exploration \ budget}$$

The $\$_{ancillary}$ refers to the sum of money allocated to nonmineral industry purposes, including investments in stocks, money in savings accounts, logging operations, etc. The $\$_{development}$ comprises those dollars required for development of prospects. The $\$_{extraction}$ comprises the more-or-less fixed sum that must be spent to keep producing mines in operation to extract ore found previously.

Thus the exploration budget must be set in relation to other factors. For example, if the company owned several promising properties needing development, if the stock market were prospering, and if no red-hot exploration tips were known, the exploration budget might be very small. On the other hand, if the stock market looked bad, if no properties to be developed existed, and if several red-hot prospects were available, the budget might be large. If the company policy were to do no exploration but to buy only established finds, the budget would be small; if the company were essentially a speculative one that sold its finds at once, the budget would be large. An ultraconservative mining company might go into the refining business and leave the mining to others. Thus the exploration budget is set by decisions reached subjectively, depending on the company's immediate prospects, its long-range policies, and perhaps the whim of the company president on the day of decision.

Once set, the exploration budget may be allocated among prospects in many ways. A financial consideration is shown by table 14.1, in which ten hypothetical properties are ranked according to the ratio of mean net value to the

TABLE 14.1. RANKING ACCORDING TO EXPLORATION–PROFIT RATIO OF TEN HYPOTHETICAL PROPERTIES

Property number	Exploration profit ratio, μ_P	Mean net value, μ_P (millions of dollars)	Total cost of exploration, C (millions of dollars)
1	5	10	2
2	5	5	1
3	2	2	1
4	1.5	3	2
5	1.33	4	3
6	1.28	9	7
7	1.2	12	10
8	0.67	2	3
9	0	0	7
10	− 0.17	− 1	6

cost of exploration, a ratio named the exploration-profit ratio by Slichter (1960). [The mean net value of a property is defined as the value of the property multiplied by the probability that it will be found minus the exploration cost (sec. 14.3).] Once compiled, the table may be used in two situations, depending on whether the exploration budget is fixed or flexible. If the budget is fixed, those properties with the highest exploration ratio are chosen, with the provision that the cost must be in discrete units. For instance, for an exploration budget of $13 million, properties 1 to 4 and 6 would be chosen, yielding an expected value of $29 million; property 5 is not chosen because the total cost would exceed the budget; nor are the properties chosen that total $13 million with the highest expected values, namely, 7, 1, and 2, because they yield an expected value of only $27 million. If the exploration budget is flexible, all those properties are explored for which the exploration-profit ratio is greater than zero, noting that a safety factor may have to be included. According to this rule, property 9 is not explored, as the company would be trading dollars; and property 10 is not explored, for which money would actually be lost.

The expected values of properties whose chances of attaining the estimated value are different may be compared. Table 14.2 contrasts two fictitious

Table 14.2. Comparison of expected values of two fictitious properties with different chances of attaining the estimated value

Value of deposit (thousands of dollars)	Probability of attaining value (%)	Expected value, $E(p)$ (thousands of dollars)
	95	95
100	90	90
	85	85
	3.5	157.5
4500	2	90
	0.5	22.5

deposits, with estimated values of $100,000 and $4,500,000. Since the first deposit has a probability of 90 percent of having the estimated value and the second has a probability of 2 percent, the expected values of both are $90,000. However, because a small change in the small probability has a large effect on the present value, a great deal of money can be either made or lost. Thus the small prospect is much less risky than the large one.

Decision Making in Theoretical and Applied Geology

Decision making is playing an ever-increasing role in both theoretical and applied geology. In theoretical geology, the impact is less because decision making, in the context of this chapter, has a strong economic input. However, it has a place in the administration of a scientific organization, such as a governmental geological survey, and in the administration of large scientific projects. For instance, it has been suggested that more decision making would have helped in the Mohole project.

In applied geology, especially in the natural-resources industries, for the many decisions that are constantly being made, operations research plays a large role today that will undoubtedly be expanded in the future in ways that are suggested in the rest of this chapter. In progressive companies, the guidance of decision making by operations research has increased rapidly in the past few years. Table 14.3, which is based on an article by Weiss (1968), indicates the scope in one large company of electronic computing, most of which has a strong operations-research connection.

14.2 SOME PROCEDURES OF OPERATIONS RESEARCH

In this section we outline three procedures of operations research that have proved useful in many industries and are likely to be valuable in geology and

Table 14.3. Scope of electronic computing in the Scientific and Engineering Computing Center of the Kennecott Copper Corporation*

I. *Current computer uses (mainly in operations research)*
    A. In exploration, including development
        Minimum economic grade and tonnage analyses
        Analyses of prospect data—geophysical, geochemical, and drilling
        Ore reserve calculations
        Mineral property evaluations based on alternative mining plans
    B. In mining
        Mineralization inventory modeling
        Long-range mine planning
        Short-range mine planning
        Production planning and scheduling
        Allocation of equipment, men, and materials
    C. In beneficiation
        Process simulation and modeling
        Data acquisition
        Process optimization by linear programing and other operations research techniques
II. *Future computer uses (mainly in operations research)*
    Information retrieval systems for already documented information and current field data
    Analysis of individual projects from more of an over-all corporate point of view
    Storage of mining data from a total systems point of view, allowing rapid and enlarged decision making
    Comparison of operating decisional information with "after-the-fact production statistics"
    Increase corporate planning and control, particularly in optimum location, capacities, and number of facilities; resource evaluation; market forecasts, resource acquisition, capital allocation
    Decision making with mathematical models for large-scale systems modeling the corporation's entire operation

---

* After Weiss, 1968.

in the mineral industries. We do not detail the mathematics and statistics for any of these procedures but instead provide a general account of them. References are given to works that give full understanding, and computer programs are available to implement the procedures. The principles behind the various algorithms are simple although their implementation may require difficult programing.

In order to evaluate any activity and to study its various manifestations, a measure of effectiveness must be devised as a yardstick against which alternatives can be compared. Some measures are production rate, cost, and profit. Once an appropriate measure is found, alternatives can be examined to find one that is optimum in terms of minimizing or maximizing the measure. Because different measures lead to different solutions, one must decide for a particular problem which measure is appropriate. This thorny issue is the hardest problem in evaluation, and once it has been grasped, it is only necessary to decide whether a budget or a required result is to be optimized.

In scientific and military work, a cost-effectiveness measure is often useful. As the name implies, such a measure evaluates the cost of an action against its effectiveness in obtaining a required result, whether this result is a cure for cancer or the killing of enemy troops. One must define rather precisely what is meant by effectiveness and then choose the procedure that will either maximize it for a given cost or achieve the required effectiveness at a minimum cost, the solutions being in general different. In business, the measure of effectiveness is usually profit; this measurement is less controversial than cost effectiveness, although today goals other than profits are increasingly a part of corporate policies, such as locating a factory in a slum as a social-action measure.

After these introductory remarks, we now proceed to the actual methods.

Linear Programing

Of several mathematical methods, linear programing is the one most widely used to find that combination of activities that minimizes or maximizes some mathematical function of them. For instance, several alternative raw materials whose available amounts and costs vary may be combined to make a finished product at minimum cost, as is explained for cement manufacture in section 14.7. Or prospecting teams drawn from a personnel pool of given size may be assigned to survey an area of maximum size divided into several subareas. Or, in theoretical geology, the partitioning of chemical elements among minerals in a suite of rocks may be studied—such as magnesium and iron among pyroxenes, amphiboles, and garnets in a suite of metamorphic rocks.

This subsection begins with an explanation of the principle of linear programing through a geometrical interpretation of a simple fictitious illustration, followed by a discussion of a mine problem, also fictitious. As in the application of any mathematical technique, the formulation of a physical problem in mathematical terms causes most of the difficulty for the user. Once a problem is formulated, computer programs are readily available to perform the calculations. The necessary mathematical techniques, not detailed in this book, are explained in several books, two of the best being those by Churchman and others (1957) and Hillier and Lieberman (1967).

We introduce linear programing through a type of problem that is covered in all the textbooks; the numerical values are those of Churchman and others (1957, p. 330), who explain the mathematics in detail. So that the problem will have a physical interpretation, assume that a coal company wishes to maximize its profit from producing two types of coal. Coal from seam A yields a profit of \$2 per ton, and coal from seam B yields \$5 per ton. The maximum daily available supply of coal is 4000 tons from seam A and 3000 tons from seam B. Cleaning coal from seam B takes twice as long as cleaning coal from seam A; and, if all the coal cleaned were from seam A, the company could process 8000 tons a day. How much coal from each seam should the company mine to maximize its profits?

This problem may be stated in mathematical terms. Let x be the number of thousand tons mined from seam A, y the number of thousand tons mined from seam B, and z the profit in thousands of dollars. The restrictions on mining and processing capacities lead then to the three inequalities

$$0 \leqslant x \leqslant 4,$$

$$0 \leqslant y \leqslant 3,$$

and

$$x + 2y \leqslant 8.$$

The objective is to maximize the value of z in the equation

$$z = 2x + 5y.$$

The previous mathematical expressions are in ordinary algebra, although solving inequalities may not be as familiar to the reader as solving linear equations. Figure 14.4 represents the algebraic statements geometrically. The first inequality states that the value of x must lie between the lines $x = 0$ and $x = 4$; the second inequality states that the value of y must lie between the lines $y = 0$ and $y = 3$; and the third inequality states that the required x, y-value must lie on or below the line $x + 2y = 8$. These lines define the

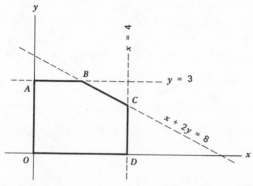

Fig. 14.4. Geometric representation of fictitious linear programing problem about mining coal.

polygon $OABCD$ in the figure. Any one of the infinite number of points within this polygon satisfies the three inequalities about the available supply of coal from seams A and B and the cleaning plant capacity.

Next the particular point must be selected to maximize the value of z in the equation

$$z = 2x + 5y,$$

z being the profit, in thousands of dollars, of processing x tons of coal from seam A and y tons of coal from seam B. If a value of z is chosen, the equation is that of a straight line with a slope of $-2/5$. If another value of z is chosen, another straight line parallel to the first is obtained; and, if the second value of z is larger than the first, the second line lies above the first, as figure 14.5 shows.

The problem then is to find the particular line that corresponds to the largest z-value and that also contains at least one point in or on the boundary of

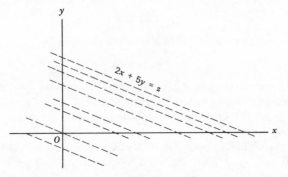

Fig. 14.5. Graph of equation $z = 2x + 5y$ for linear-programing problem.

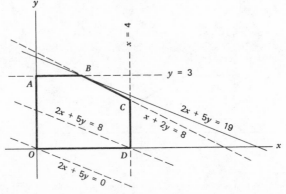

Fig. 14.6. Graphical solution to the linear-programing problem.

the polygon $OABCD$. In figure 14.6, several straight lines with the equation

$$z = 2x + 5y$$

are plotted in order to indicate that the solution is given by the coordinates $(2, 3)$ of point B, which, when substituted in the equation, yield the result

$$x = 2(2) + 5(3) = 19.$$

The meaning is that, subject to the restrictions imposed by the inequalities, the company can make a maximum profit of \$19,000 a day by processing 2000 tons from seam A and 3000 tons from seam B. Churchman and others (1957) explain the detailed solution to this problem by a procedure named the simplex method. Notably, a geometric solution, while easy for the two-dimensional case, is difficult if not impossible for more than two coal seams, although the algebraic solution follows the same principles.

In the simplex method, each inequality is replaced by an equation by adding a *dummy variable*. The resulting set of equations generally has more variables than equations, so that the solutions to the equations are not unique. For the coal illustration, the equations corresponding to the inequalities are

$$x + D_1 = 4;$$

$$y + D_2 = 3,$$

and

$$x + 2y + D_3 = 8,$$

where x and y are defined as before, and D_1, D_2, and D_3 are three dummy variables. Thus there are three equations with five variables. The simplex method is an algebraic procedure for solving these equations. When they are solved, D_1 is the amount of coal from seam A that is available but not mined (equal to 2000 tons); D_2 is the amount of coal from seam B that is available but not mined (equal to 0 tons); and D_3 is the production capacity available but not used (equal to 0 tons per day).

The second and last illustration of linear programing is a more realistic fictitious problem devised by Redmon (1964). Suppose that a lead-zinc ore is extracted from an underground mine in which on a particular day ore is available from ten stopes. Table 14.4 shows the grades, gross values per ton (including the value of copper and precious metals), production cost (Redmon's man-hour-per-ton costs converted to dollars at the rate of \$2.75), net values per ton (gross values minus production costs), and maximum amounts of ore available. Although 515 tons of ore in all is available, mill capacity is only 400 tons, and the problem is to select these 400 tons from several or all of the ten stopes, subject to the restrictions that the ore must average 2 percent or more lead and 4 percent or less zinc. Three different solutions obtained by different objective functions are given and compared.

TABLE 14.4. FICTITIOUS MINE-PRODUCTION SCHEDULING PROBLEM SOLVED BY LINEAR PROGRAMING*

Item	Pb (%)	Zn (%)	Gross value per ton ($)	Production cost ($)	Net value per ton ($)	Ore available (tons)	Minimum production cost solution		Maximum gross or net value of ore solution	
							Tons taken	Tons left	Tons taken	Tons left
Stope 1	2.50	4.75	24.05	7.975	16.075	25	5.0	20.0	25.0	0
,, 2	2.75	3.00	22.30	6.050	16.250	40	40.0	0	40.0	0
,, 3	1.75	1.50	16.45	3.575	12.875	90	66.6	23.4	0	90.0
,, 4	3.00	3.50	21.75	5.500	16.250	60	60.0	0	60.0	0
,, 5	2.00	4.00	23.80	6.875	16.925	30	30.0	0	30.0	0
,, 6	1.00	5.25	21.30	4.950	16.350	75	33.4	41.6	75.0	0
,, 7	1.50	2.50	18.75	4.125	14.625	85	85.0	0	60.0	25.0
,, 8	2.00	3.25	19.60	4.675	14.925	80	80.0	0	80.0	0
,, 9	2.75	4.50	24.90	8.800	16.100	20	0	20.0	20.0	0
,, 10	0.75	7.75	33.90	18.150	15.750	10	0	10.0	10.0	0
Totals						515	400	115	400	115
Means, ore available	1.93	3.37	20.56	5.345	15.215					
Means, ore taken, minimum production cost solution	2.00	3.05	20.00	4.865	15.135					
Means, ore taken, maximum gross or net value of ore solution	2.00	3.85	21.60	5.820	15.779					

* After Redmon, 1964.

TABLE 14.5. EQUATIONS FOR LINEAR PROGRAMING PROBLEM*

Restriction equations:

$$x_1 + x_2 + x_3 + x_4 + x_5 + x_6 + x_7 + x_8 + x_9 + x_{10} = 400$$

$$24.05x_1 + 22.30x_2 + 16.45x_3 + 21.75x_4 + 23.80x_5 + 21.30x_6 + 18.75x_7 + 19.60x_8 + 24.90x_9 + 33.90x_{10} - x_{11} = 8000$$

$$2.50x_1 + 2.75x_2 + 1.75x_3 + 3.00x_4 + 2.00x_5 + 1.00x_6 + 1.50x_7 + 2.00x_8 + 2.75x_9 + 0.75x_{10} - x_{12} = 800$$

$$4.75x_1 + 3.00x_2 + 1.50x_3 + 3.50x_4 + 4.00x_5 + 5.25x_6 + 2.50x_7 + 3.25x_8 + 4.50x_9 + 7.75x_{10} + x_{13} = 1600$$

$$x_1 + x_{14} = 25$$

$$x_2 + x_{15} = 40$$

$$x_3 + x_{16} = 90$$

$$x_4 + x_{17} = 60$$

$$x_5 + x_{18} = 30$$

$$x_6 + x_{19} = 75$$

$$x_7 + x_{20} = 85$$

$$x_8 + x_{21} = 80$$

$$x_9 + x_{22} = 20$$

$$x_{10} + x_{23} = 10$$

Objective equation:

$$7.975x_1 + 6.050x_2 + 3.757x_3 + 5.500x_4 + 6.875x_5 + 4.950x_6 + 4.125x_7 + 4.675x_8 + 8.800x_9 + 18.150x_{10} = Z$$

* After Redmon, 1964.

In the first solution, which is the one detailed by Redmon, the objective function to be minimized is labor cost, subject to the further restriction that the average value of the ore must be at least $20 per ton. Table 14.5 shows the equations for this problem, and table 14.4 shows the solution obtained when the equations are solved by the simplex method. Both the mean lead value of 2 percent and the gross value of ore of $20 are exactly the minimum allowable figures; however, the zinc does not impose a restriction because its grade is 3.05 percent rather than the allowable maximum of 4 percent. No ore is taken from stopes 9 and 10, which are the highest-cost ones, because the minimum lead grade required can be attained by taking 5 tons from the next-highest-cost stope, which is No. 1; however, only the 5 tons needed to make the required grade are taken from this stope.

Solutions 2 and 3 maximize the gross or net values of the ore regardless of the production cost, and both these solutions (table 14.4) are the same for these particular data although generally they would not be. As in solution 1, the mean lead value of this production is exactly the minimum 2 percent allowable; however, the gross value rises to $21.60 per ton because the zinc grade increases from 3.05 to 3.85 percent. This increase is achieved at the expense of raising production cost from $4.865 to $5.820 per ton. The outstanding differences from solution 1 are that no ore is taken from the lowest-cost stope, No. 3, because its grade is low; and the remaining 25 tons above the required 400 tons are left in stope 7, which is the next-lowest-cost stope but also one with a low net value of ore.

Redmon's problem, introduced to show the linear programing method, could doubtless be solved without any particular trouble even if one had never heard of linear programing, but for a real-world case a mathematical method like the simplex method is needed, and much money and effort have been saved in such problems. One example from the mineral industries is a cement-manufacture problem in section 14.7; another is a blast-furnace blending problem (Churchman and others, 1957, pp. 385–387); and many other examples are given in the periodical literature.

Critical-Path Scheduling

Scheduling problems arise whenever several jobs must be performed to complete a project. A familiar example is the construction of a house: the foundation must be poured before the frame is erected; then either the plumbers, electricians, or bricklayers can be called in, but preferably not all at the same time; etc. If prompt completion of the house or other project is advantageous, the question arises how to schedule the various jobs most efficiently. In this subsection the method of operations research named *critical-path scheduling* for solving this question is reviewed. Because, in practice, scheduling problems are solved by *heuristic* (valuable for stimulating or conducting empirical research but unproved or incapable of proof—Webster) rather than by analytic methods, this discussion stresses practical applications rather than explores the mathematical procedures involved, which basically are versions of linear programing.

Critical-path scheduling is treated in several books, for instance those by Moder and Phillips (1964) and by Shaffer and others (1965). The acronyms CPM and PERT are given to particular versions. The discussion in this book is derived from a short course presented in 1968 in the Department of Industrial Engineering and Operations Research of the University of California, Berkeley, under the direction of W. S. Jewell. We thank Professor Jewell for allowing us to use this material prior to its publication in his book.

Consider the fictitious illustration of bringing into production an open-pit mine; the following *activities* (jobs) need to be done with the specified *durations* (times to perform), the activities and durations being chosen arbitrarily to paraphrase a problem by Jewell.

	Activity	Time in months
A.	Construct concentrator	7
B.	Build crusher	8
C.	Build blending plant	6
D.	Prepare site	3
E.	Build water line to concentrator	5
F.	Let plant construction bids	3
G.	Strip pit overburden	6
H.	Prepare pit for production	3
I.	Truck ore to blending plant	2

Although all these activities must be done before the project is completed, some depend on others; for instance, ore cannot be trucked to the blending plant until it is built, and the pit cannot be prepared for production until the overburden is stripped. In detail, the *precedence* (sequential arrangements) is as follows: Activity A, construct concentrator, must follow the completion of activity D, site preparation, and activity E, build water line to concentrator. Activities B, build crusher, and C, build blending plant, can begin together or not, but only after D, site preparation, has been completed. The project is started with F, let plant construction bids, and G, strip pit overburden, as the first possible activities, with activities D, site preparation, and E, build water line to concentrator, starting only after the bids, F, are let. Activities H, prepare pit for production, and I, truck ore to blending plant (in that order), start after the pit overburden is stripped (G), except that the ore cannot be trucked (I) until the blending plant (C) is completed.

To sort out the scheduling requirements concealed in the flood of words in the previous paragraph, a graphical representation is helpful, which for a simple problem can be constructed in one of several forms without formal methodology. Figure 14.7 is a *project*, with time flowing generally from left to

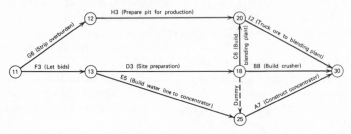

Fig. 14.7. Activity network for fictitious illustration of bringing an open-pit mine into production.

right. The different activities are represented by *arcs*, which are directed line segments, meeting at *nodes*. Nodes are junction points representing *event times*, which are simply clock times at which activities begin and end. For instance, node 13 denotes the end of activity F, which must be completed before activities D and E can start. Two nodes, 11 and 30, have special significance, as they denote the start and finish of the entire project of opening up the mine. A few conventions are followed in drawing networks. The numbering of the nodes is arbitrary, except that for some computer programs they must be numbered in ascending order of work flow from left to right. The lengths of the arcs are not proportional to the activity times, and the diagrams, being schematic, can be drawn in many different ways. The arc labeled "dummy" with zero time associated with it is drawn to show the relation that both activities D and E need to be completed before activity A starts. Precedence relations must be kept straight.

The data for this problem are of three kinds. First are the *activities*, the tasks or jobs to be performed; second are the *precedence* relations, orderings of activities; and third are *activity durations*, the times required to complete each activity. Because the minimum total time to complete the project must be found, the longest necessary path through the network is required; but paradoxically this path, called the *critical path*, must be as short as possible. It may be found by inspecting the simple network of figure 14.7, and it is F, E, A; namely, let bids, build water line to concentrator, and construct concentrator. The minimum time for completing the project is 15 months, and no time can be left between the end of one of the activities on the critical path and the start of the next. Although in this problem there is only one critical path, in other problems there may be more than one—all having the same duration.

Sometimes one or more activities may be started before the actual beginning of a project, and others may continue after it is concluded. In figure 14.8,

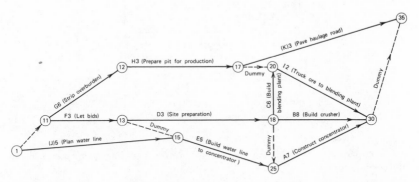

Fig. 14.8. Activity network, with two additional activities added to network of figure 14.7.

figure 14.7 is redrawn to add two new activities and four new dummy activities. These new activities, enclosed in parentheses to indicate that they are not essential to completing the project as defined, are (J) plan water line, requiring five months, and (K) pave haulage road from pit to crusher, requiring three months. It is assumed that plans for the water line can be drawn before work actually starts on the project to get it off to a good start, that little money will be lost if the board of directors decides to cancel the project, and that production can start before the haulage road is paved. Under these conditions the latest time that activity (J) can begin is two months before the start of the project, but the haulage road (K) can be completed two months before the end of the project.

Another way to represent these activities is by a bar chart, figure 14.9, which clearly shows that activities not on the critical path need not have fixed starting and ending times, but instead may have a time range through which they *float*, provided that the start and finish times of the project are observed. For instance, activity H can start at any time from month 6 to month 10 and finish three months later, provided that activities in its path, namely G and I, are started at compatible times. Table 14.6 gives the earliest and latest event times for the nodes on the network, figure 14.7.

This concludes the explanation of problems in which only time is considered. Computer programs to perform the calculations are readily available, for instance from computer manufacturers, and are essential for a network of a few hundred or thousand activities such as often occur in detailed scheduling

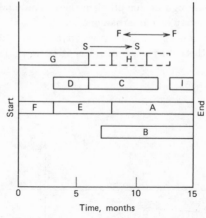

Fig. 14.9. Bar chart corresponding to fictitious illustration of bringing an open-pit mine into production.

for a large project (in the fictitious illustration of the open-pit mine if the activities were broken up).

The next problem is the modified one with cost of performing each activity considered and, in particular, a linear relation between cost and time assumed. This problem is illustrated by figure 14.10, which compares *normal* and *crash* completion times and costs; the crash costs are assumed to be equal to or larger than the normal ones. The concept is that time to complete a project

TABLE 14.6. EVENT TIMES FOR FICTITIOUS ILLUSTRATION OF BRINGING AN OPEN-PIT MINE INTO PRODUCTION CORRESPONDING TO THE ACTIVITY NETWORK, FIGURE 14.7

Event (node number)	Event times (months)	
	Earliest	Latest
11*	0	0
12	6	10
13*	3	3
18	6	7
20	9	13
25*	8	8
30*	15	15

* These events are on the critical path.

can be shortened by spending more money, perhaps by paying overtime or by purchasing new rather than second-hand equipment requiring renovation, etc.

An illustrative cost-time problem to which no physical interpretation is attached is given by figure 14.11 and table 14.7. The normal project duration is 12 months; the critical path comprises activities A, D, and G; and the normal cost is $620,000. If all activities were performed at crash times and costs, the total cost would be $1,085,000 but money would be thrown away because crashing some activities not on the critical path does not shorten the project duration.

In order to decide on the most economical method of crashing, an algorithm is used that works as follows, with modifications required for large networks:

1. Select the activity whose shortening is least expensive in cost per unit time (slope on the cost-time curve, figure 14.10) and whose shortening decreases the project time.

2. Shorten the time of this activity, either until it is fully crashed or until a new activity becomes critical.

3. Repeat steps 1 and 2 for the next best activity.

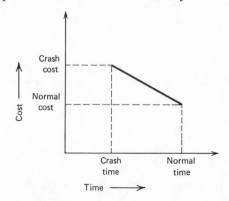

Fig. 14.10. Cost–time graph to compare normal and crash costs and times.

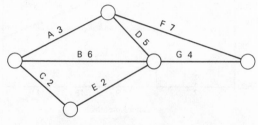

Fig. 14.11. Activity network for fictitious illustration of a cost–time problem.

4. Continue until no new activity can be shortened to decrease the project duration.

As for the time-only problem, computer programs exist to perform the tedious arithmetic.

Figure 14.12 and table 14.8 show the results of applying this algorithm to the illustrative problem. For instance, to shorten the project duration from 12 to

TABLE 14.7. DATA FOR FICTITIOUS COST–TIME PROBLEM

Activity	Time (months)		Cost (thousands of dollars)	
	Normal	Crash	Normal	Crash
A	3	2	60	100
B	6	4	140	260
C	2	1	25	50
D	5	3	100	180
E	2	2	80	80
F	7	5	115	175
G	4	2	100	240
Totals			620	1085

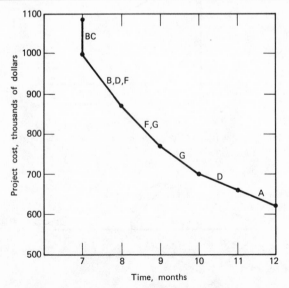

Fig. 14.12. Cost–time graph for fictitious problem of figure 14.11.

11 months, activity A is shortened at a cost of $40,000, and to shorten to 10 months, activity D is shortened one month also, at a cost of $40,000. That further shortening costs more is shown by the increasing slope to the left of the cost-time curve. The final line segment at time 7 is vertical because crashing all activities only increases project cost without a corresponding time saving; for, as a consequence of the network structure, some other activities that have been speeded up as much as possible become bottlenecks. The cost-time curve is an optimum minimum one; the meaning of the region above the curve is that a cost-time point in this region is larger in time or cost than necessary.

TABLE 14.8. ACTIVITY TIMES FOR FICTITIOUS COST–TIME PROBLEM FOR PROJECT TIMES FROM 12 TO 7 MONTHS

Activity	Project duration (months)					
	12	11	10	9	8	7
A	3*	2*	2*	2*	2*	2*
B	6	6	6*	6*	6*	5*
C	2	2	2	2	2	2
D	5*	5*	4*	4*	4*	3*
E	2	2	2	2	2	2
F	7	7	7	7*	6*	5*
G	4*	4*	4*	3*	2*	2*

* These times are for activities on the critical path.

The critical-path models introduced thus far are entirely deterministic, but corresponding probabilistic models can be formed, with costs and activity durations having frequency distributions. However, in practice, probabilistic models have seldom been productive; rather, it is better to solve problems with deterministic models and then recalculate if necessary. Thus one shipping company schedules its vessels by a critical-path method and then recalculates when one breaks down or when storms occur, because the interest is in the particular vessel or in the specific stormy part of the world rather than in the generalization implied by a frequency distribution.

Some other refinements not explained in this book may be mentioned. A project may include *divisible activities*, which are ones that can be performed at various times; for instance, a drill might be scheduled for maintenance either before or after boring a certain hole. Also, problems in manpower and resource allocation may be considered in order to handle activities that must follow directly on one another or to plan distribution of men, materials, and equipment. An example of an activity that must follow directly on another is concrete-lining a shaft, which must be done before the excavation collapses. Distribution of miners may be made from a workers' pool, and of automobiles from a car pool. Computer programs to handle these problems exist, although some of the best are proprietary.

Although some companies calculate critical paths only to satisfy a customer, particularly the U.S. Department of Defense, industry acceptance has been good if the groundwork has been laid and if the industry is one in which scheduling is important. For instance, one oil company found that critical-path scheduling saved 20 percent in refinery-shutdown costs. And a large construction company, involved in, among other things, mine and oil-refinery design and building, reduced its planning costs from 2 percent of total man-hours to $\frac{1}{2}$ percent by use of critical-path scheduling.

In summary, the outstanding advantage of critical-path scheduling is that, when drawing a network, one is forced to think out what the activities are and in what sequence they must be done. Once this is finished, the network may as well be run through a critical-path computer program, as little more work is involved. At first, networks can be drawn with times only, or with times and costs, if costs are well known, as in construction. Either formulation is likely to be useful. In doing this, remember that, for a project of long duration, the network can be drawn in detail for the first six months or a year, and in rough from then on. After more is known, the problem can be reworked, because a network should not be regarded as something cast in concrete to last forever.

Queueing

From time to time traffic jams up, lines form in front of the windows of bank tellers, people queue up at bus stops, and more automobiles break down than can be fixed at once by the available garages. All these situations are characterized by the arrival at some facility of more customers than a limited service capacity can process immediately. A highway can carry only so many cars, a bank has only a certain number of windows, a bus company has a specified number of buses, and a garage has a limited number of mechanics. In mining, the scheduling of trains and trucks from mine to crusher and the scheduling of repair facilities are common problems.

Queueing or waiting-line theory offers a well-developed method to attack these problems. The essential elements are (1) the frequency distribution of arrivals in time, (2) the frequency distribution of the service times, and (3) the queue discipline.

The first two of these elements are self-explanatory. The third element comprises rules to answer questions of the following kind: Do customers first in line get first service or are priorities established, as in a hospital emergency room, where a patient with severe injuries is treated before one with a bee sting ? Is one queue or several established where there is more than one service facility, as with several windows in a bank ?

An illustration of the application of queueing theory is an elementary problem with the simplest kinds of arrival-time distributions, service-time distributions, a single facility, and a first-come first-serve queue discipline. Suppose that ore is trucked to a crushing plant, the capacity of which must be planned and the number of trucks and crushers or unloading facilities that are needed must be specified. Such questions as the following would arise: What is the probability that an arriving truck will have to wait ? If a queue forms, what is its mean length ? If an arriving truck has to wait, how long is its mean waiting time ? If trucks were to arrive more frequently, how often would they have to arrive to make the mean waiting time equal to 10 minutes ?

In order to answer these questions, information must be obtained about the system and some assumptions must be made. Suppose that the mean time between truck arrivals at the crushing plant is 10 minutes and that the mean time to unload a truck is 5 minutes. The arrival time of trucks may be assumed to follow a Poisson distribution. If time is divided into equal intervals, say of 30 minutes, the number of trucks arriving in each interval would follow a Poisson distribution with a mean of 3. It may be further assumed that the time taken to unload the trucks follows an exponential distribution.

The questions posed may now be answered by applying formulas that are not proved in this book. For further details, and for solutions to more complicated problems, the reader is referred to Churchman and others (1957, p. 391 ff.). The solutions depend on the reciprocals of the parameters of the arrival and service times. The arrival rate λ is the reciprocal of the mean time between arrivals, and the service rate μ is the reciprocal of the mean service time. For this problem, $\lambda = 1/10 = 0.1$, and $\mu = 1/5 = 0.2$. In table 14.9 are four problems that can be solved by entering these parameter values in the appropriate formulas. The first two lines show that the probability of a queue forming is 50 percent and that, if a queue forms, its mean length is two trucks. The mean time that a truck must wait, if it has to wait, is calculated in line 3 to be 10 minutes. This time is not the mean waiting time of all trucks, because 50 percent are unloaded immediately without waiting. In line 4 is given the waiting time of all trucks, whether or not a queue forms. If this time is set equal to 10 minutes and the value of $\mu = 0.2$ is substituted, the formula can be solved for the value of λ that causes this condition to occur. The value of λ obtained is 0.133, whose reciprocal is 7.5 minutes, which means that trucks would have to arrive every 7.5 minutes on the average for the mean waiting time to be 10 minutes.

The explained problem is one of the simplest. For realistic problems involving complicated arrival distributions and numerous service centers with complex service distributions and detailed queue disciplines, simulation methods are needed to obtain solutions. Queueing theory is a powerful planning tool of operations research that has been extensively used by organizations including the Port of New York Authority, the American Telephone and Telegraph Co., and many government agencies.

Table 14.9. Four queueing problems

Line	Item	Formula	Numerical values
1	Probability that truck must wait	λ/μ	$0.1/0.2 = 0.5 = 50\%$
2	Average queue length, given that there is a queue	$\mu/(\mu - \lambda)$	$0.2/(0.2 - 0.1) = 2$ trucks
3	Average waiting time in queue, given that there is a queue	$1/(\mu - \lambda)$	$1/(0.2 - 0.1) = 10$ minutes
4	Average waiting time of all trucks whether or not there is a queue	$\lambda/\mu(\mu - \lambda)$	$\lambda/0.2(0.2 - \lambda) = 10$ minutes $\lambda = 0.133$

14.3 PRE-SEARCH STAGE

In the pre-search stage before field work begins, facts are marshaled and some decisions about targets must be made. Perhaps the first is to decide whether the search is for any and all natural resources or is restricted to one or a few substances.

For instance, a public utility may want to find coal to generate electricity and has no interest in gold. Or a metal-mining company may not look for oil. Of course, if these organizations find gold or oil, they will not ignore it, but they may sell it to someone else, and they will not stress the likelihood of finding gold or oil in their planning lest the complications stall action. A contrasting situation is found in a governmental survey of an area where the purpose is an over-all resource appraisal, as in an underdeveloped country. Here the search may be for everything; but, simply because the explorers, whether from the country or from the United Nations, will not know all details about all commodities and industries, a complete and thorough economic evaluation cannot be expected. Rather than achieving a specific objective, perhaps a source of cement-making materials sufficient for a certain mill, as in a focused exploration program, it would be satisfactory to report that potential raw materials have been found in large enough amounts and to recommend further specific studies.

Once decisions have been made about commodities, whether the search is general or directed toward one or a few commodities, one may decide in what part of the world the objective is likely to be found, if the search is for only one or a few substances. Also, assumptions can be made about expected sizes and numbers of targets in the chosen area, making use of pertinent information on target-size distributions (sec. 12.3). One must also decide what constitutes a discovery. For example, in searching for diamonds, one wants to know something about the size and shape characteristics of known diamond deposits of the world, and where the same geologic environments might be repeated in the unexplored countries.

A Model to Organize Thinking in the Pre-search Stage

Once some specific targets and areas to search are in mind, a model to organize thinking can be very helpful. An excellent example of a mathematical model for searching for petroleum devised by Drew (1967) is reviewed in this subsection, with a few changes in notation for consistency with this book.

Drew's purpose was to develop a statistical model to describe exploration on a grid for petroleum and to test the validity of the model by applying it to known oil fields. He states that the economic feasibility of the grid-drilling method of exploration can be expressed by the generalized equation

$$P = f(V, A, S, O, C),$$

where P is the net value of oil and gas that would be obtained if the entire search area were grid-drilled, the net value being defined as the value of the oil and gas minus only the cost of drilling, not minus other costs such as exploitation; V is the values of the individual targets in the search area; A is the areas, in surface projection, of the targets in the search area; S is the spatial distribution of the targets in the search area; O is the distribution of the orientations of the targets in the search area; and C is the cost of grid drilling the entire search area.

Drew transformed this generalized equation into the operational equation

$$\mu_P = \sum_{i=1}^{n} p_i v_i - C,$$

where μ_P is the mean net value of oil and gas found, p_i is the probability of finding the ith target, v_i is the value of the ith target, C is the cost of drilling the search area, and the summation is made over the number of targets. The terms in the two equations are related as follows:

Term in operational equation	Term in generalized equation
μ_P	P
p_i	$f(A, S, O)$
v_i	V
C	C

To evaluate his operational equation, Drew made the following four assumptions:

1. The target areas, as seen in surface projection, are ellipses.

2. The centers of the individual targets are uniformly distributed throughout the search area, so that the chance that a target center will be located at a particular point is the same for all points in the area.

3. The orientations of the long axes of the ellipses representing the targets are distributed uniformly, so that all values of the angle between the long axis of the target and the north direction are equally probable.

4. An oil or gas field will be recognized if it is hit by only one exploratory drill hole.

The first three assumptions make the solution of the operational equation mathematically tractable; the last assumption is geological.

By a computer program, Drew solved the operational equation for various grid spacings for 15 test areas in the United States and for the continental United States, excluding Alaska. His results for west Texas are listed in table 14.10 and graphed in figure 14.13. Column 1 gives eight grid spacings that Drew arbitrarily selected, column 2 the corresponding number of drill holes, and the remaining columns the corresponding costs and values in billions of dollars. Column 3 is drilling costs, reckoned at $100,000 a well by a geologist familiar with costs in west Texas; column 4 is mean value, which is the value of

Table 14.10. Simulated results of grid drilling West Texas (billions of dollars)*

Grid spacing, miles (1)	Number of holes (2)	Drilling cost (3)	Mean value, $\sum_{i=1}^{n} p_i v_i$ (4)	Mean net value, μP (5)	Absolute value (6)	Absolute net value, col. 6 minus col. 3 (7)
10	600	0.06	12.07	12.01	0.34	0.28
5	2,400	0.24	18.79	18.55	6.00	5.76
4	3,800	0.38	20.33	19.95	11.92	11.44
3	6,700	0.67	21.98	21.31	14.04	13.37
2.5	9,600	0.96	22.78	21.82	16.59	15.63
2	15,000	1.50	23.49	21.99	19.00	17.50
1.5	26,700	2.67	24.02	21.35	20.84	18.17
1	60,000	6.00	24.42	18.42	23.29	17.29

* After Drew, 1967.

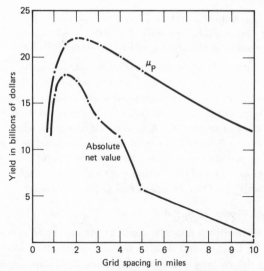

Fig. 14.13. West Texas profit curves (after Drew, 1967).

the fields that, on the average, drilling at the particular grid spacing would obtain; column 5 is mean net value, μ_P, obtained from the previous equation; and columns 6 and 7 correspond to columns 4 and 5, except that they are for absolute value and absolute net value for those fields so large that they would necessarily be found regardless of their areal distribution relative to the starting point of the grid. The number of holes corresponds to an area for the west Texas oil field of 60,000 square miles, not the area of 100,000 square miles recorded in table 1 and figure 1 of Drew's (1967) paper (Drew, personal communication, 1968).

The graph, figure 14.13, shows that, at the optimum grid spacing of 2.0 miles, the average net value μ_P is at a maximum of \$21.99 billion, with a return ratio of 14.7. The optimum grid spacing at which the absolute net value ABS(P) is a maximum is 1.5 miles, with an absolute net value of \$18.17 billion and a return ratio of 6.8. Corresponding values can be obtained from the table and curve if capital available for drilling is less than the amount needed to drill at the optimum grid spacing.

Drew states that

The grid-drilling method of exploration [is] . . . a successful strategy (assuming current prices) when applied to each of fifteen test areas in the United States and to the United States as a whole. The optimum grid spacings for the fifteen areas range from 0.75 to 3.75 miles. The optimum grid spacing for the entire United States is 3.5 miles. The [average] . . . and absolute net values realized by drilling this 3.5 mile optimum grid over the entire United States are 65 and 15 billion dollars, respectively. The cost of drilling the grid is estimated to be 25 billion dollars. These net values would have been realized solely from the oil and gas found; if all the other mineral resources likely to occur in the United States were included in the analysis, the yield would be substantially greater.

Drew's evaluation of his model is based, necessarily, on well-explored areas, but the exploration geologist is more interested in unexplored or partially explored areas. However, the results show what could be expected in an area of similar geology; for instance, Alaska should be similar to the Rocky Mountain states. The drilling of an area like Alaska would start with a large grid spacing, and as information developed, the optimum spacing could be found. Although refinements are easy to suggest, and Drew himself points out some, such as including development and production costs and other variables in the model, Drew has certainly made an excellent and provocative start.

Decisions about Money to Spend

Two kinds of decisions must be made about money to spend: how much money in all to budget, and how to make a preliminary allocation of this money among the various facets of exploration. Little that is definite can be said about this subject, because costs change rapidly and are different for different exploration entities.

Table 14.11 gives representative time and cost requirements of four exploration stages, reported for exploration programs of Bear Creek Mining Co. and other companies by Bailly (1967). While these costs are bound to change with time even within one company, the relative costs should stay about the same. These costs are basically minimum ones because they result from success at each stage of exploration, which is seldom achieved. The table suggests that, as a rough relationship, for every dollar spent in the first stage of regional appraisal, about $10 is spent in either the detailed reconnaissance or the detailed surface work and about $100 is spent in the final stage of detailed physical sampling in three dimensions of the mineral deposit.

Allocation of money among different methods of exploration, for instance geological field work, exploration geophysics, and exploration geochemistry, depends so intimately on the environment that it is not likely to be amenable to decision making in the context of operations research or statistics. However, as a safeguard, cost-benefit studies should be made, to avoid overlooking any potential exploration methods.

Go–No-Go Decision

The last step in the pre-search stage is first to rethink the decisions that have been made in the light of last-minute changes and then to decide either to go ahead or to cancel out the proposed exploration project. New targets from new discoveries, demand for new metals, sudden changes in metal prices, etc. may require new decisions.

TABLE 14.11. REPRESENTATIVE TIME AND COST REQUIREMENTS OF FOUR EXPLORATION STAGES*

Item	Four stages of exploration			
	1. Regional exploration	2. Detailed reconnaissance	3. Detailed surface investigation	4. Detailed physical sampling in three-dimensions
Exploration for a porphyry copper deposit				
Area considered, sq. miles	1,000–100,000	10–100	10–50	3–20
Cumulative cost, dollars	5,000–100,000	10,000–75,000	50,000–150,000	500,000–4,000,000
Cumulative time, months	1–12	13–18	18–24	24–60
Exploration for a blind stratiform lead, zinc deposit				
Area considered, sq. miles	1,000–100,000	10–50	5–10	$\frac{1}{2}$–10
Cumulative cost, dollars	5,000–500,000	50,000–1,000,000	250,000–1,500,000	300,000–3,000,000
Cumulative time, months	1–12	13–20	20–30	30–60
Exploration for a massive sulfide deposit				
Area considered, sq. miles	1,000–100,000	10–50	1–10	$\frac{1}{10}$–2
Cumulative cost, dollars	5,000–500,000	20,000–600,000	100,000–1,000,000	300,000–3,000,000
Cumulative time, months	1–6	7–15	16–24	24–60

* After Bailly, 1967.

Also, some other factors previously noted in Chapter 12 must be examined, including taxes, political climate, and any changes in the physical or economic environment. After all this is done, activity progresses to the exploration stage, which is discussed in the next section.

14.4   EXPLORATION STAGE

In the exploration stage, decisions of two kinds must be made, those based on the purpose of the exploration and those based on the geology of the area to be explored. First, methods of physical sampling must be chosen (Chapters 7 and 15); second, methods of exploration must be chosen (Chapter 12).

Choice of Methods of Physical Sampling

The methods of physical sampling are selected according to their cost effectiveness. In table 14.12 are given costs of some physical sampling methods, according to Bailly (1964); although these costs change with time, they provide a rough measure, and their relative sizes change slowly. Any of these normal costs may change greatly according to availability of equipment and the area where they are to be used. For instance, in 1968, the monthly cost of a seismograph crew was about $45,000 in the Rocky Mountain region of the western

TABLE 14.12. Costs of some physical sampling methods*

	Physical sampling		
	$ Cost per foot		
Drilling	To 1000 ft	1000–2000 ft	2000–3500 ft
Wagon drill	0.50– 1.25		
Churn drill	2.00– 8.00		
Rotary/diamond noncoring	0.75– 7.00	1.50– 9.00	3.00–12.00
Diamond coring	2.50–10.00	3.50–12.00	5.00–20.00
Rotary coring	3.00– 8.00	4.00–10.00	5.00–15.00
Casing	2.50– 4.00	2.50– 5.00	2.50– 6.00
	$ Cost	*Per*	
Pitting	3– 50	Cubic yard	
Trenching	2– 25	Cubic yard	
Shaft sinking	200–10000	Foot	

* After Bailly, 1964.

United States, but about 3 times that figure in most of Alaska, and about 4 times that figure on the north coast of Alaska. Similarly, in measuring effectiveness, some methods of exploration may be wholly unsuitable for a certain environment; for instance, conventional diamond drilling in unconsolidated materials or churn drilling for angle holes.

TABLE 14.13. NUMBER OF DEFLECTIONS DESIRABLE FOR DIFFERENT RATIOS OF THE COST OF A DEFLECTION TO THE COST OF A NEW HOLE FOR DATA LIKE THOSE FROM SOUTH AFRICAN GOLD MINES

$\dfrac{\text{Cost of a deflection}}{\text{Cost of a new hole}} \times 100$	Number of deflections desirable
100%	0
75%	0
50%	0
25%	1
10%	3

Once some physical methods of sampling are chosen, single methods can be examined in more detail. Table 14.13 compares the number of diamond-drill hole deflections that are desirable for different ratios of the cost of a deflection to the cost of a new hole. For instance, if the cost of a deflection is 10 percent of that of a new hole, three deflections are desirable before a new hole is bored. This table was calculated from the South African deflection data (sec. 7.4), and several statistical assumptions had to be made (Koch and Link, 1969). Although the table does not necessarily apply to data other than these from South Africa, it illustrates calculations that can be made for other physical methods of sampling and geological environments.

If some assumptions are made, physical methods of sampling can be compared according to their cost effectiveness. One comparison is illustrated in table 14.14, which contrasts five different methods of searching for a single target at 500 feet of depth; each method has a given cost and probability of detecting the target at each trial. The costs, probabilities, and target designation are all fictitious and arbitrary, although they are intended to correspond roughly to reality. For instance, the first line of the table states that, in searching with wagon-drill holes for a particular target, the probability of recognition from one intersection is 10 percent, the cost of one hole is $500, and therefore the average cost of discovery from one intersection is $5000

($500/10 percent). The separate intersections are assumed to have independent probabilities of finding the target, except for the diamond-drill-hole deflections, where some dependence is assumed (sec. 7.4).

The table shows that, if only a single hole is to be drilled, the average cost of discovery is the same for all of the first three methods. If only enough money is available for a wagon-drill hole, it would necessarily be drilled, although a churn drill affords a larger chance of success. If $3000 were to be spent, a diamond drill would be used because the average cost of discovery is the least. If diamond drilling were technically impossible and if a probability of discovery of about 65 percent were desired, a churn drill would be chosen in preference to a wagon drill because the desired probability could be obtained

TABLE 14.14. COMPARISON OF PHYSICAL METHODS OF SAMPLING ACCORDING TO THEIR COST EFFECTIVENESS

Exploration method	Number of intersections	Probability of detection	Cost	Average cost per discovery (dollars)
	1	0.10	500	5,000
	2	0.19	1,000	5,260
	3	0.27	1,500	5,560
	4	0.34	2,000	5,880
Wagon-drill holes	5	0.41	2,500	6,100
	6	0.47	3,000	6,380
	7	0.52	3,500	6,730
	8	0.57	4,000	7,020
	9	0.61	4,500	7,380
	10	0.65	5,000	7,690
	1	0.30	1,500	5,000
	2	0.51	3,000	5,880
Churn-drill holes	3	0.66	4,500	6,820
	4	0.76	6,000	7,890
	5	0.83	7,500	9,040
Diamond-drill hole	1	0.5	2,500	5,000
with deflections	2	0.7	2,750	3,930
	3	0.8	3,000	3,750
Bucket-drill holes	1	0.90	20,000	22,000
Shaft	1	0.99	100,000	101,000

at a smaller average cost per discovery and a smaller minimum cost. In none of these cases is the cost of a bucket drill or shaft warranted, unless a shaft were needed anyway, perhaps for ventilating other workings or because the target was known to be nearby.

This fictitious case illustrates that even the simplest problem soon gets complicated, and solutions can be obtained only for clearly specified objectives. It also shows what sort of information must be obtained and evaluated for a real problem.

Choice of Methods of Exploration

The first choice to be made in selecting physical methods of exploration is whether discovery is the sole aim or whether evaluation is also an aim. In practice these two objectives are often confused. Once a target is found, the expense of its evaluation can be much better afforded; the best procedure is first to find it. First things first. For instance, in drilling for uranium deposits in the Colorado Plateau area of the United States, drilling in some districts was done on a 500-foot grid, which was much too close for efficient discovery and wasted many holes, because these programs confused discovery with evaluation.

In choosing among exploration methods, one must compare the costs of exploration with the probabilities that discoveries will be made. Generally, the choice is between grid drilling (sec. 8.6) and response-surface drilling (sec. 12.5). A decision between these two methods depends (1) on whether a given area is to be explored because it is a concession or is inaccessible so that men and equipment should be moved in only once and (2) on whether there are clues to start with about target locations. Table 14.15 contrasts these two cases.

Finally, if grid drilling is selected, the first holes should be drilled on the grid at the intersections where chances of discoveries are thought to be the best, provided that this practice is not overdone at excessive expense of moving equipment, etc.

14.5   EVALUATION STAGE

In this section we make no attempt to review all the calculations and decisions that must be made, for to do so would fill another book. Rather, we focus attention on some of those that are related to the geology of the mineral deposit rather than to primarily engineering matters such as the size of trucks or cost of electric power, and we restrict the treatment to computations and decisions that may be aided by statistics and operations-research techniques. The amount of aid that can be given is increasing rapidly as operations

Table 14.15. Contrast of response-surface and grid sampling

Response-surface sampling	Grid sampling
Incomplete areal coverage. One response variable or several positively correlated variables yielding one general variable are expected to be present	Complete areal coverage. Required because several response variables of interest are expected to be present in different areas
Only one maximum is identified at a time	Positive responses may occur at more than one place
A response evaluation by a geologist or other skilled person is needed at every step until discovery or termination	Response evaluation is not necessary until the end of the program, and then only if something is found. A response may not be a valuable substance
If, after the sampling program starts, the model is deemed inappropriate, and samples are taken on a grid, a few (about 2–6) sample points may be wasted	Most sample points are probably wasted
The area is believed to contain a prize that can readily be recognized, once found	The reverse is believed
More information in general is at hand prior to sampling	Little or no information to guide sampling

researchers become more familiar with industry problems and as managers become more aware of operations research; the field is one of the most interesting and potentially rewarding in both applied statistics and applied geology.

Evaluation of natural resources depends on economics as well as on geology. The reader is referred to an economics textbook: Samuelson's (1967) is excellent for definitions of the economic terms found in the valuation literature, and Baumol (1965) clearly explains the connection of economics and operations research.

Evaluation of a nonrenewable natural resource must take into account the peculiarity that the asset to be evaluated is one that is ultimately consumed, because the residual value of the land and of the concentrating plant or oil-field apparatus is usually small. McKinstry (1948, p. 462) writes succinctly that "the valuation, then, comes down to estimating what the earnings of a mine throughout its future life are worth today. Three factors enter into such an estimate: (1) the amount that the mine will earn each year, (2) the number of years that it will continue to produce, and (3) the present value of these future earnings." The same holds true for an oil field or other nonrenewable natural resource.

For an evaluation, two kinds of factors must be considered, those that are inherent in the natural resource and those that are not. In table 14.16 these two kinds of factors are listed for contrast. Some of them were previously reviewed in the discussion of the physical and economic environment of an ore body sketched in figure 13.1.

TABLE 14.16. COMPARISON OF TWO KINDS OF FACTORS FOR EVALUATION OF A NATURAL RESOURCE: THOSE INHERENT AND THOSE EXTERNAL

Inherent factors	External factors
Tonnage of ore	Labor supply
Grade of ore	Equipment supply and repair services
Mineralogy of ore (amenability to metallurgical extraction, etc.)	Land costs
Physical properties of ore (rock strength and other properties that affect minability)	Power availability
	Transportation
	Housing availability
Geological setting of the ore body or ore bodies (physical properties of the wall rock that affect minability, depth and kind of overburden, etc.)	Medical services
	Taxes and subsidies
	Government regulations
	Market
Water supply and problems	Physical environment (caving of surface, competing land uses, conservation, pollution, weather, terrain, etc.)
Altitude	

Several good books detail evaluation methods. For mining, the reader is referred to McKinstry (1948, pp. 459–502), who has a geologist's viewpoint, and to Parks (1957), Raymond (1959), Truscott (1962, pp. 175–249), and Jones (1968), who have an engineer's viewpoint. For petroleum, not discussed explicitly in this book, the reader is referred to Kaufman (1963) and Hughes (1967). Valuation in general is discussed for the engineer and general businessman in an excellent book by Taylor (1964).

Raymond (1959, p. 136) lists steps for determining the value of a new mine (table 14.17) and also an operating mine; McKinstry (1948, p. 491) provides a similar table. Items 1 and 2 of table 14.17 are discussed in Chapter 13. In the rest of this section, the other items are discussed to which statistics or operations research can lend insight.

Besides the fundamental distinction that they work wasting assets, natural-resource extraction enterprises differ from other industries in a number of ways that affect valuation. First, the enterprise cannot be moved but must function where the resource is located, unlike a shoe factory that can be built near a market. Because of this lack of mobility, factors such as taxes, governmental stability, transportation, and availability of labor may be critical. Also, the enterprise is basically able to produce one product only, although the stage to which this product is carried may range from that of no processing, as with some coal mines, to a great deal of processing, as with some Swedish iron mines integrated from the mine to the final stainless-steel cutlery. Therefore, if demand for this product evaporates, the enterprise cannot be converted, as

Table 14.17. Steps for determining the value of a new mine*

1. Calculate the ore reserve and indicate grade or quality under the following classifications (this requires a preliminary estimate of costs and determination of mine cutoff grade):
   (a) Measurable ore
   (b) Speculative ore
2. Estimate recoverable ore, taking into consideration such factors as mine dilution, mine losses, and cost of making ore available
3. From study of flowsheets and metallurgical tests, calculate the treatment losses or metallurgical recovery
4. Estimate rate of production as determined from the mine potential and sales possibilities, as well as limitations, such as availability of power and water
5. Divide reserves by annual production to obtain life of property or operations
6. Using recovery and treatment factors, calculate total yield of salable product. Calculate the "smelter settlement value" of the ore or concentrate, or salable products
7. Estimate average sales price per annum and total average sales volume and total annual gross revenue
8. Estimate cost of sales (per ton basis), labor, materials and supplies, and overhead
9. Estimate selling (marketing), administrative, and central office costs
10. Subtract cost of sales from sales income to get gross profit
11. Subtract selling and administrative expenses from gross profit to get profit before depletion allowance—also any interest payment
12. Subtract depletion and depreciation allowances to obtain basis for computing income tax
13. Determine the income taxes
14. Estimate the total annual net profit after taxes
15. Set up work sheet to show estimated cash flow, including payments of such items as interest, principal on loans, and tax allowances for period of operations

TABLE 14.17.—CONTINUED

16. Consider special risks and hazards to operation and consider a reasonable rate of return on investment, or discount factor to be used
17. Estimate ultimate speculative tonnage that may be expected in addition to the measured reserves
18. Discount the total net income to obtain present value of future net income over the life of the operations, including any residue earnings at end of life of the property
19. From the total present value of measured reserves and residue salvage deduct the first cost of plant facilities, and working capital (total new capital requirements), to obtain the value of the mineral property and any assets thereon
20. If production is deferred, apply a discount factor for period of deferment
21. Compare earnings against investment with those current in similar enterprises

* After Raymond, 1959, p. 136.

can a factory, to making something else. Typically, although not always, there is a high ratio of invested capital to sales, and partly for this reason natural-resource enterprises usually have long lives, measured in at least a few decades, compared to short-term enterprises like World's Fairs. Behre and Arbeiter (1959, pp. 43–79) offer some further thoughts on the special nature of the mineral industries.

What is Ore?

"Ore . . . is that part of a geologic body from which the metal or metals that it contains may be extracted profitably," wrote Lindgren (1933, p. 13). This definition, admirable in its generality, is, like so many other geological concepts, difficult to express in quantitative terms susceptible to statistical analysis.

McKinstry (1948, p. 461) puts the problem clearly when he writes that

. . . despite its solid appearance, ore may be elusive . . . what is ore today may not be ore tomorrow. Millions of tons of copper-bearing rock were ore in 1930 when the price of copper was 18 cents a pound, but ceased to be ore in 1932 when the price fell to 5 cents. Much of it was ore again in 1942 when the price was 12 cents.

Since ore has no assignable value that is independent of the cost of mining and treating it, the only rational basis of valuation is earning power. Yet it is not *present* earning power that determines the value but rather the outlook for future earnings. For this reason every valuation is in the nature of a prophecy rather than a factual inventory. The factors that enter into such a prophecy are many. Some of them are technical, involving geology, mining methods, and metallurgy, but of fully equal importance are questions of economics and politics, both national and international.

In two penetrating articles, Carlisle (1954a, 1954b) reviews the mathematical and economic relations in the definition of ore. While in this book we cannot examine in detail the ramifications into which this problem soon leads, we outline some salient points. Carlisle's papers and the books already cited provide more details.

Table 14.18 and figure 14.14 present a fictitious illustration by Carlisle of a mineral deposit that will yield from 10,000 to 150,000 tons of metal, depending on the definitions of ore. In column 1 of the table is given tons of metal recovered, in 10,000-ton increments. In column 2 is given the fixed cost per ton, assumed to be $100,000 divided by the number of tons, comprising costs such as exploration, development, and other "overhead" items that are essentially fixed regardless of the size of production. In column 3 are given

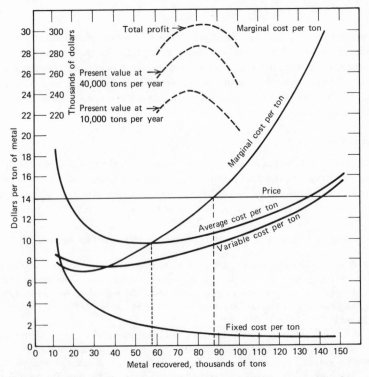

Fig. 14.14. Graph of fictitious illustration of alternative levels of recovery for a mine (after Carlisle, 1954b, p. 91).

TABLE 14.18. ALTERNATIVE LEVELS OF RECOVERY FOR A MINE*

Tons of metal recovered (1)	Fixed cost per ton (2)	Variable cost per ton (3)	Average cost per ton (4)	Total cost of production (5)	Marginal cost per ton (6)	Profit per ton (7)	Total profit (8)
10,000	10.00	8.00	18.00	180,000		-4.00	
20,000	5.00	7.50	12.50	250,000	7.00	1.50	30,000
30,000	3.33	7.25	10.58	317,400	6.74	2.42	72,600
40,000	2.50	7.30	9.80	392,000	7.46	4.20	168,000
50,000	2.00	7.55	9.55†	477,500	8.55	**4.45**	222,500
60,000	1.67	7.90	9.57	574,200	9.67	4.43	265,800
70,000	1.43	8.35	9.78	684,600	11.04	4.22	295,400
80,000	1.25	8.90	10.15	812,000	**12.74**	3.85	**308,000**
90,000	1.11	9.50	10.61	954,900	14.29	3.39	305,100
100,000	1.00	10.15	11.15	1,115,000	16.01	2.85	285,000
110,000	0.91	10.90	11.81	1,299,100	18.41	2.19	240,900
120,000	0.83	11.40	12.23	1,467,600	21.64	1.77	212,400
130,000	0.77	12.75	13.52	1,757,600	29.00	0.48	62,400
140,000	0.72	13.90	14.62	2,047,000	28.94	-0.62	
150,000	0.67	15.20	15.87	2,380,500	33.35	-1.87	

* After Carlisle, 1954b, p. 91.
† Boldface type indicates locations of maxima or minima.

variable costs, which are assumed to decline at first because of economies in mass production and then rise as costs of treating lower-grade ores, stripping, etc. increase. Summing the fixed and variable costs yields the average cost in column 4, which, multiplied by the tonnage, gives the total cost of production in column 5. In column 6 is the marginal cost per ton, which is the cost of producing 10,000 more tons of metal. In column 7 is the profit per ton calculated on the assumption that the selling price of the metal is $14 per ton, and in column 8 is the total profit obtained by multiplying the tons by profit per ton.

Carlisle's fictitious illustration shows that ore may be defined by either of two criteria: (1) average cost per ton of metal is lowest and average profit is highest at a production of about 55,000 tons, or (2) total profit is largest at about 87,000 tons, and marginal cost per ton becomes equal to the selling price.

Thus the definition of ore depends in an elusive way on the relation between the recoverable value of the metals in the ore and various economic factors. In practice, one is generally forced back to the evaluation of grade and amount as the only firm place to rest an analysis, although the concept of grade itself is indefinite, as was noted in section 13.2, when several kinds of grade were defined. To relate grade to recoverable value of metals is difficult, especially for multimetal ores. For instance, the Frisco mine in Mexico recovers gold, silver, lead, copper, zinc, and fluorspar in three different concentrates that are sold to several buyers under various contracts. Without electronic computers it would be impossible to evaluate ore, stope by stope, with changing prices, costs, and contracts. With computers, evaluation would be practical although not trivial (sec. 14.6).

Interest Rates and Valuation

Interest calculations are essential for the valuation of nonrenewable natural resources because, as mentioned at the beginning of this section, a non-renewable resource is eventually exhausted, and the operation closes down. Within the time it operates, the capital investment must be returned and the investor must be rewarded for risking his money. Future earnings, the basis for valuation, must be discounted to present time at a selected interest rate. Even though no money is necessarily borrowed and no earnings are necessarily reinvested at this or any other interest rate, an interest rate must be assumed in order to compare different prospective investments, for money has a time value. The sum of $10,000 earned 10 years in the future is worth only $5584 today at 6 percent interest. A mine that will earn $50,000 a year for 10 years is clearly worth more than one that will earn the same sum for only five years— but not twice as much, again because money has a time value. For these and many other reasons, moderate changes in the selected interest rate are often far more important to the final valuation placed on a property than details of the grade and tonnage.

If one borrows $20,000 to buy a house at $5\frac{1}{2}$ percent interest for 15 years, he pays $2000 a year or a total of $30,000 over the life of the loan. Alternatively, $20,000 is the cost of an annuity at the same interest rate, for the same time, and in the same amount. The $20,000 is the *present value* of the payments or of the annuity under the stated conditions; in the context of mine valuation, the annuity corresponds to the annual income of the mine.

TABLE 14.19. PRESENT VALUES AT 6 PERCENT INTEREST OF ALTERNATIVE LEVELS OF RECOVERY FOR A MINE*

| Life of mine (years) | Present value at 6% with various yearly production rates | | | |
	10,000 tons	20,000 tons	30,000 tons	40,000 tons
1		28,302	68,491	158,491
$1\frac{1}{4}$		37,933	93,709	208,652
$1\frac{1}{2}$		101,268	181,337	248,247
$1\frac{3}{4}$		131,098	209,556	**275,108†**
2	27,501	154,006	243,659	282,344
$2\frac{1}{4}$		177,491	265,817	277,926
$2\frac{1}{2}$		201,078	271,116	258,301
$2\frac{3}{4}$		218,959	**274,655**	217,423
3	64,687	236,828	271,844	189,248
4	145,534	**266,813**	183,997	
5	187,452	240,107		
6	212,919	174,072		
7	**235,577**			
8	239,077			
9	230,578			
10	209,763			
11	172,723			
12	148,393			
13	42,493			
14				
15				

* After Carlisle, 1954b, p. 90.
† Boldface type indicates locations of maxima.

For Carlisle's fictitious illustration, table 14.19 lists present values at 6 percent interest with various yearly production rates. This table and figure 14.14 show that (1) the present value rises as the rate of production rises, because the income is received faster, and (2) regardless of rate of production, present value rises to a maximum and then declines, a variation which suggests that, if the aim is to maximize present value, some ore is better left in the ground (however, see Carlisle and the authors that he cites for qualifications to this conclusion). In terms of the house-buying analogy, the faster money is paid back, the larger the loan that the borrower can obtain. Carlisle (1954b) writes further about the material in table 14.19 and figure 14.14:

A little reflection will show that, with positive interest rates, the maximum-present-value level is always higher than the level at which profit per unit of time is maximized, but lower than the level at which total profit neglecting its distribution in time is maximized. With moderate discount rates it falls closer to the upper level.

Another situation is depicted in table 14.20, which is for a fixed production rate of 10,000 tons a year and for different interest rates. This table shows that, with the given production, the maximum present value is between seven and eight years of life, but that present value drops sharply as the interest rate increases. For instance, at an interest rate of 6 percent, operating for five years yields a present value of $187,452, which is larger than the maximum

TABLE 14.20. PRESENT VALUES AT DIFFERENT INTEREST RATES FOR FIXED PRODUCTION RATE OF 10,000 TONS A YEAR FOR A MINE*

Life of mine (years)	Present value at fixed 10,000 production rate and various interest rates				
	6%	8%	10%	12%	16%
1					
2	27,501	26,750	26,033	25,352	24,078
3	64,687	62,366	60,183	58,124	54,351
4	145,534	139,108	133,136	127,567	117,524
5	187,452	177,675	168,690	160,414	145,706
6	212,919	200,172	188,584	178,024	159,548
7	**235,577†**	**219,710**	**205,446**	**192,592**	**170,429**
8	239,077	221.244	205,394	191,253	167,229
9	230,578	211,770	195,230	180,626	156,160
10	209,763	191,238	175,121	161,031	137,746
11	172,723	156,344	142,243	130,036	110,126
12	148,393	133,389	120,602	109,641	91,989
13	42,493	37,938	34,096	30,833	25,643
14					
15					

* After Carlisle, 1954b, p. 90.
† Boldface type indicates locations of maxima.

present value of $170,429, requiring two more years of operation at the 16 percent rate. This contrast illustrates the point that interest rates can be much more significant in valuation than engineering and geological factors such as amount or grade of ore. In terms of the home-buying analogy, as interest rates rise, the amount the borrower receives for a fixed monthly payment declines.

Thus far, only simple compound interest at a single interest rate has been discussed. However, an alternative scheme using two interest rates for mine valuation was invented in the last century and is named the Hoskold formula after the man who computed the necessary tables. Clear explanations are given by Parks (1957, pp. 190–195) and by McKinstry (1948, pp. 479–490). Hoover (1909, pp. 42–43) gives an excellent account of the principle involved; he writes,

As every mine has a limited life, the capital invested in it must be redeemed during the life of the mine. It is not sufficient that there be a bare profit over working costs. In this particular, mines differ wholly from many other types of investments, such as railways. In the latter, if proper appropriation is made for maintenance, the total income to the investor can be considered as interest or profit; but in mines, a portion of the annual income must be considered as a return of capital. Therefore, before the yield on a mine investment can be determined, a portion of the annual earnings must be set aside in such a manner that when the mine is exhausted the original investment will have been restored. If we consider the date due for the return of the capital as the time when the mine is exhausted, we may consider the annual instalments as payments before the due date, and they can be put out at compound interest until the time for restoration arrives. If they be invested in safe securities at the usual rate of about 4 percent, the addition of this amount of compound interest will assist in the repayment of the capital at the due date, so that the annual contributions to a sinking fund need not themselves aggregate the total capital to be restored, but may be smaller by the deficiency which will be made up by their interest earnings. Such a system of redemption of capital is called "amortization."

Today the Hoskold formula is less used in practice than simple compound interest at a single rate. Raymond (1959, p. 132) writes,

Among technical mining men, probably because most mining schools have emphasized the technique, the Hoskold formula is widely used. Within the industrial and financial sectors of business, however, the straight-discount method is generally used. The Hoskold formula is more complex mathematically and is not easily applicable to a varying earning rate. More important, Hoskold applies a risk interest rate on the total original investment during the life of the property similar to that of a perpetuity. The straight-discount concept assumes that the investor is entitled to interest on the unrecouped capital and that the interest so accumulated should not be used to augment a sinking-fund total. This concept is comparable to the practice used in the nonextractive industries.

One more item must be mentioned, the valuation of present value of a mine whose income will not start until some time in the future. Assume that a mine will start production and therefore start yielding income five years from now. Then the present value at the start of production is calculated in the usual way, and the resulting figure is discounted to today. Usually the same interest rate is used as was used for the original present-value calculation.

In this subsection the ramifications of interest rates in valuation have been sketched. McKinstry (1948) gives more information, and Parks (1957) provides graphs of interest-rate calculations, many problems worked out in detail, and a complete range of tables. With electronic computers, interest-rate calculations are easily programed to yield valuations at several interest rates and with different premises without the trouble of consulting tables.

Methods of Valuation

The previous discussion explains several alternative methods of valuation of natural resources. Total value of future earnings, net annual profit, or present value can be used. Two other methods of valuation are payback and discounted cash flow.

Payback is the length of time, usually measured in years, required to return the capital investment. Other things being equal rapid payback is desirable because, once capital is returned, the risk of an investment is reduced. For instance, in an investment in a politically unstable country, payback might be the variable to maximize. However, as Jones (1968, p. 1004) points out, different projects can have the same paybacks, but one can be much more desirable in another measure of value.

Discounted cash flow is a method of valuation that has become much used in recent years. The *discounted cash flow* rate of return is defined to be the interest rate r in the formula

$$P = \frac{S_1}{(1+r)^1} + \frac{S_2}{(1+r)^2} + \cdots + \frac{S_i}{(1+r)^i},$$

where P is the principal (capital investment), S_i is the cash income per year or other period (S_i can be positive, zero, or negative), n is the number of years or other periods, and r is the interest rate.

The meaning of the calculations in the previous formula may be explained through table 14.21, which is explained in detail by Groundwater (1967). Assume (line 1) that an investor lends $1000 at the beginning of a certain year designated 1 (the same time as the end of year 0), in a loan to be repaid in five years with equal installments and with interest at 10 percent on the unpaid balance. In line 2 the interest to be paid is given, which is $100 at the end of the first year, declining to $24.0 at the end of the fifth year, and totaling $319.0.

TABLE 14.21. FICTITIOUS ILLUSTRATION OF DISCOUNTED CASH FLOW*

Line	Item	Dollars at end of year indicated						Totals
		0	1	2	3	4	5	
1	Amount of investment outstanding	1000	836.2	656.0	457.8	239.8	0	
2	Interest at 10% on amount of investment outstanding		100.0	83.6	65.6	45.8	24.0	319.0
3	Dividends paid		263.8	263.8	263.8	263.8	263.8	1319.0
4	Cash flow	−1000	263.8	263.8	263.8	263.8	263.8	319.0
5	Discounted cash flow (10% interest)	−1000	239.8	218.0	198.2	180.2	163.8	0

* After Groundwater, 1967, pp. A70–A71.

When the sum, $1319.0, of interest and principal is divided by 5, the yearly payment on the loan, named "dividends paid" in line 5, is found to be equal to $263.8. From these figures, the cash flow (line 4) can be calculated; it is equal to $-$1000 at the end of year 0, and is equal to $263.8 for each of the remaining five years. As the name implies, cash flow is simply the transfer of money in and out of the account. When the positive cash flows are discounted back to the end of the year 0 at 10 percent interest, the values in line 5 are obtained. Summing across line 5 yields a total of $0, which is the criterion in the previous formula for setting the rate of interest in the discounted cash flow. In other words, an interest rate of 10 percent is obtained from the initial flow of $-$1000 and from the subsequent five equal flows of $263.8. Had the flows been equal to another value, say inward flows of $300 instead of $263.8, another interest rate would have been obtained, equal to 15.3 percent for inward flows of $300. In general, the equation may be solved conveniently by iteration with an electronic computer; the inward flows need not be equal.

A discussion of the implications of discounted cash flow would expand this book too much. Jones (1968) gives a brief review, and Groundwater (1967) gives an excellent discussion with special reference to petroleum. Johnson and Bennett (1968) provide a hypothetical application to an open-pit gold mine, implemented by a computer program. For making discounted cash-flow calculations, a computer is nearly essential; therefore this method of valuation did not make much headway until recent years.

Besides these methods of valuation, many others exist. How is the geologist concerned with valuation to choose among them? Usually, in this day of computers, he will use more than one, as once the basic facts are collected, the task is not formidable. In the next subsection a model that touches on this subject is discussed.

A Simple Valuation Model

In valuing a mine or prospect, it is instructive to construct a model, to change the parameters, and to observe the resulting changes in the values of the variables. Often 30 to 40 runs implemented by an electronic computer on a simple model will provide the information needed for decisions. A model may be purely deterministic, or it may include one or more stochastic elements.

In this subsection we illustrate the procedure for a simple model that has been slightly modified from one published by Soderberg (1959) and later discussed by Bader (1965). Although the costs by now are out of date, this model is suitable for simple discussion and, having been devised and discussed by engineers for the Kennecott Copper Co., reflects good industrial practice.

Table 14.22 sets forth Soderberg's model, broken down into the three major categories of basic, cost, and income data. The model is for a potential open-pit copper mine, and the total amount of rock available is set equal to 350 million

TABLE 14.22. VALUATION MODEL FOR A FICTITIOUS OPEN-PIT COPPER MINE*

Line	Items and units	Numerical values
	Basic data, various units	
1	Ore, tons and grade	100,000,000 tons, 0.90%
2	Low grade ore, tons and grade	40,000,000 tons, 0.55%
3	Average ore, tons and grade	140,000,000 tons, 0.80%
4	Waste	210,000,000 tons
5	Stripping ratio	1.50:1
6	Annual production of ore	5,000,000 tons
7	Annual production of copper	70,400,000 lbs
8	Daily production of ore	16,000 tons
9	Daily production of waste	24,000 tons
10	Daily total production	40,000 tons
	Cost data, dollars	
11	Mine cost	20,144,000
12	Mill cost	24,000,000
13	Total plant cost	44,144,000
14	Life in years	28
15	Mining cost, per ton	0.450
16	Stripping cost, per ton	0.675
17	Milling and general cost, per ton	1.250
18	Total operating cost, per ton	2.375
19	Total operating cost, per pound of copper	0.169
20	Treatment cost, per pound of copper	0.059
21	Grand total operating cost, per pound of copper	0.229
22	Price of copper, per pound	0.300
23	Profit, per pound	0.071
24	Annual income	5,021,000
25	Amortization	1,576,571
26	Annual income minus amortization	3,444,429
27	Depletion	1,722,214
28	Taxable income	1,722,214
29	Tax rate	52%
30	Tax	895,551

TABLE 14.22—*continued*

Line	Items and units	Numerical values
	Income data, dollars	
31	Net annual profit	4,125,449
32	Total profit	115,512,559
33	Payout, years	10.7
34	Interest rate and present value factor	6%, 13.406
35	Present value	55,306,441
36	Total plant cost	44,144,000
37	Net present worth	11,162,441

* After Soderberg, 1959, p. 58.

tons. At a cutoff grade of 0.5 percent copper, the distribution of ore and low-grade ore is shown in lines 1 and 2, providing 140 million tons of ore with an average grade of 0.8 percent. The model is constructed for a fixed production of 5 million tons of ore a year, a copper price of 30 cents a pound, a tax rate of 52 percent, and an interest rate of 6 percent. Under these conditions, the net present worth of the mine is about $11 million. In the next paragraph, which others are advised to skip, the detailed calculations are outlined for readers who want to reproduce the table.

The values in the following lines of the table are read as input to the computer program: 1, 2, 6, 12, 15, 17, 22, 29, 34. Tons and grade of average ore (line 3) are calculated from the data on ore, and low-grade ore is read for lines 1 and 2. Waste (line 4) is equal to 350 million tons, which is the total amount of rock available, minus the tonnage assigned to ore in line 3. Stripping ratio (line 5) is the ratio of waste tonnage to ore tonnage. Annual production of copper in pounds (line 7) is obtained by a linear equation fitted to the data in Soderberg's table to allow for losses in mining and milling. Daily production of ore in tons (line 8) is obtained from annual production by dividing it by 312.5. Daily production of waste in tons (line 9) is found by multiplying the daily production of ore in tons by the stripping ratio. Daily total production in tons (line 10) is the sum of the two previous lines. Mine cost (line 11) is obtained by a linear equation fitted to the data in Soderberg's table. Total plant cost (line 13) is the sum of the two previous lines. Life in years (line 14) is the tons of average ore (line 3) divided by the annual production (line 6). Stripping cost per ton (line 16) is obtained by multiplying the mining cost by the stripping ratio. Total operating cost per ton (line 18) is the sum of the previous three lines. Total operating cost per pound of copper (line 19) is equal to the total operating cost per ton of ore divided by the quotient of the annual production of copper in pounds (line 7) and the annual production of ore in tons (line 6). Treatment cost per pound of copper (line 20) varies with its grade and is obtained from a linear equation fitted to Soderberg's grade data (line 3). Grand total operating cost per pound of copper (line 21) is the sum of the two previous lines. Profit per pound (line 23) is found by subtracting line 21 from line 22. Annual income (line 24) is equal to the profit per pound (line 23) times the annual production in pounds (line 7). Amortization (line 25) is the total plant cost (line 13) divided by the life in years (line 14). Depletion (line 27) is equal to one-half the quantity annual income minus amortization (line 26). Taxable income (line 28) is equal to the annual income minus amortization (line 26) minus the depletion. Tax in dollars (line 30) is equal to the taxable income times the tax rate (line 29). The net annual profit (line 31) is the annual income (line 24) minus the tax in dollars (line 30). The total profit (line 32) is the net annual profit times the life in years (line 14). Payout in years (line 33) is the total plant cost (line 13) divided by the net annual profit (line 31). The present-value factor (line 34) is calculated from the interest rate by the standard formula. Present value (line 35) is found by multiplying the present-value factor by the net annual profit (line 31). Finally, the total plant cost (lines 13 and 36) is subtracted from the present value (line 35) to yield the net present worth (line 37).

TABLE 14.23. VALUATION MODEL FOR A FICTITIOUS OPEN-PIT COPPER MINE RUN SIX TIMES AT DIFFERENT INTEREST RATES, PRICES OF COPPER, AND CUTOFF GRADES*

Price of copper (ct./lb)	Interest rate (%)	Net annual profit (thousands of dollars)		Total profit (thousands of dollars)		Net present worth (thousands of dollars)	
		0.5% cutoff	0.4% cutoff	0.5% cutoff	0.4% cutoff	0.5% cutoff	0.4% cutoff
30	6	4,125	3,424	115,513	133,517	11,162	12,461
	7	,,	,,	,,	,,	5,927	6,695
	8	,,	,,	,,	,,	1,447	1,949
40	6	9,335	7,952	261,381	310,122	81,003	80,155
	7	,,	,,	,,	,,	69,157	66,763
	8	,,	,,	,,	,,	59,018	55,739

* After Soderberg, 1959, p. 58.

Table 14.23 gives the results of running the model six times at different interest rates, different prices of copper, and different cutoff grades. At the fixed production rate of 5 million tons of ore a year, life of the mine depends only on cutoff grade, and is 28 years for the cutoff of 0.5 percent and 39 years for the cutoff of 0.4 percent. Net annual profit and total profit depend only on cutoff and price of copper. Regardless of the copper price, net annual profit is higher but total profit is lower at the higher cutoff of 0.5 percent than at the lower cutoff of 0.4 percent. Net present worth is sensitive to the price of copper and to the interest rate but, except for a few entries, insensitive to the cutoff grade. The case of 0.5 percent cutoff in the first line is that of table 14.22; the case of 0.5 cutoff in the second line yields nearly the same result as Soderberg's (1959, p. 58), because 7 percent compound interest discounted to the present is nearly equivalent to a Hoskold factor of 6 and 3 percent.

Although table 14.23 gives the principal results from only six cases, they are enough to show what in practice is given decision makers. The model could be refined for different copper prices, cutoff grades, and values of the other variables to obtain the detailed information needed.

The deterministic model portrayed up to this point can be modified to a stochastic one. In order to demonstrate the procedure, it is shown for only one variable, although several variables could be taken, either singly or together. The variable modified is copper price, which certainly will not remain fixed over the life of the mine, having in the last 20 years ranged from a low of 19 cents a pound to a high of at least 70 cents a pound. The investigation was carried out for the 0.5 percent cutoff and the net present-worth measure of effectiveness (line 37 in table 14.22), although it could have been carried out for any other chosen measure of effectiveness.

We performed a simulation by fixing the mean copper price at 30 cents and by allowing the actual price to vary randomly from year to year about that figure. For each year, a copper price was calculated from random normal numbers with a standard deviation of 2 cents. For each of the 28 years of production, a price was simulated and annual income discounted to the present; the sum of these 28 present-worth values minus the total plant cost is the net present worth of the mine.

Figure 14.15 gives histograms for interest rates of 6 and 8 percent. Each distribution is for 100 values of net present worth for the 28 years of mining, with the copper price allowed to vary randomly. Most strikingly, the figure shows that net present worth varies widely. For the 6 percent interest rate, the net present worth ranges from $4 million to $22 million; for the 8-percent interest rate, it ranges from minus $6 million to $7 million. Thus profit can vary considerably from the usual value if the price fluctuates in this fashion. We assumed a mild fluctuation, and in real life the fluctuation and resulting variability in net present worth are likely to be more extreme.

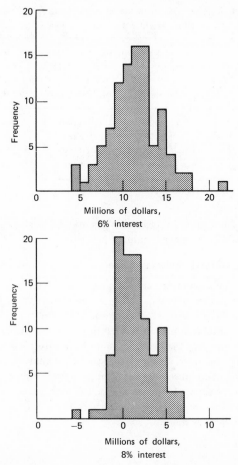

Fig. 14.15. Histograms of net present worths with interest rates of 6 and 8 percent for stochastic model of an open-pit copper mine (after Soderberg, 1959).

Table 14.24 tabulates the simulated results for four different interest rates. Comparison with table 14.23 shows that the mean results are little different from those obtained by making the calculation for a constant copper price of 30 cents. However, these mean results are not guaranteed, as shown by the upper and lower 5 percent points of the observed frequency distributions, which are given in the last two columns of the table. With a 6 percent interest rate, although the mean net worth is about $11 million, the actual net present worth may be expected to range between $6 million and $16 million, that is, to be from one-half to one-and-a-half times the expected amount. This

Table 14.24. Valuation model for a fictitious open-pit copper mine, modified with copper price as a stochastic element[*]

Interest rate (%)	Net present value (thousands of dollars)		
	Mean	Lower 5% point of distribution	Upper 5% point of distribution
6	11,309	6,000	16,000
8	1,352	− 2,000	5,000
10	− 5,798	− 10,000	− 2,000
12	− 11,309	− 14,000	− 8,000

[*] After Soderberg, 1959, p. 58.

fluctuation in value caused by price changes may be large and is one of those risks most difficult to evaluate in mining practice.

Decision Making in the Evaluation Stage

Finally, after the calculations are completed, one can make decisions. Then several properties are compared to decide which, if any, are to be mined. With a model like Soderberg's, some stochastic and some deterministic elements are present, and these different elements must be examined. A *sensitivity analysis* (Forrester, 1961, pp. 268–276) can be made to isolate those elements which change the results the most, although the stochastic analysis already provides a guide. The principal requirement is to compare several properties that are in effect laid out on the carpet for inspection on the same bases, and to think hard about what factors are important to the corporate entity or governmental unit for which the decision is made.

14.6 EXTRACTION STAGE

In the extraction stage, operations research can be applied in many ways in order to control the three substages of development, mining, and beneficiation. In this section only a few illustrations are given of the many decisions to be made in the extraction stage. Hazen (1968) provides an excellent summary of some recent papers on this subject; another good source of information is the transactions of the symposia on statistics, operations research, and computers in the mineral industries, symposia held annually (with a few exceptions) since 1961 on a rotating basis at the University of Arizona, Colorado School of Mines. Stanford University, and Pennsylvania State University, For operations research in petroleum extraction, not treated in this book, the reader is referred to Kaufman (1963), to the cited symposia transactions, and to publications of the Kansas Geological Survey.

In preparing for the extraction stage, probably the key decision is whether the mining will be done on the surface or underground, and (particularly if done underground) whether the mining will be highly selective, for instance cut-and-fill stoping, or nonselective, for instance block caving. Although at present the trend is more and more toward open-pit mining because of the economies offered by heavy equipment and smaller labor requirements, Howard (1968) suggests that in the future this trend will be reversed as value of surface land becomes too great to allow for its disruption, and underground selective mining will again come to the fore. Howard believes that underground selective mining will require a better knowledge of rock properties than presently is available, probably necessitating computer analysis of data given by sensors obtaining information before the rock face is broken, and also requiring better materials handling organized through a total systems approach.

In order to choose and implement a mining method, one must view a mining operation as a system because all the components (mining method, equipment items, operation cycles, etc.) interact with one another. Then a mathematical model can be devised to represent this system. Often selection of the size and boundaries of the system is the most difficult decision.

Once the system is understood and a mining plan is devised, the next aim is to begin production as soon as possible because the capital invested in land acquisition and in plant construction provides no return until the day the product is first sold. In table 14.25 Peters (1966) lists the pre-production intervals for a number of mines. He writes (1966, pp. 63–64) that mines with two years or less of pre-production interval "are characterized by near-optimum physical conditions; the ore is not complex and has a high unit value. Technologic conditions are favorable . . . economic conditions are such that a market is assured and only short-term financing is required." He states further that the average pre-production interval is from two to five years. In the longer interval of from five to seven years are "many of the large scale mines with low unit value ore and complex processing." Finally, mines requiring more than seven years "are characterized by difficult physical conditions, large techno-logic problems, and complex economic problems." In this last category are Western Deep Levels, South Africa (8 years), and Schefferville, Quebec, Canada (12 years). In order to reduce the pre-production interval, critical-path scheduling (sec. 14.2) may be useful, particularly a cost-time formulation including an allowance for penalty costs of reduced present value caused by a late start.

An important aid to planning for the extraction stage may be a computer-

TABLE 14.25. PRE-PRODUCTION INTERVALS FOR A NUMBER OF MINES*

Mine	Initial production (tons per day)	Year operations commenced	Type	Pre-production time (years)
Mines with 2 years or less pre-production interval				
Boriana, Arizona, tungsten	50	1931	Underground	1
Rio de Oro, New Mexico, uranium	60	1956	Underground	1
William, Arizona, copper	75	1940	Underground	1
Daisy, Arizona, copper	100	1954	Underground	1
Palawan, Philippines, mercury	150	1955	Open pit	2
Weedon, Quebec, copper	250	1952	Underground	2
Radon, Utah, uranium	250	1956	Underground	$1\frac{1}{2}$
Madsen Red Lake, Ontario, gold	300	1938	Underground	1
Mineral Hill, Arizona, copper	350	1953	Underground	1
Walton, Nova Scotia, barite	400	1941	Open pit	1
Black Rock, Australia, copper	1,000	1963	Open pit	1
Lac Dufault, Quebec, copper-zinc	1,300	1964	Underground	2
Getchell, Nevada, gold	1,500	1962	Open pit	2
Brynnor, British Columbia, iron	3,000	1962	Open pit	2
Kidd Creek, Ontario, copper-zinc-silver	6,000 (Est.)	1966	Open pit	2
Castle Dome, Arizona, copper	10,000	1943	Open pit	2
Yerington, Nevada, copper	12,000	1953	Open pit	2

Mines with 2 to 5 years pre-production interval

	1,000 tons/shift			
Con. Disc. Yellowknife Gold, N.W.T.	100	1950	Underground	4
Berens River, Ontario, gold-silver	225	1939	Underground	3
Quebec Copper, Quebec	450	1954	Underground	4
Pamour, Ontario, gold	750	1936	Underground	3
Locust Cove, Virginia, gypsum	1,000 tons/shift	1964	Underground	4
Mary Kathleen, Australia, uranium	1,120	1958	Open pit	4
Gunnar Beaverlodge, Sask., uranium	1,250	1955	Open pit	3
Sunro Copper, British Columbia	1,500	1962	Underground	$2\frac{1}{2}$
Campbell Chibougamau, Quebec, copper	1,700	1955	Underground	4
Brookfield, Nova Scotia, limestone	1,800	1965	Open pit	4
Northgate, Ireland, lead-zinc-silver	2,000	1965	Open pit	3
Carlin, Nevada, gold	2,000	1965	Open pit	3
Bagdad, Arizona, copper	2,500	1944	Underground	3
Indian Creek, Missouri, lead	2,500	1954	Underground	4
Coalinga Asbestos, California	2,500	1962	Open pit	3
Zeballos, British Columbia, iron	3,200	1962	Open pit	3
Fletcher, Missouri, lead	5,000 (Est.)	1966	Underground	3
National Gypsum, Nova Scotia	6,000	1955	Open pit	4
Endako, British Columbia, molybdenum	10,000	1965	Open pit	$2\frac{1}{2}$
Esperanza, Arizona, copper	12,000	1959	Open pit	4

Table 14.25—*continued*

Mine	Initial production (tons per day)	Year operations commenced	Type	Pre-production time (years)
Mines with 5 to 7 years pre-production interval				
Trojan, Zambia, nickel	130	1965	Underground	5
Eldorado Beaverlodge, Sask., uranium	500	1953	Underground	5
Wedge, New Brunswick, copper-zinc-lead	750	1962	Underground	6
Gaspe Copper, Quebec, copper	1,500	1955	Underground	7
Gunnar Beaverlodge, Sask., uranium	1,650	1957	Underground	5
Lynn Lake, Manitoba, nickel	2,000	1954	Underground	7
H. Young, Tennessee, zinc	2,500	1955	Underground	6
Pima, Arizona, copper	3,000	1957	Open pit	5
Viburnum, Missouri, lead	4,000	1960	Underground	5
Thompson, Manitoba, nickel	6,000	1961	Underground	5
Silver Bell, Arizona, copper	7,500	1954	Open pit	6
Bomi Hills, Liberia, iron	10,000	1951	Open pit	6
Quebec Cartier, Quebec, iron	10,000	1961	Open pit	5
Pea Ridge, Missouri, iron	12,000	1964	Underground	7
Mineral Park, Arizona, copper	12,000	1964	Open pit	5
Lavender, Arizona, copper	12,000	1954	Open pit	6
Cleveland, Ohio, salt	15,000	1965	Underground	7
Palabora, South Africa, copper	500,000 tons/year	1963	Open pit	6
Weipa, Australia, bauxite	33,000	1966	Open pit	7

* After Peters, 1966, pp. 63–64.

based analysis of alternative types of extraction. An example is the pit-design analysis implemented by computer, an analysis already discussed in sec. 13.2. Periodic review may change the definition of ore and even the method of mining (Erickson, 1968).

In the development substage new controls must be set up. A decision must be made about physical methods of sampling following the principles already stated (sec. 14.4). Also, the number of samples to be taken and their spacing must be determined. For this determination, the purpose of the sampling must be examined. If grade control for selective mining of precious metal ore is required, sampling is clearly needed; on the other hand, if base-metal ore whose grade can be "eyeballed" is mined, few if any samples may be needed. The purpose of sampling is too seldom clearly in mind: the process is often determined merely by tradition. Similarly, the decision on the pattern of samples follows arguments similar to those in the exploration stage (sec. 14.4), but on the smaller scale of the individual mine. Decisions on procedures for blocking ore depend again on the particular purpose.

An example of how development might be controlled is provided by our investigation at the City Deep mine, South Africa, of a suggested strategy for selective mining. In this mine the gold is clustered in small ore shoots, as the locations of the 39 out of 503 development points accounting for 50 percent of the total inch-pennyweight gold content show (fig. 5.3). This clustering provides a guide to the best places at which to begin stoping outward from the development workings. As an illustration of the consequences of following the proposed guide, we made calculations on the stope samples that correspond to the grade of ore actually mined. Many strategies might be adopted. If all of the background mineralization that averages 68.79 inch-pennyweights (table 16.2) is of payable grade, all of the reef should be stoped. The strategy explained here is chosen only to illustrate a procedure.

The suggested strategy is to stope outward 50 feet from the 39 high-value points, then to re-evaluate the ore with the additional assay data obtained from the stoped ground. The distance of 50 feet was taken arbitrarily to give a 100-foot-wide block, which was assumed to be the minimum size economical to mine. The number of 39 points was arbitrarily taken to correspond to 50 percent of the value. Table 14.26 shows the results that would have been obtained in mining through following the first step of this scheme. Only 16 percent of the total 1000-foot-square area would have been mined, to yield 26 percent of the total value. After this area had been mined, in order to continue the strategy, the data would have been re-evaluated with the use of the stope samples to determine where mining should have been continued another 50 feet outward. Thus it would have been possible to extract most of the gold without mining out the entire reef had such selective mining been feasible from the requirements of ground support, etc.

Table 14.26. Results of suggested strategy for stoping in the City Deep mine, South Africa

Portion of reef	Mean gold content (in.-dwt)	Percent of total area	Percent of total value
Reef within 50 feet of high-value points	316	16	26
Remainder of reef	167	84	74

In the beneficiation stage, operations research can aid in process control. Such procedures as blending ores to provide constant mill feed and crushing to specified sizes may increase the efficiency of mills. Many types of servo-mechanisms can be installed to take corrective action when the many beneficiation processes that can readily be instrumented go outside specified limits. Hazen (1968) cites several interesting recent applications.

14.7   MANUFACTURING

Operations research and other decision-making techniques are often used in manufacturing; many instances are reported in the literature cited in the first section of this chapter. In this book only one example is given, that of an industry where geology plays a role, the making of portland cement by the Riverside Cement Co., as described by Nalle and Weeks (1960) and Nalle (1962). Other information on operations research in cement manufacturing appears in articles by Tashiro and others (1964) and Weeks (1964).

Through use of linear programing, the Riverside Cement Co. is able to use impure and less expensive raw materials and thus make better use of natural resources from a conservation standpoint, as well as to utilize varying raw materials available for purchase at different times. The application of linear programing is possible only because the many calculations, simple in principle but tedious to compute, are performed by electronic computers.

Portland cement (Clausen, 1960) is a product with definite physical and chemical properties, manufactured to specifications that become more rigid every year. The specifications are set by local, state, and federal governmental units and by large private users. Cement raw materials are selected to supply the three principal components (lime, silica, and iron), which are available in about 30 different raw materials. The industry is highly competitive, and proper use of raw materials is essential for success.

The Riverside Cement Co. manufactures cement, more than 95 percent of which is composed of six oxides (Al_2O_3, SiO_2, Fe_2O_3, CaO, Na_2O, K_2O), water, and carbon dioxide. The cement contains these oxides in specified percentages within narrow tolerance limits. These oxides can be obtained from twelve raw materials, each of which contains all of them, but in different proportions; these raw materials, listed in table 14.27, vary from time to time in cost and availability, and only some of them are used to make any one batch of cement.

When the oxide composition and total tonnage of a batch of cement are specified, seven linear equations or inequalities with twelve unknowns can be written —one linear equation for each oxide and one unknown in each equation for the required amount, which may be zero, of that raw material. If one of the twelve raw materials is in short supply or unavailable, inequalities may be added to the system. For example, if only 4000 pounds of a material containing 10 percent Al_2O_3 is available, at most 400 pounds of Al_2O_3 can be supplied from this source. If none of this material at all is available, one unknown is removed from the system of equations. Thus, for the twelve raw materials, up to 84 (7×12) inequalities can in principle be added to the system. The objective function is the sum of the products of the amount of each raw material to use, multiplied by the cost of that material. The linear-programing solution (sec. 14.2) finds

TABLE 14.27. Tons of material used by the Riverside Cement Co. to make 20,000 tons of clinker*

| | | | Solution number | | |
Material	Relative cost	1	2	3	4
1 Limestone A	50				13,248
2 Limestone B	20				
3 Limestone C	6	23,422	23,452	21,352	
4 Siliceous limestone A	6				
5 Siliceous limestone B	6			3,052	10,796
6 Magnesian limestone	6	2,592	3,028	3,232	3,340
7 Shale A	6	1,469			
8 Shale B	6		616	1,944	1,894
9 Shale C	20				
10 Shale D	6	2,573	2,866		
11 Clay	89	371	522	992	992
12 Iron ore	90	395	396	363	336
Total relative cost		1.00	1.05	1.20	3.51

* After Nalle and Weeks, 1960, p. 1004.

the minimum-cost solution stating how much of each raw material to use, given all of the specifications and restrictions.

Table 14.27 gives Nalle's solutions for four different situations: the first, in which all twelve raw materials are available in unrestricted amounts; the second, when material 7 is unavailable; the third, when materials 7 and 10 are unavailable; and the fourth, when materials 2, 3, 7, and 10 are unavailable. The first situation, with all materials available in unrestricted amounts, naturally yields the lowest cost solution, the relative cost being 1.00 and six of the twelve raw materials being used.

In the remaining three situations, with more and more restrictions on the availability of raw materials, costs rise. In the second situation, with shale A unavailable, it is necessary to substitute some shale B, more of shale D, and magnesian limestone—all at the same relative cost of 6. However, the total cost increases, because 151 tons of expensive clay must be added although 73 tons of cheap material is subtracted, for a net gain in raw material of 78 tons. In the third situation, with shales A and D unavailable, even more of the expensive clay must be used, although 32 tons less of expensive iron ore is needed. The total amount of raw material needed increases by 113 tons and the amount of clay by 631 tons. Finally, in the fourth situation, limestones B and C and shales A and D are unavailable; therefore more siliceous limestone B is used, but not too much can be tolerated; consequently the difference must be made up with the high-cost limestone A, driving up the cost to 3.51, a presumably unacceptably high figure, suggesting that the company geologist had better hunt up another source of cheap limestone.

Operations-research techniques are used by some companies to schedule the distribution of cement as well as to control its manufacture. For instance, the Atlantic Cement Co. manufactures cement at a plant on the Hudson River near Albany, New York, and schedules its distribution by barge to terminals along the Atlantic coast from Maine to Florida by the critical-path method (sec. 14.2).

14.8 SURVEYS OF LARGE AREAS

A geological or natural-resources survey of a continent, subcontinent, or other large area may be made by a government, a business corporation, or another entity. For surveying a large area, operations-research methods are used to select suitable techniques from those exploration techniques explained in Chapter 12, especially in section 12.4 on the exploration stage. Some other problems peculiar to the survey of large areas arise and are discussed in this section, which is largely based on the work of J. C. Griffiths and his students at Pennsylvania State University.

In the survey of a large area, regardless of how it is made, an essential requirement is complete coverage, whether many or a few detection devices are used and whether all or a few commodities are sought. For instance, after World War II, a program searched the United States for one substance, uranium, by a wide variety of methods, including surface geology, drilling, and ground and aerial geophysics. Alternatively, one method may be used, as when an aeromagnetic survey is made to find magnetic anomalies that may reveal regional structure, an iron-ore deposit, an amphibolite body, or something else. In the Soviet Union surveys have been made by large teams of amateur prospectors, providing wide coverage, although the efficiency of search is uncertain.

In surveying a large area, the governing principle is that at first the least expensive beneficial method should be used, for example, airborne geophysics or ground geochemistry, the aim being to eliminate unpromising areas early in the program, especially if a time penalty is involved.

Exploration of a Large Area by Grid Drilling

Griffiths (1966, 1967) proposes that large areas be explored by grid drilling. He writes (1966, p. 189),

Exploration for natural resources has followed traditional trial-a.:d-error procedures from the original prospector and burro to the latest airborne detection devices, and each kind of resource tends to be associated with an individual type of search program. If the real-world objective of "making a profit" is substituted for the resource as the ultimate target of endeavor, then a systems approach, in which the kind of target is immaterial and all targets are equally welcome, appears to be most efficacious.

From the values and distributions of targets (sec. 12.3) in a well-explored area, one can estimate what is likely to be found elsewhere. For oil and gas reservoirs in the United States, Lahee (1962) compiled statistics for new field wildcat wells drilled from 1944 to 1962 with and without technical advice; Griffiths and Drew (1964) plotted these data (fig. 14.16) and fitted linear regression lines. About these results, Griffiths (1966) writes,

. . . wildcat wells drilled on the basis of technical advice . . . consistently yield a 1 : 8 success ratio over a period of 18 years, 1944 to 1962. . . . Wildcat wells drilled without advice were not nearly so successful and were much less consistent, varying over the same period from 1 : 12 to 1 : 39. The fact that the success ratio of 1 : 8 has been so consistent throughout this relatively long period is attributed to improvements in technology offsetting the increasing difficulty and increasing expense of exploration. It also suggests that a breakthrough to a superior success ratio is unlikely from technological improvements alone and that some entirely new approach will be required to prevent the inevitable decline foreshadowed by the increasingly rare and more difficult-to-locate targets.

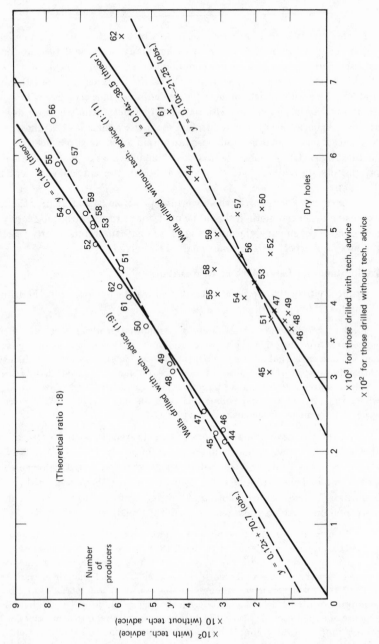

Fig. 14.16. New field wildcat wells drilled from 1944 to 1962 with and without technical advice (after a plot by Griffiths and Drew, 1964; data compiled by Lahee, 1962).

Based on these considerations and on models more or less like those discussed in section 12.3, Griffiths and Drew suggest that the United States be drilled on a grid spacing of 20 miles (1966, p. Q-1). The exact spacing is more or less arbitrary and might be modified if this program is adopted.

Expected Results of Grid Drilling the United States

As a specific example of how their proposed strategy would work, Griffiths and Drew (1966, p. Q-1) consider the consequences of drilling the continental United States, except Alaska. They write that

> . . . for example, given a continental area like the U.S.A. of 3000 by 1000 miles in size and a spacing of say 20 miles, the grid would require some 7500 wells; supposing each well was drilled to 15,000 feet at say $250,000 each, then the cost of the program would be some $1,875 \times 10^6$. Any target exceeding 28 miles in extent would be found by this grid with probability one, and such large targets yield a return of several billions of dollars so that the program appears commercially feasible.

Griffiths and Drew believe that this program would find five major oil fields in the United States to add to the 20 already known, and that the value of the new fields would be about $5 billion, well above the $1,875 billion cost of drilling and the costs of land acquisition and development. In addition, they expect that other valuable natural resources would be found.

Besides the natural resources found directly, a major gain from the suggested program is that the geology would be defined to the proposed depth of 15,000 feet with the resolution of a 20-mile grid. Through this systematic drilling, rock formations and geologic structure would be known much better. Areas whose geology is too complicated to be reasonably well defined on this scale would be identified; a continental map on this generalized scale would be obtained. Although one may quibble with Griffiths' (1966, p. 208) reference to the plan as an "equal information network," since different amounts of information are gained from such a region of simple geology as the mid-continent region compared with one of complicated geology as crystalline rocks in New England, the gain in geologic knowledge would certainly be large.

To justify Griffiths' proposal, one might try to answer the question: What is the value of the natural resources underlying one square mile of the earth's crust ? Griffiths and others (1965) made a preliminary analysis of this question for the state of Kansas, where reasonably good data are available. The following conclusions were developed by this study:

1. If the accumulated dollar value of all mineral industries' production in Kansas from 1880 to 1963 is divided by the total area of the state, the average value per square mile is $100,000. If production of commodities that could be produced at present is projected for the 20-year period 1960 to 1980, an

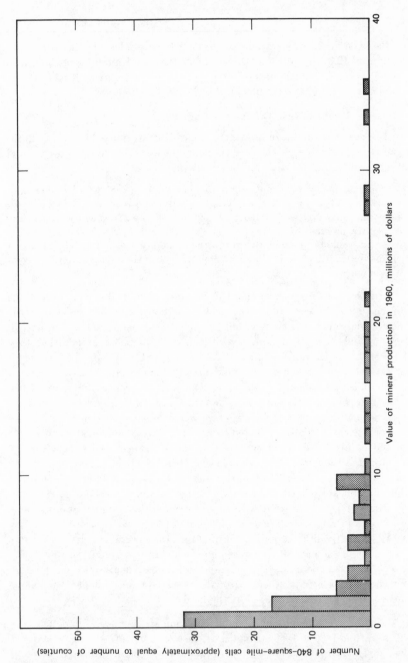

Fig. 14.17. Histogram of mineral production for 1960 of average-size counties in Kansas (data from Griffiths and others, 1965).

additional value per square mile of $120,000 is added, for a total value of $220,000.

2. The county-by-county variability is large, as demonstrated by figure 14.17, which is a histogram of production for the year 1960 per county of average size (840 square miles) based on a $14 \times 7 = 98$ grid superimposed on Kansas, so that one grid point corresponds to one of the 105 counties, except for a few that are exceptionally large or small.

3. The variability stems from at least two sources: some counties have not been thoroughly explored, and some counties are richer than others in natural resources.

4. An index of how much has been spent on exploration should be prepared for each county, for Kansas an easy task because most exploration is sub-surface, and the basis would be cost per foot of wildcat wells per county divided by cubic miles to explore above the Precambrian basement.

5. Although the location of the Precambrian suboutcrop is fairly well known, little is known about the geology of the Precambrian basement. As a model, one might take a well-explored section of the Canadian shield of as nearly similar lithology as possible. The basement is a major discontinuity for resource appraisal as well as for geology.

REFERENCES

Ackoff, R. L., 1962, Scientific method—optimizing applied research decisions: New York, John Wiley & Sons, 464 p.

Ackoff, R. L., and Rivett, Patrick, 1963, A manager's guide to operations research: New York, John Wiley & Sons, 101 p.

Bader, J. W., 1965, Mineral property evaluation with computers, *in* Proc. 5th Ann. Symp. on Computer Applications in the Mineral Industry: Tucson, Ariz., Univ. of Arizona, p. DD-1–DD-17.

Bailly, P. A., 1964, Methods, costs, land requirements, and organization in regional exploration for base metals: Preprint of paper presented at mtg. Am. Inst. Mining Engineers, Alaska sec., Fairbanks, March 18–21, 1962, 31 p.

———, 1967, Mineral exploration and mine developing problems: Statement presented to the Public Lands Law Conf., Univ. of Idaho, Boise, 44 p.

Baumol, W., Jr., 1965, Economic theory and operations analysis: Englewood Cliffs' N.J., Prentice-Hall, 586 p.

Behre, C. H., Jr., and Arbeiter, N., 1959, Distinctive features of the mineral industries, *in* Robie, E. H., ed., Economics of the mineral industries: New York, Am. Inst. Mining Engineers, p. 43–79.

Brooks, D. B., 1966, Strip mine reclamation and economic analysis: Nat. Resources Jour., New Mexico Univ. School of Law, v. 6, p. 13–14.

Carlisle, Donald, 1954a, The economics of a fund resource with particular reference to mining: Am. Econ. Rev., v. 44, p. 595–616.

———, 1954b, Economic aspects of the definition of ore: Trans. Inst. Mining Metallurgy, v. 64, p. 89–99.

Churchman, C. W., Ackoff, R. L., and Arnoff, E. L., 1957, Introduction to operations research: New York, John Wiley & Sons, 645 p.

Clark, R. W., 1962, The rise of the boffins: London, Phoenix House, 253 p.

Clausen, C. F., 1960, Cement materials, in Gillson, J. L., ed., Industrial minerals and rocks: New York, Am. Inst. Mining Engineers, p. 203–232.

Coyle, R. G., 1969, Review of the literature on operational research in the mining industry: Trans. Inst. Mining Metallurgy, v. 78, no. 746, p. A1–A9.

Crutchfield, J. A., Kates, R. W., and Sewell, W. R. D., 1967, Benefit cost analysis and the National Oceanographic Program: Nat. Resources Jour., New Mexico Univ. School of Law, v. 7, p. 361–375.

Drew, L. J., 1967, Grid-drilling exploration and its application to the search for petroleum: Econ. Geology, v. 62, p. 698–710.

Duckworth, Eric, 1962, A guide to operational research: London, Methuen & Co., 128 p.

Eckstein, Otto, 1958, Water resource development: Cambridge, Mass., Harvard Univ. Press, 300 p.

Erickson, J. D., 1968, Long-range pit open planning: Mining Eng., v. 20, p. 75–78.

Forrester, J. W., 1961, Industrial dynamics: New York, John Wiley & Sons, 464 p.

Griffiths, J. C., 1966, Exploration for natural resources: Jour. Operations Res. Society Am., v. 14, p. 189–209.

———, 1967, Mathematical exploration strategy and decision-making, in Proc. 7th World Petroleum Cong., p. 599–604.

Griffiths, J. C., and Drew, L. J., 1964, Simulation of exploration programs for natural resources by models: Colorado School Mines Quart., v. 59, no. 4, pt. A, p. 187–206.

Griffiths, J. C., and Drew, L. J., 1966, Grid spacing and success ratios in exploration for natural resources, in Proc. Symp. and Short Course on Computers and Operations Res. in Mineral Industries: Mineral Industries Expt. Sta., Pennsylvania State Univ., Spec. Pub. 2-65, p. Q-1–Q-24.

Griffiths, J. C., Koch, G. S., Jr., Meeves, Henry, and Hornbaker, Allison, 1965, Estimating regional unit value, in 1965 Mineral industries seminar, Economic analysis in the mineral industries: Lawrence, Kans., State Geol. Survey of Kansas, pp. 27–32, 74–101.

Groundwater, T. R., 1967, Role of discounted cash flow methods in the appraisal of capital projects: Trans. Inst. Mining Metallurgy, v. 76, p. A67–A82.

Hazen, S. W., Jr., 1968, Operations research, a growing force in the mineral industries: Mining Eng., v. 20, p. 88–90.

Hertz, D. B., and Eddison, R. T., eds., 1964, Progress in operations research, v. 2: New York, John Wiley & Sons, 455 p.

Hillier, F. S., and Lieberman, G. J., 1967, Introduction to operations research: San Francisco, Calif., Holden-Day, 639 p.

Hoover, H. C., 1909, Principles of mining: New York, McGraw-Hill, 193 p.

Howard, T. E., 1968, Outlook for the future in mining technology: Paper presented at the 2nd Internat. Surface Mining Conf. and Fall Mtg. of the Soc. Mining Engineers of Am. Inst. Mining Engineers, Minneapolis.

Hufschmidt, M. M., and Fiering, M. B., 1966, Simulation techniques for design of water-resource systems: Cambridge, Mass., Harvard Univ. Press, 212 p.

Hughes, R. V., 1967, Oil property valuation: New York, John Wiley & Sons., 298 p.

Johnson, E. E., and Bennett, H. J., 1968, An engineering and economic study of a gold mining operation: U.S. Bur. Mines, Inf. Circ. 8374, 53 p.

Jones, Cyril, 1968, Economic analysis for mine ventures and projects, *in* Pfleider, E. P., ed., Surface mining: New York, Am. Inst. Mining Engineers, p. 997–1013.

Kaufman, G. M., 1963, Statistical decision and related techniques in oil and gas exploration: Englewood Cliffs, N.J., Prentice-Hall, 307 p.

Koch, G. S., Jr., and Link, R. F., 1969, A statistical analysis of some data from deflected diamond-drill holes, *in* Weiss, Alfred, ed., A decade of digital computing in the mineral industry: New York, Am. Inst. Mining Engineers, p. 497–504.

Lahee, F. H., 1962, Statistics of exploratory drilling in the United States, 1945–1960: Am. Assoc. Petroleum Geologists, 135 p.

Lindgren, Waldemar, 1933, Mineral deposits: New York, McGraw-Hill, 894 p.

Maass, Arthur, et al., 1962, Design of water resource systems: Cambridge, Mass., Harvard Univ. Press, 620 p.

McCloskey, J. F., and Coppinger, J. M., eds., 1956, Operations research for management, v. 2: Baltimore, The Johns Hopkins Press, 535 p.

McKinstry, H. E., 1948, Mining geology: Englewood Cliffs, N.J., Prentice-Hall, 680 p.

Moder, J. J., and Phillips, C. R., 1964, Project management with CPM and PERT: New York, Reinhold, 283 p.

Nalle, P. B., 1962, Ore blending and process control, *in* Symposium on mathematical techniques and computer applications in mining and exploration: Tucson, Ariz., College of Mines, Univ. of Arizona, v. 2, sec. 02, 14 p.

Nalle, P. B., and Weeks, L. W., 1960, The digital computer—applications in mining and process control: Mining Eng., v. 12, p. 1001–1004.

Parks, R. D., 1957, Examination and valuation of mineral property: Reading, Mass., Addison-Wesley, 507 p.

Peters, W. C., 1966, The pre-production interval of mines: Mining Eng., v. 18, no. 8, p. 63–64.

Raymond, L. C., 1959, Valuation and mineral property, *in* Robie, E. H., ed., Economics of the mineral industries: New York, Am. Inst. Mining Engineers, p. 131–162.

Redmon, D. W., 1964, Determining mine-production schedules by linear programing: U.S. Bur. Mines, Rept. Inv. 6441, 48 p.

Samuelson, P. A., 1967, Economics: New York, McGraw-Hill, 821 p.

Shaffer, L. R., Ritter, J. B., and Meyer, W. L., 1965, The critical-path method: New York, McGraw-Hill, 224 p.

Slichter, L. B., 1960, The need of a new philosophy of prospecting: Mining Eng., v. 12, p. 570–576.

Soderberg, Adolph, 1959, Elements of long range open pit planning: Mining Cong. Jour., v. 45, no. 4, p. 54–58.

Tashiro, E., Nakagawa, T., and Hasegawa, J., 1964, Chichibu Cement Company's new plant at Kumagaya, Japan, is now under real time computer control: Colo. Sch. Mines Quart., v. 59, no. 4, pt. A, p. 465–492.

Taylor, G. A., 1964, Managerial and engineering economy: Princeton, N.J., D. van Nostrand Co., 433 p.

Truscott, S. J., 1962, Mine economics: London, Mining Pubs., 471 p.

Weeks, L. W., 1964, Computer control in cement plant operations, *in* Computers in the mineral industries, pt. 1: Stanford, Calif., School of Earth Sciences, Stanford Univ., p. 306–311.

Weiss, Alfred, 1968, Application and impact of computers in the mining industry: Paper presented at annual mtg., Am. Inst. Mining Engineers, New York.

Chapter 15

Specialized Geological Sampling

In this chapter three more sampling methods are explained in addition to those in Chapter 7, and a study of several methods of sampling a mine is outlined. The sampling methods are sampling of broken rock, sampling of placers, and sampling in exploration geochemistry. Like Chapter 7, this chapter is applied rather than theoretical, and the methods are specific to particular problems. The purpose is to discuss practical sampling problems not only for those who use these specific techniques but also for those who must devise techniques for other sampling problems.

Sampling of broken rock is related to chip sampling (sec. 7.5), because a large sample of chips may contain as much material as a sample of broken rock. In applied geology there are many occasions for sampling broken rock, which may be broken ore in stopes, chutes, or cars; millheads and other mill products; the rock from dumps and tailing piles; rock broken in road cuts; or rock from quarries. In theoretical geology broken rock is sampled less often. More occasions arise now, however, than in the past, with the recognition that the composition of a large rock body may sometimes be better determined with a sample larger than one or a few hand specimens, and with the increasing availability of the necessary mechanical equipment.

Although placer sampling is a logical subdivision of sampling of broken rock, it is explained separately in section 15.2, because it is so specialized. Many placers are geologically stratified, and the item of interest, such as gold or diamonds, is always concentrated in a sample-preparation process that is often the most influential part of the sampling plan.

15.1 SAMPLING OF BROKEN ROCK

For many years sampling of broken rock, particularly by "grab" samples, which suggests a haphazard picking-up of rocks, was in disrepute in geology and also in the mining industry. But in recent years, for many applications, the disadvantages of not breaking out a sample of rock in place have been more than compensated for by the advantages of having a large bulk of material in a sample of broken rock and the small cost of collection, always provided that the sample of broken rock can be satisfactorily reduced in volume prior to chemical or other analysis.

In sampling broken rock, the exact method of sample collection has a critical influence on the statistical analysis. Broken rock or ore is usually sampled by arbitrarily picking out one or a few pieces. Although sometimes an effort is made to take pieces within a certain size range, more often pieces are taken that the geologist or sampler finds convenient in size or location. Sometimes a system is introduced. For instance, in sampling open cars, one practice is to lay down a net on top of each car and sample at the points where the strings cross.

Instead of sampling one or a few pieces of rock, many pieces may be collected in a bulk sample. If large bulk samples are required, road-construction equipment may be needed, including a backhoe, trucks, and power equipment for crushing and splitting. A backhoe can excavate a trench 15 to 20 feet deep in loose material or can obtain rock or soil from a road cut. The material taken out can be processed through a truck-mounted crushing plant set up at the site or nearby, and a fraction can then be split out (sec. 8.4).

Two types of sampling of broken rock, not discussed in this book, are sampling of millheads and other mill products and sampling of dumps. Sampling of mill products is usually done by mechanical devices that are used throughout the manufacturing and chemical industries. Duncan (1962) explains the statistical problems involved in coal sampling, a representative example of sampling of mill products. Sampling of dumps presents special problems, well discussed for chromite dumps by Wilson (1946), who states that the main problem is to establish a reliable relationship between tonnage and grade (sec. 13.5).

Theory and practice of sampling broken rock are far apart. We discuss first the practical methods and then the theory.

Practical Methods

The practical methods of sampling broken rock depend on taking advantage of the sizing of rock fragments, because the different sizes nearly always contain different proportions of the constituent of interest. For example, figure 15.1 shows the percentage of copper in ore of three sizes versus percent-

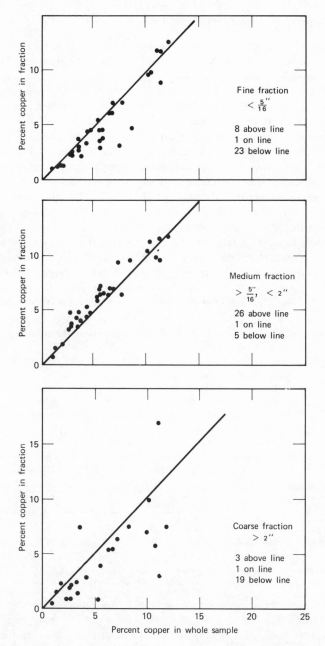

Fig. 15.1. Percentage of copper in ore of three sizes versus percentage in the whole sample at the Nchanga copper mine, Zambia (after MacKensie, 1966, p. 370).

age in the whole sample from the Nchanga, Zambia, copper mine. As size increases, the sample bias changes from negative to positive and back to negative, and the variability increases.

The sampling can be stratified in either of two ways to take advantage of the sizing of rock fragments. The first way is (1) to classify the broken rock according to size in order to determine, by chemical or other analysis, the percentage of a certain constituent present in each size fraction, and then (2) to combine these percentages, weighting by the tonnages of the different classes sampled. The second way is always to sample one class of convenient size and adjust the observations by multiplying by a constant.

To illustrate these procedures, we apply them to data for gold, silver, copper, and zinc from the Quemont mine, Quebec, Canada (McLachlan and others, 1954), for which grades and weights of different size-fractions were available. An arbitrary number was assigned to each mesh size in the Tyler screen series (from 4 inches to 200 per inch), and a straight line was fitted by linear regression (sec. 9.2) to relate this number to the grade for that mesh size. From the weight and grade of ore in each mesh size, the over-all average for the ore could be calculated. Using this weighted average in the equation for the straight line, we could solve for the mesh size whose ore was most nearly the average, to yield the sizes in column 2 of table 15.1. If one size near the middle is taken,

TABLE 15.1. CALCULATION OF GRADE FOR SIZED FRACTIONS OF ORE FROM THE QUEMONT MINE, QUEBEC, CANADA

			Grades		
Metal and units (1)	Calculated size of rock to sample (2)	Constant for rock of size 5 (3)	Of rock of size 5 (4)	Estimated from rock of size 5 (5)	From weighted average of crusher discharge (6)
Gold, ounces/ton	6.1	0.005	0.14	0.145	0.14
Silver, ounces/ton	4.3	− 0.02	1.06	1.04	0.87
Copper, percent	4.9	− 0.005	1.53	1.525	1.29
Zinc, percent	3.6	− 0.1	2.9	2.8	2.18

namely 5, corresponding to a mesh size of 1 to $1\frac{1}{2}$ inches, the adjustment constants in column 3 can be obtained. Applying these constants to the observed grades of ore of size 1 to $1\frac{1}{2}$ inch (column 4) yields the estimates in column 5. In column 6 are the best estimates based on all the ore. For all metals, the predictions are somewhat high, and examination of the original data shows that, for these particular data, 1- to $1\frac{1}{2}$-inch size ore is a little richer than average.

For these Quemont data the method works fairly well; in practice many more data should be studied for a mine, ore body, or other geological entity. For these data, the straight lines were fitted so that the grades of each fraction were related to actual mesh sizes and to areas of screen openings, as well as to the arbitrarily assigned numbers. The arbitrary numbers happened to be best for these particular data.

Appraisal of Broken-Ore Sampling

As with any kind of sampling, the best way to appraise broken-ore sampling is to compare it with another, preferably better, method of sampling. The procedure may be demonstrated for some data from the Giant Yellowknife gold mine, Northwest Territory (Dadson and Emery, 1968, p. 10), used to compare sampling of broken ore in cars with chip sampling. Table 15.2

TABLE 15.2. COMPARISON OF SAMPLING BROKEN ORE IN CARS WITH CHIP SAMPLING, GIANT YELLOWKNIFE GOLD MINE, NORTHWEST TERRITORY

| Items compared | Degrees of freedom | Calculated t-values | | $t_{10\%}$ |
		Grade	Gold content	
All stopes	25	-0.09	1.76	1.708
B-shaft stopes	20	0.75	1.76	1.725
C-shaft stopes	4	-0.81	1.49	2.132

compares by a paired t-test these two kinds of sampling for 26 stopes and for two subdivisions of stopes according to location. Gold grades are in remarkably good agreement. Disregarding weighting by tonnage, the grade estimated for all stopes from chip sampling is only slightly lower than that estimated from broken-ore sampling (negative calculated t-value). On the other hand, the gold contents estimated from chip sampling in all stopes and in those from shaft B are significantly larger than those from car sampling.

Theoretical Studies

Various investigators have approached the general problem of geological sampling through theoretical studies of sampling broken ores. Starting with the distributions of individual particles assumed to be all of the same size, they are led to binomial and multinomial distributions (sec. 6.1), whence they consider compound binomial distributions and other complicated results for particles of various sizes. Becker (1964–1968, 1968) and Gy (1968) have been active in these studies in recent years.

Although these studies are of considerable intrinsic interest, their predicted results seldom agree with practice. The investigators necessarily assume that the material studied is basically homogeneous. However, it has been shown repeatedly that for most rocks, even those nearly homogeneous, the amount of small-scale variability (which is the sort that theoretical studies of particle statistics could hope to predict) is negligible compared to the amount of large-scale variability. Therefore the large-scale variability that theory has assumed to be zero dominates practical sampling.

Extending an already-involved theory in order to consider large-scale variability is extremely difficult, because even the simplest models will complicate the theory even further, and sophisticated models lead to a theory that is mathematically intractable.

Fortunately, empirical experience alone can provide a guide to action for any situation. This experience need not be extensive and can be generalized from a local level to a more extensive geographical area, since minerals and rocks display characteristic types of variability. The empirical method, which embodies taking a statistical sample that is properly randomized over a correct target or sample population, furnishes workable answers to real sampling problems, without the obstacle of overcoming purely mathematical difficulties that are inherent in the theoretical approach.

15.2   PLACER SAMPLING

For particular kinds of geological deposits and sampling requirements, many specialized sampling methods have been devised. One of these, placer sampling, is discussed in this section to illustrate the kinds of problems that can arise in specialized sampling and some procedures for dealing with such problems. Another reason for this discussion of placer sampling is its importance in applied geology and its potential, at present not realized, in theoretical geology.

There are several specialized aspects of placer sampling. First is the interesting viewpoint, firmly held by engineers who have dominated this field rather than geologists, that the sampling method should closely resemble the mining method, so that only the material that can be recovered in mining will be sampled. Thus, in sampling placer gold, samplers aim for a concentration method like that used on a dredge, and the fine-sized gold is lost just as it is lost from the dredge. Second, the geological samples must be concentrated before the grade is determined in order to improve accuracy and because in most placers the unconcentrated gravel does not contain enough gold (or other material of interest) from which to determine grade by available analytical methods. Third, since placer engineers traditionally use many arbitrary "safety" factors at different places in their calculations, the sources of variability at different stages in the sampling operation are masked and it is difficult or impossible to evaluate the different sources of variability. Malozemoff (1939, p. 47) writes,

A prevalent and highly questionable practice of evaluating the placer recoveries is that of comparing the amount of gold recovered by actual operation with that determined to be present in the ground by preliminary sampling. This yields recovery percentages varying widely from 50 to 200 percent or more—figures that really evaluate nothing, and at best, when they are proudly claimed to be just more than 100 percent, they merely give tribute to the conservative empirical wisdom of the man who sampled the ground, who made a good guess at the proper allowance for losses in the tailings of the projected dredging operation.

Placer exploitation, a specialized art in little contact with other fields of geology and engineering, has been described in detail in numerous works. Good accounts, with references to the voluminous older literature, are provided by Griffith (1960), Graves (1939), Romanowitz et al. (1970), and Truscott (1962), and in the proceedings of a Symposium on Opencast Mining, Quarrying, and Alluvial Mining (1965).

Types and Exploitation of Deposits

Lindgren (1933, pp. 213–251) and the authors cited above describe the geological environments of placer deposits. The discussion in this book is restricted to deposits in water-laid gravels, mostly those deposited in streams and found either in present-day stream channels or on benches, but including beach gravels as well. The valuable substances in most placers are gold, cassiterite, or diamonds; but many other minerals that are chemically inert and not too friable are also recovered from placers.

Today practically all placers are mined by dredges, an extremely low-cost method. Thus the grade of the valuable substances may be extremely low; at costs prevailing today in the United States, one can dredge gravel containing 25 cents a yard in gold, equivalent to 0.33 gram or 0.017 cubic centimeter of gold per cubic meter, which is only 17 parts per billion by volume. Previously, mining was also done by spraying high-pressure streams of water at the gravel face (hydraulicking) or with the aid of other devices, such as sluice boxes, all of these depending on gravity concentration under running water.

Knowledge of how the valuable mineral or minerals are distributed vertically or horizontally in a placer may help in devising a sampling plan. With regard to vertical distribution, figure 15.2 shows an increase downward of gold concentration in a terrace gravel in the Klondike area, Yukon Territory, Canada, with 86.6 percent of the gold being within 6 feet of bedrock, where the grade reaches 450 parts per million or 735 cents per cubic yard. However, another concentration of gold was found about 45 feet above bedrock, localized above

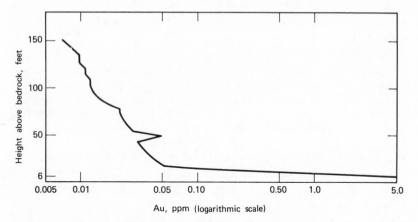

Fig. 15.2. Increase in gold downward in high terrace bench gravels in the Yukon Territory, Canada (data from McConnell, 1907, p. 220).

a *false bedrock*, which is a clayey or other compact stratum that the gold cannot settle through. A sampling plan should take advantage of such a vertical distribution, once it is recognized.

Horizontally, a change in mineral composition may be found. At any one place laterally in a channel, composition is rather constant, presumably because of mixing in sedimentation. This fact is of economic importance when mining gold, because native gold can contain from a trace to 50 percent silver, in addition to base metals. One or two determinations from a given placer are all that are said to be needed (Mertie, 1940). For cassiterite placers, the tin content of the cassiterite is also said to be constant within a given placer. However, the composition of individual minerals can vary laterally if a long-stream channel is involved, as Ferguson (1916) found for placer gold at Manhattan, Nevada. In figure 15.3 Ferguson's data are plotted with fitted linear regression lines (sec. 9.2), which are evidently suitable because the variation with distance is nearly linear (the correlation coefficient, r, is nearly equal to $+1$ for gold and -1 for silver). Gold content of the nuggets (fineness) increases downstream, and silver content decreases because silver and base metals are leached out of the nuggets.

Obtaining Samples

Placer samples are obtained from drill holes or from pits, shafts, or drifts excavated by hand or by machinery. Unless caissons are sunk, an uncommon procedure because of the excessive cost, drills are always used for sampling below the water table; above the water table any of the sampling methods listed can be used.

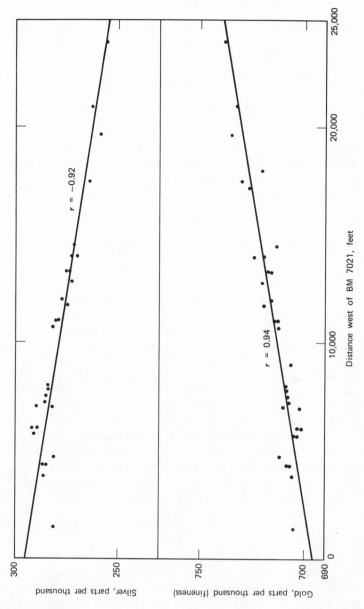

Fig. 15.3. Variation in gold and silver fineness with distance downstream (westward) from source, Manhattan, Nevada (after Ferguson, 1916).

Holes are usually bored with a churn drill. When this method of drilling is used to sample placer deposits, a hollow pipe or casing, usually 4 to 6 inches in diameter, is driven by successive blows into the ground to be sampled; any boulders too large to enter the casing are displaced or broken. The ground inside the casing is then broken by repeated blows with a weighted, chisel-shaped bit that is lifted and dropped by gravity, breaking the ground beneath it. The cuttings are lifted to the surface by a bailing device. The advantage of churn drilling is that the casing prevents contamination from the walls; however, if the water at the bottom of the casing is under pressure, sand may run into the hole there. Volume calculations, explained in the cited references, are made to determine whether sand has run in and also to keep track of the tonnage-volume factors involved (sec. 13.3).

Concentrating Samples and Estimating Grades

Because placer samples are always concentrated as an essential part of the sampling process, this subject is explained here, although it also relates to section 8.4 on variability in sample preparation. Samples must be concentrated because the proportion of valuable minerals in them is otherwise too small for an accurate determination of grade, because splitting is not satisfactory when grain size ranges from a micron up to coarse boulders, and because the sampling method is intended to duplicate the recovery process in mining.

The simplest concentrating device is the gold pan, with which an experienced man can recover nearly all of the valuable mineral or minerals, the test being for another worker to pan the reject material. The gold pan works on the principles of gravity and centrifugal separation. Not only for concentrating minerals from placers but for making heavy mineral concentrates from disintegrated rocks in general, the gold pan is a valuable replacement for or adjunct to heavy (high-density) liquids. Besides, gold panning is more fun than fishing. Mertie (1954) and Theobald (1957) have written about the gold pan as a geological tool and about heavy mineral recoveries that can be attained.

Heavy minerals may be recovered in a variety of other ways, including sluice boxes, vibrating riffles, screens, and jigs (see previously cited references). Of particular use to geologists who wish to recover heavy minerals from soils, saprolites, and gravel beds is the Gold Saver, a machine manufactured by the Denver Equipment Company that will fit in a pickup truck and concentrate heavy minerals out of about 2 cubic yards of materials per hour through a rotating cylindrical screen (a scrubbing trommel) and a vibrating riffle.

These recovery devices utilize methods similar to those used by commercial placer-mining recovery plants. Only material is saved that can be recovered by these plants. For example, gold entrapped in sulfide or other minerals is not evaluated because it could not be recovered by the mining methods conventionally used for placers.

Data Analysis

First, statistical analysis of data conventionally obtained is discussed, and then some suggestions are made for other data that might profitably be collected to afford material for more incisive analyses.

Data from churn-drill holes may be compared with data from test pits sunk at the same places. Table 15.3 illustrates the procedure with data from a

TABLE 15.3. COMPARISON OF ESTIMATES OF GRADE FROM CHURN-DRILL HOLES AND TEST PITS IN A CALIFORNIA GOLD PLACER*

Depth of hole (ft)	Grade estimate (ppm)†	
	From churn drilling	From test pit
12	0.042	0.071
25	0.372	0.383
29	0.103	0.071
25	0.101	0.547
15	0.175	0.157
14	0.150	0.279
17.5	0.055	0.077
15	0.144	0.134
Mean	0.143	0.215
Standard deviation	0.103	0.174
Coefficient of variation	0.72	0.81

* Data from Janin, 1918, p. 30.

† Converted from cents per cubic yard in original paper, using an approximate specific gravity of 1.8.

California gold placer; a paired t-test of the grades of the two types of samples yields a value of -1.28, which is outside the critical region defined by the two-tailed tabled value of 1.895 with 7 degrees of freedom at the 10 percent significance level. [To put them in units familiar to geologists, the original values in cents per cubic yard have been converted to parts per million, although cents per cubic yard is the practical unit for placer mining (sec. 13.3).] Gardner (1921) compares many drill and pit samples of placer gold, with results similar to those in table 15.3, although Gardner's t-values tend to be negative and inside the critical region (for example, $t = -4.87$, with 14 degrees of freedom, for another California deposit). In any area, one might expect higher-grade samples from pits than from drill holes, because of better gold recovery from the pits, and a lower coefficient of variation, because of the larger sample volume. Most of the few pit samples that we have studied are higher in grade, but no consistent pattern is shown by the coefficients of variation.

Placer sampling is usually evaluated by comparing estimated grade with recovery in mining, but Malozemoff (1939) points out that this comparison is not meaningful because, unlike a concentrator at a hard-rock mine, a dredge in a placer mine has no adequate provision to determine how much of the valuable mineral is lost in the tailings. In fact, many placer deposits have been redredged two, three, or even more times, with profitable recovery each time. Therefore, as table 15.4 shows for data for hydraulic and bulldozer mining of

TABLE 15.4. COMPARISON OF ESTIMATES AND MINED GRADE FOR
DREDGING IN THE YUKON*

	Grade (ppm)†		Percent of estimated grade mined
Year	Estimated	Mined	
1952	0.177	0.097	54.9
1953	0.194	0.537	277.3
1954	0.191	0.256	133.9
1955	0.359	0.686	191.0
1956	0.347	0.279	80.4
1957	0.221	0.194	87.8
1958	0.194	0.230	118.6
1960	0.267	0.352	131.6
Mean	0.244	0.323	
Standard deviation	0.073	0.193	

* Data from McFarland, 1965, p. 194.

† Converted from cents per cubic yard in original paper, using an approximate specific gravity of 1.8.

placers in the Yukon, Canada, recoveries can be very different from estimated grades, ranging from 54.9 percent to 277.3 percent in this example. The table also shows a much higher standard deviation for the mined than for the sampled grade, which first increases and then gradually declines, except for the last year. One might be tempted to compare the two kinds of data by a paired t-test, but this would not be relevant because (1) the estimated grades are evidently not calculated on a consistent basis each year (for instance, the 1958 estimate is identical with the 1957 production) and (2) the quantities are more nearly parameters than statistics, as only one quantity is obtained each year.

Data obtained in placer sampling can also be compared to learn whether or not they are internally consistent. When a panner has concentrated in his pan a sample obtained from drilling or test pitting, he estimates the number of particles (or "colors" if he is panning gold) and their weight; then, eventually, the mineral is weighed. From studying records of this procedure from a number of places, we have the impression that some panners are proficient in estimating but that arithmetic blunders occur in recording and in processes like averaging.

Interpretation of sample data from placers could be much improved, but first one would need to know distributions of particles, which today seem to be unknown or at least unpublished. For instance, figure 15.4 shows schematically

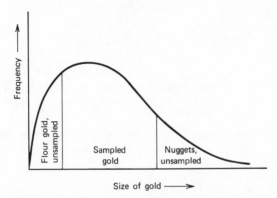

Fig. 15.4. Schematic frequency distribution of size of gold grains in placer deposits.

that, in the usual practice at present, fine-grained gold, not recovered in the gold pan, is not sampled, nor are nuggets, which are not included in the sample in order to provide a "safety factor." Lack of inclusion of nuggets explains partly why recoveries of more than 100 percent of the estimated gold are not uncommon (table 15.4). Table 15.5, listing the weight of gold in different size fractions, is characteristic of the scanty published data and suggests that about half the value is in the coarsest fractions. Unfortunately, because these data are insufficiently detailed, one cannot draw any more definite conclusions.

15.3 SAMPLING IN EXPLORATION GEOCHEMISTRY

Exploration geochemistry comprises the application of geochemical methods to search for valuable metals, petroleum, and other substances. In this section the purpose of exploration geochemistry, the nature of the data, and some

Table 15.5. Weight of gold in different size fractions from a California placer deposit*

Size fraction (screen mesh, inches)	Sample number				
	1	2	3	4	5
60 and coarser	60.85	47.25	56.3	55.67	70.23
60 to 100	12.16	16.4	17.92	20.54	9.7
100 to 120	2.13	2.67	3.57	3.72	2.7
120 to 150	22.94	2.88	2.47	2.06	1.27
150 and finer	—	30.8	19.73	18.01	16.09
Total	98.08	100.00	99.99	100.00	99.99

* After Lindgren, 1911, p. 221.

methods of data analysis are taken up in turn. The emphasis is on some special sampling problems that may benefit from statistical treatment.

Purpose of Exploration Geochemistry

The authors of the standard English-language textbook in the field, Hawkes and Webb (1962, p. 1), state the definition and purpose of geochemistry in mineral exploration as follows:

> Geochemical prospecting for minerals, as defined by common usage, includes any method of mineral exploration based on systematic measurement of one or more chemical properties of a naturally occurring material. The chemical property measured is most commonly the trace content of some element or group of elements; the naturally occurring material may be rock, soil, gossan, glacial debris, vegetation, stream sediment, or water. The purpose of the measurements is the discovery of abnormal chemical patterns, or *geochemical anomalies*, related to mineralization.

In the search for ore bodies with exploration geochemistry, the geological principle exploited is that many ore bodies are surrounded by a halo of certain chemical elements, including some of those present in the ore body, whose abundance is above the average in the country rocks of the area. Because the intensity of such a halo varies with details of the geology, degree of erosion, and temporal effects such as the season, and because detection of a halo depends on the details of the geochemical method used, the relative rather than the absolute sizes of the observations define the halo. Moreover, as a further consequence of the uncertainty of the data, they often suggest that an ore body is present when one is not, just as large magnetic observations may suggest the presence of magnetite when none is present.

Thus the statistical problem in geochemical exploration becomes one of how to gather, analyze, and interpret data so that the false-alarm rate is not excessively high, but also so that the chance of missing an ore body is small. In this section we discuss techniques for dealing with some of these problems. Many unsolved statistical problems remain.

Two conceptual approaches to geochemical exploration are possible. In the first approach trend-surface analysis (chapter 9) is used to seek a maximum, since, as a geochemical survey approaches the object sought, the values of some trace elements should gradually increase in size. In the second approach, *anomalies*, which are geographic clusters of unusually high (or, rarely, unusually low) values, define the halos. This second approach, which is generally more useful, is the only one considered in this section.

In finding anomalies, the relative rather than the absolute sizes of the observations are important; therefore, unlike most geological sampling, the primary interest is not in the mean, and the bias, which may be introduced through transforming observations, is not objectionable (sec. 6.3). Anomalies are found by setting up *threshold values* based on *orientation surveys* of unmineralized areas. Geographically clustered values that are higher (or rarely lower) than the threshold values define the anomalies. It is important to recognize that, at least in the present state of the art, the sizes of ore bodies are unrelated to the sizes of the anomalies (Miesch, 1969).

In the past, exploration geochemists have paid much attention to analytical methods for more accurate determination of chemical elements and other substances, less attention to problems of field sampling, and little attention to statistical analysis of the data. But in order to find anomalies that are small in size, small in value, or poorly defined geographically, good sampling and data analysis are required. That the need for statistical analysis is recognized by some investigators was remarked on at a recent International Geochemical Exploration Symposium by Erickson (1969), Boyle (1969), Webb (1969), and others.

Nature of Data from Exploration Geochemistry

The conceptually simplest procedure is to collect the data on a grid, which is usually square. At each station, one or more geological samples—which may be soil, rock, water, or other material—are taken. They are then chemically analyzed, sometimes at the station but usually in the laboratory. The largest expense incurred is in reaching the field area, and most of the remaining expense is in occupying the station, an activity that includes recording its location, laying out a compass line to the next station, etc., as well as collecting the sample. Therefore, taking several samples spaced, say, a few feet apart, rather than only one sample at a station, costs little more and the added information may be large.

Figures 15.5 and 15.6 are frequency distributions of typical geochemical data, representing the logarithms of copper observations in background and anomalous areas in Zambia (Hawkes and Webb, 1962). Because these distributions of logarithms are approximately normal, the original observations follow a skewed distribution that is more or less lognormal.

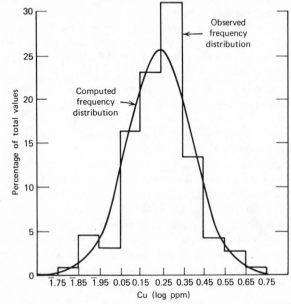

Fig. 15.5. Histogram of readily soluble copper in 216 observations of background stream sediment from Zambia (after Hawkes and Webb, 1962, p. 25).

Tennant and White (1959) studied several sets of geochemical data from different areas by plotting them on logarithmic probability paper (sec. 6.4). Because most of the data sets define curved rather than straight lines, Tennant and White concluded that more than one lognormal distribution is present, but other explanations are possible (sec. 6.5).

In exploration geochemistry, the use of highly precise and accurate analytical methods is usually sacrificed in order to reduce cost and time. Several problems accordingly may arise. *Truncated* and *censored* data (sec. 8.5) are common, and many data, particularly those from spectroscopic and some colorimetric analyses, are *grouped* in frequency classes.

Variability and bias tend to be high, both because of the analytical methods used and because they are often introduced in sampling. Many sampling problems arise, stemming from the geological history and particle size of different soils, from changes in stream flow at different times of year, from glacial materials, and so on.

The high variability of geochemical data is reflected in high coefficients of variation. Values as large as 4 and 5 are not uncommon (Miesch, 1968, personal communication); these are among the highest found in geology. Table 15.6 lists some coefficients of variation calculated from extensive data published by Newman (1962) on distribution of trace elements in sedimentary rocks of the Colorado Plateau, by Garrett and Nichol (1967) on distribution of trace elements in stream sediments in Sierra Leone, and by Earll (1965) on distribution of lead and zinc in soil samples from the Winston mining district, Montana.

Miesch (1969) makes several pertinent remarks about sampling in exploration geochemistry. He states that, because both the sampling and analytical methods are rapid and low in cost, randomization is important to detect analytical blunders, and that replication of the sampling is at least as important as replication of the analyses. Costs of replicating either of these steps are relatively low, and therefore this work should be done if any doubt exists about whether it is necessary.

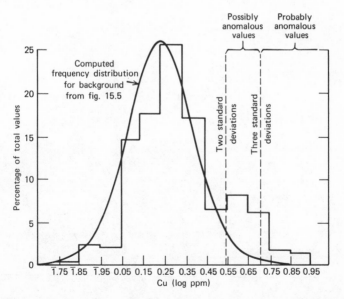

Fig. 15.6. Histogram of readily soluble copper in 825 observations of stream sediment from both background and anomalous areas, Zambia. Data from same field problem as figure 15.5 (after Hawkes and Webb, 1962, p. 30).

Table 15.6. Some coefficients of variation for data from exploration geochemistry

Source of data	Newman (1962)	Garrett and Nichol (1967)	Earll (1965)
Analytical method	Spectrographic and colorimetric	Spectrographic and colorimetric	Colorimetric
Chemical element			
Ag	4.3		
Al	1.7		
B	0.8		
Ba	1.9		
Be	5.7		
Ca	2.1		
Co	1.8	1.1	
Cr	1.2	4.3	
Cu	1.9	0.7	
Fe	3.4		
Ga	1.5		
K	1.2		
La	2.6		
Mg	1.3		
Mo	3.8	0.6	
Mn	1.7	1.6	
Na	1.5		
Nb	3.4		
Ni	1.4	1.2	
Pb	4.6	0.9	3.0
Sc	1.8		
Sn		1.4	
Sr	4.6		
Ti	1.0	0.7	
V	1.8	0.7	
Y	1.9		
Yb	1.3		
Zn		0.6	1.6
Zr	1.8		

Data Analysis

To analyze geochemical data, one must first decide what is an *anomalous observation*. Current practice is to make an orientation survey, to calculate the mean and standard deviation of the observations, which are logarithms of the original chemical determinations, and then to define any observation more than two standard deviations above the mean as possibly anomalous and any one more than three standard deviations above the mean as certainly anomalous. If the parameters of the background population sampled in the orientation survey were known, the highest 2.5 percent of observations would thus be defined as possibly anomalous and the highest 0.13 percent as certainly anomalous. But because only the statistics are known, the actual percentages of anomalous observations are not these exact values; however, the discrepancies are not critical. The demarcation point for defining an anomalous observation may be a serious question, but not one that is pursued in this book.

Once the orientation survey is completed, the next step is to make the exploration survey, to identify any anomalous observations, and to plot them on a map. Then the *anomalies* are established. Here again problems of definition arise, as for the observations themselves. In present practice, anomalies are defined purely subjectively. More objective statistical methods to define them exist, and others could be invented. One difficult problem is to decide how many anomalous observations are needed to define an anomaly. One is usually reluctant to define an anomaly with only a single anomalous observation, unless the sampling and analytical work are sufficiently replicated, because of the large sampling and analytical variability of geochemical data.

An anomaly is usually defined by the clustering of the certainly and possibly anomalous observations. Whether or not clustering exists can be investigated through the Poisson distribution (as was done for meteorite data in section 6.1). The area of exploration is divided into about 100 squares, and the number of points in each square is counted (the total number of points is unimportant). If the points follow the Poisson distribution, they may be randomly distributed, and no real anomalies may exist; otherwise the evidence suggests clustering, and the two or three cells with the largest number of anomalous points may be defined as anomalies.

Another potential technique for investigating the existence of a postulated anomaly is to compare the perimeter of the smallest convex areas of anomalous points (fig. 15.7) with the average perimeter of the same number of randomly distributed points. Statistical tables needed for this test have been calculated (Schuenemeyer, Lienert, and Koch, 1971).

The identification of an anomaly is good positive evidence that one exists, but lack of identification may mean only that an anomaly is too small to be detected with the method used or that the analytical or sampling methods were not working on the day that a particular area was surveyed. If a coefficient of variation is extremely high, say 5 or above, the observations may be so erratic that an anomaly could not be distinguished from random geographic clustering of points. Another problem is that thresholds are sometimes different in different parts of the test area. This problem, discussed by Horsnail and others (1969), is illustrated by figure 15.8.

Although the discussion in this section applies only to single elements, exploration geochemists more often analyze for several elements and become involved with multivariate problems that may be investigated by techniques explained in Chapter 10. For example, if an anomaly is based on either of two or more metals, say on both high cobalt and high nickel derived from a cobalt-nickel ore, only one anomaly may be detected in some of the samples, perhaps because of the occurrence of an element in an insoluble mineral or the leaching out of an element. Then a joint anomaly based on the presence of either or both elements may be defined. Garrett and Nichol (1967) and Nichol and others (1969) discuss some of the problems involved in analyzing several elements.

Finally, in analyzing data from exploration geochemistry, as in analyzing any other kind of data, one is likely to obtain better results if he has some kind of a model in mind. Dahlberg (1969, p. 195) devised a statistical model relating

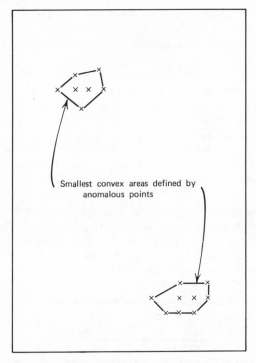

Fig. 15.7. Smallest convex areas of anomalous points.

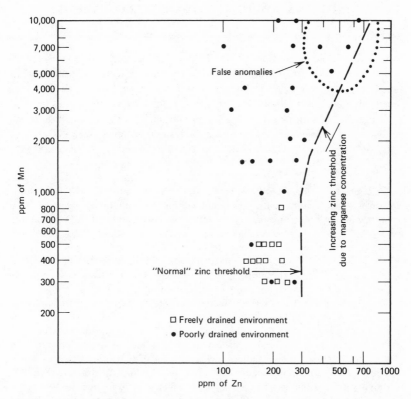

Fig. 15.8. Changing threshold value for zinc (after Horsnail and others, 1969, p. 320).

geochemical prospecting data to the geologic attributes of an area of native copper mineralization in Pennsylvania. His description follows:

For explaining variation in stream sediment trace metal data as a function of some regional properties, a model expressed by the general equation

$$T = f(L, H, G, C, V, M, e)$$

has been utilized. T denotes element concentration, L represents lithological influences, H denotes drainage system effects, G represents geological features, C denotes cultural (man-made) influences and V stands for types of natural cover. M represents the effects of mineral bodies and e denotes error plus effect of additional factors not explicitly defined in the model.

Dahlberg converted this general model into an operational model through a multiple-regression technique combined with factor analysis.

15.4 AN INVESTIGATION TO APPRAISE SEVERAL SAMPLING METHODS

In a careful investigation at the Kilembe mine, Uganda, Davis (1962) compared several of the sampling methods discussed in this chapter and in Chapter 7. A review of part of Davis' study ties together some of the ideas previously developed, discusses work whose results are pertinent to sampling in theoretical as well as applied geology, and shows still another use of the single-degree-of-freedom version of the analysis of variance (sec. 5.9).

About the geological setting of the Kilembe mine, Davis (pp. 146–147) writes,

The deposit is situated in the rugged eastern foothills of the Ruwenzori mountains in western Uganda, with those mine workings which are considered in this account flanking both sides of a steep valley between 4500 and 5100 feet above sea level. The pre-Cambrian Kilembe series . . . consists of about 3000 feet thickness of metamorphosed layered rocks which are generally accepted as mainly altered sediments, sandwiched between acid gneisses. Economic sulfide mineralization follows the upper portion of the quartzo-feldspathic amphibolites . . . and the overlying "Ore Zone Granulite." . . . Average ore contains about 6.5 percent of chalcopyrite, which is the sole primary copper mineral, and about 10.5 percent of the iron sulfides pyrite . . . and minor pyrrhotite, which carry 1 to 2 percent cobalt in solid solution. . . . This stratiform ore, while preserving uniformity in the broad sense, is far from uniform in detail. The ore zone is structurally complex on all scales owing to a set of folds and dragfolds being superimposed at a high angle across an earlier set of folds. . . . Horizontal widths of ore vary from a few feet to many scores of feet, due to folding, faulting, and intensity of mineralization. . . . Sampling had therefore to contend with details of rapidly changing copper and cobalt content in association with varying hardness due to rock type, fracturing, and shearing.

During the exploration phase at Kilembe, Davis sampled for copper in ten crosscuts, three of which were subdivided. In order to simplify the presentation of his findings in table 15.7, only the copper data for the undivided crosscuts from the following sources are presented:

1. Referee value, accepted as most accurate, obtained by processing part of the broken ore from each crosscut through a pilot mill (described in detail by Davis, 1962, p. 155 ff.).

2. Grab samples, each consisting of either one or two shovelfuls from cars of broken development rock, comprising about 0.7 percent of the rock broken.

3. Muck samples, obtained from the broken rock at the face by taking shovelfuls at different points in the muck pile at intervals throughout the mucking operation, comprising about 0.3 percent of the rock broken.

Table 15.7. Results of several types of sampling in the Kilembe copper mine, Uganda*

Crosscut number	Referee value	Grab samples		Muck samples	Channel samples		Drill core	
		2 shovelfuls/car	1 shovelful/car		East wall	West wall	First half	Second half
1	2.260	2.24	2.31	2.16	2.20	2.61	2.94	2.82
2	0.795	0.81	0.77	0.76	0.95	0.60	1.01	0.83
3	1.825	1.78	1.84	1.79	1.76	1.71	2.16	1.92
4	1.395	1.22	1.25	1.57	0.97	0.92	1.13†	1.13
5	1.825	1.89	1.89	2.17	1.97	1.60	2.08	1.94
6	1.305	1.29	1.29	1.26	1.29	0.88	1.50	1.38
7	1.240	1.18	1.16	1.32	1.37	1.19	1.43†	1.43
8	1.820	1.84	1.84†	2.03	1.65	3.15	2.22	1.77
9	3.508	3.87	3.87†	3.68	3.83	3.22	3.16	3.16
10	1.324	1.37	1.37†	1.81	1.05	1.69	1.43	1.79

* Date from Davis, 1962.
† No data. Values assumed by authors of this book.

4. Channel samples, cut along both walls, about 3 pounds of rock per foot of length, carefully subdivided according to rock type.

5. Drill core, AX size, split with a Longyear core splitter.

In table 15.8 are presented the results of an analysis of variance incorporating single-degree-of-freedom tests of the data in table 15.7. Practically all of the variability is among crosscuts rather than among methods; in fact the agreement among methods, with nearly all the calculated F-values being less than 1, is so good that one is suspicious. Perhaps we have ignored something essential in Davis' procedure and our analysis is inappropriate because there is some way of further reducing or breaking up the residual variability.

Table 15.8. Analysis of variance to compare several types of sampling in the Kilembe copper mine, Uganda

Source of variation	Sum of squares	Degrees of freedom	Mean square	F	$F_{10\%}$
Among-crosscuts	43.787556	9	4.86528	66.98	1.74
Among-methods	0.329386	7	0.04706	0.65	1.82
Between halves of drill core	0.049506	1	0.04951	0.68	2.79
Between east and west wall channels	0.014045	1	0.01404	0.19	2.79
Between one and two shovel/car grab samples	0.000714	1	0.00071	0.00	2.79
Referee vs. drill core	0.115809	1	0.11581	1.59	2.79
Referee vs. channel	0.000004	1	0.00000	0.00	2.79
Referee vs. grab	0.003937	1	0.00394	0.05	2.79
Referee vs. muck pile	0.078500	1	0.07850	1.08	2.79
Residual variability	4.212518	58	0.07263		

REFERENCES

Becker, R. M., 1964–1968, Some generalized probability distributions with special reference to the mineral industries (in 5 pts.): U.S. Bur. Mines Rept. Inv. 6329, 53 p.; 6552, 101 p.; 6598, 79 p.; 6627, 57 p.; 6768, 60 p.

————, 1968, A multistage probability model of sample reduction in the mineral industries: U.S. Bur. Mines Rept. Inv. 7177, 121 p.

Boyle, R. W., 1969, Panel discussion: Education of the exploration geochemist, *in* Internat. Geochem. Explor. Symp. Proc.: Colo. Sch. Mines Quart., v. 64, no. 1, Jan. 1969, p. 28–29.

Chan, S. S. M., 1969, Suggested guides for exploration from geochemical investigation of ore veins at the Galena mine deposits, Shoshone county, Idaho, *in* Internat. Geochem. Explor. Symp. Proc.: Colo. Sch. Mines Quart., v. 64, no. 1, Jan. 1969, p. 139–168.

Dadson, A. S., and Emery, D. J., 1968, Ore estimation and grade control at the Giant Yellowknife mine, *in* Ore reserve estimation and grade control: Canadian Inst. Mining Metall., spec. v. 9, p. 215–226.

Dahlberg, E. C., 1969, Use of model for relating geochemical prospecting data to geological attributes of a region, South Mountain, Pennsylvania, *in* Internat. Geochem. Explor. Symp. Proc.: Colo. Sch. Mines Quart., v. 64, no. 1, Jan. 1969, p. 195–216.

Davis, G. R., 1962, Results of comparative sampling methods at Kilembe mine, Uganda: Trans. Inst. Mining Metall., v. 72, p. 156–157.

Duncan, Acheson J., 1962, Bulk sampling: problems and lines of attack: Technometrics, v. 4, no. 3, p. 319–344.

Earll, F. N., 1965, Economic geology and geochemical study of Winston mining district, Broadwater county, Winston, Montana: Montana Bur. Mines & Geology Bull. 41, 56 p.

Erickson, R. L., 1969, U.S. Geological Survey program in geochemical exploration research, *in* Internat. Geochem. Explor. Symp. Proc.: Colo. Sch. Mines Quart., v. 64, no. 1, Jan. 1969, p. 237–244.

Ferguson, H. G., 1916, Placer deposits of the Manhattan district, Nevada: U.S. Geol. Survey Bull. 640-J, p. 163–193.

Gardner, C. W., 1921, Drilling results and dredging returns: Eng. Mining Jour., v. 112, no. 117, p. 646–648; no. 118, p. 688–692.

Garrett, R. G., and Nichol, Ian, 1967, Regional geochemical reconnaissance in eastern Sierra Leone: Inst. Mining Metall. Trans. Sec. B, v. 76, p. B97–B112.

Graves, Thomas A., 1939. The examination of placer deposits: New York, Richard R. Smith, 158 p.

Griffith, S. V., 1960, Alluvial prospecting and mining: London, Pergamon Press, 245 p.

Gy, Pierre, 1968, Theory and practice of sampling broken ores, *in* Ore reserve estimation and grade control: Canadian Inst. Mining Metall., spec. v. 9, p. 5–10.

Hawkes, H. E., and Webb, J. S., 1962, Geochemistry in mineral exploration: New York, Harper & Row, 401 p.

Horsnail, R. F., Nichol, Ian, and Webb, J. S., 1969, Influence of variations in secondary environment on the metal content of drainage sediments, *in* Internat. Geochem. Explor. Symp. Proc.: Colo. Sch. Mines Quart., v. 64, no. 1, Jan. 1969, p. 307–322.

Institution of Mining and Metallurgy (London), 1965, Symposium on opencast mining, quarrying, and alluvial mining: London, published by the institution, 772 p.

Janin, Charles, 1918, Gold dredging in the United States: U.S. Bur. Mines Bull. 127, p. 180–192.

Lindgren, Waldemar, 1911, The Tertiary gravels of the Sierra Nevada of California: U.S. Geol. Survey Prof. Paper 73, 226 p.

――――, 1933, Mineral deposits: New York, McGraw-Hill, 894 p.

McConnell, R. G., 1907, Report on gold values in the Klondike high level gravels: Canada Geol. Survey Rept. 979.

McFarland, W. H. S., 1965, Operations of the Yukon Consolidated Gold Corporation, *in* Symposium on opencast mining, quarrying, and alluvial mining: London, Inst. Mining Metall., p. 189–195.

MacKensie, J. H., 1966, Evaluation, sampling, and forecasting grades of ore at Nchanga, *in* Symposium on mathematical statistics and computer applications in ore valuation: South African Inst. Mining Metall., p. 358–371.

McLachlan, C. G., Bennett, M. J. S., and Coleman, R. L., 1954, The Quemont milling operation: Canadian Inst. Mining Metall. Bull., v. 46, p. 386–401.

Malozemoff, P., 1939, Testing for tailing losses in placer mining: Eng. Mining Jour., v. 140, no. 9, p. 47–52.

Mertie, J. B., Jr., 1940, Placer gold in Alaska: Washington Acad. Sci. Jour., v. 30, no. 3, p. 93–124.

――――, 1954, The gold pan; a neglected geological tool: Econ. Geology, v. 49, p. 639–651.

Miesch, A. T., 1969, Experimental design in geochemical exploration (abs.), *in* Internat. Geochem. Explor. Symp. Proc.: Colo. Sch. Mines Quart., v. 64, no. 1, Jan. 1969, p. 512.

Newman, W. L., 1962, Distribution of elements in sedimentary rocks of the Colorado Plateau—a preliminary report: U.S. Geol. Survey Bull. 1107-F, p. 337–445.

Nichol, Ian, Garrett, R. G., and Webb, J. S., 1969, The role of some statistical and mathematical methods in the interpretation of regional geochemical data: Econ. Geology, v. 64, no. 2, p. 204–220.

Romanowitz, C. M., Bennett, H. J., and Dare, W. L., 1970, Gold placer mining—placer evaluation and dredge selection: U.S. Bur. Mines Inf. Circ. 8462, 56 p.

Schuenemeyer, J. H., Lienert, C. E., and Koch, G. S., Jr., 1971, Delineation of clustered points in two dimensions by measuring perimeters of convex hulls: U.S. Bur. Mines Rept. Inv., in press.

Tennant, C. B., and White, M. L., 1959, Study of the distribution of some geochemical data: Econ. Geol., v. 54, p. 1281–1290.

Theobald, P. K., Jr., 1957, The gold pan as a quantitative geologic tool: U.S. Geol. Survey Bull. 1071-A, 54 p.

Truscott, S. J., 1962, Mine Economics: London, Mining Pubs., Ltd., 471 p.

Webb, J. S., 1969, Panel discussion: Education of the exploration geochemist, *in* Internat. Geochem. Explor. Symp. Proc.: Colo. Sch. Mines Quart., v. 64, no. 1, Jan. 1969, p. 41–44.

Wilson, N. W., 1946, Notes on the estimation of tonnage and grade of some chromite dumps: Inst. Mining Metall. Trans., v. 56, p. 341–355.

Chapter 16

Gold and the Lognormal Distribution

In this chapter we discuss the statistical analysis of gold assay data in relation to the lognormal frequency distribution, which is considered in section 6.2. We again stress the development of taste and judgment in the application of statistics to real problems.

The most important purpose of this chapter is to show the complexities that arise in the analysis of real data, which are rarely if ever as well behaved as ideal data. Moreover, many other substances of geological interest, besides gold, seldom follow the normal distribution for whose analysis well-defined statistical methods are available. The second purpose of this chapter is to demonstrate how different statistical approaches can be put to work on one particular problem. The third purpose is to tie together some of the many references to gold data made throughout this book.

Millions of gold assays have been performed, many estimates have been made of mean grades, and the distribution of gold has undoubtedly been studied more than that of any other trace element. Much information of great scientific as well as commercial interest can be derived from the study of gold assay data.

Most of the knowledge of the statistics of gold distribution is due to the pioneering work of two investigators in South Africa, H. S. Sichel and D. G. Krige, who have written numerous papers, cited in the works listed at the end of this chapter. This chapter is drawn in part from a paper by Link, Koch, and Schuenemeyer (1971).

TABLE 16.1. COEFFICIENTS OF VARIATION FOR GOLD ASSAYS FROM SEVERAL DEPOSITS

Reference	Ore deposit	C
3	Homestake Saddle Project, Nevada	0.33
1	Loraine, South Africa (Basal Reef)	0.80
1	Freddies, South Africa	0.86
1	President Steyn, South Africa	0.87
1	Hartes, South Africa	0.89
1	Harmony, South Africa	0.95
1	Buffels, South Africa	0.98
1	Welkom, South Africa	0.99
1	Virginia, South Africa	0.99
1	Vaal Reefs, South Africa	1.02
2	Getchell mine, Nevada (wagon-drill hole cuttings)	1.03
1	Welgedacht, South Africa	1.05
1	Rand Leases, South Africa (main reef leader)	1.07
1	Stilfontein, South Africa	1.20
3	Fresnillo mine, Zacatecas, Mexico (2137 vein)	1.24
1	Crown, South Africa	1.25
2	Getchell mine, Nevada (diamond-drill hole cores)	1.25
1	Blyvooruitzicht, South Africa	1.27
1	Western Holdings, South Africa	1.28
1	Kinross Area, South Africa (3 mines)	1.29
1	President Brand, South Africa	1.30
1	City Deep, South Africa	1.35
1	Merriespruit, South Africa	1.35
1	E.R.P.M., South Africa	1.52
3	Knob Hill mine, Washington	1.54
3	Carlin mine, Nevada	1.58
1	Free State Geduld, South Africa	1.81
1	East Geduld, South Africa	1.92
3	Frisco mine, Chihuahua, Mexico (Brown vein)	2.24
1	AMEX Joint Venture Project, Crescent Valley, Nevada	2.68
1	Loraine, South Africa (B reef)	2.81
1	Rand Leases, South Africa (Black Bar)	2.93
3	Homestake mine, Lead, South Dakota (30-g samples)	3.96
3	Nome, Alaska, marine placers	4.10
3	Columbia Pit, California, Tertiary placers	5.10

References: 1—Krige, 1960.
2—Koch and Link, 1970.
3—Various Bureau of Mines' published and unpublished papers.

The coefficients of variation of gold data are characteristically among the highest found in geological data, and are as extreme as those found in any other discipline. Table 16.1 compares coefficients of variation, C, for gold with those for base metals. Most of those for gold fall in the range above 1.2, where Finney's efficiency curve (fig. 6.7) suggests the use of a logarithmic transformation.

Most gold ore contains only a minute amount of gold. A deposit with 1 troy ounce (or 20 pennyweights) of gold per short ton contains only 34 parts per million (or 34 grams per metric ton, or 0.0034 percent) of gold. Moreover, these 34 parts per million are rarely if ever spread evenly throughout the ore, but are instead sporadically distributed. As an example, assume that the gold in a 1-ounce-per-ton deposit is in the form of small but easily seen particles, 1 cubic millimeter in size, corresponding to cubes 0.04 inch on a side. Each particle would then weigh 0.019 gram, and a metric ton of ore would contain only about 1800 particles; only 6 out of every million particles 1 cubic millimeter in size would be gold. In reality, the amount of gold in most ore is less than 1 ounce per ton, and many mines work ore with a grade of 5 to 10 parts per million.

16.1 EXAMPLE DATA FROM THE CITY DEEP MINE, SOUTH AFRICA

For the first example of gold data that are lognormally distributed, we return to data from the City Deep mine, South Africa (Koch and Link, 1966), already introduced in section 5.4. Figure 16.1 is a plot on logarithmic probability paper of cumulative frequency distributions for the 503 development values and the 1536 stope inch-pennyweight values. Because both sets of values lie nearly but not exactly on a straight line, these observations are distributed approximately, not perfectly, lognormally. A method to make these observations lognormal by adding a constant to each is explained in section 16.4.

The outstanding fact about gold distribution in the 1000-foot-square block of the City Deep mine that was studied is that most of the gold is from only a few of the sampled points. A plot of cumulative metal content, figure 16.2, shows that fewer than 10 percent of the stope samples account for more than 50 percent of the total value.

The cumulative frequency distribution and the plot of cumulative metal content display the value distribution of the gold but provide no information about its geographic distribution. The geography of gold distribution in the 1000-foot-square block is shown by two kinds of maps. The first one, figure 5.3, locates the 39 out of 503 development points that account for 50 percent of the total inch-pennyweight gold content. Because the high-value points in the

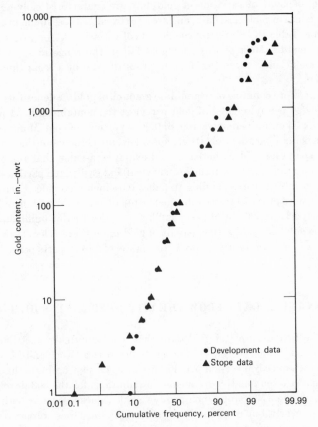

Fig. 16.1. Cumulative frequency distribution for inch-pennyweight gold observations from development workings and stopes, City Deep mine, South Africa.

development workings are clustered rather than scattered, a guide to planning the extraction is offered (sec. 14.6). The second one, figure 16.3, is a contour map of inch-pennyweight values from the stopes. To prepare this map, a 50-foot-square grid was superimposed on the 1000-foot-square block, and average location and inch-pennyweight values were calculated for each of the 400 squares (each one 20 feet on a side). This contoured map of the averaged data shows an irregular pattern of mineralization.

To illustrate that low-grade gold, which may be termed "background mineralization," is spread rather evenly through the reef, we prepared table 16.2 and figure 16.4, portraying the geographic distribution of the lowest 80 percent of the inch-pennyweight values. The table shows that this 80 per-

TABLE 16.2. SUMMARY ASSAY DATA FROM 80 PERCENT OF SAMPLE POINTS YIELD-
ING LOWEST INCH-PENNYWEIGHT VALUES (PRESUMED TO REPRESENT BACKGROUND
DATA) IN THE CITY DEEP MINE, CENTRAL WITWATERSRAND, SOUTH AFRICA

Unit	Stope data			Development data		
	Mean	s	C	Mean	s	C
Dwt	3.93	5.82	1.48	6.48	9.99	1.54
In.	20.58	9.20	0.45	17.78	9.91	0.56
In.-dwt	68.79	71.12	1.03	91.27	106.32	1.16

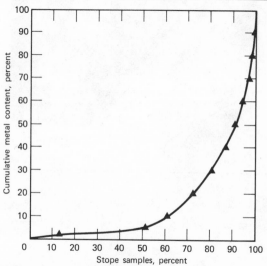

Fig. 16.2. Relation of cumulative gold content to percentage of stope samples, City
Deep mine, South Africa.

cent averages only 68.79 inch-pennyweights, although the mean sample width
is the same as for all of the points, within the limits of statistical fluctuation.
The figure shows that the low-value background mineralization is quite uniform over the area when the high-value observations are removed.

In section 5.4 the stope and development data from the City Deep mine are
compared by a one-way analysis of variance (table 5.10), which leads to the
conclusion that they come from two different populations. The difference
between the 273-inch-pennyweight estimate of grade from the development
workings and the 192-inch-pennyweight estimate from the stopes is so large
that at first we thought a bias must have been introduced by the sampling
and assaying or by a mistake in transcribing the data.

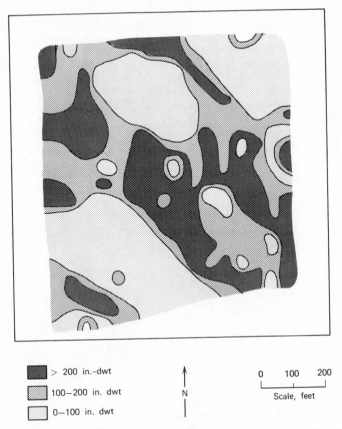

> 200 in.–dwt

100–200 in. dwt

0–100 in. dwt

↑
N
|

0   100   200
|____|____|
Scale, feet

Fig. 16.3. Contour map of inch-pennyweight gold values, based on 20 by 20 grid, City Deep mine, South Africa.

However, when the locations of the 39 development points accounting for 50 percent of the total gold content are plotted (fig. 5.3), nine of the highest observations are found to lie at the northeast end of the northernmost drift. These nine observations were then removed in order to demonstrate the great effect that even so few observations can have on the standard deviation, and the statistics were recalculated, yielding the results in table 16.3. Next the one-way analysis of variance from table 5.11 was repeated for the new data, leading to the acceptance of the hypothesis H_0 that the two population means are equal (table 16.4). In geological terms, this example of development sampling affords a reliable estimate of the grade of ore mined in the stopes, provided that these nine observations are disregarded.

TABLE 16.3. SUMMARY SAMPLE DATA FROM 1000-FOOT-SQUARE BLOCK, CITY DEEP MINE, CENTRAL WITWATERSRAND, SOUTH AFRICA

Unit	Stope data, 1536 samples			Development data, 503 samples			Development data, 494 samples		
	Mean	s	C	Mean	s	C	Mean	s	C
Dwt	12.00	31.31	2.61	22.58	71.33	3.16	15.56	32.84	2.11
In.	20.84	9.47	0.45	18.20	10.29	0.57	18.32	10.22	0.56
In.-dwt	192.40	374.70	1.95	273.54	607.77	2.22	223.84	440.75	1.97

These calculations were performed to demonstrate that a few sample values can have a profound influence on the estimates of grade, and not to suggest that in practice certain values should be removed. They show that exclusion of fewer than 2 percent of the development observations that were concentrated in one place changed the estimated grade about 50 inch-pennyweights, or 20 percent.

Alternatively, the stope and development data can be compared by forming confidence intervals. Table 16.5 shows that the confidence interval for the inch-pennyweight mean difference does not contain zero for the complete statistical

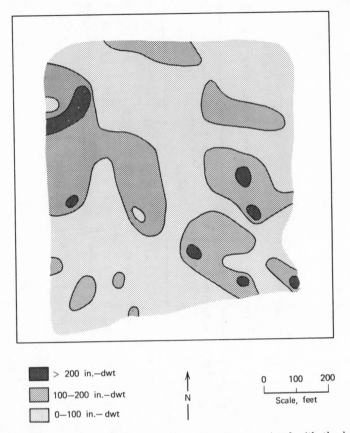

> 200 in.—dwt

100—200 in.—dwt

0—100 in.— dwt

N

0   100   200

Scale, feet

Fig. 16.4. Contour map of inch-pennyweight gold values associated with the lowest 80 percent of the total inch-pennyweight gold content, based on a 20 by 20 grid, City Deep mine, South Africa.

TABLE 16.4. ONE-WAY ANALYSES OF VARIANCE TO COMPARE DEVELOPMENT AND STOPE GOLD CONTENT IN THE CITY DEEP MINE, SOUTH AFRICA

A. All observations included

Source of variation	Sum of squares	Degrees of freedom	Mean square	F	$F_{10\%}$
Between stope and development	2,489,747	1	2,489,747	12.7	2.71
Within groups	400,963,342	2,037	196,647		

B. With the 9 highest development observations removed

Source of variation	Sum of squares	Degrees of freedom	Mean square	F	$F_{10\%}$
Between stope and development	369,624	1	369,624	2.40	2.71
Within groups	311,766,232	2,028	153,731		

TABLE 16.5. CONFIDENCE INTERVALS FOR THE INCH-PENNYWEIGHT MEAN DIFFERENCE WITH AND WITHOUT NINE HIGH-VALUE POINTS REMOVED FOR DATA FROM THE CITY DEEP MINE, SOUTH AFRICA

Source of development data	Unit	Observed average difference between stope and development data	10% Confidence interval based on raw data	
			Lower limit	Upper limit
All data	Dwt	10.6	6.8	14.4
	Inches	−2.64	−3.46	−1.82
	In.-dwt	81	44	119
Data with 9 high points removed	Dwt	3.6	0.9	6.3
	Inches	−2.52	−3.34	−1.70
	In.-dwt	31	−2	65

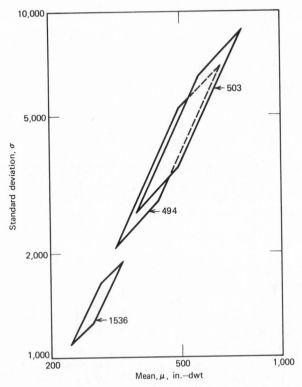

Fig. 16.5. Confidence regions in μ, σ-space for lognormally distributed gold observations from the City Deep mine, Central Witwatersrand, South Africa.

sample of 503 development points but does contain zero for the sample with the nine points removed. These data can also be compared by graphing the joint-confidence intervals for μ and σ, computed by Mood's method (fig. 16.5).

That the data may be reconciled by removing nine observations is evidence that accurate prediction from development to stopes requires more observations than the 503 available here because, with such skewed distributions, the coefficients of variation are large. These data also reveal other problems. Although the width measurements have a small coefficient of variation (table 16.3), a bias exists between those from the development workings (mean of 18.20) and those from the stopes (mean of 20.84). Clearly, because almost all of the 1000-foot-square block was stoped, the stope samples were cut across greater widths. This bias could cause problems unless the cut was wide enough in both the stopes and the development workings to sample all of the gold.

16.2 EXAMPLE DATA FROM THE HOMESTAKE MINE, LEAD, SOUTH DAKOTA

Data from the Homestake gold mine (Koch and Link, 1967, 1971) were introduced previously. In section 3.4 random sampling of 1-inch-long diamond-drill cores from Homestake is explained; in section 5.6 a nested analysis of variance (table 5.15) for those data is given; and in section 5.10 this analysis of variance is reanalyzed through the single-degree-of-freedom interpretation (table 5.34). In this section we re-examine the previous data and relate them to a second set of Homestake data.

For convenience, the previous Homestake data are called the 1-inch data, corresponding to the length of core sampled. Similarly, the second Homestake data are called the 1-foot data, their source being 1-foot lengths of EX diamond-drill core from holes drilled in the regular course of the company's operations. The core was sampled only within the ore-bearing Homestake formation.

As for the City Deep data, most of the Homestake sampled gold is in only a few of the samples, as is clearly displayed by sorting the assay values according to their sizes and then plotting cumulative metal grade against the number of observations. For the 219 observations of the 1-inch data, the graph (fig. 16.6) shows that the 50 percent of the observations with the lowest grade contain almost no gold, whereas only 3 percent of the highest grade observations, six in number, contain one-half the gold. For the 900 observations of the 1-foot data, the graph in figure 16.7 shows that, as before, 50 percent of the observations with the lowest grade contain almost no gold, whereas 6.3 percent of the highest observations, 57 in number, contain one-half the gold. Table 16.6 summarizes statistics computed from these two sets of data.

Although the sample means of the two sets of observations are different, whether estimated by the arithmetic mean or from logarithms, the confidence

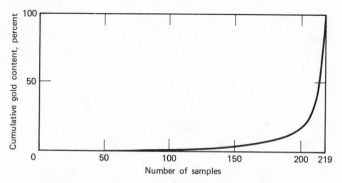

Fig. 16.6. Relation of cumulative gold content to the number of observations for 1-inch data from the Homestake mine, Lead, South Dakota.

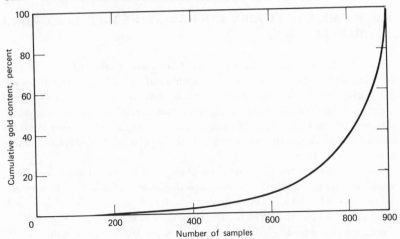

Fig. 16.7. Relation of cumulative gold content to number of observations for 1-foot data from the Homestake mine, Lead, South Dakota.

regions overlap. It is interesting to consider how these seemingly inconsistent results may be reconciled. The diamond-drill holes were bored in an ore body selected as typical by the Homestake geologists. The company assay data from this part of the mine indicate ore with a grade that is near the current mine average of about 11.3 parts per million (0.33 ounce per short ton); yet the average grade in each of the five holes is higher. The company assay records and the geology of the crosscuts suggest that the holes may have been collared in ore of above-average grade. However, the confidence limits do not clearly indicate whether the five holes in fact cut ore of higher than average grade, or whether the holes cut run-of-mine ore and the high sampled grades are due merely to statistical fluctuations. The mined grade of ore corresponds more closely to the sampled grade of the 900 observations.

The 1-inch data reveal the extreme variability in gold mineralization in the Homestake mine. This variability is that naturally present on the smallest scale that can be found by fire assay of samples weighing about 29.167 grams (1 assay ton). With samples obtained by splitting cores of larger volumes, as in the 1-foot data, variability is less.

Clustering of Gold in the Homestake Mine

Gold particles may be distributed throughout a rock body in many ways. They may be distributed at random; in this case the chance of a gold particle being present at one place is as good as the chance of its being present elsewhere. Or the gold particles may be distributed in a well-defined pattern, of which specified grid points on a three-dimensional grid is a far-fetched illus-

Table 16.6. Summary data calculated from two sets of gold assay data from the Homestake mine, Lead, South Dakota

	Type of data			
Item	Gold (ppm)		Gold (logarithms)	
219 observations on 1-in.-long EX cores				
Mean	$\overline{w}$	42.70	$\overline{u}$	1.036
Variance	s_w^2	28,613	s_u^2	5.6555
Standard deviation	s_w	169	s_u	2.378
Geometric mean	$e^{\overline{u}}$	2.817		
Multiplying factor for geometric mean	$\Psi_n\left(\tfrac{1}{2}s_u^2\right)$	16.115		
Estimate of mean from logarithms	m	45.4		
900 observations on 1-ft.-long EX cores				
Mean	$\overline{w}$	7.59	$\overline{u}$	0.578
Variance	s_w^2	327	s_u^2	3.024
Standard deviation	s_w	18	s_u	1.739
Geometric mean	$e^{\overline{u}}$	1.783		
Multiplying factor for geometric mean	$\Psi_n\left(\tfrac{1}{2}s_u^2\right)$	4.516		
Estimate of mean from logarithms	m	8.1		

tration. The gold in the Homestake mine, however, is evidently distributed in clusters rather than in either of these other two ways. An examination of this clustering develops principles that can be applied to other geological data that are clustered.

Because the gold is clustered, the standard-error-of-the-mean law, which depends on a random distribution, does not apply. Hence more mine samples are required than this law specifies in order to secure enough that contain gold to provide a reliable estimate of the gold content. Most of the gold is contained in the clusters, although their boundaries are ill defined. Even though the clusters can, in principle, be tied to mineralogy and rock types, this association is evidently too subtle to take into account in mine sampling. Clustering is therefore an awkward evil that must be accepted and provided for. After establishing the degrees of clustering, we suggest an interpretation and propose a treatment.

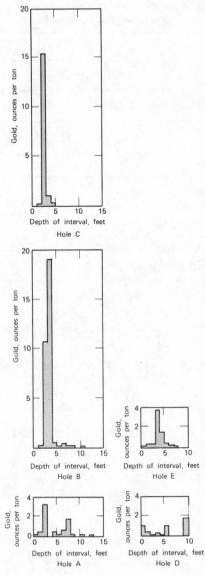

Fig. 16.8. Mean grade of gold at 1-foot intervals, calculated for the 1-inch data from the Homestake mine, Lead, South Dakota.

Clustering of the Homestake gold is demonstrated by figure 16.8, which graphs the variation in gold content in the five holes for the 1-inch data. The fact that distinct segments contain much higher than average gold demonstrates, in one dimension, the clustering present in three dimensions.

Table 16.7 investigates the consequences of clustering by comparing some statistics calculated from the 1-foot data with those calculated from other data. The first part of the table is for 1- to 5-foot intervals constructed by combining the original assays of the 1-foot-long cores. The first two columns list the number of observations, which is the original 900 for the 1-foot intervals, and then one-half of 900, or 450, for the 2-foot intervals, in which the assays are combined in pairs, and so on. The third column gives the mean, which is 7.60 regardless of how the assays are combined; and the fourth column gives the variance.

For the 1-foot intervals, the variance is 327.21 or about 320. If the standard-error-of-the-mean law applied to these data, the variance for the 2-foot intervals would be σ^2/n, which is $320/2 = 160$. However, the observed variance, 218.48, is larger than 160. This fact shows that the assays, rather than being distributed independently and randomly, are correlated so that a high assay tends to have neighboring high assays on either side. The 3- to 5-foot intervals display the same effect. For the 5-foot intervals, the observed variance of 129.91 is about twice the variance of 64, which is predicted by the standard-error-of-the-mean law.

TABLE 16.7. ESTIMATES OF MEANS AND VARIANCES FROM VARIOUS TYPES OF DATA FROM THE HOMESTAKE MINE, LEAD, SOUTH DAKOTA

| | | Gold (ppm) | | |
Source of estimate	Number of observations	Mean	Variance	Variance based on σ^2/n law
Experimental data:				
One-foot intervals	900	7.60	327.21	320
Two-foot intervals	450		218.48	160
Three-foot intervals	300		165.77	107
Four-foot intervals	225		141.44	80
Five-foot intervals	180		124.91	64
Standard mine data, nearby drill holes:				
Group 1	295	4.01	63.83	
Group 2	30	7.90	141.91	
Experimental data, 1-inch cores:	219	42.71	28,613	

The next two rows of table 16.7 present data from EX diamond-drill cores prepared in the company's standard way at 5-foot intervals. Group 1 is for 295 five-foot intervals. Group 2 is a subgroup of group 1, chosen to have a mean about equal to that of the 1-foot experimental data. The variances for these 5-foot cores are about 17 times the mean, as is the variance for the simulated 5-foot intervals constructed by combining the 1-foot data; for group 2, the variance is nearly identical with that of the simulated 5-foot intervals. These facts suggest that mixing of the 5-foot cores during sample preparation (sec. 8.4) was fairly good. The last row gives calculations made on the 1-inch data.

In order to examine whether the variance for the 1-inch data is consistent with the other variances in this table, the sample-volume variance relation (sec. 7.4) may be applied. The variance is inversely proportional to the sample volume, with the ratio of mean to variance assumed to be constant. There are two starting points. The first point is based on the 1-foot data; if the variance of 327.21 in line 1 is multiplied by 12, the resulting variance is 3927, which would be appropriate for 1-inch data with the same mean. Because the means are different, if variance is proportional to the mean, 3927 must be multiplied by the ratio of the two means, $42.71/7.60 = 5.62$, to yield a product of about 22,000. Alternatively, if the calculations are repeated for the 5-foot standard data, the product obtained is about 40,000. These values imply that the effective sample volume of the 1-foot data taken in 5-foot intervals is about 2 feet rather than 5 feet. Another implication is that the effective sample volume of 1-foot data taken at 4-foot intervals is about 5 feet, because 141.91 (the variance for the group 2 standard mine data) is about equal to 141.44 (the variance for the experimental data at 4-foot intervals).

16.3   OTHER EXAMPLE GOLD DATA

Other gold data that we have studied exhibit a spectrum of variability from distributions like those for City Deep and Homestake, which are approximately lognormal or even more skewed, to distributions that are less skewed. A complete range of variability that can be related to geological type is found. Among the data with the least variability are those from the Getchell mine, Humboldt County, Nevada, where the gold is submicroscopic in size, and spread more evenly throughout the ore than visible gold usually is. Figure 16.9, a plot of cumulative frequency for 956 representative Getchell assays, substantially deviates from lognormality.

Augmenting High Observations

Because in most gold deposits nearly all geological samples contain little or

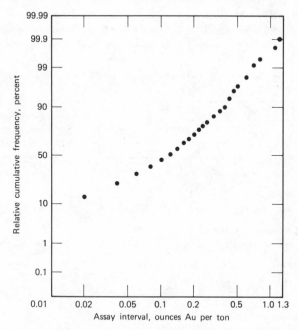

Fig. 16.9. Plot on lognormal probability graph paper of 956 gold assays from the Getchell mine, Humboldt County, Nevada.

no gold, as a consequence of the extremely skewed frequency distribution, usually enough samples are taken to estimate the mean of the low-grade ore, but seldom are enough taken to estimate the mean of the high-grade ore. In statistical terms, the variance is not homoscedastic from place to place. The same problem arises generally in geologic sampling wherever the substance sampled is erratically distributed. It would therefore be desirable to devise a plan for sampling more of the high-grade ore than would be sampled with ordinary systematic or random sampling methods. The rock to be sampled more intensively could be identified in one of several ways: by scanning the core or other sample with some quick-measurement device, such as one depending on x-ray or neutron activation, by making additional assays when an initial assay finds high-grade rock, etc.

One method, which we call *augmenting*, for obtaining more samples of high-grade ore is to make one or more additional assays of all material that yields a high-grade assay. For the Homestake 1-foot data, we investigated the statistical properties of estimates obtained by augmenting high assays by two additional assays from the 1-foot cores on either side of the high-assay samples, or, alternatively, by four additional assays, two on each side. In this scheme, the original assays were augmented if they were higher than one of the specified values or cutoff grades (sec. 13.2) listed in table 16.8. These values were chosen so that the percentage of assays that were augmented ranged from 81 to 6.

TABLE 16.8. STATISTICS CALCULATED FROM AUGMENTED OBSERVATIONS, HOME-STAKE MINE, LEAD, SOUTH DAKOTA*

Line	Item	Specified assay (ppm)				
		0.25	1.00	5.80	16.60	29.90
1	Percent of assays to augment	81	60	28	12	6
2	Estimated percentage bias, augmenting by 2 assays	98	96	86	81	83
3	Estimated percentage bias, augmenting by 4 assays	78	77	68	68	74
4	Estimated variances of unbiased estimates, augmenting by 2 assays	0.85	0.80	0.63	0.55	0.52
5	Estimated variances of unbiased estimates, augmenting by 4 assays	0.72	0.66	0.50	0.44	0.50

* This table is condensed from more detailed ones by Koch and Link (1971). The specified assays are not integer values because they were computed originally in other units.

The principal difficulty with augmenting is that the means obtained by augmenting are biased with respect to the unaugmented means. Lines 2 and 3 of table 16.8 indicate the average percentage biases that are introduced; the tabled values are the percentages of the unweighted means corresponding to the augmented means. For instance, if assays above 16.60 parts per million were augmented by 4, the estimated grade would be divided by 0.68 to remove the bias. Bias is greatest for augmenting by 4 and with specified assays of 5.80 and 16.60 parts per million. If these biases held for the entire mine, they could readily be eliminated, but unfortunately the bias will vary with the grade of ore, its distribution in the mine, and other unforeseen factors; therefore it would be unsafe to use these tabled results for the mine as a whole.

However, one may determine which cutoff grade and augmentation scheme yield the minimum variance. For this determination, the variances of unbiased estimates were calculated for each of the cutoff grades and both of the augmenting schemes. These variances, listed in lines 4 and 5 of table 16.8, show that augmenting by 4 assays at a cutoff grade of 16.60 ppm provides a minimum estimated variance. The results do not depend closely on the cutoff grade, but are somewhat sensitive to the number of assays used to augment.

16.4 DEPARTURES OF GOLD OBSERVATIONS FROM THE LOGNORMAL MODEL

For some gold assay data, Krige (1960) postulates that the observations are too small to be distributed lognormally by an amount that is constant or nearly so, regardless of the sizes of the individual observations. If gold observations depart from the lognormal model, lognormal theory is not entirely satisfactory, because the estimated arithmetic means are too large unless adjustments are made. Figure 16.10 and table 16.9, which display a lognormal distribution with two different constants added, illustrate that adding either a positive or a negative constant to originally lognormal observations produces nonlognormal observations. On the other hand, if all of the original observations are too small by the same constant amount, adding the proper constant

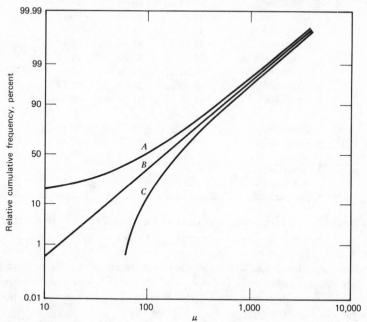

Fig. 16.10. Distortion of a lognormal distribution obtained by adding positive and negative constants.

TABLE 16.9. FREQUENCY TABLE FOR THREE DISTRIBUTIONS OBTAINED BY ADDING CONSTANTS OF -50, 0, AND $+50$ TO A LOGNORMAL DISTRIBUTION WITH $\alpha = 5$, $\beta^2 = 1$

	Relative cumulative frequency (%)		
Class interval	$A(-50)$	$B(0)$	$C(+50)$
0–10	18.77	0.43	0
10–20	23.20	2.07	0
20–40	27.01	5.26	0
30–40	31.15	9.55	0
40–50	35.32	14.38	0
50–60	38.76	18.77	0.43
60–70	42.03	23.20	2.07
70–80	45.10	27.01	5.26
80–90	47.80	31.15	9.55
90–100	50.75	35.32	14.38
100–200	70.30	62.02	50.75
200–300	80.71	76.46	70.30
300–400	86.60	83.93	80.71
400–500	90.47	88.75	86.60
500–600	93.11	91.88	90.47
600–700	95.00	94.09	93.11
700–800	96.14	95.70	95.00
800–900	96.98	96.69	96.14
900–1000	97.68	97.34	96.98
1000–2000	99.55	99.51	99.49
2000–3000	99.89	99.88	99.87
3000–4000	99.94	99.94	99.93
4000–5000	99.99	99.98	99.98
5000–6000	100.00	100.00	100.00

will make their distribution lognormal, as happens if curve A in the figure is changed to curve B. Then lognormal theory can be applied. Suggesting a physical explanation for too-small observations, Krige writes that a constant amount of gold might have been leached from the ore after its formation; but, as he points out, whether or not a physical explanation is known, if the postulated phenomenon occurs, improved estimates of means can be obtained by adding the constant.

Krige (1960, p. 236) discusses various ways to compute the constant and concludes that the preferred method is as follows:

Using logarithmic-probability paper, the basic distribution of z [the logarithm of the original observation] is first plotted, and then the distributions of $(z + \alpha)$ with the constant α being increased in stages until the resultant plot yields the most satisfactory straight line fit. For all practical purposes this procedure will be found adequate, as the fit is evidently not very sensitive to small variations in the value of α either side of the optimum value.

After the new observations are formed by adding a constant to each of the original observations, logarithms of the new observations are taken, and a mean and variance are calculated by using lognormal theory. Finally, the constant is subtracted from the mean of the new observations in order to obtain the desired mean, which is the arithmetic mean of the original observations calculated by lognormal theory with the use of the constant. The desired variance is of course the same, because subtracting the constant does not change the variance of the new observations.

Table 16.10 summarizes data on 28 distributions of gold assay data from South African mines, originally published by Krige (1960), who gives the names of the mines and reefs, the reef areas, frequency distributions, and other data. The table lists four kinds of means. The first three are the arithmetic mean, the lognormal mean computed without a constant, and the lognormal mean computed with a constant. The fourth is calculated from mine-production data, supplied by Krige (1967, personal communication), who comments about them as follows:

Production data reasonably representative of the specific mine section covered by the distributions were available only for the nine mines shown. In.-dwt of gold actually accounted for on surface in gold recovered and in slime residues dumps are tabulated corresponding approximately to ore from these sections of the mine represented by the distribution shown. The difference between this value and the mean of the distribution is due to (a) bias in normal cutting of underground samples (generally negative), (b) gold actually lost in mining operations (negative), and (c) improvements (positive) due to selectively mining only ore above the accepted pay limit (significant mainly for mines numbers 27 and 28).

The table shows that the mined grade averages 90.3 percent of the sampled grade. Finally, the column farthest to the right gives coefficients of variation, which are approximate because they are calculated from standard deviations obtained from classed data in Krige's frequency tables rather than from the original observations.

What can be learned from table 16.10? First, the means estimated from logarithms of observations formed by adding a constant are nearly the same as the arithmetic means, if the appropriate constants are added. Although in principle a constant is chosen, as Krige (1960) explains, in order to make a distribution as nearly lognormal as possible, in practice subjective judgment is

TABLE 16.10. COMPARISON OF FOUR KINDS OF MEANS FOR 28 DISTRIBUTIONS OF GOLD ASSAY DATA

Distribution number	Number of observations	Arithmetic mean (in.-dwt)	Lognormal mean (in.-dwt)		Constant (in.-dwt)	Mine production grade (in.-dwt)	Percent of predicted grade*	Coefficient of variation
			Without a constant	With a constant				
1	216	308	406	304	40			1.29
2	3,600	477	478	472	5			1.92
3	4,200	90	91	91	10			1.03
4	1,648	373	396	366	10			1.52
5	9,599	301	366	304	80			1.25
6	2,477	307	667	299	90			1.35
7	1,003	240	293	236	30			1.07
8	964	259	320	320	0			2.93
9	2,840	126†	129†	125	2			1.40
10	28,334	680	681	681	0	630	93	1.15
11	25,474	341	344	339	7	283	83	1.20
12	22,500	441	467	446	20	396	89	0.89
13	1,000	448	449	444	20			1.08

14	14,920	537	550	534	30			0.98
15	25,112	477	497	478	25			1.02
16	568	142	147	143	30			0.80
17	517	210‡	242	204	4			2.81
18	158	299	310	296	60			0.86
19	12,238	1,230	1,244	1,229	12			1.31
20	17,726	934	989	931	35			1.28
21	33,031	319	358	319	50	254	80	0.99
22	14,590	1,103	1,283	1,113	95	981	92	1.19
23	740	1,026	1,234	1,019	60			1.58
24	19,687	499	505	499	4	415	83	0.87
25	44	362	445	356	25			1.59
26	8,436	559	610	559	70	492	88	0.95
27	197	219	244	225	20	248	110	0.99
28	1,000	261	339	257	55	245	95	1.35

* Ratio of mine production mean to lognormal mean with a constant.

† Pennyweights instead of inch-pennyweights for this mine only.

‡ Based on value of all development to December 1959.

required. Another way to choose the constant is to select one that makes the mean calculated from lognormal theory equal to the arithmetic mean, which must be unbiased regardless of the form of the distribution, provided only that the sample is random. Most of the means estimated by adding a constant are a fraction of a percentage point below the arithmetic means. These constants may in fact have been added as a method for reducing the means, because the mined grades are even further below the arithmetic means. As expected, low coefficients of variation are associated with large constants, because, if a large constant must be added, the lower part of the original distribution disappears and the ratio of standard deviation to mean (coefficient of variation) must therefore be smaller.

Figure 16.11 plots on logarithmic probability graph paper the development values of ore from 16 South African mines. The means should be reliable because they are estimated from many observations for each mine and are distributed approximately lognormally, suggesting the hypothesis that all of the mines work ore whose gold-assay values belong to a common lognormal population. Restated, this hypothesis says that the distribution of gold in these mines that work several reefs constitutes one enormous lognormal population, and the differences in mean grade from mine to mine represent statistical fluctuation on a large scale. For the entire postulated population, the mean grade based on lognormal theory is 485 inch-pennyweights (the arithmetic mean is 490 inch-pennyweights), which may approximate the mean

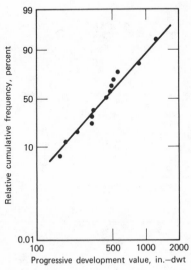

Fig. 16.11. Plot on lognormal probability graph paper of development values of gold ore from 16 South African mines.

grade of the reefs, although the data are not weighted by the sizes of the mines.

We conducted a sampling experiment in order to investigate the consequences of adding a constant to observations before applying the methods of estimation appropriate for lognormal data, in full knowledge of the fact that it was inappropriate to add a constant. In the experiment, 1000 samples of size 10 were taken from a population that was modified from a lognormal population with an arithmetic mean of 245 and a variance of logarithms of 1. This modification was made by subtracting 50 from all those observations larger than 50 and by replacing all those observations smaller than 50 with the observation 1. The resulting distribution, with an arithmetic mean of 196, is of the kind postulated by Krige for gold observations that were originally lognormal but from each of which a constant amount of gold was removed, with the smallest observations, which the assayer would record as trace or zero, being replaced by the arbitrary small number 1.

Table 16.11. Statistics calculated in a sampling experiment to test the effect of adding a constant to a modified lognormal population

Constant	Average of 1000 estimated means, assuming lognormality	Estimated bias	Percent of 90 percent confidence intervals that do not contain the population mean		Variance of logarithms
			Lower	Upper	
0	346	150	10.9	0.1	4.28
5	229	33	5.4	0.2	2.33
10	210	14	5.2	0.8	1.81
25	193	−3	5.2	2.7	1.18
50	186	−10	5.5	5.1	0.78
100	185	−11	5.7	8.6	0.47
500	189	−7	8.0	15.1	0.09
1000	191	−5	8.5	16.3	0.04

Table 16.11 gives the results of this experiment. The first column lists the constants; the second column lists the arithmetic means of the 1000 estimates of the population mean; the third lists the estimated biases obtained by subtracting the means in column 2 from the theoretical mean of 196; the fourth lists the percentages of confidence intervals that do not contain the population mean; and the fifth lists the estimated variances of the logarithms. Because the confidence intervals, calculated for each sample of size 10 by Mood's method, are nominal 90 percent intervals, about 10 percent should fail to contain the population mean. The table shows that, whether or not a constant is needed,

adding too large a constant does not on the average lead to much bias; but if a constant is needed, adding too small a constant leads to intolerable bias. Similarly, the confidence intervals are fairly reliable unless the constants are either too small or extremely large.

16.5 SUMMARY

From these studies, which are detailed in the original paper (Link and others, 1971), we draw the following conclusions about the statistical analysis of gold and similar trace element data:

1. The logarithmic transformation should be used cautiously, if at all. There are two principal reasons for this. (a) A large bias can be introduced, especially if the data are truncated or if the distribution is inappropriate. In theory this bias can be removed by adding a constant, but in practice the constant is difficult to estimate unless many data are available, or unless the observations in question are drawn from a distribution whose characteristics are well known. (b) Calculated by either Mood's or Sichel's method, the lower confidence limits are about the same, whether or not a constant is added. The issue of whether or not to add a constant therefore becomes unimportant if Mood's method is used and if the investigator is interested mainly in the lower confidence limit.

2. If the coefficient of variation is less than 1.2, the mean ordinarily should not be estimated from logarithms, because the possible gain in efficiency is less than the risks of (a) introducing bias if the lognormal distribution is inappropriate and (b) choosing a constant of the wrong size.

3. Even if the coefficient of variation is less than 1.2, Mood's method may provide the best estimate of the lower confidence limit, because Student's t-method may give too low a limit, especially if the number of observations is small.

4. If a constant is added, it is better to choose one too large rather than one too small, because this choice will yield an estimated arithmetic mean that is less biased on the average. However, if the constant is very much too large, the confidence limits become too narrow because the variance of logarithms becomes too small.

5. The principal difficulty, especially for gold assay data, is that a few high assays tend to dominate the statistical analyses, regardless of the kinds of analyses that are made. Extreme care must be taken, especially when estimating grade of ore near the cutoff grade (sec. 13.2), because one does not want to overestimate and thereafter develop a mine that can operate only at a loss;

on the other hand, one does not want to underestimate and turn down a mine on which a profit could be made.

REFERENCES

Koch, G. S., Jr., and Link, R. F., 1966, Some comments on the distribution of gold in a part of the City Deep mine, Central Witwatersrand, South Africa, *in* Symposium on mathematical statistics and computer applications in ore valuation: South African Inst. Mining Metall., p. 173–189.

Koch, G. S., Jr., and Link, R. F., 1967, Gold distribution in diamond-drill core from the Homestake mine, Lead, South Dakota: U.S. Bur. Mines Rept. Inv. 6897, 27 p.

Koch, G. S., Jr., and Link, R. F., 1970, A statistical interpretation of sample assay data from the Getchell mine, Humboldt County, Nevada: U.S. Bur. Mines Rept. Inv. 7383, 23 p.

Koch, G. S., Jr., and Link, R. F., 1971, Sampling gold ore by diamond drilling in the Homestake mine, Lead, South Dakota: U.S. Bur. Mines Rept. Inv., in press.

Krige, D. G., 1960, On the departure of ore value distributions from the lognormal model in South African gold mines: Jour. South African Inst. Mining Metall., v. 61, p. 231–244.

Link, R. F., Koch, G. S., Jr., and Schuenemeyer, J. H., 1971, Statistical analysis of gold assay and other trace-element data: U.S. Bur. Mines Rept. Inv., in press. (*Note:* Entitled "The lognormal frequency distribution in relation to gold assay data" in references in Volume 1 of this book.)

ELECTRONIC COMPUTERS AND GEOLOGY

The use of electronic computers is almost essential to the effective statistical analysis of voluminous geological data. In Part VI of this book, which is Chapter 17, we explain methods of exploiting computer capabilities—methods that have proven effective for us in several years of scientific computing. These methods may help the geologist to keep control of the situation when dealing with computer scientists, programers, and operators, and with the computers themselves.

Chapter 17

Electronic Computers and Geology

What is different about a computer? An electronic computer can perform calculations heretofore performed by clerks using first pencil and paper, then slide rules and desk calculators, and finally tabulating machines. Calculating by computer differs only in degree from the kind of calculating done in the past; the number of calculations is larger but the operations performed are essentially the same. However, the true significance of electronic computers is that they can perform calculations that before were out of the question because time requirements would have been too great, or because the necessary procedures could not possibly have been effectively organized. Moreover, computers perform calculations at such high speeds that the results are available in time to be immediately put to use to control manufacturing processes, satellites, artillery, etc.

17.1 COMPUTER APPLICATIONS

Two types of computing, business and scientific, may be distinguished. Although the computing machines may be the same for both, the philosophy behind the two types is different. In business computing relatively simple arithmetic and algebraic operations are applied to usually large volumes of data. Typically the data are processed periodically, perhaps daily, monthly, quarterly, or yearly, to perform such functions as informing management, preparing bills for customers, keeping accounts, and making out pay checks. Once set up, procedures are followed for as long as possible, because, naturally, the people involved wish to have their monthly bills appear on the same date in the same form and to compare them year by year on the same basis. In scientific computing, on the other hand, complex mathematical operations are typically performed on data that may or may not be voluminous. Usually a particular calculation is made only one or a few times, although the mathematical methods basic to scientific calculations, such as elementary arithmetic, matrix inversions, and computing of statistics, may be applied many times in various combinations. Even when a procedure is done repeatedly, as for guiding satellites, it is often modified after each use, in order to encompass previous results.

The cardinal principle in computing is to organize one's ideas and work. Without organization, a large penalty is usually paid for inefficiency, and indeed no useful results may be obtained. Although organization imposes the disadvantage that more time may be needed to get started than for hand calculating, where one can pick up a slide rule and make a quick division, the not insignificant advantage accrues that one thinks about the problem long enough to decide at least tentatively what sort of calculations are likely to yield insight.

Electronic computers work on either real or simulated data, examples of which are given in sections 9.8 and 14.5. In simulation, fictitious data are obtained, usually by a random process, and are then manipulated to simulate numerical results of processes that may occur in the real world. For instance, when rocks are subjected to stress, they may yield by faulting, folding, or some other kind of deformation. As these processes are dynamic, observed field data, which portray essentially only the frozen results of the experiment, may be wholly inadequate for a meaningful statistical analysis. Through simulation, fictitious data can be obtained and manipulated to investigate the natural process. The procedure is not unlike that applied in clay-cake and other physical-model experiments; and, as in these experiments, the results suggest possible natural processes rather than uniquely specifying a single process. Simulation may also provide insight and suggest new field measurements or new field areas that may yield more information through a designed experiment. Simulation of geologic data is treated by many authors, particularly by Harbaugh and Merriam (1968) and Harbaugh and Bonham-Carter (1970).

Two broad classes of electronic computers, analog and digital, are in common use. Analog computers work on the same principle as a slide rule. Although they are suitable for some specialized applications, especially the solving of differential equations, they are not nearly as versatile for most geologic problems as digital computers, and therefore are not considered in this book. Digital computers manipulate numbers as discrete items, that is, as whole numbers. The basic unit in a digital computer is a switch, which may be either on or off. Even though the fundamental arithmetic in most modern digital computers is done in the binary rather than in the decimal number system, the results can be provided in any number system desired, and the details of binary arithmetic need not concern the geologist.

A digital computer can be considered to consist of two parts, classified according to their functions, each part generally housed in a separate cabinet or cabinets. One part, the *central processor*, consists of three components: a *control unit*, which operates the computer under the direction of a computer program; *memory*, which stores the computer program and the data being worked on at any particular time; and the *arithmetic unit*, which adds, subtracts, multiplies, divides, and makes equality comparisons. The other part of the computer, the function of which is to get data into and out of the central processor, consists of one or more pieces of *input-output* (I/O) equipment. Depending on how data are fed into the computer, the input devices read different types of media (described subsequently in sec. 17.3), including punched cards, punched paper tape, magnetic tape, video screen, and direct records such as canceled checks. The principal output devices produce data on the same media as the input devices and also on a high-speed printer or a video screen (sec. 17.3). The computer is controlled from microsecond to microsecond by a *computer program* (sec. 17.2), and on a gross basis, for instance, for turning the computer on and off, by the *console*, which is the manually operated panel or table with the switches and colored lights that have been exhibited enough on television commercials to need no discussion.

A computing system consists of the physical machinery, named *hardware*; a monitor or operating system, which is a special kind of computer program to perform housekeeping functions inside the computer; an ever-growing and changing library of written instructions for performing particular operations on specific data, named the *software*; and the people involved in the operation and in maintenance.

Prior to the development of electronic computers, the most advanced methods available to process large volumes of data were tabulating machines, whose input was punched cards and whose output was either punched cards or typed or printed paper. Tabulating machines can code, sort, and calculate, and although, in comparison to electronic computers, they are slow and limited in scope, they are still useful today for many purposes. Of particular value is the International Business Machines IBM 101 electronic statistical machine, which will add, count, and sort data on punched cards. For many computing jobs, if electronic computers are either unavailable or difficult to gain access to because of procedural problems, the IBM 101 machine may be very suitable.

Computing is discussed in this chapter in more or less the order in which a problem is handled by an investigator with an electronic computer. In section 17.2 computer programing is discussed with particular emphasis on the computer language FORTRAN, which is the language used in most scientific programing. Organization of computing is also explained. In section 17.3 data acquisition, storage, and retrieval are considered. Data acquisition requires the organization of a method to get the data and record it, either in the field or in the laboratory. Data storage involves preserving the data, and data retrieval involves getting them back so that they can be processed by computer. In section 17.4, editing data to get them right initially and to maintain their integrity is discussed. Data editing includes not only the editing, copyreading, and proofreading done in newspaper or book publishing, but also other functions peculiar to digital computing. Numerical methods for data analysis are appraised in section 17.5. Ways to display the results of calculations are taken up in section 17.6; because a large volume of results can readily be obtained by digital computer, it is important to consider carefully how best to display them. In section 17.7 computing strategy is discussed to tie together the previous sections. Finally, in section 17.8, the technology of digital computers and future trends in the industry are reviewed.

In the pages ahead, as we describe a road to scientific computing that we have found successful, we also describe the pitfalls in that road. In fact, we list so many pitfalls that the road may appear obscured, but a smooth journey may be achieved if a few precautions are observed at each stage. These we try to describe in order to assist the reader who is about to start on a computing job. The reader will learn from his own mistakes but may find it less painful to study ours first.

Of great assistance to the geologist in this day of large, complex computers are the people who work closely with them: programers, operators, and computer-center directors. They continually assist users with computing work, and the difficulty that seems monstrous to the geologist may be the kitchen mouse to them. But their knowledge extends only so far into geology and into the particular mathematical or statistical procedure to be applied; therefore the geologist must be sure that his work reaches the place where their competence begins.

17.2 COMPUTER PROGRAMING

It is convenient to distinguish three kinds of computer programing: (1) programing for operating systems or monitors, (2) compiler-language programing, and (3) applications programing. In programing for operating systems or monitors, the concern is for the efficient operation and interaction of the computer's many electrical and mechanical components. In compiler-language programing, programs are devised to translate other programs written in easy-to-use languages into instructions that the computer understands. In applications programing, specific problems of users are written in a language that the other two kinds of programing translate to a form that can be solved with the aid of a machine. Because the only kind of programing that most geologists interested in computing need to understand is applications programing, it is the only kind discussed in any detail in this book; henceforth, whenever computer programing is mentioned in the text, it will be applications programing that is meant.

The function of applications programing is to translate a mathematical equation into a form understood by a computer through a compiler program. This function is conceptually simple but not trivial, because it is like translating, from one language to another, technical information of two civilizations at different stages of development—for instance, translating a book on electronics from English to Latin. Applications programing may or may not be easy. Although some mathematical expression of the variables in question must be accomplished before starting to program, a complete mathematical formulation of a problem may be unnecessary. Instead, preliminary programing may be done and some trial results obtained in order to provide insight into the mathematical form suitable for the final program.

A computer program is simply a complete list of what, how, and the order in which computing is to be done—a list that is written in a language the computer understands. A computer program is analogous to a list of instructions written for a simple-minded operator of a desk calculator or a slide rule. Nothing can be left to the electronic computer's imagination or discretion, as it completely lacks these qualities.

A computer program is written in a language that contains the same elements as a familiar language such as English or algebra, namely, words, symbols, numbers, grammar, and syntax. There are a variety of computer languages for different purposes. Although in principle a computer program can be written in a language that the computer understands directly (a so-called machine language), in practice machine languages are seldom used today because they are difficult to learn and tedious and time-consuming to use. They were the only languages available for the first computers. Today so-called higher-order or compiler languages have been devised that are easier to use because their grammar and syntax are easier, and the words, symbols, and numbers are closely related to those familiar in English and algebra.

The principal languages for application programing are FORTRAN, ALGOL, COBOL, Programing Language I, and APL. As a result of competition, all of the computer manufacturers have devised different versions of these languages and have revised them periodically. Other special-purpose languages have been written by companies that concentrate on this work. Therefore no computer accepts all versions of all languages, although today

FORTRAN, ALGOL, and COBOL are more or less interchangeable among machines. Special programs named translators and simulators have been written that enable languages to be read, with various losses of efficiency, by machines for which they were not devised. For details, the reader is referred to representatives of the manufacturers, but is advised to take with a large grain of salt what he may hear.

What the compiler does is to translate the higher-level language into one understood by the machine. Each version of a language has a different compiler for each machine. Compiler efficiency varies and should be investigated for large problems or for programs to be run many times. However, for scientific computing, compiler efficiency is seldom critical.

Most programs for scientific computing, both published and unpublished, are written in FORTRAN. A few are written in ALGOL (especially those from Stanford University); a few are written in COBOL, which was devised for business computing; and very few have yet been written in Programing Language I or APL. The only language considered in this book is FORTRAN. Because newer versions of these programing languages are being written continually with simpler grammar and more words, human-computer interaction becomes easier year by year.

FORTRAN

Table 17.1 is an extremely simple FORTRAN program that demonstrates some attributes of the language, in particular the "Do loop," which is probably its most powerful feature. Each line of the table is one FORTRAN statement

TABLE 17.1. SIMPLE FORTRAN PROGRAM TO DEMONSTRATE SOME ATTRIBUTES OF THE LANGUAGE

Line	Statement	Type of statement
1	SUM = 0.0	Housekeeping
2	DO 10 N=1, 100	Iterative instruction
3	READ 1, W	Input instruction
4	SUM = SUM + W	Algebraic
5	WBAR = SUM/N	Algebraic
6	10 PRINT 2, N, SUM, WBAR	Output instruction
7	1 FORMAT (F10.0)	Housekeeping
8	2 FORMAT (I10,2F15.5)	Housekeeping
9	END	Housekeeping

that provides an instruction to the computer. Those instructions that are commands for action are executed by the computer in the order listed. The purpose of this computer program is to take 100 numbers, one at a time, and compute the sum and mean each time a new number is added. The table shows that the FORTRAN statements look very much like a combination of English and algebra.

This program may be explained line by line. In line 1 an algebraic variable named SUM is defined and is set equal to zero. This is a housekeeping statement, analogous to clearing a register on a desk calculator. Line 2 is the statement "DO 10 N = 1, 100," an iterative instruction that commands the computer to repeat 100 times the sequence of instructions from line 2 down to the statement in line 6. The purpose of the integer 10 is to identify the statement in line 6 of the program, a statement that is preceded by this same number, which is merely an arbitrary method of identification. The phrase "N = 1, 100" means that the commands in the Do loop are to be repeated exactly 100 times, which is analogous to repeating a certain operation 100 times on a desk calculator or on a slide rule. The instructions might be given to a desk-calculator operator to "Repeat the listed sequence of operations exactly 100 times."

A digression from the explanation of this program is introduced here to discuss the Do loop, because this is one of the most powerful statements in the FORTRAN language. Through use of the Do loop in the illustrative program, a computer is instructed to repeat a series of arithmetic operations exactly 100 times. Thus statements of the form

$$x = \sum w_i,$$

where i ranges from M to N, are easy to write. The power and compactness of this statement in many operations of mathematics and statistics are obvious.

This discussion now returns to the program illustrated in table 17.1. The instruction in line 3 is an input instruction that means that a variable named "W" should be read according to format 1, which appears in line 7. The format statement simply explains the place or "form" in which the number is to be obtained. Similarly, a desk-calculator operator might be told to read a number from column 1 in a certain table. The statement "SUM = SUM + W" in line 4 means that the variable "SUM" is to be set equal to its previous value, plus the number read in line 3. This is an algebraic statement similar to, but not exactly like, one in algebra, where the equality sign implies equivalence, whereas in FORTRAN it implies replacement. The initial value of "SUM" having been set equal to zero, the sum of W is accumulated by this command through iteration, just as numbers are accumulated in the register of a desk calculator. Line 5 contains the command that the new algebraic variable named "WBAR" is to be set equal to "SUM" divided by "N," another algebraic statement; this statment is exactly like one in algebra. The command in line 6 is an output statement directing the computer how to display the result obtained in the calculations; the command states that, in format 2, the three variables "N," "SUM," and "WBAR" are to be printed. Lines 7 and 8 are housekeeping statements specifying the formats designed 1 and 2 by a shorthand, which is equivalent to specifying columns in a table, the number of decimal points to be kept, etc. Finally, the statement "END" in line 9 simply tells the computer that the instructions are concluded. Table 17.2 gives the first and last lines of sample output produced by this program on a General Electric Company GE-225 computer.

TABLE 17.2. SAMPLE HIGH-SPEED PRINTER OUTPUT PRODUCED BY THE FORTRAN PROGRAM OF TABLE 17.1

N	SUM	WBAR
1	1.42550	1.42550
2	2.93410	1.46705
3	5.87050	1.95683
4	6.50460	1.62615
5	6.89100	1.37820
Lines 6 to 95 omitted		
96	123.71170	1.28866
97	125.07900	1.28947
98	126.12540	1.28699
99	127.66270	1.28952
100	130.14180	1.30142

Advantages of the FORTRAN language are many. The most important one is the correspondence of FORTRAN to algebra and English. However, the language has disadvantages, although they can be surmounted. For instance, step 5 in the illustrative program of table 17.1 is grammatically inadmissible in most versions of FORTRAN. Moreover, if fewer or more than 100 numbers to be summed are supplied, trouble results; the kind of trouble depends on the computing system, but the results are liable to be more unexpected than those obtained by even a naive desk-calculator operator, who should at least have sense enough to ask why 100 numbers were not provided. Safeguards, which can be incorporated in the computer program to obviate these difficulties, would be provided if this particular program were to be used many times, or without close supervision.

The FORTRAN language is explained in many books; McCracken (1961) has written one of the most widely used and certainly one of the best. The elements of FORTRAN can be learned through about six hours of course instruction or by reading McCracken's or another book. The geologist concerned with digital computing is well advised to invest this time in order to learn to write simple programs and in order to communicate more effectively with a programer. Complicated programing is best left to a professional programer, who should be able to organize computations better, to discern special cases or loopholes in requested analyses, and to perform at least some numerical analysis. Familiarity with FORTRAN enables the geologist to suggest to the programer efficient ways to perform the required operations as well as formats for convenient data input and clear display of output.

Organization of Programing

Programs represent capital investment, since the average output of a computer programer is the surprisingly small number of ten debugged (error-eliminated) FORTRAN statements per day. Completed and working programs should be documented so that this expense is not wasted. Documentation requires a description of what a program does, directions for input and output, sample input and accompanying output, instructions about what to do in case of trouble, and a list of the algebraic variables, together with as much explanation of them as is necessary. Documentation should also include a flow chart, which is a schematic pictorial diagram of the logical steps and their interrelationships. Fortunately, computer programs to make semifinished flow charts are becoming available to make the job easier. Documentation should be internally consistent and should apply to the documented program, not to some earlier or later version.

Organization of programing includes assembling and cataloging a library of programs. The library consists of programs written within the particular user organization (perhaps the geology department of a university or a petroleum research organization) and programs collected from manufacturers, organizations of computer users, friends, and the literature. The outstanding source of published programs in geology is the Kansas Geological Survey, which periodically publishes lists of its programs (some 50 in number at the end of 1970). Programs written within the user organization can be presumed to have worked at least at some time. Programs acquired from other sources are termed "canned" programs in the industry and may or may not work on the user's computer, even if they actually worked on their author's computer. However, even if they do not work, their flow charts and algebraic methods may be worth having.

Besides the programing manuals that provide primers to the computer languages, several books on numerical methods designed specifically for various languages have been written. Those by McCracken and Dorn (1964) and Ralston and Wilf (1960, 1967) are particularly recommended.

Computer programs should be interchangeable from machine to machine, and from model to model of the same manufacturer, but not always are. In fact, we know of programs that work successfully only in one computer installation and will not work elsewhere even though the computer systems are represented as identical. Therefore one should not be surprised if a program does not work on a machine on which it was not tested. This disconcerting condition, which reflects the complexity of computers and the newness of the industry, may be cured someday.

17.3 DATA ACQUISITION, STORAGE, AND RETRIEVAL

The acquisition, storage, and retrieval of data comprise the housekeeping chores of obtaining, keeping, and using data. The close attention of the geologist interested in the analysis of the data is required. The work cannot be left to other people because, being less interested than the investigator himself, they may neglect the constant, careful attention needed to preserve the data and keep them in order.

In digital computing the words *field*, *item*, and *record* have taken on technical meanings. A *field* is one or more spaces or columns used to enter a specific piece of information, such as a spectrographic analysis or a fossil name. A logical grouping by the user of fields is an *item*, which may correspond to a part of a record, one record, or several records. A *record*, as distinguished from an item, is a physical unit set by the configuration of a particular piece of computer gear; it may be one punched card, one line on a high-speed printer, one line on a typewriter, or a variable-length block of data on magnetic tape separated from the next record by a gap of specified length.

Data may be recorded in either the field or the laboratory. Most field recording is done on paper, which is either in loose sheets or bound in notebooks, the former being easier to lose initially but simpler to process later. Because data recorded on paper must be processed into a machine-readable form, their collection should be arranged in a manner suitable for transfer to machine-readable records, as discussed below. Although unnecessary, specially printed forms may facilitate this transfer, and a knowledge of input formats (table 17.1) for FORTRAN, or other language used, is helpful.

Other methods of field recording are generally less practical but may sometimes be advantageous. In mark sensing, often used by students in examinations, a special pencil deposits a graphite mark that conducts electricity. The marks are unreliable, however, because not enough graphite may be deposited, and they are difficult to erase if the marker changes his mind. Optical scanning, in which data are read optically, is today suitable at best only for clean environments. A third method of collection of field data is by a device named the Porta Punch, in which holes are made in pre-perforated cards by a special punch. But because the tiny rectangles of punched-out cardboard tend to adhere to the card and later fill in some of the holes again, we do not recommend this method.

Laboratory recording includes all the recording methods recommended for field use, optical scanning, and some additional methods. Data collected by seismic and other geophysical surveys, including those in field laboratories mounted in trucks, may be digitized directly on cards or on magnetic tape for machine input. Rock analyses, whether made chemically, by x-ray fluorescence, or by some other method, and crystallographic data can be digitized directly. Also useful are coordinate readers, which resemble a drafting machine or pantograph. When the stylus of the reader is placed on a map point, the point's coordinates are automatically recorded in machine-readable form. There are other methods of laboratory recording too numerous to mention.

If machine-readable records are not made initially, they must be produced from the original data. Verification of the records is essential, and planning will substantially reduce cost. The most common machine-readable record by far is the Hollerith punch card, named for its inventor and familiar to all Americans in the form of the $7\frac{3}{8}$ by $3\frac{1}{4}$-inch check or subscription form with the instructions "Do not fold, spindle, or mutilate." The Hollerith card has 80 columns and 12 rows for recording numerical or alphabetic information by punched holes and is read by transmitting either an electrical or a light signal through the holes. Advantages of punched cards are that they cost little, can be easily punched on widely available machines, and can be readily verified. Verification is done by an operator rekeying the original data on a machine named a verifier, which emits a signal if the character keyed does not match the punch on the card. If more than one operator is available, a different one verifies than did the original punching.

Unfortunately, since not all punch cards in use today are standardized, some tiresome details that can lead to trouble will be mentioned. Most punch cards used in scientific data processing are nominally of the size given in the last paragraph and are punched with rectangular holes. Generally, one or more of the corners of the cards are clipped off to provide a ready check on whether any cards in a deck are backward or upside down. Other cards, less widely used, are of different sizes and may have round rather than rectangular holes. Another variation, which cannot be read by some machines, is to clip one corner of the cards and to round the others. Also, if one or both of the lower corners of the cards are clipped, some card readers will not accept them. Good-quality cards are worth paying more for because poor-quality cards may deviate so much in size, thickness, and resistance to warping that they will jam some machines.

Other, less frequently used kinds of machine-readable records are also available. Punched paper tape, resembling adding-machine tape, is punched by a machine similar to a card punch. Its disadvantages are that it is difficult to verify because mistakes can be corrected only by punching a whole new tape, it is awkward to use, and it is easily torn. Magnetic tape, like that used in tape sound recorders, has the advantage that mistakes can be corrected during verification, but it is rather expensive and is less flexible than Hollerith cards. Through optical or magnetic scanning, either punch cards or magnetic tape can be made; however, so far as we are aware, these methods are not at present used for geological data.

Data Storage

In organizing data storage, the governing factors are the amount of data involved, how often and how fast the data must be recovered from storage, the kind of equipment available, and the use to be made of data after recovery.

Data are frequently stored on Hollerith cards. A good upper limit for one data file is 20,000 cards, beyond which a file starts to become unwieldy and other methods of storage should be considered. Advantages of card storage are that cards are inexpensive, easy to manipulate, and reasonably durable—although for long-term storage a controlled climate is required. Punched paper tape is undesirable for data storage because it is fragile and deteriorates rapidly. Magnetic tape is more durable than punched cards although more expensive, and there is no natural upper limit to the amount of tape that can be stored. But magnetic tapes are more difficult to manipulate than cards because the data are arranged linearly; moreover, duplicate tapes are mandatory because the tapes can be accidentally erased.

Two additional methods of storage are the disc file and the data cell. Both have the advantage that large amounts of data can be stored and quickly retrieved. Because disc files are expensive, they have been used primarily to maintain large files that receive almost daily use. However, if a disc file is available in a computer system, voluminous scientific data, such as sedimentary petrographical data from a large basin or basalt analyses of the world, might well be stored on a disc and put into the disc file periodically for updating, as the individual discs themselves are not expensive. The advantage of a disc over a tape is that the data are directly accessible in a few milliseconds as opposed to the time-consuming sequential search involved in retrieving data that are stored linearly on a tape. A data cell is physically different from a disc file but has the same advantages over magnetic tape, with access rate slower than for the disc file but at a lower cost.

Data Retrieval

For small amounts of data on punch cards, retrieval is easy because the cards can be sorted on a sorting machine and because the information punched in the card can also be printed on it and read by eye. For data stored on tapes or discs, retrieval is more difficult, because the user must either invest time enough to understand programing details or resign himself to output in a rigid format. In summary, organization is relatively informal for data on cards but becomes more and more formal for data recorded in other ways as the data file grows.

In the oil industry, where voluminous data from many wells are maintained and must be retrieved, data retrieval becomes even more important. These problems are further discussed by Dillon (1967), with references to the principal published papers.

17.4   EDITING DATA

Editing in electronic computing comprises all of the functions ordinarily performed in editing newspapers and books, including copyreading and proofreading.

While editing is essential in all disciplines, it is overwhelmingly important in scientific computing because mistakes can vitiate all results. For instance, in a trend-surface analysis (sec. 9.4), if the average north and east coordinates are 10,000 with a standard deviation of 200, and if there are 10,000 data points, one blank card read as a data card by mistake can have an effect on the resulting equation that is roughly equivalent to that of all the other data points. Potentially the situation is much more dangerous than the erroneous payroll check occasionally produced in business data processing, since the latter error will be quickly detected. If one out of 10,000 payroll checks is blank, the employee will be upset, and if the check is for $10,000,000, the bank cashier will presumably think twice before honoring it. Other common problems that arise in data editing include losing large numbers of data cards or other records—for instance, all those for a certain year or a certain month, or those having nonnumeric data in numeric fields.

Good editing requires the will to do it well and proper organization. Among magazines, *The New Yorker* is noted in publishing circles for careful editing and a vanishingly small proportion of typographical errors. Indeed, it delights in printing typographical errors from other publications. Scientific editing can be as good.

Initial Editing

The purpose of initial editing is to get a correct data file; this work is the longest and most tiresome part of many computing jobs. The principles are few but essential: avoid hand copying, verify repeatedly, assign proper identification, and make numerical checks for erroneous data.

Hand copying should never be done when ingenuity can find a way to avoid it, because mistakes are inevitable. Data collection should be planned to make hand copying unnecessary. Even if data are obtained whose collection was unplanned, an attempt should be made to transcribe them into machine-readable form without hand copying. For transcribing difficult-to-read records, key punching is usually the only feasible way. Many key-punch operators welcome the challenge to punch difficult copy as a change from routine. For instance, we have had data punched directly from large and illegible mine-assay maps and from old envelopes covered with bloody fish scales. When key-punching data, the U.S. Census deliberately introduces redundancy and tests for errors by appropriate wiring of the key punch.

All transcribed data must be verified, preferably by a different person, because if the same key-punch operator does both transcription and verification, she is almost certain to make some of the same mistakes twice. At best, verification does not ensure perfect transcription. If the character "8" is mispunched "3," by chance alone the verifying operator may again punch a "3," even though she intends to punch an "8;" if the character is difficult to read, the chance for error is increased. If machine verification is impossible, transcribed records must be proofread. Although proofreading can be excellent, as in *The New Yorker*, more often it is poor, but nonetheless better than nothing.

Both the original and the machine-readable records must be properly identified. The original records perhaps consist of several notebooks, a pile of maps, or loose sheets of paper. Whatever the original records are, they should be regarded as documents and assigned document numbers in sequence. These numbers should be stamped with a numbering stamp, not written by hand, because, if they are written by hand, inevitably numbers will be repeated or skipped. Thus all numbers in some range—say 50 to 59, or 101 to 200—may inadvertently be left out, or two number 243s put in. On machine-readable records, an essential part of the identification is serial numbers, which must also be assigned by machine. The last few columns on punched cards may be reserved for serial numbers. Other typical identification for machine-readable records comprises geographic location, date of information, name of collector name of chemical element, and name of variable. If data are on cards, the data format may be preprinted on them.

The essential last step in initial editing is checking for errors. In checking, nothing helps so much as familiarity with the subject matter, ingenuity, and bitter experience. The programer cannot be expected to know that trilobites are more nearly 2 centimeters than 2 meters long, that the distance between outcrops is more likely to be 10 miles than 10 inches, and that trace-element determinations are usually reported in parts per million or per billion than in percentages. Some or all of the following checks may be made in approximately the order listed:

1. Check for duplicate data. In a spirit of helpfulness, your two graduate students may have each handed in, on different days, the same data notebook for key-punching. We once wasted a week on a set of duplicate data carefully disguised by the sender through the clever ruse of changing mine names and drill-hole numbers.

2. Check to see if numeric fields contain numbers. If they contain letters or other miscellaneous characters, like asterisks or dollar signs, something is wrong.

3. For each numeric field, sort the data in order of size, and count the records. If the data are on punched cards, sort also on columns in which the proper character is a decimal point in order to identify mispunched or un-punched decimal points and any other of several common mistakes. List the sorted data. Calculate reasonable bounds for each variable, and see if all of the observations are within them (for example, reasonable bounds for iron in analyses of magnetite ore are from zero to 74 percent). Check the largest and smallest observations to see if they are reasonable. FUNOP (sec. 6.4) and FUNOM (Tukey, 1962) may be useful.

4. Restore the data to their original sequence. Verify that the serial numbers are in sequence, with none omitted or duplicated, and that the number of records is correct.

5. Make frequency distributions and histograms for all variables to reveal clumps in the data, which may be caused by the analyst or recorder preferring certain digits (for instance, a chemist may tend to record observations as either odd or even numbers or prefer some numbers such as 10 or 20), and to provide clues to other errors such as doubled-up observations (for instance, twice the expected observations for one year and none for the next because of a coding mistake).

6. Repeat some or all of the above steps for any obvious derived variables. For instance, original observations might be collar and bottom elevations of drill holes. One derived variable is hole depth, equal to collar elevation minus bottom elevation, which should be a positive number for holes drilled vertically downward.

7. For all variables, compute the mean and standard deviation.

We have used these editing schemes to advantage for data files of a few thousand or a few tens of thousands of records. For larger data files, other schemes have been suggested. One investigator recommends reading records into the computer in small groups, say of nine records each, computing statistics for each group, and comparing these statistics with those calculated for successive groups. Another worker recommends checking a 10 percent sample of the data to identify major changes and errors introduced over a period of time.

Despite great care, errors are bound to occur, and the geologist needs to be constantly on guard against them. In an interesting and entertaining article, Coale and Stephan (1962) explain how knowledge of the subject matter and careful detective work enabled them to solve "the Case of the Indians and the Teen-age Widows." They write,

An examination of tables from the 1950 U.S. Census of Populations, and of the basic Persons punch card, shows that a few of the cards were punched one column to the right of the proper position in at least some columns. The result is that numbers reported in certain rare categories—very young widows and divorcees, and male Indians 10–14 or 20–24—were greatly exaggerated. These errors occurred in spite of a careful checking program, and illustrate the necessity for users to view data concerning rare categories with special caution.

Maintaining Data Integrity

The first rule in maintaining data integrity is to label everything completely and carefully. Otherwise it is difficult if not impossible to remember the meaning of the data—to discover in what format and conventions the data were recorded, for what purpose the data file was assembled, and to whom it belongs. When this happens, a great deal of hand waving is indulged in; various people connected with the installation from previous years are consulted; and the end result is usually that a potentially valuable data file must be thrown out.

Some specific ways to maintain data integrity are the following:

1. Keep a master file. This file is preserved and used only as a source for making working files, or as a reference for correcting the files.

2. Keep one or more working files. If a file is on punched cards, a working file is highly desirable because cards can always be mutilated by the card reader, lost by users, or misplaced. If files are on other media, it is absolutely essential to keep master files that are not used in working on the data because it is so very easy to erase the other media by mistake, particularly magnetic tape. Working files should be checked for data integrity periodically and whenever trouble occurs—such as after a day in which card-reader malfunctions have been unusually large, operators have been especially harrassed, handfuls of cards have been dropped on the floor, etc. To verify the reliability of a file, recompute the number of records (for a punched file, the number of cards), the mean for all variables, and the standard deviation. If all of these agree with the original computations made in initial editing, there is a good chance that data integrity has been maintained.

3. When data integrity has been lost, first find out the cause of the difficulty without using the master file. The villain may be the computer, which could erase the master tape exactly as it erased the working tape. After the computer

is found to be not at fault, the master file should be duplicated to make one or more new working files.

17.5   NUMERICAL METHODS

In ordinary operation, electronic computers perform most calculations in about eight significant figures, or in 16 figures in the "double-precision" mode of operation, in which twice the usual number of significant figures are carried in the calculation. Because the calculations are done with only a small number of significant figures, severe accuracy problems can arise from cumulative round-off errors. These problems may be especially severe when matrices are inverted for such operations as trend-surface analyses (Chap. 9) and when calculations are made of eigenvalues and eigenvectors (Chap. 10). As an example of round-off errors, table 17.3 gives the result of adding a number to itself 100 times on a General Electric Company GE-225 computer.

Table 17.3. Result of adding a number to itself 100 times for five numbers

Number	Result of adding the number to itself 100 times	
	Correct result	Computer result
0.01	1.00	0.99999998
0.02	2.00	1.99999997
0.03	3.00	3.00000002
0.04	4.00	3.99999993
0.05	5.00	4.99999998

One must always be concerned about numerical accuracy. Even if a program has been run many times on a computer, results should always be inspected to see that they are reasonable. Either the computer or the program can fail to work properly and yet display no obvious symptoms. On several different computers, we have run the same problem on the same computer on different days with different results. Everyone with computing experience has had this sort of trouble.

For many problems, random numbers are required. Those in the RAND tables (1955) should be used whenever possible because they are the best genuine random numbers published, having been tested repeatedly and found valid. The RAND numbers must be read into the computer, however, and, to avoid this trouble various methods have been devised to generate pseudo-random numbers within a computer. Sometimes these pseudo-random numbers may be used with acceptable results, but one should be cautious when using them since evidence has accumulated that patterns can occur, especially in multivariate problems.

Among the best of many books that explain numerical methods are those by McCracken and Dorn (1964), Ralston and Wilf (1960, 1967), Milne (1949), and Hamming (1962).

17.6 DISPLAY OF RESULTS OF CALCULATIONS

The principal media for displaying the results of calculations are paper and the video screen. Information is recorded on paper by a high-speed printer, a graphical plotter, or a typewriter. The most widely used of these media is the high-speed printer, which is therefore emphasized in this section.

Today more than 10 percent of all printing in the United States is computer printout. The Livermore Laboratory of the University of California alone prints the equivalent of 300 average-size novels a week. The large volume is attained because digital computers can calculate and output results very rapidly; high-speed printers list at the rate of several hundred lines per minute. Thought must be given to the display of the results of calculations, or the output will be so repulsive that no one will read it.

Organization of results does not imply abbreviation; on the contrary, voluminous output is generally desirable because unsuspected interesting results are often found. Tukey (1965) advocates obtaining computer output measured in "side-feet," that is, the height of the pile of computer-printed paper rather than the length of the paper when unfolded. There is no place in effective computing for an effort to save paper; it is much the cheapest part of the computing process and should be used freely.

All output should be carefully labeled. This stricture also applies to intermediate output, like card decks, that is to be listed. If labeling is done by humans such as computer operators, clerks, or the investigator, it will be put off until tomorrow and then forgotten; the labeling should therefore be done by machine. One must be sure that the labeling matches the output on any particular page.

It is advisable to print out intermediate results, which may be unexpectedly interesting in their own right or may indicate, if the final results are wrong, how to modify the program to get correct results next time. It is also helpful to print out some or all of the original data to verify that the computations were applied to the investigator's data rather than to other data that were lying around the computer room.

Several principles of organizing data output are worth mentioning. Fixed decimal-point formats should be used rather than exponential (floating-point) formats whenever possible, because numbers printed in this way are much easier to read. Lines should be skipped between pages; otherwise, if the pages get out of register, valuable information printed on the perforations may become unreadable when the sheets are torn apart. Results should be printed on the same page as the labels, because the pages are invariably torn apart by accident or design, and then the unlabeled pages may be unidentifiable.

There are two basic types of data display, tabular and graphical, and several media on which to display the two types. Most results are displayed on paper that is outputted by a high-speed printer. The widths of the standard industry print lines are either 120 or 132 characters, each requiring $\frac{1}{10}$ inch. The paper is either in rolls that can be automatically cut into desired lengths after printing or in pages folded together in accordion pleats. Using this medium, the data can be displayed in a table or graph that is 12 or 13.2 inches wide and of indefinite length. In the past, paper was also printed by computer-driven typewriters, but they are unreliable and slow and are little used at present except for subordinate output. Paper is also printed by a Teletype, which has a print line of 72 characters, each requiring a $\frac{1}{10}$-inch space. The advantage of Teletype printing is that the machine can serve for both input and output even though it is remote from the computer. But Teletype printing is slow and therefore undesirable if much input or output is required. Paper, up to 4 feet wide, is also driven by automatic graphical plotters, which are expensive and slow. They are driven with the aid of "canned" programs supplied by their manufacturers.

Another mode of display is the video screen, which may be of two types. The first reproduces information essentially as does a television screen, for example, for process control in industry. Disadvantages are that the number of characters per picture is limited, generally to a 32 by 32-inch grid, and that the information is temporary. In principle, selective information retrieval is also possible if there is too little time and paper with which to recover all of the large amount of information that is stored in the computer memory. The second type makes high-quality graphs that can be photographed for a permanent record.

17.7   COMPUTING STRATEGY

Unless the investigator has a clear purpose or purposes in mind, nothing good will result from computing. In scientific computing, little has ever been accomplished by accumulating large tables of data without giving previous thought to what sort of information may be extracted from them. This is not to say that objectives should not be changed or modified as a project develops, but some tentative goals should be set before data are collected. For example, nothing is gained by collecting all of the world's granodiorite analyses on punched cards unless one has some idea of what can be done with this information.

The strategies suggested in this section have proven effective for us throughout some 30 man-years of scientific computing in both basic and applied research. We hope that some of these methods, many of which are similar to or the same as those used by others who do scientific computing, will be helpful to the reader.

The first principle is to establish tentative objectives. Perhaps species in two fossil populations are to be compared, ore reserves are to be calculated faster and more accurately, or differentiation in a batholith is to be studied by examining rock types. All these problems are relatively simple, can be well defined, and are suitable for computer study. Seldom is anything accomplished by attempting a "complete" computerized study of a large geologic area, of zircon analyses of the world, or the like.

Preliminary computing should be done after some of the data have been collected or simulated. Several of the methods listed in table 17.4 may be applied for the initial summarization, along with others that seem desirable. Those at the top of the table are more standard than those at the bottom.

TABLE 17.4. SOME METHODS FOR INITIAL SUMMARIZATION OF DATA

Frequency distributions
Probability plots on normal or lognormal probability graph paper
Lists sorted on different variables
Scatter diagrams
Standard summary statistics: mean, standard deviation, coefficient of variation, correlation coefficients, confidence intervals
Summarization in one-, two-, or three-dimensional cells
Computer-printed contour maps
Cumulative sums

Analyses of variance
Regression analyses, including trend-surface analyses
Inspection of residuals from analyses of variance and regression analyses
Transformations
FUNOR, FUNOM
Computation of standard variables such as tonnage-volume factor and normative minerals

---

Frequency distributions summarize the basic information about variability in the data and identify any unusual clumps. For instance, a frequency distribution of grain-size measurements may show whether they are from a single or a mixed distribution (sec. 6.5). Lists of each variable, sorted according to its size, with the corresponding observations for the other variables, will show how the different variables are related to the one sorted. For instance, a list of the Fresnillo assay data that is sorted on silver reveals that most of the high-grade assays are from the upper mine levels and are associated with low lead assays (sec. 6.5). A probability plot on either normal or lognormal probability paper (sec. 6.4) may indicate that the data follow one of these well-understood theoretical frequency distributions. Plotting data on a scatter diagram, easily done by computer, will quickly reveal any strong linear or curvilinear correlations between variables.

If data assigned to geographic locations are numerous, with, say, a few hundred or a few thousand or more data points, summarizing the observations in equal-sized cells (in either one, two, or three dimensions) will show whether the number of points and the local variability in each cell are approximately the same. If the data are distributed in two dimensions, perhaps on a surface map or a cross-section, a contour map, prepared by linear interpolation among points (sec. 9.9), can provide a clear initial summary of the data. Standard summary statistics—mean, standard deviation, coefficient of variation, correlation coefficient, and confidence intervals—should also be computed. These statistics can be compared with those expected from theory or those characteristic of similar types of data. Similarity may stem from chemical elements with comparable inherent variability, from geological environments, from fossil populations, or from many other attributes that experience and theory suggest. For some kinds of data, particularly assays and chemical analyses, cumulative sums show which of the observations account for most of the sum.

Some or all of the more specialized analyses in the bottom part of the table may also be informative. Analyses of variance (Chap. 5) may define the sources of variability; trend-surface (Chap. 9) and other forms of regression analysis (Chap. 10) may reveal that two or more variables are related. Residuals (sec. 9.8) from analyses of variance and regression analyses often show interesting patterns. Transformations (Chap. 6) may be valuable if some variable other than that measured is more natural. FUNOP, FUNOM, and FUNOR (sec. 6.4) are applicable with some problems. Finally, any standard derived variables should be computed, examples of which are the normative, von Wolff, and CPIW values for minerals. Special programs for calculating these values were published by McIntyre in 1963.

One should not stop with the methods in table 17.4, which are drawn arbitrarily from our experience. Those in other fields should use the "standard" methods traditional to their field. The investigator should, moreover, always be on the lookout for other ideas suggested by his ingenuity, reading, or conversations with other workers.

It is neither necessary nor desirable to have all the data in hand before starting a study. A pilot investigation may provide valuable insight that can guide data collection for a more complete investigation. Common practice in questionnaire-type surveys is to pretest the questions on a selected sample before questioning the entire group of people. In geology, one or a few outcrops or formations may be sampled before the study is extended to the entire area of interest.

Thus the preliminary computing serves as a guide—perhaps to obtaining additional data but certainly for testing the procedures to be used for the final data analysis.

For some reason obscure to us, when individuals or organizations first embark on computing, they generally want a master computer program that will analyze all data of all kinds for all purposes. Unfortunately, such over-mechanization is not the road to good computing. Instead it is much better to form a library of relatively simple programs that can be applied to data of all sorts to yield specific results. Thus the data analyst needs a set of programs, just as the carpenter needs a set of tools, each with its particular function, and, just as the carpenter's tools are related to each other so that one tool begins to work the wood where another leaves off, computer programs should be organized so that output from one serves as input for another.

Once the computing is done and output has been obtained, its reliability must be assessed, in one or both of two ways. *Assume that trouble is normal and be prepared to cope with it.*

The first way to assess reliability is to exercise common sense. Does the mean lie between the highest and lowest numbers in the data set, and are the largest and smallest observations small multiples of the standard deviation? Is the frequency distribution skewed and, if so, should it be, in accordance with theory and experience? Are the correlation coefficients less than 1? The second way to evaluate reliability is to apply geological knowledge. Is the average feldspar content of the granite less than 95 percent, and is the thickness of the limestone bed less than 1000 feet?

As a simple example of how computing strategy may be applied, we briefly mention a typical problem once put to us. A 1000 by 1000 grid of points was to be laid out on a topographic base. The elevation was to be measured at each point and the results of natural erosional processes were to be simulated. When one looks at this problem in terms of the 1 million (1000 × 1000) points, it seems formidable. The computer need handle only a small part of the problem at any one time, however, once the task is understood in terms of traversing an array of points row by row or column by column. As illustrated by this example, a general knowledge of the computer operation may be necessary in order to plan a successful strategy.

17.8 TECHNOLOGY OF DIGITAL COMPUTERS

A computer system is composed of the following elements: the computer, the software, the programers, the operators, the operating rules, and maintenance. Computer systems are changing rapidly today. The trend is toward larger and more complex systems. The first digital computers built during World War II were designed, constructed, and operated by the same group of men, all of whom were intimately acquainted with the peculiarities of the individual machines. In the 1950s most computers were still small enough to be operated on an "open-shop" basis, which means that the investigators and programers could themselves run the computers directly. With the advent of larger computers, which are both more complex and more costly to operate, the tendency has been toward "closed-shop" computing, with the computers run by special operators. The advantage is that these operators are more familiar with the routine working of the machine than are technically trained men, but the disadvantage is that, because their function is essentially clerical, they make the sorts of mistakes expected of clerks, such as introducing programs and data upside down or backward and failing to take appropriate measures when unexpected, nonroutine things happen.

Operating rules can be very simple and therefore of little concern or, on the other hand, so complex that for the user they are the dominant element of the computer system. An analogy will clarify this contention. Organizations whose employees use automobiles have policies about them. At one extreme, each employee or department requiring an automobile is assigned one or more cars or else the employee is paid for operating his private automobile. From the viewpoint of the organization, this is a simple operating rule, although the employee may need to keep records to justify being reimbursed by his employers. In contrast is the scheme of keeping automobiles in a car pool and assigning them to individuals as needed. Here relatively complicated operating rules are required, because forms must be filled out to obtain the vehicles, priorities must be established, reservations must be made, etc. The advantage may be better utilization of the equipment. On the other hand, the operating rules may become so onerous that employees defer traveling or circumvent the rules by taking airplanes or by hiring a rental car, the operating rules for which are purposely made simple by the car-rental companies.

Operating rules in computing cause the same sorts of problems as in automobile usage. Having worked in computer systems with both simple and complicated operating rules, we have found that our most productive computing has been done under systems with simple rules. Even then our most insightful computing has been done on weekends, when most operating rules are suspended. The most important operating rule sets "turnaround time," which is the elapsed time from when a job is handed in to when results are obtained. If this time is measured in minutes, the user can wait for his results and readjust the remainder of his program and data according to the results already obtained. If this turnaround time is an hour or two, the user can walk back to the computer while his idea is still somewhat fresh in mind. However, if the turnaround time is overnight, chances are that the inspiration that suggested the computing has dissipated and the steps to be taken next, provided that the preliminary program works, have been forgotten. And if the turnaround time lengthens to much more than a day, only routine computing is accomplished, because new ideas are so time-consuming to implement that the user becomes discouraged.

If, then, operating rules for complicated computing systems must be complex, what is the best solution for the user doing scientific computing? We believe that it is to do most computing on a system with simple operating rules and then, after a program and data are properly working and organized, to transfer the entire problem, if need be, to a large computer.

In recent years, multiple-use computers used for several functions have become increasingly popular. For instance, a computer may do scientific computing on the first shift, do business computing on the second shift, and be rented to an outside organization for the third shift. Multiple-use computing can be satisfactory and efficient, but when the inevitable breakdowns in equipment and systems occur and schedules must be readjusted, priority is liable to be given to preparing such items as payroll checks, that are of interest to everyone, in preference to scientific computing.

Time sharing is a kind of computing that has become operational, commercially rather than experimentally, only in the past few years. In this form of computing, satellite input-output devices, mostly Teletypes, are connected to a central computer, which may be thousands of miles away. The central computer can work efficiently, at least in principle, on problems relayed by the satellite stations.

Hardware Technology

The current trend in hardware technology is toward larger, faster, and more expensive computers, with which their users often communicate through time-sharing terminals. Acquisition of a computer, whether by purchase or rental, is likely to require investment of much time as well as money, because not only the hardware must be considered but the computing system as a whole. For instance, one university recently installed a large computer whose potential is not nearly realized because the output equipment has been entirely unable to keep up with the central processor. This difficulty results from a failure to anticipate the ratio of computing to the output time that would be required by the computer's users.

Thus the acquisition of a computer is highly specific to individual situations, and general advice is difficult if not impossible to give. Mathews (1968) has written a good general article, and the periodical literature and the manufacturers can provide detailed additional information.

Software Technology

For most users in all disciplines, and certainly for most geologists, software technology is far more important than hardware technology. Since it affects primarily size and speed, hardware technology has little or no impact on most users and is of great concern only for those with exceedingly large volumes of routine data to handle, such as insurance companies, the Internal Revenue Service, and the Census Bureau.

As computing systems become larger and more complex, software technology changes, from the user's point of view, for better or for worse. The changes for the worse are several. First, the user must cope with an operating system, which is the over-all program that schedules the various computer components for different jobs with various computational problems, priorities, and lengths. A new one every year or so is to be expected, because manufacturers and computer-center directors delight in devising them. With each new system, new rules and conventions must be learned, and programs that worked before may no longer operate. Another bad change is that, as systems get bigger, closed shops become inevitable, turnaround time increases, and the user is farther and farther removed from the actual computing process—all of which makes less efficient the interaction between the human and the computer. Moreover, really major catastrophes are possible. For instance, if the data are on tapes, they may all be erased by mistake.

However, the changes for the better may outweigh those for the worse. Programing aids, for instance error messages, get better all the time. More advanced languages, which are library-linked subroutines, are becoming more available and more efficient. In these languages, a single statement equals a list of FORTRAN (or other equivalent language) statements. For instance, one statement can program a computer to perform a factor analysis or compute correlation coefficients. Some of these more advanced languages are P-STAT, SNAP, and SNOBOL. Real-time applications are possible, and more efficient computer use can be attained through multiprograming, in which several programs are run on the computer at one time. The computer then selects those that are to be handled in various parts of the central processor at any one time. These changes are all to the good, especially for users who dislike dealing directly with computers.

Perhaps the most important change is the trend toward time sharing, which is nearly the same as having a small computer like those introduced in the 1950s. With time sharing, computing has come full circle, back to the days when human-computer interaction was relatively easy, with the large added advantages of more memory for larger problems, better programing languages, and improved diagnostic aids. Although time sharing still has substantial problems, it offers a great potential for the future.

REFERENCES

Coale, A. J., and Stephan, F. F., 1962, The case of the Indians and the teen-age widows: Jour. Am. Statistical Assoc., v. 57, no. 298, p. 338–347.

Dillon, E. L., 1967, Information storage and retrieval systems: Seventh World Petroleum Cong. Proc., v. 2, p. 555–560.

Hamming, R. W., 1962, Numerical methods for scientists and engineers: New York, McGraw-Hill, 432 p.

Harbaugh, J. W. and Bonham-Carter, Graeme, 1970, Computer simulation in geology: New York, John Wiley & Sons, 575 p.

Harbaugh, J. W., and Merriam, D. F., 1968, Computer applications in stratigraphic analysis: New York, John Wiley & Sons, 259 p.

Mathews, M. V., 1968, Choosing a scientific computer for service: Science, v. 161, July 5 issue, p. 23–27.

McCracken, D. D., 1961, A guide to FORTRAN programming: New York, John Wiley & Sons, 88 p.

McCracken, D. D., and Dorn, W. S., 1964, Numerical methods and FORTRAN programming: New York, John Wiley & Sons, 457 p.

McIntyre, D. B., 1963, FORTRAN II program for norm and von Wolff computations: Claremont, Calif., Pomona College Dept. of Geology, Tech. Rept. 14.

Milne, W. E., 1949, Numerical calculus: Princeton, N.J., Princeton Univ. Press, 393 p.

Ralston, A., and Wilf, H. S., 1960, Mathematical methods for digital computers, v. 1: New York, John Wiley & Sons, 293 p.

Ralston, A., and Wilf, H. S., 1967, Mathematical methods for digital computers, v. 2: New York, John Wiley & Sons, 287 p.

RAND Corporation, The, 1955, A million random digits with 100,000 random deviates: Glencoe, Ill., Free Press, 200 p.

Tukey, J. W., 1962, The future of data analysis: Ann. Mathematical Statistics, v. 33, p. 1–67.

————, 1965, The technical tools of statistics: Am. Statistician, v. 19, p. 23–27.

APPENDIX

Table A.1. Percentiles of the distribution of r when $\rho = 0$*

N	$r_{95\%}$	$r_{97.5\%}$	$r_{99\%}$	$r_{99.5\%}$	$r_{99.95\%}$	N	$r_{95\%}$	$r_{97.5\%}$	$r_{99\%}$	$r_{99.5\%}$	$r_{99.95\%}$
5	.805	.878	.934	.959	.991	20	.378	.444	.516	.561	.679
6	.729	.811	.882	.917	.974	22	.360	.423	.492	.537	.652
7	.669	.754	.833	.875	.951	24	.344	.404	.472	.515	.629
8	.621	.707	.789	.834	.925	26	.330	.388	.453	.496	.607
9	.582	.666	.750	.798	.898	28	.317	.374	.437	.479	.588
10	.549	.632	.715	.765	.872	30	.306	.361	.423	.463	.570
11	.521	.602	.685	.735	.847	40	.264	.312	.366	.402	.501
12	497	576	.658	.708	.823	50	.235	.279	.328	.361	.451
13	.476	.553	.634	.684	.801	60	.214	.254	.300	.330	.414
14	.457	.532	.612	.661	.780	80	.185	.220	.260	.286	.361
15	.411	.514	.592	.641	.760	100	.165	.196	.232	.256	.324
16	.426	.497	.574	.623	.742	250	.104	.124	.147	.163	.207
17	.412	.482	.558	.606	.725	500	.074	.088	.104	.115	.147
18	.400	.468	.543	.590	.708	1000	.052	.062	.074	.081	.104
19	.389	.456	.529	.575	.693	∞	0	0	0	0	0
N	$-r_{5\%}$	$-r_{2.5\%}$	$-r_{1\%}$	$-r_{0.5\%}$	$-r_{0.05\%}$	N	$-r_{5\%}$	$-r_{2.5\%}$	$-r_{1\%}$	$-r_{0.5\%}$	$-r_{0.05\%}$

* From Dixon, W. J., and Massey, F. J., Jr., 1969, Introduction to statistical analysis: New York, McGraw-Hill, p. 569.

Table A.2. Values of the quantity $\left\{ \left(\dfrac{1}{0.1} \right)^{1/(n-1)} - 1 \right\}$, n from 3 to 100

N	0	1	2	3	4	5	6	7	8	9
0	—	—	—	2.162	1.154	0.778	0.585	0.468	0.389	0.334
10	0.292	0.259	0.233	0.212	0.194	0.179	0.166	0.155	0.145	0.136
20	0.129	0.122	0.116	0.110	0.105	0.101	0.096	0.093	0.089	0.086
30	0.083	0.080	0.077	0.075	0.072	0.070	0.068	0.066	0.064	0.062
40	0.061	0.059	0.058	0.056	0.055	0.054	0.053	0.051	0.050	0.049
50	0.048	0.047	0.046	0.045	0.044	0.044	0.043	0.042	0.041	0.040
60	0.040	0.039	0.038	0.038	0.037	0.037	0.036	0.036	0.035	0.034
70	0.034	0.033	0.033	0.032	0.032	0.032	0.031	0.031	0.030	0.030
80	0.030	0.029	0.029	0.028	0.028	0.028	0.027	0.027	0.027	0.027
90	0.026	0.026	0.026	0.025	0.025	0.025	0.025	0.024	0.024	0.024
100	0.024									

TABLE A.3. SIGNIFICANCE POINTS OF R_0 AT THE 10 PERCENT LEVEL*

0	—	—	—	2.45	2.85	3.19	3.50	3.78	4.05	4.30
10	4 54	4.76	4.97	5.18	5.38	5.57	5.75	5.93	6.10	6.27
20	6.44	6.60	6.75	6.90	7.05	7.20	7.34	7.48	7.62	7.76
30	7.89	8.02	8.15	8.28	8.40	8.52	8.65	8.77	8.88	9.00
40	9.12	9.23	9.34	9.45	9.56	9.67	9.78	9.88	9.99	10.09
50	10.19	10.30	10.40	10.50	10.59	10.69	10.79	10.89	10.98	11.08
60	11.17	11.26	11.35	11.45	11.54	11.63	11.72	11.80	11.89	11.98
70	12.07	12.15	12.24	12.32	12.41	12.49	12.57	12.66	12.74	12.82
80	12.90	12.98	13.06	13.14	13.22	13.30	13.38	13.46	13.53	13.61
90	13.68	13.76	13.84	13.91	13.99	14.06	14.13	14.21	14.28	14.36
100	14.42									

* Calculated by J. H. Schuenmeyer, using Watson's method.

TABLE A.4. ONE-HALF WIDTH OF A 95 PERCENT TWO-SIDED CONFIDENCE INTERVAL FOR γ

	γ									
n	0.25	0.50	0.75	1.00	1.25	1.50	1.75	2.00	2.25	2.50
10	0.122	0.278	0.492	0.773	1.128	1.557*	2.063*	2.646*	3.305*	4.042*
20	0.084	0.193	0.343	0.541	0.792	1.095	1.453	1.865	2.331*	2.852*
30	0.068	0.157	0.279	0.441	0.645	0.893	1.185	1.521	1.902	2.327
40	0.059	0.135	0.241	0.381	0.558	0.773	1.025	1.317	1.646	2.015
50	0.052	0.121	0.215	0.341	0.499	0.691	0.917	1.177	1.472	1.802
60	0.048	0.110	0.196	0.311	0.455	0.630	0.837	1.075	1.344	1.645
70	0.044	0.102	0.182	0.288	0.421	0.583	0.775	0.995	1.244	1.522
80	0.041	0.095	0.170	0.269	0.394	0.546	0.724	0.930	1.164	1.424
90	0.039	0.090	0.160	0.254	0.371	0.514	0.683	0.877	1.097	1.342
100	0.037	0.085	0.152	0.240	0.352	0.488	0.648	0.832	1.041	1.274
150	0.030	0.069	0.124	0.196	0.288	0.398	0.529	0.679	0.849	1.040
200	0.026	0.060	0.107	0.170	0.249	0.345	0.458	0.588	0.736	0.900
300	0.021	0.049	0.088	0.139	0.203	0.282	0.374	0.480	0.601	0.735
400	0.018	0.042	0.076	0.120	0.176	0.244	0.324	0.416	0.520	0.637
500	0.016	0.038	0.068	0.107	0.157	0.218	0.290	0.372	0.465	0.569
1000	0.012	0.027	0.048	0.076	0.111	0.154	0.205	0.263	0.329	0.403

* Lower confidence limit = 0.

TABLE A.5. ONE-HALF WIDTH OF A 90 PERCENT TWO-SIDED CONFIDENCE INTERVAL
FOR γ

| n | \multicolumn{10}{c}{γ} |
	0.25	0.50	0.75	1.00	1.25	1.50	1.75	2.00	2.25	2.50
10	0.102	0.233	0.413	0.649	0.946	1.307	1.732	2.221*	2.774*	3.393*
20	0.071	0.162	0.288	0.454	0.665	0.919	1.219	1.565	1.957	2.394
30	0.057	0.132	0.234	0.370	0.541	0.749	0.994	1.277	1.596	1.953
40	0.049	0.114	0.202	0.320	0.468	0.648	0.861	1.105	1.382	1.691
50	0.044	0.101	0.181	0.286	0.419	0.580	0.770	0.988	1.236	1.512
60	0.040	0.092	0.165	0.261	0.382	0.529	0.702	0.902	1.128	1.380
70	0.037	0.086	0.153	0.241	0.354	0.490	0.650	0.835	1.044	1.278
80	0.035	0.080	0.143	0.226	0.331	0.458	0.608	0.781	0.977	1.195
90	0.033	0.075	0.134	0.213	0.312	0.432	0.573	0.736	0.921	1.127
100	0.031	0.071	0.127	0.202	0.296	0.410	0.544	0.698	0.873	1.069
150	0.025	0.058	0.104	0.165	0.241	0.334	0.444	0.570	0.713	0.873
200	0.022	0.050	0.090	0.143	0.209	0.289	0.384	0.494	0.617	0.756
300	0.018	0.041	0.073	0.116	0.171	0.236	0.314	0.403	0.504	0.617
400	0.015	0.036	0.064	0.101	0.148	0.205	0.272	0.349	0.437	0.534
500	0.014	0.032	0.057	0.090	0.132	0.183	0.243	0.312	0.390	0.478
1000	0.010	0.023	0.040	0.064	0.093	0.129	0.172	0.221	0.276	0.338

* Lower confidence limit = 0.

Partial List of Symbols

r	sample correlation coefficient	16
	interest rate	312
R	mode of eigenvalue analysis	126
	vector resultant	140
	ratio	154
s_e^2	sample variance of the random fluctuation e	11
	residual mean square	65
s_u^2	eigenvalue	124
S_i	cash income per period	312
SS	sum of squares	11
t	Student's t-statistic	21
T	tonnage	252
T^2	Hotelling's T^2-statistic	102
u	a transformed observation	104
u_i	rotated axis	122
V	volume	252
w	observation or dependent variable	4
	numerator of a ratio	154
$\overline{w}$	sample mean	11
$\hat{w}$	estimated value of w	8
x	east coordinate	4
	denominator of a ratio	154
$\bar{x}$	mean of x	11
x_i	ith independent variable	4
	an open array of variables	169
y	north coordinate	4
y_i	a closed array of variables	169

Greek Letters

α	constant parameter in regression analysis	4
	intercept	8
	percentage risk of type I erro	17
β	slope	8
β_i	ith regression coefficient	4
	population coefficient of variation	165

θ	azimuth of a directional observation	132
μ	population mean	102
γ	population correlation coefficient	16
σ	population standard deviation	16
σ^2	population variance	16
σ_e^2	population variance of the random fluctuation e	11
ϕ	plunge of a linear regression surface in three dimensions	38
	inclination of a directional observation	132

Index

A CATALOGUE OF
SELECTED DOVER BOOKS
IN ALL FIELDS OF INTEREST

A CATALOGUE OF SELECTED DOVER
BOOKS IN ALL FIELDS OF INTEREST

CELESTIAL OBJECTS FOR COMMON TELESCOPES, T. W. Webb. The most used book in amateur astronomy: inestimable aid for locating and identifying nearly 4,000 celestial objects. Edited, updated by Margaret W. Mayall. 77 illustrations. Total of 645pp. 5⅜ x 8½.
20917-2, 20918-0 Pa., Two-vol. set $9.00

HISTORICAL STUDIES IN THE LANGUAGE OF CHEMISTRY, M. P. Crosland. The important part language has played in the development of chemistry from the symbolism of alchemy to the adoption of systematic nomenclature in 1892. ". . . wholeheartedly recommended,"—Science. 15 illustrations. 416pp. of text. 5⅝ x 8¼. 63702-6 Pa. $6.00

BURNHAM'S CELESTIAL HANDBOOK, Robert Burnham, Jr. Thorough, readable guide to the stars beyond our solar system. Exhaustive treatment, fully illustrated. Breakdown is alphabetical by constellation: Andromeda to Cetus in Vol. 1; Chamaeleon to Orion in Vol. 2; and Pavo to Vulpecula in Vol. 3. Hundreds of illustrations. Total of about 2000pp. 6⅛ x 9¼.
23567-X, 23568-8, 23673-0 Pa., Three-vol. set $26.85

THEORY OF WING SECTIONS: INCLUDING A SUMMARY OF AIR-FOIL DATA, Ira H. Abbott and A. E. von Doenhoff. Concise compilation of subatomic aerodynamic characteristics of modern NASA wing sections, plus description of theory. 350pp. of tables. 693pp. 5⅜ x 8½.
60586-8 Pa. $7.00

DE RE METALLICA, Georgius Agricola. Translated by Herbert C. Hoover and Lou H. Hoover. The famous Hoover translation of greatest treatise on technological chemistry, engineering, geology, mining of early modern times (1556). All 289 original woodcuts. 638pp. 6¾ x 11.
60006-8 Clothbd. $17.50

THE ORIGIN OF CONTINENTS AND OCEANS, Alfred Wegener. One of the most influential, most controversial books in science, the classic statement for continental drift. Full 1966 translation of Wegener's final (1929) version. 64 illustrations. 246pp. 5⅜ x 8½. 61708-4 Pa. $3.00

THE PRINCIPLES OF PSYCHOLOGY, William James. Famous long course complete, unabridged. Stream of thought, time perception, memory, experimental methods; great work decades ahead of its time. Still valid, useful; read in many classes. 94 figures. Total of 1391pp. 5⅜ x 8½.
20381-6, 20382-4 Pa., Two-vol. set $13.00

YUCATAN BEFORE AND AFTER THE CONQUEST, Diego de Landa. First English translation of basic book in Maya studies, the only significant account of Yucatan written in the early post-Conquest era. Translated by distinguished Maya scholar William Gates. Appendices, introduction, 4 maps and over 120 illustrations added by translator. 162pp. 5⅜ x 8½.
23622-6 Pa. $3.00

THE MALAY ARCHIPELAGO, Alfred R. Wallace. Spirited travel account by one of founders of modern biology. Touches on zoology, botany, ethnography, geography, and geology. 62 illustrations, maps. 515pp. 5⅜ x 8½.
20187-2 Pa. $6.95

THE DISCOVERY OF THE TOMB OF TUTANKHAMEN, Howard Carter, A. C. Mace. Accompany Carter in the thrill of discovery, as ruined passage suddenly reveals unique, untouched, fabulously rich tomb. Fascinating account, with 106 illustrations. New introduction by J. M. White. Total of 382pp. 5⅜ x 8½. (Available in U.S. only)   23500-9 Pa. $4.00

THE WORLD'S GREATEST SPEECHES, edited by Lewis Copeland and Lawrence W. Lamm. Vast collection of 278 speeches from Greeks up to present. Powerful and effective models; unique look at history. Revised to 1970. Indices. 842pp. 5⅜ x 8½.   20468-5 Pa. $8.95

THE 100 GREATEST ADVERTISEMENTS, Julian Watkins. The priceless ingredient; His master's voice; 99 44/100% pure; over 100 others. How they were written, their impact, etc. Remarkable record. 130 illustrations. 233pp. 7⅞ x 10 3/5.   20540-1 Pa. $5.00

CRUICKSHANK PRINTS FOR HAND COLORING, George Cruickshank. 18 illustrations, one side of a page, on fine-quality paper suitable for watercolors. Caricatures of people in society (c. 1820) full of trenchant wit. Very large format. 32pp. 11 x 16.   23684-6 Pa. $5.00

THIRTY-TWO COLOR POSTCARDS OF TWENTIETH-CENTURY AMERICAN ART, Whitney Museum of American Art. Reproduced in full color in postcard form are 31 art works and one shot of the museum. Calder, Hopper, Rauschenberg, others. Detachable. 16pp. 8¼ x 11.
23629-3 Pa. $2.50

MUSIC OF THE SPHERES: THE MATERIAL UNIVERSE FROM ATOM TO QUASAR SIMPLY EXPLAINED, Guy Murchie. Planets, stars, geology, atoms, radiation, relativity, quantum theory, light, antimatter, similar topics. 319 figures. 664pp. 5⅜ x 8½.
21809-0, 21810-4 Pa., Two-vol. set $10.00

EINSTEIN'S THEORY OF RELATIVITY, Max Born. Finest semi-technical account; covers Einstein, Lorentz, Minkowski, and others, with much detail, much explanation of ideas and math not readily available elsewhere on this level. For student, non-specialist. 376pp. 5⅜ x 8½.
60769-0 Pa. $4.00

THE COMPLETE BOOK OF DOLL MAKING AND COLLECTING, Catherine Christopher. Instructions, patterns for dozens of dolls, from rag doll on up to elaborate, historically accurate figures. Mould faces, sew clothing, make doll houses, etc. Also collecting information. Many illustrations. 288pp. 6 x 9. 22066-4 Pa. $4.00

THE DAGUERREOTYPE IN AMERICA, Beaumont Newhall. Wonderful portraits, 1850's townscapes, landscapes; full text plus 104 photographs. The basic book. Enlarged 1976 edition. 272pp. 8¼ x 11¼. 23322-7 Pa. $6.00

CRAFTSMAN HOMES, Gustav Stickley. 296 architectural drawings, floor plans, and photographs illustrate 40 different kinds of "Mission-style" homes from The Craftsman (1901-16), voice of American style of simplicity and organic harmony. Thorough coverage of Craftsman idea in text and picture, now collector's item. 224pp. 8⅛ x 11. 23791-5 Pa. $6.00

PEWTER-WORKING: INSTRUCTIONS AND PROJECTS, Burl N. Osborn. & Gordon O. Wilber. Introduction to pewter-working for amateur craftsman. History and characteristics of pewter; tools, materials, step-by-step instructions. Photos, line drawings, diagrams. Total of 160pp. 7⅞ x 10¾. 23786-9 Pa. $3.50

THE GREAT CHICAGO FIRE, edited by David Lowe. 10 dramatic, eye-witness accounts of the 1871 disaster, including one of the aftermath and rebuilding, plus 70 contemporary photographs and illustrations of the ruins—courthouse, Palmer House, Great Central Depot, etc. Introduction by David Lowe. 87pp. 8¼ x 11. 23771-0 Pa. $4.00

SILHOUETTES: A PICTORIAL ARCHIVE OF VARIED ILLUSTRA-TIONS, edited by Carol Belanger Grafton. Over 600 silhouettes from the 18th to 20th centuries include profiles and full figures of men and women, children, birds and animals, groups and scenes, nature, ships, an alphabet. Dozens of uses for commercial artists and craftspeople. 144pp. 8⅜ x 11¼. 23781-8 Pa. $4.00

ANIMALS: 1,419 COPYRIGHT-FREE ILLUSTRATIONS OF MAM-MALS, BIRDS, FISH, INSECTS, ETC., edited by Jim Harter. Clear wood engravings present, in extremely lifelike poses, over 1,000 species of animals. One of the most extensive copyright-free pictorial sourcebooks of its kind. Captions. Index. 284pp. 9 x 12. 23766-4 Pa. $7.50

INDIAN DESIGNS FROM ANCIENT ECUADOR, Frederick W. Shaffer. 282 original designs by pre-Columbian Indians of Ecuador (500-1500 A.D.). Designs include people, mammals, birds, reptiles, fish, plants, heads, geometric designs. Use as is or alter for advertising, textiles, leathercraft, etc. Introduction. 95pp. 8¾ x 11¼. 23764-8 Pa. $3.50

SZIGETI ON THE VIOLIN, Joseph Szigeti. Genial, loosely structured tour by premier violinist, featuring a pleasant mixture of reminiscenes, insights into great music and musicians, innumerable tips for practicing violinists. 385 musical passages. 256pp. 5⅝ x 8¼. 23763-X Pa. $3.50

TONE POEMS, SERIES II: TILL EULENSPIEGELS LUSTIGE STREICHE, ALSO SPRACH ZARATHUSTRA, AND EIN HELDEN-LEBEN, Richard Strauss. Three important orchestral works, including very popular *Till Eulenspiegel's Marry Pranks,* reproduced in full score from original editions. Study score. 315pp. 9⅜ x 12¼. (Available in U.S. only)
23755-9 Pa. $7.50

TONE POEMS, SERIES I: DON JUAN, TOD UND VERKLARUNG AND DON QUIXOTE, Richard Strauss. Three of the most often performed and recorded works in entire orchestral repertoire, reproduced in full score from original editions. Study score. 286pp. 9⅜ x 12¼. (Available in U.S. only)
23754-0 Pa. $7.50

11 LATE STRING QUARTETS, Franz Joseph Haydn. The form which Haydn defined and "brought to perfection." (*Grove's*). 11 string quartets in complete score, his last and his best. The first in a projected series of the complete Haydn string quartets. Reliable modern Eulenberg edition, otherwise difficult to obtain. 320pp. 8⅜ x 11¼. (Available in U.S. only)
23753-2 Pa. $6.95

FOURTH, FIFTH AND SIXTH SYMPHONIES IN FULL SCORE, Peter Ilyitch Tchaikovsky. Complete orchestral scores of Symphony No. 4 in F Minor, Op. 36; Symphony No. 5 in E Minor, Op. 64; Symphony No. 6 in B Minor, "Pathetique," Op. 74. Bretikopf & Hartel eds. Study score. 480pp. 9⅜ x 12¼. 23861-X Pa. $10.95

THE MARRIAGE OF FIGARO: COMPLETE SCORE, Wolfgang A. Mozart. Finest comic opera ever written. Full score, not to be confused with piano renderings. Peters edition. Study score. 448pp. 9⅜ x 12¼. (Available in U.S. only) 23751-6 Pa. $11.95

"IMAGE" ON THE ART AND EVOLUTION OF THE FILM, edited by Marshall Deutelbaum. Pioneering book brings together for first time 38 groundbreaking articles on early silent films from *Image* and 263 illustrations newly shot from rare prints in the collection of the International Museum of Photography. A landmark work. Index. 256pp. 8¼ x 11.
23777-X Pa. $8.95

AROUND-THE-WORLD COOKY BOOK, Lois Lintner Sumption and Marguerite Lintner Ashbrook. 373 cooky and frosting recipes from 28 countries (America, Austria, China, Russia, Italy, etc.) include Viennese kisses, rice wafers, London strips, lady fingers, hony, sugar spice, maple cookies, etc. Clear instructions. All tested. 38 drawings. 182pp. 5⅜ x 8.
23802-4 Pa. $2.50

THE ART NOUVEAU STYLE, edited by Roberta Waddell. 579 rare photographs, not available elsewhere, of works in jewelry, metalwork, glass, ceramics, textiles, architecture and furniture by 175 artists—Mucha, Seguy, Lalique, Tiffany, Gaudin, Hohlwein, Saarinen, and many others. 288pp. 8⅜ x 11¼. 23515-7 Pa. $6.95

THE AMERICAN SENATOR, Anthony Trollope. Little known, long unavailable Trollope novel on a grand scale. Here are humorous comment on American vs. English culture, and stunning portrayal of a heroine/villainess. Superb evocation of Victorian village life. 561pp. 5⅜ x 8½.
23801-6 Pa. $6.00

WAS IT MURDER? James Hilton. The author of *Lost Horizon* and *Goodbye, Mr. Chips* wrote one detective novel (under a pen-name) which was quickly forgotten and virtually lost, even at the height of Hilton's fame. This edition brings it back—a finely crafted public school puzzle resplendent with Hilton's stylish atmosphere. A thoroughly English thriller by the creator of Shangri-la. 252pp. 5⅜ x 8. (Available in U.S. only)
23774-5 Pa. $3.00

CENTRAL PARK: A PHOTOGRAPHIC GUIDE, Victor Laredo and Henry Hope Reed. 121 superb photographs show dramatic views of Central Park: Bethesda Fountain, Cleopatra's Needle, Sheep Meadow, the Blockhouse, plus people engaged in many park activities: ice skating, bike riding, etc. Captions by former Curator of Central Park, Henry Hope Reed, provide historical view, changes, etc. Also photos of N.Y. landmarks on park's periphery. 96pp. 8½ x 11.
23750-8 Pa. $4.50

NANTUCKET IN THE NINETEENTH CENTURY, Clay Lancaster. 180 rare photographs, stereographs, maps, drawings and floor plans recreate unique American island society. Authentic scenes of shipwreck, lighthouses, streets, homes are arranged in geographic sequence to provide walking-tour guide to old Nantucket existing today. Introduction, captions. 160pp. 8⅞ x 11¾.
23747-8 Pa. $6.95

STONE AND MAN: A PHOTOGRAPHIC EXPLORATION, Andreas Feininger. 106 photographs by *Life* photographer Feininger portray man's deep passion for stone through the ages. Stonehenge-like megaliths, fortified towns, sculpted marble and crumbling tenements show textures, beauties, fascination. 128pp. 9¼ x 10¾.
23756-7 Pa. $5.95

CIRCLES, A MATHEMATICAL VIEW, D. Pedoe. Fundamental aspects of college geometry, non-Euclidean geometry, and other branches of mathematics: representing circle by point. Poincare model, isoperimetric property, etc. Stimulating recreational reading. 66 figures. 96pp. 5⅝ x 8¼.
63698-4 Pa. $2.75

THE DISCOVERY OF NEPTUNE, Morton Grosser. Dramatic scientific history of the investigations leading up to the actual discovery of the eighth planet of our solar system. Lucid, well-researched book by well-known historian of science. 172pp. 5⅜ x 8½.
23726-5 Pa. $3.00

THE DEVIL'S DICTIONARY. Ambrose Bierce. Barbed, bitter, brilliant witticisms in the form of a dictionary. Best, most ferocious satire America has produced. 145pp. 5⅜ x 8½.
20487-1 Pa. $1.75

HISTORY OF BACTERIOLOGY, William Bulloch. The only comprehensive history of bacteriology from the beginnings through the 19th century. Special emphasis is given to biography-Leeuwenhoek, etc. Brief accounts of 350 bacteriologists form a separate section. No clearer, fuller study, suitable to scientists and general readers, has yet been written. 52 illustrations. 448pp. 5⅝ x 8¼. 23761-3 Pa. $6.50

THE COMPLETE NONSENSE OF EDWARD LEAR, Edward Lear. All nonsense limericks, zany alphabets, Owl and Pussycat, songs, nonsense botany, etc., illustrated by Lear. Total of 321pp. 5⅜ x 8½. (Available in U.S. only) 20167-8 Pa. $3.00

INGENIOUS MATHEMATICAL PROBLEMS AND METHODS, Louis A. Graham. Sophisticated material from Graham *Dial*, applied and pure; stresses solution methods. Logic, number theory, networks, inversions, etc. 237pp. 5⅜ x 8½. 20545-2 Pa. $3.50

BEST MATHEMATICAL PUZZLES OF SAM LOYD, edited by Martin Gardner. Bizarre, original, whimsical puzzles by America's greatest puzzler. From fabulously rare *Cyclopedia*, including famous 14-15 puzzles, the Horse of a Different Color, 115 more. Elementary math. 150 illustrations. 167pp. 5⅜ x 8½. 20498-7 Pa. $2.50

THE BASIS OF COMBINATION IN CHESS, J. du Mont. Easy-to-follow, instructive book on elements of combination play, with chapters on each piece and every powerful combination team—two knights, bishop and knight, rook and bishop, etc. 250 diagrams. 218pp. 5⅜ x 8½. (Available in U.S. only) 23644-7 Pa. $3.50

MODERN CHESS STRATEGY, Ludek Pachman. The use of the queen, the active king, exchanges, pawn play, the center, weak squares, etc. Section on rook alone worth price of the book. Stress on the moderns. Often considered the most important book on strategy. 314pp. 5⅜ x 8½. 20290-9 Pa. $3.50

LASKER'S MANUAL OF CHESS, Dr. Emanuel Lasker. Great world champion offers very thorough coverage of all aspects of chess. Combinations, position play, openings, end game, aesthetics of chess, philosophy of struggle, much more. Filled with analyzed games. 390pp. 5⅜ x 8½. 20640-8 Pa. $4.00

500 MASTER GAMES OF CHESS, S. Tartakower, J. du Mont. Vast collection of great chess games from 1798-1938, with much material nowhere else readily available. Fully annotated, arranged by opening for easier study. 664pp. 5⅜ x 8½. 23208-5 Pa. $6.00

A GUIDE TO CHESS ENDINGS, Dr. Max Euwe, David Hooper. One of the finest modern works on chess endings. Thorough analysis of the most frequently encountered endings by former world champion. 331 examples, each with diagram. 248pp. 5⅜ x 8½. 23332-4 Pa. $3.50

SECOND PIATIGORSKY CUP, edited by Isaac Kashdan. One of the greatest tournament books ever produced in the English language. All 90 games of the 1966 tournament, annotated by players, most annotated by both players. Features Petrosian, Spassky, Fischer, Larsen, six others. 228pp. 5⅜ x 8½. 23572-6 Pa. $3.50

ENCYCLOPEDIA OF CARD TRICKS, revised and edited by Jean Hugard. How to perform over 600 card tricks, devised by the world's greatest magicians: impromptus, spelling tricks, key cards, using special packs, much, much more. Additional chapter on card technique. 66 illustrations. 402pp. 5⅜ x 8½. (Available in U.S. only) 21252-1 Pa. $3.95

MAGIC: STAGE ILLUSIONS, SPECIAL EFFECTS AND TRICK PHO-TOGRAPHY, Albert A. Hopkins, Henry R. Evans. One of the great classics; fullest, most authorative explanation of vanishing lady, levitations, scores of other great stage effects. Also small magic, automata, stunts. 446 illustrations. 556pp. 5⅜ x 8½. 23344-8 Pa. $5.00

THE SECRETS OF HOUDINI, J. C. Cannell. Classic study of Houdini's incredible magic, exposing closely-kept professional secrets and revealing, in general terms, the whole art of stage magic. 67 illustrations. 279pp. 5⅜ x 8½. 22913-0 Pa. $3.00

HOFFMANN'S MODERN MAGIC, Professor Hoffmann. One of the best, and best-known, magicians' manuals of the past century. Hundreds of tricks from card tricks and simple sleight of hand to elaborate illusions involving construction of complicated machinery. 332 illustrations. 563pp. 5⅜ x 8½. 23623-4 Pa. $6.00

MADAME PRUNIER'S FISH COOKERY BOOK, Mme. S. B. Prunier. More than 1000 recipes from world famous Prunier's of Paris and London, specially adapted here for American kitchen. Grilled tournedos with anchovy butter, Lobster a la Bordelaise, Prunier's prized desserts, more. Glossary. 340pp. 5⅜ x 8½. (Available in U.S. only) 22679-4 Pa. $3.00

FRENCH COUNTRY COOKING FOR AMERICANS, Louis Diat. 500 easy-to-make, authentic provincial recipes compiled by former head chef at New York's Fitz-Carlton Hotel: onion soup, lamb stew, potato pie, more. 309pp. 5⅜ x 8½. 23665-X Pa. $3.95

SAUCES, FRENCH AND FAMOUS, Louis Diat. Complete book gives over 200 specific recipes: bechamel, Bordelaise, hollandaise, Cumberland, apricot, etc. Author was one of this century's finest chefs, originator of vichyssoise and many other dishes. Index. 156pp. 5⅜ x 8. 23663-3 Pa. $2.50

TOLL HOUSE TRIED AND TRUE RECIPES, Ruth Graves Wakefield. Authentic recipes from the famous Mass. restaurant: popovers, veal and ham loaf, Toll House baked beans, chocolate cake crumb pudding, much more. Many helpful hints. Nearly 700 recipes. Index. 376pp. 5⅜ x 8½. 23560-2 Pa. $4.00

"OSCAR" OF THE WALDORF'S COOKBOOK, Oscar Tschirky. Famous American chef reveals 3455 recipes that made Waldorf great; cream of French, German, American cooking, in all categories. Full instructions, easy home use. 1896 edition. 907pp. 6⅝ x 9⅜.   20790-0 Clothbd. $15.00

COOKING WITH BEER, Carole Fahy. Beer has as superb an effect on food as wine, and at fraction of cost. Over 250 recipes for appetizers, soups, main dishes, desserts, breads, etc. Index. 144pp. 5⅜ x 8½. (Available in U.S. only)   23661-7 Pa. $2.50

STEWS AND RAGOUTS, Kay Shaw Nelson. This international cookbook offers wide range of 108 recipes perfect for everyday, special occasions, meals-in-themselves, main dishes. Economical, nutritious, easy-to-prepare: goulash, Irish stew, boeuf bourguignon, etc. Index. 134pp. 5⅜ x 8½.
23662-5 Pa. $2.50

DELICIOUS MAIN COURSE DISHES, Marian Tracy. Main courses are the most important part of any meal. These 200 nutritious, economical recipes from around the world make every meal a delight. "I . . . have found it so useful in my own household,"—*N.Y. Times.* Index. 219pp. 5⅜ x 8½.   23664-1 Pa. $3.00

FIVE ACRES AND INDEPENDENCE, Maurice G. Kains. Great back-to-the-land classic explains basics of self-sufficient farming: economics, plants, crops, animals, orchards, soils, land selection, host of other necessary things. Do not confuse with skimpy faddist literature; Kains was one of America's greatest agriculturalists. 95 illustrations. 397pp. 5⅜ x 8½.
20974-1 Pa. $3.95

A PRACTICAL GUIDE FOR THE BEGINNING FARMER, Herbert Jacobs. Basic, extremely useful first book for anyone thinking about moving to the country and starting a farm. Simpler than Kains, with greater emphasis on country living in general. 246pp. 5⅜ x 8½.
23675-7 Pa. $3.50

A GARDEN OF PLEASANT FLOWERS (PARADISI IN SOLE: PARADISUS TERRESTRIS), John Parkinson. Complete, unabridged reprint of first (1629) edition of earliest great English book on gardens and gardening. More than 1000 plants & flowers of Elizabethan, Jacobean garden fully described, most with woodcut illustrations. Botanically very reliable, a "speaking garden" of exceeding charm. 812 illustrations. 628pp. 8½ x 12¼.   23392-8 Clothbd. $25.00

ACKERMANN'S COSTUME PLATES, Rudolph Ackermann. Selection of 96 plates from the *Repository of Arts,* best published source of costume for English fashion during the early 19th century. 12 plates also in color. Captions, glossary and introduction by editor Stella Blum. Total of 120pp. 8⅜ x 11¼.   23690-0 Pa. $4.50

MUSHROOMS, EDIBLE AND OTHERWISE, Miron E. Hard. Profusely illustrated, very useful guide to over 500 species of mushrooms growing in the Midwest and East. Nomenclature updated to 1976. 505 illustrations. 628pp. 6½ x 9¼. 23309-X Pa. $7.95

AN ILLUSTRATED FLORA OF THE NORTHERN UNITED STATES AND CANADA, Nathaniel L. Britton, Addison Brown. Encyclopedic work covers 4666 species, ferns on up. Everything. Full botanical information, illustration for each. This earlier edition is preferred by many to more recent revisions. 1913 edition. Over 4000 illustrations, total of 2087pp. 6⅛ x 9¼. 22642-5, 22643-3, 22644-1 Pa., Three-vol. set $24.00

MANUAL OF THE GRASSES OF THE UNITED STATES, A. S. Hitchcock, U.S. Dept. of Agriculture. The basic study of American grasses, both indigenous and escapes, cultivated and wild. Over 1400 species. Full descriptions, information. Over 1100 maps, illustrations. Total of 1051pp. 5⅜ x 8½. 22717-0, 22718-9 Pa., Two-vol. set $12.00

THE CACTACEAE,, Nathaniel L. Britton, John N. Rose. Exhaustive, definitive. Every cactus in the world. Full botanical descriptions. Thorough statement of nomenclatures, habitat, detailed finding keys. The one book needed by every cactus enthusiast. Over 1275 illustrations. Total of 1080pp. 8 x 10¼. 21191-6, 21192-4 Clothbd., Two-vol. set $35.00

AMERICAN MEDICINAL PLANTS, Charles F. Millspaugh. Full descriptions, 180 plants covered: history; physical description; methods of preparation with all chemical constituents extracted; all claimed curative or adverse effects. 180 full-page plates. Classification table. 804pp. 6½ x 9¼. 23034-1 Pa. $10.00

A MODERN HERBAL, Margaret Grieve. Much the fullest, most exact, most useful compilation of herbal material. Gigantic alphabetical encyclopedia, from aconite to zedoary, gives botanical information, medical properties, folklore, economic uses, and much else. Indispensable to serious reader. 161 illustrations. 888pp. 6½ x 9¼. (Available in U.S. only) 22798-7, 22799-5 Pa., Two-vol. set $11.00

THE HERBAL or GENERAL HISTORY OF PLANTS, John Gerard. The 1633 edition revised and enlarged by Thomas Johnson. Containing almost 2850 plant descriptions and 2705 superb illustrations, Gerard's *Herbal* is a monumental work, the book all modern English herbals are derived from, the one herbal every serious enthusiast should have in its entirety. Original editions are worth perhaps $750. 1678pp. 8½ x 12¼. 23147-X Clothbd. $50.00

MANUAL OF THE TREES OF NORTH AMERICA, Charles S. Sargent. The basic survey of every native tree and tree-like shrub, 717 species in all. Extremely full descriptions, information on habitat, growth, locales, economics, etc. Necessary to every serious tree lover. Over 100 finding keys. 783 illustrations. Total of 986pp. 5⅜ x 8½. 20277-1, 20278-X Pa., Two-vol. set $10.00

AMERICAN BIRD ENGRAVINGS, Alexander Wilson et al. All 76 plates. from Wilson's *American Ornithology* (1808-14), most important ornithological work before Audubon, plus 27 plates from the supplement (1825-33) by Charles Bonaparte. Over 250 birds portrayed. 8 plates also reproduced in full color. 111pp. 9⅜ x 12½. 23195-X Pa. $6.00

CRUICKSHANK'S PHOTOGRAPHS OF BIRDS OF AMERICA, Allan D. Cruickshank. Great ornithologist, photographer presents 177 closeups, groupings, panoramas, flightings, etc., of about 150 different birds. Expanded *Wings in the Wilderness*. Introduction by Helen G. Cruickshank. 191pp. 8¼ x 11. 23497-5 Pa. $6.00

AMERICAN WILDLIFE AND PLANTS, A. C. Martin, et al. Describes food habits of more than 1000 species of mammals, birds, fish. Special treatment of important food plants. Over 300 illustrations. 500pp. 5⅜ x 8½. 20793-5 Pa. $4.95

THE PEOPLE CALLED SHAKERS, Edward D. Andrews. Lifetime of research, definitive study of Shakers: origins, beliefs, practices, dances, social organization, furniture and crafts, impact on 19th-century USA, present heritage. Indispensable to student of American history, collector. 33 illustrations. 351pp. 5⅜ x 8½. 21081-2 Pa. $4.00

OLD NEW YORK IN EARLY PHOTOGRAPHS, Mary Black. New York City as it was in 1853-1901, through 196 wonderful photographs from N.-Y. Historical Society. Great Blizzard, Lincoln's funeral procession, great buildings. 228pp. 9 x 12. 22907-6 Pa. $7.95

MR. LINCOLN'S CAMERA MAN: MATHEW BRADY, Roy Meredith. Over 300 Brady photos reproduced directly from original negatives, photos. Jackson, Webster, Grant, Lee, Carnegie, Barnum; Lincoln; Battle Smoke, Death of Rebel Sniper, Atlanta Just After Capture. Lively commentary. 368pp. 8⅜ x 11¼. 23021-X Pa. $8.95

TRAVELS OF WILLIAM BARTRAM, William Bartram. From 1773-8, Bartram explored Northern Florida, Georgia, Carolinas, and reported on wild life, plants, Indians, early settlers. Basic account for period, entertaining reading. Edited by Mark Van Doren. 13 illustrations. 141pp. 5⅜ x 8½. 20013-2 Pa. $4.50

THE GENTLEMAN AND CABINET MAKER'S DIRECTOR, Thomas Chippendale. Full reprint, 1762 style book, most influential of all time; chairs, tables, sofas, mirrors, cabinets, etc. 200 plates, plus 24 photographs of surviving pieces. 249pp. 9⅞ x 12¾. 21601-2 Pa. $6.50

AMERICAN CARRIAGES, SLEIGHS, SULKIES AND CARTS, edited by Don H. Berkebile. 168 Victorian illustrations from catalogues, trade journals, fully captioned. Useful for artists. Author is Assoc. Curator, Div. of Transportation of Smithsonian Institution. 168pp. 8½ x 9½. 23328-6 Pa. $5.00

THE SENSE OF BEAUTY, George Santayana. Masterfully written discussion of nature of beauty, materials of beauty, form, expression; art, literature, social sciences all involved. 168pp. 5⅜ x 8½. 20238-0 Pa. $2.50

ON THE IMPROVEMENT OF THE UNDERSTANDING, Benedict Spinoza. Also contains *Ethics, Correspondence,* all in excellent R. Elwes translation. Basic works on entry to philosophy, pantheism, exchange of ideas with great contemporaries. 402pp. 5⅜ x 8½. 20250-X Pa. $4.50

THE TRAGIC SENSE OF LIFE, Miguel de Unamuno. Acknowledged masterpiece of existential literature, one of most important books of 20th century. Introduction by Madariaga. 367pp. 5⅜ x 8½.
20257-7 Pa. $3.50

THE GUIDE FOR THE PERPLEXED, Moses Maimonides. Great classic of medieval Judaism attempts to reconcile revealed religion (Pentateuch, commentaries) with Aristotelian philosophy. Important historically, still relevant in problems. Unabridged Friedlander translation. Total of 473pp. 5⅜ x 8½. 20351-4 Pa. $5.00

THE I CHING (THE BOOK OF CHANGES), translated by James Legge. Complete translation of basic text plus appendices by Confucius, and Chinese commentary of most penetrating divination manual ever prepared. Indispensable to study of early Oriental civilizations, to modern inquiring reader. 448pp. 5⅜ x 8½. 21062-6 Pa. $4.00

THE EGYPTIAN BOOK OF THE DEAD, E. A. Wallis Budge. Complete reproduction of Ani's papyrus, finest ever found. Full hieroglyphic text, interlinear transliteration, word for word translation, smooth translation. Basic work, for Egyptology, for modern study of psychic matters. Total of 533pp. 6½ x 9¼. (Available in U.S. only) 21866-X Pa. $4.95

THE GODS OF THE EGYPTIANS, E. A. Wallis Budge. Never excelled for richness, fullness: all gods, goddesses, demons, mythical figures of Ancient Egypt; their legends, rites, incarnations, variations, powers, etc. Many hieroglyphic texts cited. Over 225 illustrations, plus 6 color plates. Total of 988pp. 6⅛ x 9¼. (Available in U.S. only)
22055-9, 22056-7 Pa., Two-vol. set $12.00

THE ENGLISH AND SCOTTISH POPULAR BALLADS, Francis J. Child. Monumental, still unsuperseded; all known variants of Child ballads, commentary on origins, literary references, Continental parallels, other features. Added: papers by G. L. Kittredge, W. M. Hart. Total of 2761pp. 6½ x 9¼.
21409-5, 21410-9, 21411-7, 21412-5, 21413-3 Pa., Five-vol. set $37.50

CORAL GARDENS AND THEIR MAGIC, Bronsilaw Malinowski. Classic study of the methods of tilling the soil and of agricultural rites in the Trobriand Islands of Melanesia. Author is one of the most important figures in the field of modern social anthropology. 143 illustrations. Indexes. Total of 911pp. of text. 5⅝ x 8¼. (Available in U.S. only)
23597-1 Pa. $12.95

THE PHILOSOPHY OF HISTORY, Georg W. Hegel. Great classic of Western thought develops concept that history is not chance but a rational process, the evolution of freedom. 457pp. 5⅜ x 8½.   20112-0 Pa. $4.50

LANGUAGE, TRUTH AND LOGIC, Alfred J. Ayer. Famous, clear introduction to Vienna, Cambridge schools of Logical Positivism. Role of philosophy, elimination of metaphysics, nature of analysis, etc. 160pp. 5⅜ x 8½. (Available in U.S. only)   20010-8 Pa. $1.75

A PREFACE TO LOGIC, Morris R. Cohen. Great City College teacher in renowned, easily followed exposition of formal logic, probability, values, logic and world order and similar topics; no previous background needed. 209pp. 5⅜ x 8½.   23517-3 Pa. $3.50

REASON AND NATURE, Morris R. Cohen. Brilliant analysis of reason and its multitudinous ramifications by charismatic teacher. Interdisciplinary, synthesizing work widely praised when it first appeared in 1931. Second (1953) edition. Indexes. 496pp. 5⅜ x 8½.   23633-1 Pa. $6.00

AN ESSAY CONCERNING HUMAN UNDERSTANDING, John Locke. The only complete edition of enormously important classic, with authoritative editorial material by A. C. Fraser. Total of 1176pp. 5⅜ x 8½.
                        20530-4, 20531-2 Pa., Two-vol. set $14.00

HANDBOOK OF MATHEMATICAL FUNCTIONS WITH FORMULAS, GRAPHS, AND MATHEMATICAL TABLES, edited by Milton Abramowitz and Irene A. Stegun. Vast compendium: 29 sets of tables, some to as high as 20 places. 1,046pp. 8 x 10½.   61272-4 Pa. $14.95

MATHEMATICS FOR THE PHYSICAL SCIENCES, Herbert S. Wilf. Highly acclaimed work offers clear presentations of vector spaces and matrices, orthogonal functions, roots of polynomial equations, conformal mapping, calculus of variations, etc. Knowledge of theory of functions of real and complex variables is assumed. Exercises and solutions. Index. 284pp. 5⅝ x 8¼.   63635-6 Pa. $4.50

THE PRINCIPLE OF RELATIVITY, Albert Einstein et al. Eleven most important original papers on special and general theories. Seven by Einstein, two by Lorentz, one each by Minkowski and Weyl. All translated, unabridged. 216pp. 5⅜ x 8½.   60081-5 Pa. $3.00

THERMODYNAMICS, Enrico Fermi. A classic of modern science. Clear, organized treatment of systems, first and second laws, entropy, thermodynamic potentials, gaseous reactions, dilute solutions, entropy constant. No math beyond calculus required. Problems. 160pp. 5⅜ x 8½.
                                        60361-X Pa. $2.75

ELEMENTARY MECHANICS OF FLUIDS, Hunter Rouse. Classic undergraduate text widely considered to be far better than many later books. Ranges from fluid velocity and acceleration to role of compressibility in fluid motion. Numerous examples, questions, problems. 224 illustrations. 376pp. 5⅝ x 8¼.   63699-2 Pa. $5.00

AN AUTOBIOGRAPHY, Margaret Sanger. Exciting personal account of hard-fought battle for woman's right to birth control, against prejudice, church, law. Foremost feminist document. 504pp. 5⅜ x 8½.
20470-7 Pa. $5.50

MY BONDAGE AND MY FREEDOM, Frederick Douglass. Born as a slave, Douglass became outspoken force in antislavery movement. The best of Douglass's autobiographies. Graphic description of slave life. Introduction by P. Foner. 464pp. 5⅜ x 8½.
22457-0 Pa. $5.00

LIVING MY LIFE, Emma Goldman. Candid, no holds barred account by foremost American anarchist: her own life, anarchist movement, famous contemporaries, ideas and their impact. Struggles and confrontations in America, plus deportation to U.S.S.R. Shocking inside account of persecution of anarchists under Lenin. 13 plates. Total of 944pp. 5⅜ x 8½.
22543-7, 22544-5 Pa., Two-vol. set $9.00

LETTERS AND NOTES ON THE MANNERS, CUSTOMS AND CONDITIONS OF THE NORTH AMERICAN INDIANS, George Catlin. Classic account of life among Plains Indians: ceremonies, hunt, warfare, etc. Dover edition reproduces for first time all original paintings. 312 plates. 572pp. of text. 6⅛ x 9¼.
22118-0, 22119-9 Pa.. Two-vol. set $10.00

THE MAYA AND THEIR NEIGHBORS, edited by Clarence L. Hay, others. Synoptic view of Maya civilization in broadest sense, together with Northern, Southern neighbors. Integrates much background, valuable detail not elsewhere. Prepared by greatest scholars: Kroeber, Morley, Thompson, Spinden, Vaillant, many others. Sometimes called Tozzer Memorial Volume. 60 illustrations, linguistic map. 634pp. 5⅜ x 8½.
23510-6 Pa. $7.50

HANDBOOK OF THE INDIANS OF CALIFORNIA, A. L. Kroeber. Foremost American anthropologist offers complete ethnographic study of each group. Monumental classic. 459 illustrations, maps. 995pp. 5⅜ x 8½.
23368-5 Pa. $10.00

SHAKTI AND SHAKTA, Arthur Avalon. First book to give clear, cohesive analysis of Shakta doctrine, Shakta ritual and Kundalini Shakti (yoga). Important work by one of world's foremost students of Shaktic and Tantric thought. 732pp. 5⅜ x 8½. (Available in U.S. only)
23645-5 Pa. $7.95

AN INTRODUCTION TO THE STUDY OF THE MAYA HIEROGLYPHS, Syvanus Griswold Morley. Classic study by one of the truly great figures in hieroglyph research. Still the best introduction for the student for reading Maya hieroglyphs. New introduction by J. Eric S. Thompson. 117 illustrations. 284pp. 5⅜ x 8½.
23108-9 Pa. $4.00

A STUDY OF MAYA ART, Herbert J. Spinden. Landmark classic interprets Maya symbolism, estimates styles, covers ceramics, architecture, murals, stone carvings as artforms. Still a basic book in area. New introduction by J. Eric Thompson. Over 750 illustrations. 341pp. 8⅜ x 11¼.
21235-1 Pa. $6.95